Stewart's Calculus

Second Edition

Volume I

Student Solutions Manual
Early Transcendentals

James Stewart
McMaster University

Daniel Anderson
University of Iowa

Daniel Drucker
Wayne State University

with the assistance of Lothar Redlin
The Pennsylvania State University

Terry Tiballi
North Harris County College

Brooks/Cole Publishing Company
Pacific Grove, California

Preface

I have edited this solutions manual by taking the solutions provided by Daniel Anderson, Daniel Drucker, and Terry Tiballi, comparing them with solutions that were produced by me and my colleagues at McMaster University. A further check was provided by comparison with solutions produced by several of my best calculus students whom I hired to work all the problems: Eric Bosch, Don Callfas, Aaron Childs, Loris Corazza, Ron Donaberger, Michael Gmell, Brett Goodwin, Sara Lee, Marc Riehm, Martin Sarabura, Jeff Schultheiss, Tony di Silvestro, and Joe Vetrone. Solutions to the Applications Plus exercises were checked by Garrett Etgen, David McKay, Lothar Redlin, and Saleem Watson.

I thank all of these people for contributing to the accuracy of this solutions manual.

JAMES STEWART

Brooks/Cole Publishing Company
A Division of Wadsworth, Inc.

Printed in the United States of America

10 9 8 7 6 5 4 3 2 1

ISBN 0-534-13832-2

Contents

REVIEW AND PREVIEW

EXERCISES 1

1. $|5 - 23| = |-18| = 18$

3. $|-\pi| = \pi$ since $\pi > 0$.

5. $\left|\sqrt{5} - 5\right| = -(\sqrt{5} - 5) = 5 - \sqrt{5}$ since $\sqrt{5} - 5 < 0$.

7. For $x < 2$, $x - 2 < 0$, so $|x - 2| = -(x - 2) = 2 - x$.

9.
$$|x + 1| = \begin{cases} x + 1 \text{ for } x + 1 \geq 0 \Leftrightarrow x \geq -1 \\ -(x + 1) \text{ for } x + 1 < 0 \Leftrightarrow x < -1 \end{cases}$$

11. $\left|x^2 + 1\right| = x^2 + 1$ (since $x^2 + 1 \geq 0$ for all x).

13. $2x + 7 > 3 \Leftrightarrow 2x > -4 \Leftrightarrow x > -2$; so $x \in (-2, \infty)$.

15. $1 - x \leq 2 \Leftrightarrow -x \leq 1 \Leftrightarrow x \geq -1$; so $x \in [-1, \infty)$.

17. $2x + 1 < 5x - 8 \Leftrightarrow 9 < 3x \Leftrightarrow 3 < x$; so $x \in (3, \infty)$.

19. $-1 < 2x - 5 < 7 \Leftrightarrow 4 < 2x < 12 \Leftrightarrow 2 < x < 6$; so $x \in (2, 6)$.

21. $0 \leq 1 - x < 1 \Leftrightarrow -1 \leq -x < 0 \Leftrightarrow 1 \geq x > 0$; so $x \in (0, 1]$.

23. $4x < 2x + 1 \le 3x + 2$. So $4x < 2x + 1 \Leftrightarrow 2x < 1 \Leftrightarrow x < \frac{1}{2}$, and $2x + 1 \le 3x + 2$ $\Leftrightarrow -1 \le x$. Thus $x \in [-1, \frac{1}{2})$.

25. $1 - x \ge 3 - 2x \ge x - 6$. So $1 - x \ge 3 - 2x \Leftrightarrow x \ge 2$, and $3 - 2x \ge x - 6 \Leftrightarrow 9 \ge 3x$ $\Leftrightarrow 3 \ge x$. Thus $x \in [2, 3]$.

27. $(x - 1)(x - 2) > 0$. Case (i): $x - 1 > 0 \Leftrightarrow x > 1$, and $x - 2 > 0 \Leftrightarrow x > 2$; so $x \in (2, \infty)$. Case (ii): $x - 1 < 0 \Leftrightarrow x < 1$, and $x - 2 < 0 \Leftrightarrow x < 2$; so $x \in (-\infty, 1)$. Solution set: $(-\infty, 1) \cup (2, \infty)$.

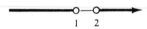

29. $2x^2 + x \le 1 \Leftrightarrow 2x^2 + x - 1 \le 0 \Leftrightarrow (2x - 1)(x + 1) \le 0$. Case (i): $2x - 1 \ge 0 \Leftrightarrow x \ge \frac{1}{2}$, and $x + 1 \le 0 \Leftrightarrow x \le -1$; which is impossible. Case (ii): $2x - 1 \le 0 \Leftrightarrow x \le \frac{1}{2}$, and $x + 1 \ge 0 \Leftrightarrow x \ge -1$; so $x \in [-1, \frac{1}{2}]$. Thus, the solution set is: $[-1, \frac{1}{2}]$.

31. $x^2 + x + 1 > 0 \Leftrightarrow x^2 + x + \frac{1}{4} + \frac{3}{4} > 0 \Leftrightarrow (x + \frac{1}{2})^2 + \frac{3}{4} > 0$. But since $(x + \frac{1}{2})^2 \ge 0$ for every real x, the original inequality will be true for all real x as well. Thus, the solution set is: $(-\infty, \infty)$.

33. $x^2 < 3 \Leftrightarrow x^2 - 3 < 0 \Leftrightarrow (x - \sqrt{3})(x + \sqrt{3}) < 0$. Case (i): $x > \sqrt{3}$ and $x < -\sqrt{3}$, which is impossible. Case (ii): $x < \sqrt{3}$ and $x > -\sqrt{3}$. Thus, the solution set is: $(-\sqrt{3}, \sqrt{3})$. [Another method: $x^2 < 3 \Leftrightarrow |x| < \sqrt{3} \Leftrightarrow -\sqrt{3} < x < \sqrt{3}$.]

35. $x^3 - x^2 \le 0 \Leftrightarrow x^2(x - 1) \le 0$. Since $x^2 \ge 0$ for all x, the inequality is satisfied when
     $x - 1 \le 0 \Leftrightarrow x \le 1$. Thus, the solution set is $(-\infty, 1]$.

37. $x^3 > x \Leftrightarrow x^3 - x > 0 \Leftrightarrow x(x^2 - 1) > 0 \Leftrightarrow x(x - 1)(x + 1) > 0$. Constructing a table:

Interval	x	x − 1	x + 1	x(x − 1)(x + 1)
x < −1	−	−	−	−
−1 < x < 0	−	−	+	+
0 < x < 1	+	−	+	−
x > 1	+	+	+	+

Since $x^3 > x$ when the last column is positive, the solution set is: $(-1, 0) \cup (1, \infty)$.

39. $1/x < 4$. This is clearly true for $x < 0$. So suppose $x > 0$. Then $1/x < 4 \Leftrightarrow 1 < 4x$
     $\Leftrightarrow \frac{1}{4} < x$. Thus, the solution set is: $(-\infty, 0) \cup (\frac{1}{4}, \infty)$.

41. Multiply both sides by x. Case (i): If $x > 0$, then $4/x < x \Leftrightarrow 4 < x^2 \Leftrightarrow 2 < x$.
     Case (ii): If $x < 0$, then $4/x < x \Leftrightarrow 4 > x^2 \Leftrightarrow -2 < x < 0$. Thus, the solution set is:
     $(-2, 0) \cup (2, \infty)$.

43. $\dfrac{2x + 1}{x - 5} < 3$. Case (i): If $x - 5 > 0$ (i.e., $x > 5$) then $2x + 1 < 3(x - 5) \Leftrightarrow 16 < x$, so
     $x \in (16, \infty)$. Case (ii): If $x - 5 < 0$ (i.e., $x < 5$) then $2x + 1 > 3(x - 5) \Leftrightarrow 16 > x$, so in
     this case $x \in (-\infty, 5)$. Combining the two cases, the solution set is: $(-\infty, 5) \cup (16, \infty)$.

45. $\frac{x^2-1}{x^2+1} \geq 0,$ since $x^2+1 > 0$ for all real x, this is equivalent to $x^2-1 \geq 0$

$\Leftrightarrow (x-1)(x+1) \geq 0.$ Case (i): $x \geq 1$ and $x \geq -1$, so $x \in [1,\infty).$ Case (ii): $x \leq 1$

and $x \leq -1$, so $x \in (-\infty,-1].$ Thus, the solution set is: $(-\infty,-1] \cup [1,\infty).$

[Another method: $x^2 \geq 1 \Leftrightarrow |x| \geq 1 \Leftrightarrow x \geq 1$ or $x \leq -1.$]

47. $C = \frac{5}{9}(F-32) \Rightarrow F = \frac{9}{5}C+32.$ So $50 \leq F \leq 95 \Rightarrow 50 \leq \frac{9}{5}C+32 \leq 95$

$\Rightarrow 18 \leq \frac{9}{5}C \leq 63 \Rightarrow 10 \leq C \leq 35.$ So the interval is $[10,35].$

49. (a) Let T represent the temperature in degrees Celsius and h the height in km.

T = 20 when h = 0 and T decreases by 10°C for every km.

Thus $T = 20-10h$ when $0 \leq h \leq 12.$

(b) From (a), $T = 20-10h \Rightarrow h = 2-T/10.$ So $0 \leq h \leq 5$

$\Rightarrow 0 \leq 2-T/10 \leq 5 \Rightarrow -2 \leq -T/10 \leq 3 \Rightarrow -30 \leq T \leq 20.$

Thus the range of temperature to be expected is $[-30,20].$

51. $|2x| = 3 \Leftrightarrow 2x = 3$ or $2x = -3 \Leftrightarrow x = \frac{3}{2}$ or $x = -\frac{3}{2}.$

53. $|x+3| = |2x+1| \Leftrightarrow x+3 = 2x+1$ or $x+3 = -(2x+1).$ In the first case $x = 2$, and

in the second case $3x = -4 \Leftrightarrow x = -\frac{4}{3}.$

55. By (6) Property 5, $|x| < 3 \Leftrightarrow -3 < x < 3$; so $x \in (-3,3).$

57. $|x-4| < 1 \Leftrightarrow -1 < x-4 < 1 \Leftrightarrow 3 < x < 5$; so $x \in (3,5).$

59. $|x+5| \geq 2 \Leftrightarrow x+5 \geq 2$ or $x+5 \leq -2 \Leftrightarrow x \geq -3$ or $x \leq -7$;

so $x \in (-\infty,-7] \cup [-3,\infty).$

61. $|2x-3| \leq 0.4 \Leftrightarrow -0.4 \leq 2x-3 \leq 0.4 \Leftrightarrow 2.6 \leq 2x \leq 3.4 \Leftrightarrow 1.3 \leq x \leq 1.7$;

so $x \in [1.3,1.7].$

63. $1 \leq |x| \leq 4.$ So either $1 \leq x \leq 4$ or $1 \leq -x \leq 4 \Leftrightarrow -1 \geq x \geq -4.$

Thus, $x \in [-4,-1] \cup [1,4].$

65. $|x| > |x-1|.$ Since $|x|, |x-1| \geq 0$; $|x| > |x-1| \Leftrightarrow |x|^2 > |x-1|^2$

$\Leftrightarrow x^2 > (x-1)^2 = x^2-2x+1 \Leftrightarrow 0 > -2x+1 \Leftrightarrow x > \frac{1}{2}$, so $x \in (\frac{1}{2},\infty).$

67. $\left|\frac{x}{2+x}\right| < 1 \Leftrightarrow \left(\frac{x}{2+x}\right)^2 < 1 \Leftrightarrow x^2 < (2+x)^2 \Leftrightarrow x^2 < 4+4x+x^2 \Leftrightarrow 0 < 4+4x$

$\Leftrightarrow -1 < x$; so $x \in (-1,\infty).$

69. $a(bx-c) \geq bc \Leftrightarrow bx-c \geq \frac{bc}{a} \Leftrightarrow bx \geq \frac{bc}{a}+c = \frac{bc+ac}{a} \Leftrightarrow x \geq \frac{bc+ac}{ab}.$

4

71. $ax + b < c \Leftrightarrow ax < c - b \Leftrightarrow x > \frac{c-b}{a}$ *(since a < 0)*.

73. $|(x + y) - 5| = |(x - 2) + (y - 3)| \leq |x - 2| + |y - 3| < 0.01 + 0.04 = 0.05$.

75. If $a < b$ then $a + a < a + b$ and $a + b < b + b$. Thus $2a < a + b < 2b$, so dividing by 2 gives: $a < \frac{a+b}{2} < b$.

77. $|ab| = \sqrt{(ab)^2} = \sqrt{a^2 b^2} = \sqrt{a^2}\sqrt{b^2} = |a||b|$

79. If $0 < a < b$ then $a \cdot a < a \cdot b$ and $a \cdot b < b \cdot b$ [using (2), Rule 3]. So $a^2 < ab < b^2$ and hence $a^2 < b^2$.

81. Observe that the sum, difference and product of two integers is always an integer. Let the rational numbers be represented by $r = \frac{m}{n}$ and $s = \frac{p}{q}$ (where m, n, p and q are integers with n and $q \neq 0$). Now, $r + s = \frac{m}{n} + \frac{p}{q} = \frac{mq + pn}{nq}$ but $mq + pn$ and nq are both integers so $\frac{mq + pn}{nq} = r + s$ is a rational number by the definition of a rational number. Similarly, $r - s = \frac{m}{n} - \frac{p}{q} = \frac{mq - pn}{nq}$ is a rational number. Finally, $r \cdot s = \frac{m}{n} \cdot \frac{p}{q} = \frac{mp}{nq}$ but mp and nq are both integers so $\frac{mp}{nq} = r \cdot s$ is a rational number by the definition of a rational number.

EXERCISES 2

1. From the Distance Formula (8) with $x_1 = 1$, $x_2 = 4$, $y_1 = 1$, $y_2 = 5$ we get the distance to be $\sqrt{(4-1)^2 + (5-1)^2} = \sqrt{3^2 + 4^2} = \sqrt{25} = 5$.

3. $\sqrt{(-1-6)^2 + (3-(-2))^2} = \sqrt{(-7)^2 + 5^2} = \sqrt{74}$.

5. $\sqrt{(4-2)^2 + (-7-5)^2} = \sqrt{2^2 + (-12)^2} = \sqrt{148} = 2\sqrt{37}$.

7. From (9), the slope is $\frac{11-5}{4-1} = \frac{6}{3} = 2$. 9. $m = \frac{-6-3}{-1-(-3)} = \frac{-9}{2}$

11. Since $|AC| = \sqrt{(-4-0)^2 + (3-2)^2} = \sqrt{(-4)^2 + 1^2} = \sqrt{17}$ and $|BC| = \sqrt{(-4-(-3))^2 + (3-(-1))^2} = \sqrt{(-1)^2 + 4^2} = \sqrt{17}$, the triangle has two sides of equal length and so is isosceles.

13. Label the points A, B, C and D respectively. Then:

$|AB| = \sqrt{(4-(-2))^2 + (6-9)^2} = \sqrt{6^2 + (-3)^2} = 3\sqrt{5}$,

$|BC| = \sqrt{(1-4)^2 + (0-6)^2} = \sqrt{(-3)^2 + (-6)^2} = 3\sqrt{5},$

$|CD| = \sqrt{(-5-1)^2 + (3-0)^2} = \sqrt{(-6)^2 + 3^2} = 3\sqrt{5},$ and

$|DA| = \sqrt{(-2-(-5))^2 + (9-3)^2} = \sqrt{3^2 + 6^2} = 3\sqrt{5}.$ So all sides are of equal length.

Moreover, $m_{AB} = \dfrac{6-9}{4-(-2)} = -\dfrac{1}{2},$ $m_{BC} = \dfrac{0-6}{1-4} = 2,$ $m_{CD} = \dfrac{3-0}{-5-1} = -\dfrac{1}{2},$

$m_{DA} = \dfrac{9-3}{-2-(-5)} = 2,$ so the sides are perpendicular. Thus, it is a square.

15. The slope of the line segment AB is $\dfrac{4-1}{7-1} = \dfrac{1}{2},$ the slope of CD is $\dfrac{7-10}{-1-5} = \dfrac{1}{2},$ the slope

of BC is $\dfrac{10-4}{5-7} = -3,$ and the slope of DA is $\dfrac{1-7}{1-(-1)} = -3.$ So AB is parallel to CD

and BC is parallel to DA. Hence ABCD is a parallelogram.

17. $x = 3$

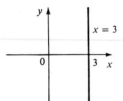

19. $xy = 0 \Leftrightarrow x = 0$ or $y = 0$

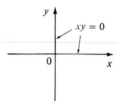

21. From (10), the equation of the line is $y - (-3) = 6(x-2)$ or $y = 6x - 15.$

23. $y - 7 = \dfrac{2}{3}(x-1)$ or $2x - 3y + 19 = 0.$

25. The slope is $m = \dfrac{6-1}{1-2} = -5,$ so the equation of the line is $y - 1 = -5(x-2)$ or
$5x + y = 11.$

27. From (11), the equation is $y = 3x - 2.$

29. Since the line passes through $(1,0)$ and $(0,-3),$ its slope is $m = \dfrac{-3-0}{0-1} = 3,$ so its
equation is $y = 3x - 3.$

31. Since $m = 0,$ $y - 5 = 0(x-4)$ or $y = 5.$

33. Putting the line $x + 2y = 6$ into slope-intercept form $y = -\dfrac{1}{2}(x) + 3,$ we see that this
line has slope $-\dfrac{1}{2}.$ So we want the line of slope $-\dfrac{1}{2}$ that passes through the point
$(1, -6):$ $y - (-6) = -\dfrac{1}{2}(x-1) \Leftrightarrow y = -\dfrac{1}{2}x - \dfrac{11}{2}$ or $x + 2y + 11 = 0.$

35. $2x + 5y + 8 = 0 \Leftrightarrow y = -\dfrac{2}{5}x - \dfrac{8}{5}.$ Since this line has slope $-\dfrac{2}{5},$ a line perpendicular to
it would have slope $\dfrac{5}{2},$ so the required line is: $y - (-2) = \dfrac{5}{2}(x - (-1)) \Leftrightarrow y = \dfrac{5}{2}x + \dfrac{1}{2}$ or
$5x - 2y + 1 = 0.$

37. $x + 3y = 0 \Leftrightarrow y = -\frac{1}{3}x$, so the slope is $-\frac{1}{3}$ and the y-intercept is 0.

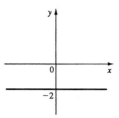

39. $y = -2$ is a horizontal line with slope 0 and y-intercept -2.

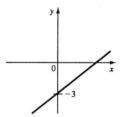

41. $3x - 4y = 12 \Leftrightarrow y = \frac{3}{4}x - 3$, so the slope is $\frac{3}{4}$ and the y-intercept is -3.

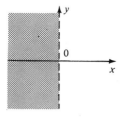

43. $\{ (x,y) \mid x < 0 \}$

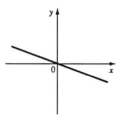

45. $\{ (x,y) \mid xy < 0 \} = \{(x,y) \mid x < 0 \text{ or } y < 0\}$

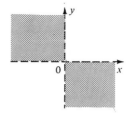

47. $\{(x,y) \mid |x| \le 2\} = \{(x,y) \mid -2 \le x \le 2\}$

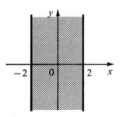

49. $\{(x,y) \mid 0 \le y \le 4, \ x \le 2\}$

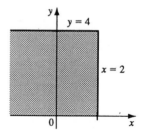

51. $\{(x,y) \mid 1 + x \le y \le 1 - 2x\}$

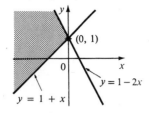

53. Let $P(0,y)$ be a point on the y-axis. The distance from P to $(5, -5)$ is
$$\sqrt{(5-0)^2 + (-5-y)^2} = \sqrt{5^2 + (y+5)^2}. \text{ The distance from P to } (1,1) \text{ is}$$
$$\sqrt{(1-0)^2 + (1-y)^2} = \sqrt{1^2 + (y-1)^2}. \text{ We want these distances to be equal:}$$
$$\sqrt{5^2 + (y+5)^2} = \sqrt{1^2 + (y-1)^2} \ \Leftrightarrow \ 5^2 + (y+5)^2 = 1^2 + (y-1)^2$$
$$\Leftrightarrow \ 25 + (y^2 + 10y + 25) = 1 + (y^2 - 2y + 1) \ \Leftrightarrow \ 12y = -48 \ \Leftrightarrow \ y = -4. \text{ So the}$$
desired point is $(0, -4)$.

55. Using the midpoint formula of Exercise 54, we get:

(a) $\left(\dfrac{1+7}{2}, \dfrac{3+15}{2}\right) = (4,9)$ (b) $\left(\dfrac{-1+8}{2}, \dfrac{6-12}{2}\right) = (\tfrac{7}{2}, -3)$.

57. $2x - y = 4 \ \Leftrightarrow \ y = 2x - 4 \ \Rightarrow \ m_1 = 2$ and $6x - 2y = 10 \ \Leftrightarrow \ 2y = 6x - 10 \ \Leftrightarrow \ y = 3x - 5$
$\Rightarrow \ m_2 = 3$. Since $m_1 \ne m_2$ the two lines are not parallel $(0.13(a))$. To find the point
of intersection: $2x - 4 = 3x - 5 \ \Leftrightarrow \ x = 1 \ \Rightarrow \ y = -2$. Thus, the point of intersection is
$(1, -2)$.

59. The slope of the segment AB is $\dfrac{-2-4}{7-1} = -1$, so its perpendicular bisector has slope 1.
The midpoint of AB is $\left(\dfrac{1+7}{2}, \dfrac{4-2}{2}\right) = (4,1)$, so the equation of the perpendicular
bisector is $y - 1 = 1(x-4)$ or $y = x - 3$.

61. (a) Since the x-intercept is a, the point $(a, 0)$ is on the line, and similarly, since the
y-intercept is b, $(0, b)$ is on the line. Hence the slope of the line is $m = \dfrac{b - 0}{0 - a} = -\dfrac{b}{a}$.
Substituting into $y = mx + b$ gives $y = -\dfrac{b}{a}x + b \Leftrightarrow y + \dfrac{b}{a}x = b \Leftrightarrow \dfrac{y}{b} + \dfrac{x}{a} = 1$

(b) Letting $a = 6$ and $b = -8$ gives $\dfrac{y}{-8} + \dfrac{x}{6} = 1 \Leftrightarrow 6y - 8x = -48$
$\Leftrightarrow 8x - 6y - 48 = 0 \Leftrightarrow 4x - 3y - 24 = 0$

63. (a) Let $d =$ distance traveled (in miles) and $t =$ time elapsed (in hours). At $t = 0$, $d = 0$
and at $t = 50$ minutes $= 50 \cdot \dfrac{1}{60} = \dfrac{5}{6}$ h, $d = 40$. Thus we have two points $(0,0)$ and
$(\frac{5}{6}, 40)$ so $m = \dfrac{40 - 0}{\frac{5}{6} - 0} = 48$ and so $d = 48t$.

(b)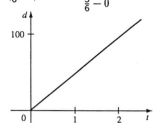

(c) The slope is 48 and represents Jason and
Debbie's speed in mi/h.

65. If L_1 and L_2 are perpendicular, $\theta = 90°$ and $\tan \theta$ is undefined. For $\dfrac{m_2 - m_1}{1 + m_1 m_2}$ to be
undefined, $1 + m_1 m_2 = 0 \Leftrightarrow m_1 m_2 = -1$.

EXERCISES 3

1. From (14), the equation is $(x - 3)^2 + (y + 1)^2 = 25$.

3. The equation has the form $x^2 + y^2 = r^2$. Since $(4, 7)$ lies on the circle, we have
$4^2 + 7^2 = r^2 \Rightarrow r^2 = 65$. So the required equation is $x^2 + y^2 = 65$.

5. $x^2 + y^2 - 4x + 10y + 13 = 0 \Leftrightarrow x^2 - 4x + y^2 + 10y = -13$
$\Leftrightarrow (x^2 - 4x + 4) + (y^2 + 10y + 25) = -13 + 4 + 25 = 16 \Leftrightarrow (x - 2)^2 + (y + 5)^2 = 4^2$.
Thus, we have a circle with center $(2, -5)$ and radius 4.

7. $x^2 + y^2 + x = 0 \Leftrightarrow (x^2 + x + \frac{1}{4}) + y^2 = \frac{1}{4} \Leftrightarrow (x + \frac{1}{2})^2 + y^2 = (\frac{1}{2})^2$.
Thus, we have a circle with center $(-\frac{1}{2}, 0)$ and radius $\frac{1}{2}$.

9. $2x^2 + 2y^2 - x + y = 1 \Leftrightarrow 2(x^2 - \frac{1}{2}x + \frac{1}{16}) + 2(y^2 + \frac{1}{2}y + \frac{1}{16}) = 1 + \frac{1}{8} + \frac{1}{8}$
$\Leftrightarrow 2(x - \frac{1}{4})^2 + 2(y + \frac{1}{4})^2 = \frac{5}{4} \Leftrightarrow (x - \frac{1}{4})^2 + (y + \frac{1}{4})^2 = \frac{5}{8}$. Thus, we have a circle with
center $(\frac{1}{4}, -\frac{1}{4})$ and radius $\dfrac{\sqrt{5}}{2\sqrt{2}} = \dfrac{\sqrt{10}}{4}$.

11. $y = -x^2$. Parabola.

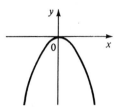

13. $x^2 + 4y^2 = 16 \Leftrightarrow \frac{x^2}{16} + \frac{y^2}{4} = 1$. Ellipse.

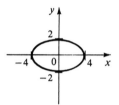

15. $16x^2 - 25y^2 = 400 \Leftrightarrow \frac{x^2}{25} - \frac{y^2}{16} = 1$. Hyperbola.

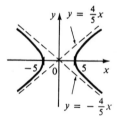

17. $4x^2 + y^2 = 1 \Leftrightarrow \frac{x^2}{\frac{1}{4}} + y^2 = 1$. Ellipse.

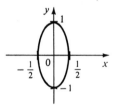

19. $x = y^2 - 1$. Parabola with vertex at $(-1, 0)$.

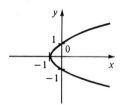

21.　$9y^2 - x^2 = 9 \Leftrightarrow y^2 - \frac{x^2}{9} = 1$.　Hyperbola.

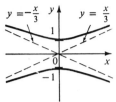

23.　$xy = 4$.　Hyperbola.

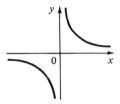

25.　$9(x-1)^2 + 4(y-2)^2 = 36 \Leftrightarrow \frac{(x-1)^2}{4} + \frac{(y-2)^2}{9} = 1$.　Ellipse centered at $(1,2)$.

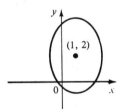

27.　$y = x^2 - 6x + 13 = (x^2 - 6x + 9) + 4 = (x-3)^2 + 4$.　Parabola with vertex at $(3,4)$.

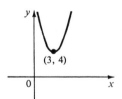

29.　$x = -y^2 + 4$.　Parabola with vertex at $(4,0)$.

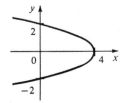

31. $x^2 + 4y^2 - 6x + 5 = 0 \Leftrightarrow (x^2 - 6x + 9) + 4y^2 = -5 + 9 = 4 \Leftrightarrow \frac{(x-3)^2}{4} + y^2 = 1.$

Ellipse centered at $(3, 0)$.

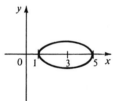

33. $y = 3x$ and $y = x^2$ intersect where $3x = x^2 \Leftrightarrow 0 = x^2 - 3x = x(x-3)$ or at $(0,0)$ and $(3, 9)$.

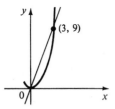

35. The parabola must have an equation of the form $y = a(x-1)^2 - 1$. Substituting $x = 3$ and $y = 3$ into the equation gives $3 = a(3-1)^2 - 1$ so $a = 1$, and the equation is $y = (x-1)^2 - 1 = x^2 - 2x$. (*Note that using the other point $(-1, 3)$ would have given the same value for a and hence the same equation.*)

37. $\{(x,y)\,|\,x^2 + y^2 \le 1\}$

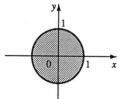

39. $\{(x,y)\,|\,y \ge x^2 - 1\}$

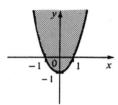

EXERCISES 4

1. $f(x) = x^2 - 3x + 2$, so $f(1) = 1^2 - 3(1) + 2 = 0$, $f(-2) = (-2)^2 - 3(-2) + 2 = 12$,
    $f(\frac{1}{2}) = (\frac{1}{2})^2 - 3(\frac{1}{2}) + 2 = \frac{3}{4}$, $f(\sqrt{5}) = (\sqrt{5})^2 - 3(\sqrt{5}) + 2 = 7 - 3\sqrt{5}$, $f(a) = a^2 - 3a + 2$,
    $f(-a) = (-a)^2 - 3(-a) + 2 = a^2 + 3a + 2$.

3. $g(x) = \frac{1-x}{1+x}$, so $g(2) = \frac{1-2}{1+2} = -\frac{1}{3}$, $g(-2) = \frac{1-(-2)}{1+(-2)} = -3$, $g(\pi) = \frac{1-\pi}{1+\pi}$,
    $g(a) = \frac{1-a}{1+a}$, $g(a-1) = \frac{1-(a-1)}{1+(a-1)} = \frac{2-a}{a}$, $g(-a) = \frac{1-(-a)}{1+(-a)} = \frac{1+a}{1-a}$.

5. $f(x) = x - x^2$, so $f(2+h) = 2 + h - (2+h)^2 = 2 + h - 4 - 4h - h^2 = -(h^2 + 3h + 2)$,
    $f(x+h) = x + h - (x+h)^2 = x + h - x^2 - 2xh - h^2$,
    $\frac{f(x+h) - f(x)}{h} = \frac{x + h - x^2 - 2xh - h^2 - x + x^2}{h} = \frac{h - 2xh - h^2}{h} = 1 - 2x - h.$

7. $f(x) = \sqrt{x}$, $0 \leq x \leq 4$.

    Machine Diagram   Arrow Diagram   Graph

    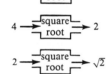      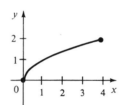

9. The range of f is the set of values of f, $\{0, 1, 2, 4\}$.

    Arrow Diagram   Graph

    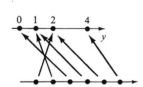   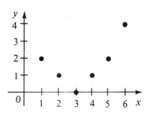

11. $f(x) = 6 - 4x$, $-2 \leq x \leq 3$. The domain is $[-2, 3]$. If $-2 \leq x \leq 3$, then
    $14 = 6 - 4(-2) \geq 6 - 4x \geq 6 - 4(3) = -6$, so the range is $[-6, 14]$.

13. $g(x) = \frac{2}{3x - 5}$. This is defined when $3x - 5 \neq 0$, so the domain of f is
    $\{x \mid x \neq \frac{5}{3}\} = (-\infty, \frac{5}{3}) \cup (\frac{5}{3}, \infty)$, and the range is $\{y \mid y \neq 0\} = (-\infty, 0) \cup (0, \infty)$.

15. $h(x) = \sqrt{2x - 5}$ is defined when $2x - 5 \geq 0$ or $x \geq \frac{5}{2}$, so the domain is $[\frac{5}{2}, \infty)$ and the
    range is $[0, \infty)$.

17. $F(x) = \sqrt{1-x^2}$ is defined when $1-x^2 \geq 0 \Leftrightarrow x^2 \leq 1 \Leftrightarrow |x| \leq 1 \Leftrightarrow -1 \leq x \leq 1$, so the domain is $[-1,1]$ and the range is $[0,1]$.

19. $f(x) = \dfrac{x+2}{x^2-1}$ is defined for all x except when $x^2 - 1 = 0 \Leftrightarrow x = 1$ or $x = -1$, so the domain is $\{\, x \mid x \neq \pm 1 \,\}$.

21. $g(x) = \sqrt[4]{x^2 - 6x}$ is defined when $0 \leq x^2 - 6x = x(x-6) \Leftrightarrow x \geq 6$ or $x \leq 0$, so the domain is $(-\infty, 0] \cup [6, \infty)$.

23. $\phi(x) = \sqrt{\dfrac{x}{\pi - x}}$ is defined when $\dfrac{x}{\pi - x} \geq 0$. So either $x \leq 0$ and $\pi - x < 0$ ($\Leftrightarrow x > \pi$), which is impossible, or $x \geq 0$ and $\pi - x > 0$ ($\Leftrightarrow x < \pi$), and so the domain is $[0, \pi)$.

25. $f(t) = \sqrt[3]{t-1}$ is defined for every t, since every real number has a cube root. The domain is the set of all real numbers.

27. $f(x) = 2$. Domain is R.

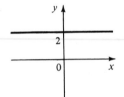

29. $f(x) = 3 - 2x$. Domain is R.

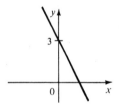

31. $f(x) = -x^2$. Domain is R.

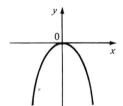

33. $f(x) = x^2 + 2x - 1 = (x^2 + 2x + 1) - 2 = (x + 1)^2 - 2$, so the graph is a parabola with vertex at $(-1, -2)$. The domain is R.

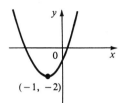

35. $g(x) = x^4$. Domain is R.

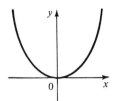

37. $g(x) = \sqrt{-x}$. Domain is $\{\, x \mid -x \geq 0 \,\} = (-\infty, 0]$.

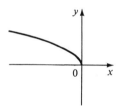

39. $h(x) = \sqrt{4 - x^2}$. Now $y = \sqrt{4 - x^2} \Rightarrow y^2 = 4 - x^2 \Leftrightarrow x^2 + y^2 = 4$, so the graph is the top half of a circle of radius 2. The domain is $\{\, x \mid 4 - x^2 \geq 0 \,\} = [-2, 2]$.

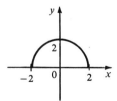

41. $F(x) = \frac{1}{x}$. Domain is $\{\, x \mid x \neq 0 \,\}$.

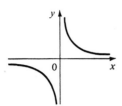

43.
$$G(x) = |x| + x = \begin{cases} 2x, & x \geq 0 \\ 0, & x < 0 \end{cases}$$
Domain is R.

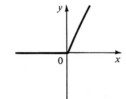

45.
$$H(x) = |2x| = \begin{cases} 2x, & x \geq 0 \\ -2x, & x < 0 \end{cases}$$
Domain is R.

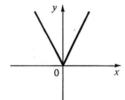

47.
$$f(x) = \frac{x}{|x|} = \begin{cases} 1, & x > 0 \\ -1, & x < 0 \end{cases}$$
Domain is $\{\, x \mid x \neq 0 \,\}$.

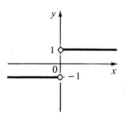

49. $f(x) = \frac{x^2-1}{x-1} = \frac{(x+1)(x-1)}{x-1}$, so for $x \neq 1$, $f(x) = x+1$. Domain is $\{\, x \mid x \neq 1 \,\}$.

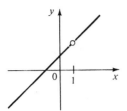

NOTE: In each of Problems 51-61, the domain is all of R.

51.
$$f(x) = \begin{cases} x+1, & x \neq 1 \\ 1, & x = 1 \end{cases}$$

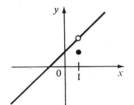

53.
$$f(x) = \begin{cases} 0, & x < 2 \\ 1, & x \geq 2 \end{cases}$$

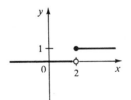

55.
$$f(x) = \begin{cases} x, & x \leq 0 \\ x+1, & x > 0 \end{cases}$$

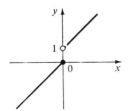

57.
$$f(x) = \begin{cases} -1, & x < -1 \\ x, & -1 \le x \le 1 \\ 1, & x > 1 \end{cases}$$

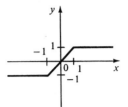

59.
$$f(x) = \begin{cases} x+2, & x \le -1 \\ x^2, & x > -1 \end{cases}$$

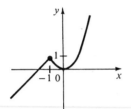

61.
$$f(x) = \begin{cases} -1, & x \le -1 \\ 3x+2, & -1 < x < 1 \\ 7-2x, & x \ge 1 \end{cases}$$

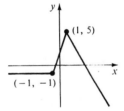

63. Yes, the curve is the graph of a function. The domain is $[-3, 2]$ and the range is $[-2, 2]$.

65. No, this is not the graph of a function since for $x = -1$ there are infinitely many points on the curve.

67. The slope of this line segment is $\dfrac{-6-1}{4-(-2)} = -\dfrac{7}{6}$, so its equation is $y - 1 = -\dfrac{7}{6}(x+2)$. The function is $f(x) = -\dfrac{7}{6}x - \dfrac{4}{3}$, $-2 \le x \le 4$.

69. $x + (y-1)^2 = 0 \Leftrightarrow y - 1 = \pm\sqrt{-x}$. The bottom half is given by the function $f(x) = 1 - \sqrt{-x}$, $x \le 0$.

71. Let the length and width of the rectangle be L and W respectively. Then the perimeter is $2L + 2W = 20$, and the area is $A = LW$. Solving the first equation for W in terms of L gives $W = \frac{20 - 2L}{2} = 10 - L$. Thus $A(L) = L(10 - L) = 10L - L^2$. Since lengths are positive, the domain of A is $0 < L < 10$.

73. Let the length of a side of the equilateral triangle be x. Then by Pythagoras' Theorem the height y of the triangle satisfies $y^2 + \left(\frac{x}{2}\right)^2 = x^2$, so that $y = \frac{\sqrt{3}}{2}x$. Thus the area of the triangle is $A = \frac{1}{2}xy$ and so $A(x) = \frac{1}{2}\left(\frac{\sqrt{3}}{2}x\right)x = \frac{\sqrt{3}x^2}{4}$, with domain $x > 0$.

75. Let each side of the base of the box have length x, and let the height of the box be h. Since the volume is 2, we know that $2 = hx^2$, so that $h = 2/x^2$, and the surface area is $S = x^2 + 4xh$. Thus $S(x) = x^2 + 4x(2/x^2) = x^2 + 8/x$, with domain $x > 0$.

77. The height of the box is x and the length and width are $L = 20 - 2x$, $W = 12 - 2x$. $V = LWx$ and so $V(x) = (20 - 2x)(12 - 2x)(x) = 4(10 - x)(6 - x)(x)$
$= 4x(60 - 16x + x^2) = 4x^3 - 64x^2 + 240x$, with domain $0 < x < 6$.

79. $f(-x) = \dfrac{1}{(-x)^2} = \dfrac{1}{x^2} = f(x)$, so f is an even function.

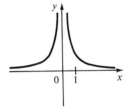

81. $f(-x) = (-x)^2 + (-x) = x^2 - x$. Since this is neither f(x) nor $-f(x)$, the function f is neither even nor odd.

83. $f(-x) = (-x)^3 - (-x) = -x^3 + x = -(x^3 - x) = -f(x)$, so f is odd.

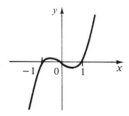

EXERCISES 5

In this problem set, "D = " stands for "the domain of the function is".

1. $f(x) = x^2 - x$, $g(x) = x + 5$. $(f+g)(x) = x^2 - x + x + 5 = x^2 + 5$, D = R.

   $(f-g)(x) = x^2 - x - (x+5) = x^2 - 2x - 5$, D = R.

   $(fg)(x) = (x^2 - x)(x+5) = x^3 + 4x^2 - 5x$, D = R. $(f/g)(x) = (x^2 - x)/(x+5)$,

   $D = \{x \mid x \neq -5\}$.

3. $f(x) = \sqrt{1+x}$, $D = [-1, \infty)$; $g(x) = \sqrt{1-x}$, $D = (-\infty, 1]$. $(f+g)(x) = \sqrt{1+x} + \sqrt{1-x}$,

   $D = (-\infty, 1] \cap [-1, \infty) = [-1, 1]$. $(f-g)(x) = \sqrt{1+x} - \sqrt{1-x}$, $D = [-1, 1]$.

   $(fg)(x) = \sqrt{1+x} \cdot \sqrt{1-x} = \sqrt{1-x^2}$, $D = [-1, 1]$. $(f/g)(x) = \sqrt{1+x}/\sqrt{1-x}$, $D = [-1, 1)$.

5. $f(x) = \sqrt{x}$, $D = [0, \infty)$; $g(x) = \sqrt[3]{x}$, $D = R$. $(f+g)(x) = \sqrt{x} + \sqrt[3]{x}$, $D = [0, \infty)$.

   $(f-g)(x) = \sqrt{x} - \sqrt[3]{x}$, $D = [0, \infty)$. $(fg)(x) = \sqrt{x} \cdot \sqrt[3]{x} = x^{5/6}$, $D = [0, \infty)$.

   $(f/g)(x) = \sqrt{x}/\sqrt[3]{x} = x^{1/6}$, $D = (0, \infty)$.

7. $F(x) = (\sqrt{4-x} + \sqrt{3+x})/(x^2 - 2)$. Domain of the numerator is

   $(-\infty, 4] \cap [-3, \infty) = [-3, 4]$. Since the denominator is 0 when $x = \pm\sqrt{2}$, the domain of

   F is $[-3, -\sqrt{2}) \cup (-\sqrt{2}, \sqrt{2}) \cup (\sqrt{2}, 4]$.

9. $f(x) = x^3$, $g(x) = 1$.   11. $f(x) = x$, $g(x) = \frac{1}{x}$.

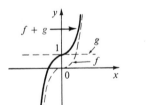 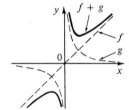

13. $f(x) = 2x + 3$, $g(x) = 4x - 1$. Since both f and g have domain and range R, so will all

    their composite functions.

    $(f \circ g)(x) = f(g(x)) = f(4x - 1) = 2(4x - 1) + 3 = 8x + 1$.

    $(g \circ f)(x) = g(f(x)) = g(2x + 3) = 4(2x + 3) - 1 = 8x + 11$.

    $(f \circ f)(x) = f(f(x)) = f(2x + 3) = 2(2x + 3) + 3 = 4x + 9$.

    $(g \circ g)(x) = g(g(x)) = g(4x - 1) = 4(4x - 1) - 1 = 16x - 5$.

15. $f(x) = 2x^2 - x$, $g(x) = 3x + 2$. D = R for both f and g, and hence for their composites

    as well.

    $(f \circ g)(x) = f(g(x)) = f(3x + 2) = 2(3x + 2)^2 - (3x + 2) = 18x^2 + 21x + 6$.

    $(g \circ f)(x) = g(f(x)) = g(2x^2 - x) = 3(2x^2 - x) + 2 = 6x^2 - 3x + 2$.

$(f \circ f)(x) = f(f(x)) = f(2x^2 - x) = 2(2x^2 - x)^2 - (2x^2 - x) = 8x^4 - 8x^3 + x.$

$(g \circ g)(x) = g(g(x)) = g(3x + 2) = 3(3x + 2) + 2 = 9x + 8.$

17. $f(x) = 1/x$, $D = \{x \mid x \neq 0\}$; $g(x) = x^3 + 2x$, $D = R$.

$(f \circ g)(x) = f(g(x)) = f(x^3 + 2x) = 1/(x^3 + 2x)$, $D = \{x \mid x^3 + 2x \neq 0\} = \{x \mid x \neq 0\}$.

$(g \circ f)(x) = g(f(x)) = g(1/x) = 1/x^3 + 2/x$, $D = \{x \mid x \neq 0\}$.

$(f \circ f)(x) = f(f(x)) = f(1/x) = \dfrac{1}{1/x} = x$, $D = \{x \mid x \neq 0\}$.

$(g \circ g)(x) = g(g(x)) = g(x^3 + 2x) = (x^3 + 2x)^3 + 2(x^3 + 2x)$

$= x^9 + 6x^7 + 12x^5 + 10x^3 + 4x$, $D = R$.

19. $f(x) = \sqrt[3]{x}$, $D = R$; $g(x) = 1 - \sqrt{x}$, $D = [0, \infty)$.

$(f \circ g)(x) = f(g(x)) = f(1 - \sqrt{x}) = \sqrt[3]{1 - \sqrt{x}}$, $D = [0, \infty)$.

$(g \circ f)(x) = g(f(x)) = g(\sqrt[3]{x}) = 1 - x^{1/6}$, $D = [0, \infty)$.

$(f \circ f)(x) = f(f(x)) = f(\sqrt[3]{x}) = x^{1/9}$, $D = R$.

$(g \circ g)(x) = g(g(x)) = g(1 - \sqrt{x}) = 1 - \sqrt{1 - \sqrt{x}}$,

$D = \{x \geq 0 \mid 1 - \sqrt{x} \geq 0\} = [0, 1]$.

21. $f(x) = \dfrac{x + 2}{2x + 1}$, $D = \{x \mid x \neq -\frac{1}{2}\}$; $g(x) = \dfrac{x}{x - 2}$, $D = \{x \mid x \neq 2\}$.

$(f \circ g)(x) = f(g(x)) = f\!\left(\dfrac{x}{x - 2}\right) = \dfrac{x/(x - 2) + 2}{2x/(x - 2) + 1} = \dfrac{3x - 4}{3x - 2}$, $D = \{x \mid x \neq 2, \frac{2}{3}\}$.

$(g \circ f)(x) = g(f(x)) = g\!\left(\dfrac{x + 2}{2x + 1}\right) = \dfrac{(x + 2)/(2x + 1)}{(x + 2)/(2x + 1) - 2} = \dfrac{-x - 2}{3x}$,

$D = \{x \mid x \neq 0, -\frac{1}{2}\}$.

$(f \circ f)(x) = f(f(x)) = f\!\left(\dfrac{x + 2}{2x + 1}\right) = \dfrac{(x + 2)/(2x + 1) + 2}{2(x + 2)/(2x + 1) + 1} = \dfrac{5x + 4}{4x + 5}$,

$D = \{x \mid x \neq -\frac{1}{2}, -\frac{5}{4}\}$. $(g \circ g)(x) = g(g(x)) = g\!\left(\dfrac{x}{x - 2}\right) = \dfrac{x/(x - 2)}{x/(x - 2) - 2} = \dfrac{x}{4 - x}$,

$D = \{x \mid x \neq 2, 4\}$.

23. $(f \circ g \circ h)(x) = f(g(h(x))) = f(g(x - 1)) = f(\sqrt{x - 1}) = \sqrt{x - 1} - 1.$

25. $(f \circ g \circ h)(x) = f(g(h(x))) = f(g(\sqrt{x})) = f(\sqrt{x} - 5) = (\sqrt{x} - 5)^4 + 1.$

27. Let $g(x) = x - 9$ and $f(x) = x^5$. Then $(f \circ g)(x) = (x - 9)^5 = F(x)$.

29. Let $g(x) = x^2$ and $f(x) = \dfrac{x}{x + 4}$. Then $(f \circ g)(x) = \dfrac{x^2}{x^2 + 4} = G(x)$.

31. Let $h(x) = x^2$, $g(x) = x + 1$ and $f(x) = \frac{1}{x}$. Then $(f \circ g \circ h)(x) = \dfrac{1}{x^2 + 1} = H(x)$.

33. Let r be the radius of the ripple in cm. The area of the ripple is $A = \pi r^2$ but, as a function of time, $r = 60t$. Thus, $A = \pi (60t)^2 = 3600 \pi t^2$.

35. We need a function g so that $f(g(x)) = 3(g(x)) + 5 = h(x) = 3x^2 + 3x + 2$
$= 3(x^2 + x) + 2 = 3(x^2 + x - 1) + 5$. So we see that $g(x) = x^2 + x - 1$.

37. The function $g(x) = x$ has domain $(-\infty, \infty)$. However the function fof for $f(x) = \frac{1}{x}$ has for its domain $(-\infty, 0) \cup (0, \infty)$ even though the rule is the same: $(f \circ f)(x) = f(\frac{1}{x}) = x$.

EXERCISES 6

1. $y = x^8$

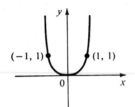

3. $y = -\frac{1}{x}$

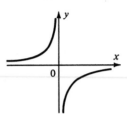

5. $y = 2 \sin x$

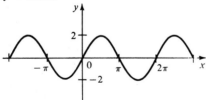

7. $y = (x-1)^3 + 2$

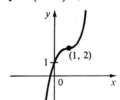

9. $y = 1 + \sqrt[6]{-x}$

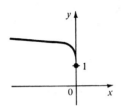

11. $y = \cos \frac{x}{2}$

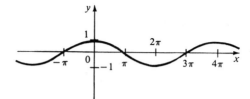

13. $y = \dfrac{1}{x-3}$

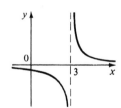

15. $y = \frac{1}{3}\sin\left(x - \frac{\pi}{6}\right)$

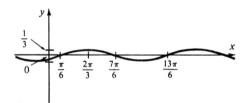

17. $y = 1 + 2x - x^2 = -(x-1)^2 + 2$

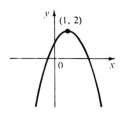

19. $y = 2 - \sqrt{x+1}$

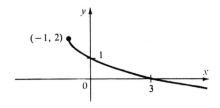

21. $y = x^2 - 2x = (x-1)^2 - 1$

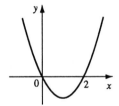

$\Rightarrow$

$y = \left|x^2 - 2x\right|.$

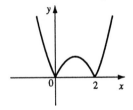

23. $y = \left|\,|x| - 1\,\right|.$

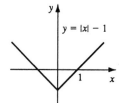

25. (a)   To obtain $y = f(|x|)$, the portion of $y = f(x)$ right of the y–axis is reflected in the y–axis.

(b) $y = \sin|x|.$

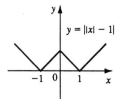

REVIEW EXERCISES

1. $2 + 5x \leq 9 - 2x \iff 7x \leq 7 \iff x \leq 1$; so $x \in (-\infty, 1]$.

3. $|x + 3| < 7 \iff -7 < x + 3 < 7 \iff -10 < x < 4$; so $x \in (-10, 4)$.

5. $x^2 - 5x + 6 > 0 \iff (x - 3)(x - 2) > 0$. Case (i): $x > 3$ and $x > 2$, so $x \in (3, \infty)$.
    Case (ii): $x < 3$ and $x < 2$, so $x \in (-\infty, 2)$. Thus, the solution set is:
    $(-\infty, 2) \cup (3, \infty)$.

7. $\sqrt{(-4-4)^2 + (-4-2)^2} = \sqrt{(-8)^2 + (-6)^2} = \sqrt{100} = 10$.

9. $(x - 2)^2 + (y - 1)^2 = 3^2 = 9$.

11. $x^2 + 2x + y^2 - 8y = -8 \iff (x^2 + 2x + 1) + (y^2 - 8y + 16) = -8 + 1 + 16$
    $\iff (x + 1)^2 + (y - 4)^2 = 9 = 3^2$. Center $(-1, 4)$, radius 3.

13. Slope $m = \dfrac{-1 - (-6)}{2 - (-1)} = \frac{5}{3}$, so $y - (-1) = \frac{5}{3}(x - 2)$ or $5x - 3y = 13$.

15. $x + 2y = 1 \iff y = -\frac{1}{2}x + \frac{1}{2}$, so the slope is $-\frac{1}{2}$. So $y - 3 = -\frac{1}{2}(x - 2)$ or $x + 2y = 8$.

17. Slope $m = \dfrac{0 + \frac{5}{2}}{\frac{3}{2} - 0} = \frac{5}{3}$. So $y = \frac{5}{3}x - \frac{5}{2}$ or $10x - 6y = 15$.

    OR: From Section 2 Exercise 61(a), with $a = \frac{3}{2}$ and $b = -\frac{5}{2}$:

    $\dfrac{x}{3/2} + \dfrac{y}{-5/2} = 1 \Rightarrow 10x - 6y = 15$.

19. $y = 8 - 2x^2$, parabola.

21. $x^2 - 4y^2 = 4 \iff \dfrac{x^2}{4} - y^2 = 1$,
    hyperbola.

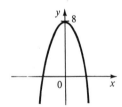

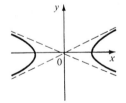

23. $2x^2 + y^2 - 16x + 30 = 0 \Leftrightarrow 2(x^2 - 8x) + y^2 = -30$

$\Leftrightarrow 2(x^2 - 8x + 16) + y^2 = -30 + 32 = 2 \Leftrightarrow 2(x - 4)^2 + y^2 = 2$

$\Leftrightarrow (x - 4)^2 + \dfrac{y^2}{2} = 1.$ Ellipse.

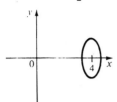

25. $f(x) = 1 + \sqrt{x - 1}$, so $f(5) = 1 + \sqrt{5 - 1} = 3$, $f(9) = 1 + \sqrt{9 - 1} = 1 + 2\sqrt{2}$,

$f(-x) = 1 + \sqrt{-x - 1}$, $f(x^2) = 1 + \sqrt{x^2 - 1}$, $\left[f(x)\right]^2 = \left[1 + \sqrt{x - 1}\right]^2 = x + 2\sqrt{x - 1}$.

27. $g(x)$ is defined unless the denominator is 0. $x^2 + x - 1 = 0$

$\Leftrightarrow x = \dfrac{-1 \pm \sqrt{1^2 - 4(-1)}}{2} = \dfrac{-1 \pm \sqrt{5}}{2}.$ Domain is $\left\{ x \mid x \neq \dfrac{-1 \pm \sqrt{5}}{2} \right\}.$

29. $f(x) = -1.$

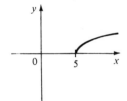

31. $g(x) = x^2 + 2.$

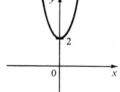

33. $h(x) = \sqrt{x - 5}.$

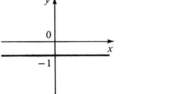

35. $y = -\sin 2x.$

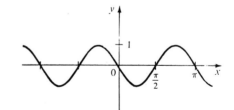

37. $y = x^2 - 2|x|$. Consider $x \geq 0 \Rightarrow y = x^2 - 2x \Leftrightarrow y = (x-1)^2 - 1$; a parabola with vertex $(1, -1)$. Since $f(-x) = (-x)^2 - 2|-x| = x^2 - 2x = f(x)$, we just reflect in the y-axis to complete the graph.

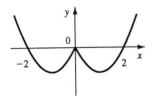

39. $f(x) = x^2$, $g(x) = x + 2$; both have domain R.

   (a) $(f+g)(x) = x^2 + x + 2$. Domain is R.

   (b) $(f/g)(x) = x^2/(x+2)$. Domain is $\{x \mid x \neq -2\}$.

   (c) $(f \circ g)(x) = f(g(x)) = f(x+2) = (x+2)^2$. Domain is R.

   (d) $(g \circ f)(x) = g(f(x)) = g(x^2) = x^2 + 2$. Domain is R.

41. Let $f(x) = \sqrt{x}$ and $g(x) = x^2 + x + 9$. Then $(f \circ g)(x) = \sqrt{x^2 + x + 9} = h(x)$.

43. $A = \pi r^2$ but $C = 2\pi r \Rightarrow r = \frac{C}{2\pi}$. Thus $A(C) = \pi \left(\frac{C}{2\pi}\right)^2 = \frac{C^2}{4\pi}$.

CHAPTER ONE

EXERCISES 1.1

1. For the curve $f(x) = 1 + x + x^2$ and the point $P(1, 3)$:

    (a)

x	Q	m_{PQ}
2	$(2, 7)$	4
1.5	$(1.5, 4.75)$	3.5
1.1	$(1.1, 3.31)$	3.1
1.01	$(1.01, 3.0301)$	3.01
1.001	$(1.001, 3.003001)$	3.001
0	$(0, 1)$	2
0.5	$(0.5, 1.75)$	2.5
0.9	$(0.9, 2.71)$	2.9
0.99	$(0.99, 2.9701)$	2.99
0.999	$(0.999, 2.997001)$	2.999

    (b) The slope appears to be 3.

    (c) $y - 3 = 3(x - 1)$ or $y = 3x$.

3. For the curve $y = \sqrt{x}$ and the point $P(4, 2)$:

    (a)

x	Q	m_{PQ}
5	$(5, 2.236068)$	0.236068
4.5	$(4.5, 2.121320)$	0.242641
4.1	$(4.1, 2.024846)$	0.248457
4.01	$(4.01, 2.002498)$	0.249844
4.001	$(4.001, 2.000250)$	0.249984
3	$(3, 1.732051)$	0.267949
3.5	$(3.5, 1.870829)$	0.258343
3.9	$(3.9, 1.974842)$	0.251582
3.99	$(3.99, 1.997498)$	0.250156
3.999	$(3.999, 1.999750)$	0.250016

    (b) The slope appears to be 1/4.
    (c) $y - 2 = (1/4)(x - 4)$ or
    $x - 4y + 4 = 0$

5. (a) At $t = 2$, $y = 40(2) - 16(2)^2 = 16$. The average velocity between times 2 and

    $2 + h$ is $\dfrac{40(2 + h) - 16(2 + h)^2 - 16}{h} = \dfrac{-24h - 16h^2}{h} = -24 - 16h$, if $h \neq 0$.

    (i) $h = 0.5$, -32 ft/s.   (ii) $h = 0.1$, -25.6 ft/s.

    (iii) $h = 0.05$, -24.8 ft/s.   (iv) $h = 0.01$, -24.16 ft/s.

    (b) The instantaneous velocity when $t = 2$ is -24 ft/s.

7. (a) The average velocity between times 0 and h is $\dfrac{s(h) - s(0)}{h} = \dfrac{h^2 + h - 0}{h} = h + 1$.

    (i) $[0, 2]$: $2 + 1 = 3$ m/s.   (ii) $[0, 1]$: $1 + 1 = 2$ m/s.

    (iii) $[0, 0.5]$: $0.5 + 1 = 1.5$ m/s.   (iv) $[0, 0.1]$: $0.1 + 1 = 1.1$ m/s.

    (b) As h approaches 0, the velocity approaches 1 m/s.

(c), (d)

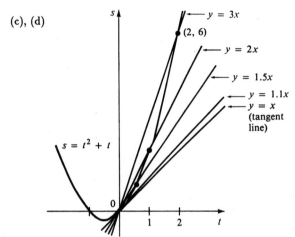

EXERCISES 1.2

1.   (a) $\lim_{x \to 1} f(x) = 3$ (b) $\lim_{x \to 3^-} f(x) = 2$ (c) $\lim_{x \to 3^+} f(x) = -2$

(d) $\lim_{x \to 3} f(x)$ doesn't exist   (e) $f(3) = 1$ (f) $\lim_{x \to -2^-} f(x) = -1$

(g) $\lim_{x \to -2^+} f(x) = -1$ (h) $\lim_{x \to -2} f(x) = -1$ (i) $f(-2) = -3$

3.   (a) $\lim_{x \to 3} f(x) = 2$ (b) $\lim_{x \to 1} f(x) = -1$ (c) $\lim_{x \to -3} f(x) = 1$

(d) $\lim_{x \to 2^-} f(x) = 1$ (e) $\lim_{x \to 2^+} f(x) = 2$ (f) $\lim_{x \to 2} f(x)$ doesn't exist

5.   (a)

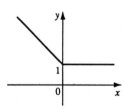

(b) (i) $\lim_{x \to 0^-} f(x) = 1$ (ii) $\lim_{x \to 0^+} f(x) = 1$ (iii) $\lim_{x \to 0} f(x) = 1.$

7.   For $f(x) = 1 - 2x^3$:

x	f(x)
1	-1
1.5	-5.75
1.9	-12.718
1.99	-14.761198
1.999	-14.976012
3	-53
2.5	-30.25
2.1	-17.522
2.01	-15.241202
2.001	-15.024012

It appears that $\lim_{x \to 2} (1 - 2x^3) = -15.$

9.   For $g(x) = \frac{x-1}{x^3-1}$:

x	g(x)
0.2	0.806452
0.4	0.641026
0.6	0.510204
0.8	0.409836
0.9	0.369004
0.99	0.336689
1.8	0.165563
1.6	0.193798
1.4	0.229358
1.2	0.274725
1.1	0.302115
1.01	0.330022

It appears that $\lim\limits_{x \to 1} \frac{x-1}{x^3-1} = 0.\bar{3}$ or $\frac{1}{3}$.

11.   For $F(x) = \frac{\frac{1}{\sqrt{x}} - \frac{1}{5}}{x-25}$:

x	F(x)
26	−0.003884
25.5	−0.003941
25.1	−0.003988
25.05	−0.003994
25.01	−0.003999
24	−0.004124
24.5	−0.004061
24.9	−0.004012
24.95	−0.004006
24.99	−0.004001

It appears that $\lim\limits_{x \to 25} F(x) = -0.004$.

13.   For $f(x) = \frac{1-\cos x}{x^2}$:

x	f(x)
1	0.459698
0.5	0.489670
0.4	0.493369
0.3	0.496261
0.2	0.498336
0.1	0.499583
0.05	0.499896
0.01	0.499996

It appears that $\lim\limits_{x \to 0} \frac{1-\cos x}{x^2} = 0.5$.

15. Let $h(x) = (1+x)^{1/x}$

x	h(x)
1.0	2.0
0.1	2.593742
0.01	2.704814
0.001	2.716924
0.0001	2.718146
0.00001	2.718268
0.000001	2.718280
0.0000001	2.718282
0.00000001	2.718282
0.000000001	2.718282

It appears that

$$\lim_{x \to 0} (1+x)^{1/x} \approx 2.71828$$

17. For $f(x) = x^2 - (2^x/1000)$:

(a)

x	f(x)
1	0.998000
0.8	0.638259
0.6	0.358484
0.4	0.158680
0.2	0.038851
0.1	0.008928
0.05	0.001465

It appears that $\lim_{x \to 0} f(x) = 0$.

(b)

x	f(x)
0.04	0.000572
0.02	-0.000614
0.01	-0.000907
0.005	-0.000978
0.003	-0.000993
0.001	-0.001000

It appears that $\lim_{x \to 0} f(x) = -0.001$.

EXERCISES 1.3

1. $$\lim_{x \to 4} (5x^2 - 2x + 3) = \lim_{x \to 4} 5x^2 - \lim_{x \to 4} 2x + \lim_{x \to 4} 3 \qquad (\textit{Limit Laws } 2 \text{ \& } 1)$$
    $$= 5 \lim_{x \to 4} x^2 - 2 \lim_{x \to 4} x + 3 \qquad (3 \text{ \& } 7)$$
    $$= 5(4)^2 - 2(4) + 3 = 75 \qquad (9 \text{ \& } 8)$$

3. $$\lim_{x \to 2} (x^2 + 1)(x^2 + 4x) = \lim_{x \to 2} (x^2 + 1) \lim_{x \to 2} (x^2 + 4x) \qquad (4)$$
    $$= \left(\lim_{x \to 2} x^2 + \lim_{x \to 2} 1 \right) \left(\lim_{x \to 2} x^2 + 4 \lim_{x \to 2} x \right) \qquad (1 \text{ \& } 3)$$
    $$= \left((2)^2 + 1 \right) \left((2)^2 + 4(2) \right) = 60 \qquad (9,7 \text{ \& } 8)$$

5.
$$\lim_{x \to -1} \frac{x - 2}{x^2 + 4x - 3} = \frac{\lim_{x \to -1} (x - 2)}{\lim_{x \to -1} (x^2 + 4x - 3)} \tag{5}$$

$$= \frac{\lim_{x \to -1} x - \lim_{x \to -1} 2}{\lim_{x \to -1} x^2 + 4 \lim_{x \to -1} x - \lim_{x \to -1} 3} \tag{2,1 \& 3}$$

$$= \frac{(-1) - 2}{(-1)^2 + 4(-1) - 3} = \frac{1}{2} \tag{8,7 \& 9}$$

7.
$$\lim_{x \to -1} \sqrt{x^3 + 2x + 7} = \sqrt{\lim_{x \to -1} (x^3 + 2x + 7)} \tag{11}$$

$$= \sqrt{\lim_{x \to -1} x^3 + 2 \lim_{x \to -1} x + \lim_{x \to -1} 7} \tag{1 \& 3}$$

$$= \sqrt{(-1)^3 + 2(-1) + 7} = 2 \tag{9,8 \& 6}$$

9.
$$\lim_{t \to -2} (t + 1)^9 (t^2 - 1) = \lim_{t \to -2} (t + 1)^9 \lim_{t \to -2} (t^2 - 1) \tag{4}$$

$$= \left[\lim_{t \to -2} (t + 1) \right]^9 \lim_{t \to -2} (t^2 - 1) \tag{6}$$

$$= \left[\lim_{t \to -2} t + \lim_{t \to -2} 1 \right]^9 \left[\lim_{t \to -2} t^2 - \lim_{t \to -2} 1 \right] \tag{1 \& 2}$$

$$= \left[(-2) + 1 \right]^9 \left[(-2)^2 - 1 \right] = -3 \tag{8,7 \& 9}$$

11.
$$\lim_{w \to -2} \sqrt[3]{\frac{4w + 3w^3}{3w + 10}} = \sqrt[3]{\lim_{w \to -2} \frac{4w + 3w^3}{3w + 10}} \tag{11}$$

$$= \sqrt[3]{\frac{\lim_{w \to -2} (4w + 3w^3)}{\lim_{w \to -2} (3w + 10)}} \tag{5}$$

$$= \sqrt[3]{\frac{4 \lim_{w \to -2} w + 3 \lim_{w \to -2} w^3}{3 \lim_{w \to -2} w + \lim_{w \to -2} 10}} \tag{1 \& 3}$$

$$= \sqrt[3]{\frac{4(-2) + 3(-2)^3}{3(-2) + 10}} = -2 \tag{8,9 \& 7}$$

13.
$$\lim_{h \to \frac{1}{2}} \frac{2h}{h + \frac{1}{h}} = \frac{\lim_{h \to \frac{1}{2}} 2h}{\lim_{h \to \frac{1}{2}} \left(h + \frac{1}{h} \right)} \tag{5}$$

$$= \frac{2 \lim_{h \to \frac{1}{2}} h}{\lim_{h \to \frac{1}{2}} h + \frac{1}{\lim_{h \to \frac{1}{2}} h}}$$ (3,1,5 & 7)

$$= \frac{2\left(\frac{1}{2}\right)}{\frac{1}{2} + \frac{1}{1/2}} = \frac{2}{5}$$ (8)

15.　(a)　$\lim_{x \to a}\left[f(x) + h(x)\right] = \lim_{x \to a} f(x) + \lim_{x \to a} h(x) = -3 + 8 = 5$

　　　(b)　$\lim_{x \to a}\left[f(x)\right]^2 = \left[\lim_{x \to a} f(x)\right]^2 = (-3)^2 = 9$

　　　(c)　$\lim_{x \to a} \sqrt[3]{h(x)} = \sqrt[3]{\lim_{x \to a} h(x)} = \sqrt[3]{8} = 2$

　　　(d)　$\lim_{x \to a} \frac{1}{f(x)} = \frac{1}{\lim_{x \to a} f(x)} = \frac{1}{-3} = -\frac{1}{3}$

　　　(e)　$\lim_{x \to a} \frac{f(x)}{h(x)} = \frac{\lim_{x \to a} f(x)}{\lim_{x \to a} h(x)} = \frac{-3}{8} = -\frac{3}{8}$

　　　(f)　$\lim_{x \to a} \frac{g(x)}{f(x)} = \frac{\lim_{x \to a} g(x)}{\lim_{x \to a} f(x)} = \frac{0}{-3} = 0$

　　　(g)　Does not exist since $\lim_{x \to a} g(x) = 0$ but $\lim_{x \to a} f(x) \neq 0$.

　　　(h)　$\lim_{x \to a} \frac{2f(x)}{h(x) - f(x)} = \frac{2 \lim_{x \to a} f(x)}{\lim_{x \to a} h(x) - \lim_{x \to a} f(x)} = \frac{2(-3)}{8 - (-3)} = -\frac{6}{11}$

17.　$\lim_{x \to -3} \frac{x^2 - x + 12}{x + 3}$ does not exist since $x + 3 \to 0$ but $x^2 - x + 12 \to 24$　as $x \to -3$.

19.　$\lim_{x \to 0} \frac{x + x^2}{x} = \lim_{x \to 0} \frac{x(1 + x)}{x} = \lim_{x \to 0} (1 + x) = 1$

21.　$\lim_{x \to -1} \frac{x^2 - x - 2}{x + 1} = \lim_{x \to -1} \frac{(x + 1)(x - 2)}{x + 1} = \lim_{x \to -1} (x - 2) = -3$

23.　$\lim_{t \to 1} \frac{t^3 - t}{t^2 - 1} = \lim_{t \to 1} \frac{t(t^2 - 1)}{t^2 - 1} = \lim_{t \to 1} t = 1$

25.　$\lim_{h \to 0} \frac{(h - 5)^2 - 25}{h} = \lim_{h \to 0} \frac{(h^2 - 10h + 25) - 25}{h} = \lim_{h \to 0} \frac{h^2 - 10h}{h} = \lim_{h \to 0} (h - 10) = -10$

27.　$\lim_{h \to 0} \frac{(1 + h)^4 - 1}{h} = \lim_{h \to 0} \frac{(1 + 4h + 6h^2 + 4h^3 + h^4) - 1}{h} = \lim_{h \to 0} \frac{4h + 6h^2 + 4h^3 + h^4}{h}$

　　$= \lim_{h \to 0} (4 + 6h + 4h^2 + h^3) = 4$

29.　$\lim_{x \to -2} \frac{x + 2}{x^2 - x - 6} = \lim_{x \to -2} \frac{x + 2}{(x - 3)(x + 2)} = \lim_{x \to -2} \frac{1}{x - 3} = -\frac{1}{5}$

31. $\lim\limits_{x \to -2} \left[\dfrac{x^2}{x+2} + \dfrac{2x}{x+2} \right] = \lim\limits_{x \to -2} \dfrac{x^2 + 2x}{x+2} = \lim\limits_{x \to -2} \dfrac{x(x+2)}{x+2} = \lim\limits_{x \to -2} x = -2$

33. $\lim\limits_{t \to 9} \dfrac{9-t}{3-\sqrt{t}} = \lim\limits_{t \to 9} \dfrac{(3+\sqrt{t})(3-\sqrt{t})}{3-\sqrt{t}} = \lim\limits_{t \to 9} (3+\sqrt{t}) = 3 + \sqrt{9} = 6$

35. $\lim\limits_{t \to 0} \dfrac{\sqrt{2-t} - \sqrt{2}}{t} = \lim\limits_{t \to 0} \dfrac{\sqrt{2-t} - \sqrt{2}}{t} \cdot \dfrac{\sqrt{2-t} + \sqrt{2}}{\sqrt{2-t} + \sqrt{2}} = \lim\limits_{t \to 0} \dfrac{-t}{t(\sqrt{2-t} + \sqrt{2})}$

$= \lim\limits_{t \to 0} \dfrac{-1}{\sqrt{2-t} + \sqrt{2}} = -\dfrac{1}{2\sqrt{2}} = -\dfrac{\sqrt{2}}{4}$

37. $\lim\limits_{x \to 9} \dfrac{x^2 - 81}{\sqrt{x} - 3} = \lim\limits_{x \to 9} \dfrac{(x-9)(x+9)}{\sqrt{x} - 3} = \lim\limits_{x \to 9} \dfrac{(\sqrt{x} - 3)(\sqrt{x} + 3)(x+9)}{\sqrt{x} - 3}$

$= \lim\limits_{x \to 9} (\sqrt{x} + 3)(x+9) = \lim\limits_{x \to 9} (\sqrt{x} + 3) \lim\limits_{x \to 9} (x+9) = (\sqrt{9} + 3)(9+9) = 108$

39. $\lim\limits_{t \to 0} \left[\dfrac{1}{t\sqrt{1+t}} - \dfrac{1}{t} \right] = \lim\limits_{t \to 0} \dfrac{1 - \sqrt{1+t}}{t\sqrt{1+t}} = \lim\limits_{t \to 0} \dfrac{(1 - \sqrt{1+t})(1 + \sqrt{1+t})}{t\sqrt{t+1}(1 + \sqrt{1+t})}$

$= \lim\limits_{t \to 0} \dfrac{-t}{t\sqrt{1+t}(1 + \sqrt{1+t})} = \lim\limits_{t \to 0} \dfrac{-1}{\sqrt{1+t}(1 + \sqrt{1+t})} = \dfrac{-1}{\sqrt{1+0}(1 + \sqrt{1+0})} = -\dfrac{1}{2}$

41. $\lim\limits_{x \to 0} \dfrac{x}{\sqrt{1+3x} - 1} = \lim\limits_{x \to 0} \dfrac{x(\sqrt{1+3x} + 1)}{(\sqrt{1+3x} - 1)(\sqrt{1+3x} + 1)} = \lim\limits_{x \to 0} \dfrac{x(\sqrt{1+3x} + 1)}{3x}$

$= \lim\limits_{x \to 0} \dfrac{\sqrt{1+3x} + 1}{3} = \dfrac{\sqrt{1} + 1}{3} = \dfrac{2}{3}$

43. $3x \le f(x) \le x^3 + 2$ for $0 \le x \le 2$, also $\lim\limits_{x \to 1} 3x = 3$ and $\lim\limits_{x \to 1} (x^3 + 2)$

$= \lim\limits_{x \to 1} x^3 + \lim\limits_{x \to 1} 2 = 1^3 + 2 = 3$. Therefore, by the Squeeze Theorem, $\lim\limits_{x \to 1} f(x) = 3$.

45. $-1 \le \sin \dfrac{1}{\sqrt[3]{x}} \le 1 \Rightarrow -\sqrt[3]{|x|} \le \sqrt[3]{x} \sin \dfrac{1}{\sqrt[3]{x}} \le \sqrt[3]{|x|}$. But $\lim\limits_{x \to 0} \sqrt[3]{|x|}$

$= \sqrt[3]{\lim\limits_{x \to 0} |x|} = \sqrt[3]{0} = 0 = \lim\limits_{x \to 0} -\sqrt[3]{|x|}$. Thus, by the Squeeze Theorem,

$\lim\limits_{x \to 0} \sqrt[3]{x} \sin \dfrac{1}{\sqrt[3]{x}} = 0$.

47. $\lim\limits_{x \to 5^+} (\sqrt{x-5} + \sqrt{5x}) = \sqrt{\lim\limits_{x \to 5^+} (x-5)} + \sqrt{\lim\limits_{x \to 5^+} 5x} = \sqrt{5-5} + \sqrt{5 \cdot 5} = 5$

49. $\lim\limits_{x \to -1.5^+} (\sqrt{3+2x} + x) = \sqrt{\lim\limits_{x \to -1.5^+} 3 + 2 \lim\limits_{x \to -1.5^+} x} + \lim\limits_{x \to -1.5^+} x$

$= \sqrt{3 + 2(-1.5)} - 1.5 = -1.5$

51. If $x < -4$, then $|x+4| = -(x+4)$, so $\lim\limits_{x \to -4^-} \dfrac{|x+4|}{x+4} = \lim\limits_{x \to -4^-} \dfrac{-(x+4)}{x+4}$

$= \lim\limits_{x \to -4^-} (-1) = -1$.

53. As $x \to 2^+$, $x - 2 \to 0$, so $\dfrac{1}{x-2}$ becomes very large and $\lim\limits_{x \to 2} \dfrac{1}{x-2}$ doesn't exist.

55. $[\![x]\!] = 9$ for $9 \le x < 10$, so $\lim\limits_{x \to 9^+} [\![x]\!] = \lim\limits_{x \to 9^+} 9 = 9$.

57. $[\![x]\!] = -2$ for $-2 \le x < -1$, so $\lim\limits_{x \to -2^+} [\![x]\!] = \lim\limits_{x \to -2^+} (-2) = -2$.

$[\![x]\!] = -3$ for $-3 \le x < -2$, so $\lim\limits_{x \to -2^-} [\![x]\!] = \lim\limits_{x \to -2^-} (-3) = -3$.

The right and left limits are different, so $\lim\limits_{x \to -2} [\![x]\!]$ does not exist.

59. $\lim\limits_{x \to 8^+} (\sqrt{x-8} + [\![x+1]\!]) = \lim\limits_{x \to 8^+} \sqrt{x-8} + \lim\limits_{x \to 8^+} [\![x+1]\!] = 0 + 9 = 9$ because $[\![x+1]\!] = 9$

for $8 \le x < 9$.

61. $\lim\limits_{x \to -2^-} \sqrt{x^2 + x - 2} = \sqrt{\lim\limits_{x \to -2^-} x^2 + \lim\limits_{x \to -2^-} x - \lim\limits_{x \to -2^-} 2} = \sqrt{(-2)^2 - 2 - 2} = 0$.

(Notice that the domain of $\sqrt{x^2 + x - 2}$ is $(-\infty, -2] \cup [1, \infty)$.)

63. $\lim\limits_{x \to 5^+} x \sqrt[6]{x^2 - 25} = \lim\limits_{x \to 5^+} x \sqrt[6]{\lim\limits_{x \to 5^+} x^2 - \lim\limits_{x \to 5^+} 25} = 5 \sqrt[6]{5^2 - 25} = 0$.

65. Since $|x| = x$ for $x > 0$, we have $\lim\limits_{x \to 0^+} \left(\frac{1}{x} - \frac{1}{|x|} \right) = \lim\limits_{x \to 0^+} \left(\frac{1}{x} - \frac{1}{x} \right) = \lim\limits_{x \to 0^+} 0 = 0$.

67. $H(t) = 1$ for $t \ge 0$, so $\lim\limits_{t \to 0^+} H(t) = \lim\limits_{t \to 0^+} 1 = 1$. $H(t) = 0$ for $t < 0$, so

$\lim\limits_{t \to 0^-} H(t) = \lim\limits_{t \to 0^-} 0 = 0$. By Theorem 1.8, $\lim\limits_{t \to 0} H(t)$ does not exist.

69. (a) $\lim\limits_{x \to -1^-} g(x) = \lim\limits_{x \to -1^-} (-x^3) = -(-1)^3 = 1$.

$\lim\limits_{x \to -1^+} g(x) = \lim\limits_{x \to -1^+} (x+2)^2 = (-1+2)^2 = 1$.

(b) Therefore $\lim\limits_{x \to -1} g(x) = 1$.

(c)

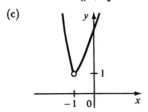

71. (a) (i) $[\![x]\!] = n - 1$ for $n - 1 \le x < n$, so $\lim\limits_{x \to n^-} [\![x]\!] = \lim\limits_{x \to n^-} (n-1) = n - 1$.

(ii) $[\![x]\!] = n$ for $n \le x < n + 1$, so $\lim\limits_{x \to n^+} [\![x]\!] = \lim\limits_{x \to n^+} n = n$.

(b) $\lim\limits_{x \to a} [\![x]\!]$ exists $\Leftrightarrow$ a is not an integer.

73. (a) (i) $\lim\limits_{x \to 1^+} \frac{x^2 - 1}{|x - 1|} = \lim\limits_{x \to 1^+} \frac{x^2 - 1}{x - 1} = \lim\limits_{x \to 1^+} (x + 1) = 2$.

(ii) $\lim\limits_{x \to 1^-} \frac{x^2 - 1}{|x - 1|} = \lim\limits_{x \to 1^-} \frac{x^2 - 1}{-(x - 1)} = \lim\limits_{x \to 1^-} -(x + 1) = -2$.

(b) No, $\lim\limits_{x \to 1} F(x)$ does not exist since $\lim\limits_{x \to 1^+} F(x) \ne \lim\limits_{x \to 1^-} F(x)$.

(c)

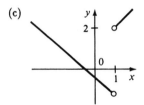

75. Since $p(x)$ is a polynomial $p(x) = a_0 + a_1x + a_2x^2 + \cdots + a_nx^n$. Thus,

$$\lim_{x \to a} p(x) = \lim_{x \to a} (a_0 + a_1x + a_2x^2 + \cdots + a_nx^n)$$

$$= a_0 + a_1 \lim_{x \to a} x + a_2 \lim_{x \to a} x^2 + \cdots + a_n \lim_{x \to a} x^n \text{ (by the Limit Laws)}$$

$$= a_0 + a_1a + a_2a^2 + \cdots + a_na^n = p(a). \text{ Thus, for any polynomial p, } \lim_{x \to a} p(x) = p(a).$$

77. Observe that $0 \le f(x) \le x^2$ for all x, and $\lim_{x \to 0} 0 = 0 = \lim_{x \to 0} x^2$. So, by the Squeeze

Theorem, $\lim_{x \to 0} f(x) = 0$.

79. Let $f(x) = H(x)$ and $g(x) = 1 - H(x)$, where H is the Heaviside function defined in

Example 7 in Section 1.2 Then $\lim_{x \to 0} f(x)$ and $\lim_{x \to 0} g(x)$ do not exist but

$$\lim_{x \to 0} [f(x) g(x)] = \lim_{x \to 0} 0 = 0.$$

81. Let $t = \sqrt[3]{1 + cx}$. Then $t \to 1$ as $x \to 0$ and $t^3 = 1 + cx \Rightarrow x = (t^3 - 1)/c$. *(If $c = 0$,*

then the limit is obviously 0.) Therefore $\lim_{x \to 0} \dfrac{\sqrt[3]{1 + cx} - 1}{x} = \lim_{t \to 1} \dfrac{t - 1}{(t^3 - 1)/c}$

$$= \lim_{t \to 1} \frac{c(t - 1)}{(t - 1)(t^2 + t + 1)} = \lim_{t \to 1} \frac{c}{t^2 + t + 1} = \frac{c}{1^2 + 1 + 1} = \frac{c}{3}. \text{ [Another method:}$$

multiply numerator and denominator by $(1 + cx)^{2/3} + (1 + cx)^{1/3} + 1.]$

EXERCISES 1.4

1. (a) $|(6x + 1) - 19| < 0.1 \Leftrightarrow |6x - 18| < 0.1 \Leftrightarrow 6|x - 3| < 0.1$

$\Leftrightarrow |x - 3| < (0.1)/6 = 1/60.$

(b) $|(6x + 1) - 19| < 0.01 \Leftrightarrow |x - 3| < (0.01)/6 = 1/600.$

3. Given $\epsilon > 0$, we need $\delta > 0$ so that if $|x - 2| < \delta$, then $|(3x - 2) - 4| < \epsilon \Leftrightarrow |3x - 6| < \epsilon$

$\Leftrightarrow 3|x - 2| < \epsilon \Leftrightarrow |x - 2| < \frac{\epsilon}{3}.$ So if we choose $\delta = \frac{\epsilon}{3}$, then $|x - 2| < \delta$

$\Rightarrow |(3x - 2) - 4| < \epsilon.$ Thus $\lim_{x \to 2} (3x - 2) = 4$ by the definition of a limit.

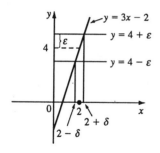

5. Given $\epsilon > 0$, we need $\delta > 0$ so that if $|x - (-1)| < \delta$, then $|(5x + 8) - 3| < \epsilon$

$\Leftrightarrow |5x + 5| < \epsilon \Leftrightarrow 5|x + 1| < \epsilon \Leftrightarrow |x - (-1)| < \frac{\epsilon}{5}$. So if we choose $\delta = \frac{\epsilon}{5}$, then

$|x - (-1)| < \delta \Rightarrow |(5x + 8) - 3| < \epsilon$. Thus $\lim_{x \to -1} (5x + 8) = 3$ by the definition of a

limit.

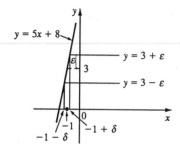

7. Given $\epsilon > 0$, we need $\delta > 0$ so that if $|x - 2| < \delta$ then $\left|\frac{x}{7} - \frac{2}{7}\right| < \epsilon \Leftrightarrow \frac{1}{7}|x - 2| < \epsilon$

$\Leftrightarrow |x - 2| < 7\epsilon$. So take $\delta = 7\epsilon$. Then $|x - 2| < \delta \Rightarrow \left|\frac{x}{7} - \frac{2}{7}\right| < \epsilon$. Thus $\lim_{x \to 2} \frac{x}{7} = \frac{2}{7}$

by the definition of a limit.

9. Given $\epsilon > 0$, we need $\delta > 0$ so that if $|x - (-5)| < \delta$ then $\left|(4 - \frac{3x}{5}) - 7\right| < \epsilon$

$\Leftrightarrow \frac{3}{5}|x + 5| < \epsilon \Leftrightarrow |x - (-5)| < \frac{5\epsilon}{3}$. So take $\delta = \frac{5\epsilon}{3}$. Then $|x - (-5)| < \delta$

$\Rightarrow \left|(4 - \frac{3x}{5}) - 7\right| < \epsilon$. Thus $\lim_{x \to -5} \left(4 - \frac{3x}{5}\right) = 7$ by the definition of a limit.

11. Given $\epsilon > 0$, we need $\delta > 0$ so that if $|x - a| < \delta$ then $|x - a| < \epsilon$. So obviously $\delta = \epsilon$

will work.

13. Given $\epsilon > 0$, we need $\delta > 0$ so that if $|x - 2| < \delta$ then $|c - c| < \epsilon$. But $|c - c| = 0$, so

this will be true no matter what δ we pick.

15. Given $\epsilon > 0$, we need $\delta > 0$ so that if $|x| < \delta$ then $\left|x^2 - 0\right| < \epsilon \Leftrightarrow x^2 < \epsilon \Leftrightarrow |x| < \sqrt{\epsilon}$.

Take $\delta = \sqrt{\epsilon}$. Then $|x - 0| < \delta \Rightarrow \left|x^2 - 0\right| < \epsilon$. Thus $\lim_{x \to 0} x^2 = 0$ by the definition of a

limit.

17. Given $\epsilon > 0$, we need $\delta > 0$ so that if $|x - 0| < \delta$ then $||x| - 0| < \epsilon$. But $||x|| = |x|$. So this is true if we pick $\delta = \epsilon$.

19. Given $\epsilon > 0$, we need $\delta > 0$ so that if $|x - (-2)| < \delta$ then $|(x^2 - 1) - 3| < \epsilon$ or upon simplifying we need $|x^2 - 4| < \epsilon$ whenever $|x + 2| < \delta$. Notice that if $|x + 2| < 1$, then $-1 < x + 2 < 1 \Rightarrow -5 < x - 2 < -3 \Rightarrow |x - 2| < 5$. So take $\delta = \min\{\epsilon/5, 1\}$. Then $|x - 2| < 5$ and $|x + 2| < \epsilon/5$, so
$$|(x^2 - 1) - 3| = |(x + 2)(x - 2)| = |x + 2||x - 2| < (\epsilon/5)(5) = \epsilon.$$
Therefore by the definition of a limit $\lim_{x \to -2} x^2 - 1 = 8$

21. (a) Suppose that $\lim_{x \to 0} (1/x) = L$ for some L. Then, given $\epsilon = 1$, there exists δ such that $0 < |x| < \delta \Rightarrow |(1/x) - L| < 1$. Thus $0 < x < \delta \Rightarrow L - 1 < \frac{1}{x} < L + 1$. This is impossible since $1/x$ takes on arbitrarily large values in any interval $(-\delta, \delta)$. In fact, $\frac{1}{x} > L + 1$ if $0 < x < \frac{1}{|L| + 1}$.

   (b) Suppose that $\lim_{x \to -4} (x + 4)^{-2} = L$ for some L. Then, given $\epsilon = 1$, there exists δ such that $0 < |x + 4| < \delta \Rightarrow |(x + 4)^{-2} - L| < 1$. Thus $0 < |x + 4| < \delta \Rightarrow L - 1 < \frac{1}{(x + 4)^2} < L + 1$. This is impossible since $(x + 4)^{-2}$ takes on arbitrarily large values in any interval $(-4 - \delta, -4 + \delta)$. In fact, $(x + 4)^{-2} > L + 1$ if $-4 < x < -4 + \frac{1}{\sqrt{L + 1}}$.

23. Given $\epsilon > 0$, we let $\delta = \min\{2, \frac{\epsilon}{8}\}$. If $0 < |x - 3| < \delta$ then $|x - 3| < 2 \Rightarrow 1 < x < 5$ $\Rightarrow |x + 3| < 8$. Also $|x - 3| < \frac{\epsilon}{8}$, so $|x^2 - 9| = |x + 3||x - 3| < 8 \cdot \frac{\epsilon}{8} = \epsilon$. Thus $\lim_{x \to 3} x^2 = 9$.

25. 1. (*Guessing a value for δ*) Given $\epsilon > 0$, we must find $\delta > 0$ such that $|\sqrt{x} - \sqrt{a}| < \epsilon$ whenever $0 < |x - a| < \delta$. But $|\sqrt{x} - \sqrt{a}| = \frac{|x - a|}{\sqrt{x} + \sqrt{a}} < \epsilon$ (*from the hint*). Now if we can find a positive constant C such that $\sqrt{x} + \sqrt{a} > C$ then $\frac{|x - a|}{\sqrt{x} + \sqrt{a}} < \frac{|x - a|}{C} < \epsilon$, and we take $|x - a| < C\epsilon$. We can find this number by restricting x to lie in some interval centered at a. If $|x - a| < a/2$, then $a/2 < x < 3a/2 \Rightarrow \sqrt{x} + \sqrt{a} > \sqrt{a/2} + \sqrt{a}$, and so $C = \sqrt{a/2} + \sqrt{a}$ is a suitable choice for the constant. So $|x - a| < (\sqrt{a/2} + \sqrt{a})\epsilon$. This suggests that we let $\delta = \min\{a/2, (\sqrt{a/2} + \sqrt{a})\epsilon\}$.

   2. (*Showing that δ works*) Given $\epsilon > 0$, we let $\delta = \min\{a/2, (\sqrt{a/2} + \sqrt{a})\epsilon\}$. If $0 < |x - a| < \delta$, then $|x - a| < a/2 \Rightarrow \sqrt{x} + \sqrt{a} > \sqrt{a/2} + \sqrt{a}$ (as in part 1). Also $|x - a| < (\sqrt{a/2} + \sqrt{a})\epsilon$, so $|\sqrt{x} - \sqrt{a}| = \frac{|x - a|}{\sqrt{x} + \sqrt{a}} < \frac{(\sqrt{a/2} + \sqrt{a})\epsilon}{(\sqrt{a/2} + \sqrt{a})} < \epsilon$.
   Therefore $\lim_{x \to a} \sqrt{x} = \sqrt{a}$ by the definition of a limit.

27. Suppose that $\lim\limits_{x \to 0} f(x) = L$. Given $\epsilon = \frac{1}{2}$, there exists $\delta > 0$ such that $0 < |x| < \delta$ $\Rightarrow |f(x) - L| < \frac{1}{2}$. Take any rational number r with $0 < |r| < \delta$. Then $f(r) = 0$, so $|0 - L| < \frac{1}{2}$, so $L \le |L| < \frac{1}{2}$. Now take any irrational number s with $0 < |s| < \delta$. Then $f(s) = 1$, so $|1 - L| < \frac{1}{2}$. Hence $1 - L < \frac{1}{2}$, so $L > \frac{1}{2}$. This contradicts $L < \frac{1}{2}$. Thus $\lim\limits_{x \to 0} f(x)$ does not exist.

EXERCISES 1.5

1. (a) f is discontinuous at $-5, -3, -1, 3, 5, 8$ and 10.

  (b) f is continuous from the left at -5 and -3, and continuous from the right at 8.
   It is continuous on neither side at $-1, 3, 5$, and 10.

3. $\lim\limits_{x \to 3} (x^4 - 5x^3 + 6) = \lim\limits_{x \to 3} x^4 - 5 \lim\limits_{x \to 3} x^3 + \lim\limits_{x \to 3} 6 = 3^4 - 5(3^3) + 6 = -48 = f(3)$.

Thus f is continuous at 3.

5. $\lim\limits_{x \to 5} f(x) = \lim\limits_{x \to 5} \left(1 + \sqrt{x^2 - 9}\right) = \lim\limits_{x \to 5} 1 + \sqrt{\lim\limits_{x \to 5} x^2 - \lim\limits_{x \to 5} 9} = 1 + \sqrt{5^2 - 9} = 5 = f(5)$.

Thus f is continuous at 5.

7. $\lim\limits_{t \to -8} g(t) = \lim\limits_{t \to -8} \dfrac{\sqrt[3]{t}}{(t + 1)^4} = \dfrac{\sqrt[3]{\lim\limits_{t \to -8} t}}{\left(\lim\limits_{t \to -8} t + 1\right)^4} = \dfrac{\sqrt[3]{-8}}{(-8 + 1)^4}$

$= -\dfrac{2}{2401} = g(-8)$. Thus g is continuous at -8.

9. For $-4 < a < 4$ we have $\lim\limits_{x \to a} f(x) = \lim\limits_{x \to a} x\sqrt{16 - x^2} = \lim\limits_{x \to a} x \sqrt{\lim\limits_{x \to a} 16 - \lim\limits_{x \to a} x^2}$

$= a\sqrt{16 - a^2} = f(a)$, so f is continuous on $(-4, 4)$. Similarly, we get $\lim\limits_{x \to 4^-} f(x) = 0 = f(4)$

and $\lim\limits_{x \to -4^+} f(x) = 0 = f(-4)$, so f is continuous from the left at 4
and from the right at -4. Thus f is continuous on $[-4, 4]$.

11. For any $a \in R$ we have $\lim\limits_{x \to a} f(x) = \lim\limits_{x \to a} (x^2 - 1)^8 = \left(\lim\limits_{x \to a} x^2 - \lim\limits_{x \to a} 1\right)^8 = (a^2 - 1)^8$
$= f(a)$. Thus f is continuous on $(-\infty, \infty)$.

13. $f(x) = \dfrac{3x^2 - 5x - 2}{x - 2}$ is discontinuous at 2

since $f(2)$ is not defined. For the graph, note

that $f(x) = \dfrac{(3x + 1)(x - 2)}{x - 2} = 3x + 1, x \ne 2$.

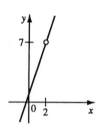

15. $f(x) = -\dfrac{1}{(x-1)^2}$ is discontinuous at 1 since f(1) is not defined.

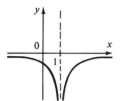

17. $\lim\limits_{x \to 1} f(x) = \lim\limits_{x \to 1} -\dfrac{1}{(x-1)^2}$ does not exist. Therefore f is discontinuous at 1.

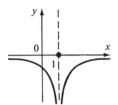

19. Since $f(x) = x^2 - 2$ for $x \neq -3$, $\lim\limits_{x \to -3} f(x) = \lim\limits_{x \to -3} (x^2 - 2) = (-3)^2 - 2 = 7$.

But $f(-3) = 5$, so $\lim\limits_{x \to -3} f(x) \neq f(-3)$. Therefore f is discontinuous at -3.

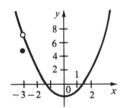

21. $f(x) = (x+1)(x^3 + 8x + 9)$ is a polynomial, so by Theorem 1.20 it is continuous on R.

23. $h(x) = \dfrac{x^2 + 2x - 1}{x + 1}$ is a rational function, so by Theorem 1.20 it is continuous on its domain, which is $\{x \mid x \neq -1\}$.

25. $g(x) = x + 1$, a polynomial, is continuous (*by Theorem* 1.20) and $f(x) = \sqrt{x}$ is continuous on $[0, \infty)$ by Theorem 1.22, so $f(g(x)) = \sqrt{x + 1}$ is continuous on $[-1, \infty)$ by Theorem 1.24. By Theorem 1.19(5), $H(x) = 1/\sqrt{x+1}$ is continuous on $(-1, \infty)$.

27. $g(x) = x - 1$ and $G(x) = x^2 - 2$ are both polynomials, so by Theorem 1.20 they are continuous. Also $f(x) = \sqrt[5]{x}$ is continuous by Theorem 1.22, so $f(g(x)) = \sqrt[5]{x-1}$ is continuous on R by Theorem 1.24. Thus the product $h(x) = \sqrt[5]{x-1}\,(x^2 - 2)$ is continuous on R by Theorem 1.19(4).

29. Since the discriminant of $t^2 + t + 1$ is negative, $t^2 + t + 1$ is always positive. So the domain of F(t) is R. By Theorem 1.20 the polynomial $(t^2 + t + 1)^3$ is continuous. By Theorems 1.22 and 1.24 the composition $F(t) = \sqrt{(t^2 + t + 1)^3}$ is continuous on R.

31. $H(x) = \sqrt{(x - 2)/(5 + x)}$. The domain is $\{x \mid (x - 2)/(5 + x) > 0\} = (-\infty, -5) \cup [2, \infty)$ by the methods of the Preview and Review Chapter. By Theorem 1.20 the rational function $(x - 2)/(5 + x)$ is continuous. Since the square root function is continuous (*Theorem* 1.22), the composition $H(x) = \sqrt{(x - 2)/(5 + x)}$ is continuous on its domain by Theorem 1.24.

33. $g(x) = x^3 - x$ is continuous on R since it is a polynomial $(1.20(a))$, and $f(x) = |x|$ is continuous on R by Example 9(a). So $L(x) = |x^3 - x|$ is continuous on $(-\infty, \infty)$ by 1.24.

35. f is continuous on $(-\infty, 3)$ and $(3, \infty)$ since on each of these intervals it is a polynomial. Also $\lim\limits_{x \to 3^+} f(x) = \lim\limits_{x \to 3^+} (5 - x) = 2$ and $\lim\limits_{x \to 3^-} f(x) = \lim\limits_{x \to 3^-} (x - 1) = 2$, so $\lim\limits_{x \to 3} f(x) = 2$. Since $f(3) = 5 - 3 = 2$, f is also continuous at 3. Thus f is continuous on $(-\infty, \infty)$.

37. f is continuous on $(-\infty, 0)$ and $(0, \infty)$ since on each of these intervals it is a polynomial.

Now $\lim\limits_{x \to 0^-} f(x) = \lim\limits_{x \to 0^-} (x - 1)^3 = -1$

and $\lim\limits_{x \to 0^+} f(x) = \lim\limits_{x \to 0^+} (x + 1)^3 = 1$.
Thus $\lim\limits_{x \to 0} f(x)$ does not exist, so f is discontinuous at 0. Since $f(0) = 1$, f is continuous from the right at 0.

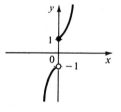

39. f is continuous on $(-\infty, -1)$, $(-1, 1)$ and $(1, \infty)$. Now $\lim\limits_{x \to -1^-} f(x) = \lim\limits_{x \to -1^-} \frac{1}{x} = -1$ and $\lim\limits_{x \to -1^+} f(x) = \lim\limits_{x \to -1^+} x = -1$, so $\lim\limits_{x \to -1} f(x) = -1 = f(-1)$ and f is continuous at -1. Also $\lim\limits_{x \to 1^-} f(x) = \lim\limits_{x \to 1^-} x = 1$ and $\lim\limits_{x \to 1^+} f(x) = \lim\limits_{x \to 1^+} \frac{1}{x^2} = 1$, so $\lim\limits_{x \to 1} f(x) = 1 = f(1)$ and f is continuous at 1. Thus f has no discontinuities.

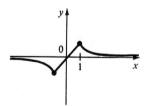

41. $f(x) = H(x) - H(x-1) = \begin{cases} 0 & \text{if } x < 0 \\ 1 & \text{if } 0 \le x < 1. \\ 0 & \text{if } x \ge 1 \end{cases}$ f is continuous on $(-\infty, 0)$, $(0, 1)$ and

$(1, \infty)$. $\lim_{x \to 0^-} f(x) = \lim_{x \to 0^-} 0 = 0$ and $\lim_{x \to 0^+} f(x) = \lim_{x \to 0^+} 1 = 1$. Since $f(0) = 1$, f is

continuous only from the right at 0. Also $\lim_{x \to 1^-} f(x) = 1$ and $\lim_{x \to 1^+} f(x) = 0$, so f is

continuous only from the right at 1.

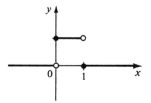

43. $f(x) = [\![2x]\!]$ is continuous except when $2x = n \Leftrightarrow x = n/2$, n an integer. In fact,

$\lim_{x \to n/2^-} [\![2x]\!] = n - 1$ and $\lim_{x \to n/2^+} [\![2x]\!] = n = f(n)$, so f is continuous only from the

right at n/2.

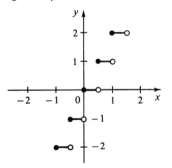

45. f is continuous on $(-\infty, 3)$ and $(3, \infty)$. Now $\lim_{x \to 3^-} f(x) = \lim_{x \to 3^-} (cx + 1) = 3c + 1$ and

$\lim_{x \to 3^+} f(x) = \lim_{x \to 3^+} (cx^2 - 1) = 9c - 1$. So f is continuous $\Leftrightarrow 3c + 1 = 9c - 1$

$\Leftrightarrow 6c = 2 \Leftrightarrow c = \frac{1}{3}$. Thus for f to be continuous on $(-\infty, \infty)$, $c = \frac{1}{3}$.

47. (a) $\lim_{x \to 1^-} f(x) = \lim_{x \to 1^-} (1 - x^2) = 0$ and $\lim_{x \to 1^+} f(x) = \lim_{x \to 1^+} (1 + x/2) = \frac{3}{2}$. Thus

$\lim_{x \to 1} f(x)$ does not exist, so f is not continuous at 1.

(b) $f(0) = 1$ and $f(2) = 2$. For $0 \le x \le 1$, f takes the values in $[0, 1]$. For $1 < x \le 2$,

f takes the values in $(1.5, 2]$. Thus f does not take on the value 1.5 (or any other

value in $(1, 1.5]$).

49. $f(x) = x^3 - x^2 + x$ is continuous on $[2, 3]$ and $f(2) = 6$, $f(3) = 21$. Since $6 < 10 < 21$, there is a number c in $(2, 3)$ such that $f(c) = 10$ by the Intermediate Value Theorem.

51. $f(x) = x^3 - 3x + 1$ is continuous on $[0, 1]$ and $f(0) = 1$, $f(1) = -1$. Since $-1 < 0 < 1$, there is a number c in $(0, 1)$ such that $f(c) = 0$ by the Intermediate Value Theorem. Thus there is a root of the equation $x^3 - 3x + 1 = 0$ in the interval $(0, 1)$.

53. $f(x) = x^4 - 3x^3 - 2x^2 - 1$ is continuous on $[3, 4]$ and $f(3) = -19$, $f(4) = 31$. Since $-19 < 0 < 31$, there is a number c in $(3, 4)$ such that $f(c) = 0$ by the Intermediate Value Theorem. Thus there is a root of the equation $x^4 - 3x^3 - 2x^2 - 1 = 0$ in the interval $(3, 4)$.

55. $f(x) = x^3 + 2x - (x^2 + 1) = x^3 + 2x - x^2 - 1$ is continuous on $[0, 1]$ and $f(0) = -1$, $f(1) = 1$. Since $-1 < 0 < 1$, there is a number c in $(0, 1)$ such that $f(c) = 0$ by the Intermediate Value Theorem. Thus there is a root of the equation $x^3 + 2x - x^2 - 1 = 0$, or equivalently, $x^3 + 2x = x^2 + 1$, in the interval $(0, 1)$.

57. (a) $f(x) = x^3 - x + 1$ is continuous on $[-2, -1]$ and $f(-2) = -5$, $f(-1) = 1$. Since $-5 < 0 < 1$, there is a number c in $(-2, -1)$ such that $f(c) = 0$ by the Intermediate Value Theorem. Thus there is a root of the equation $x^3 - x + 1 = 0$ in the interval $(-2, -1)$.

(b) $f(-1.33) \approx -0.0226$ and $f(-1.32) \approx 0.0200$, so there is a root between -1.33 and -1.32.

59. $f(x) = \begin{cases} 0 & \text{if x is rational} \\ 1 & \text{if x is irrational} \end{cases}$ is continuous nowhere. For, given any number a and any $\delta > 0$, the interval $(a - \delta, a + \delta)$ contains both infinitely many rational and infinitely many irrational numbers. Since $f(a) = 0$ or 1, there are infinitely many numbers x with $|x - a| < \delta$ and $|f(x) - f(a)| = 1$. Thus $\lim_{x \to a} f(x) \neq f(a)$. (In fact, $\lim_{x \to a} f(x)$ does not even exist.)

61. ($\Rightarrow$) If f is continuous at a, then by Theorem 1.23 with $g(h) = a + h$, we have
$$\lim_{h \to 0} f(a + h) = f\left(\lim_{h \to 0} (a + h)\right) = f(a).$$
($\Leftarrow$) Let $\epsilon > 0$. Since $\lim_{h \to 0} f(a + h) = f(a)$, there exists $\delta > 0$ such that $|h| < \delta \Rightarrow$
$|f(a + h) - f(a)| < \epsilon$. So if $|x - a| < \delta$, then $|f(x) - f(a)| = |f(a + (x - a)) - f(a)| < \epsilon$.
Thus $\lim_{x \to a} f(x) = f(a)$ and so f is continuous at a.

EXERCISES 1.6

1. $\lim\limits_{x \to \infty} \dfrac{1}{x\sqrt{x}} = \lim\limits_{x \to \infty} \dfrac{1}{x^{3/2}} = 0$ by Theorem 1.29.

3. $\lim\limits_{x \to \infty} \dfrac{5 + 2x}{3 - x} = \lim\limits_{x \to \infty} \dfrac{\frac{5}{x} + 2}{\frac{3}{x} - 1} \overset{(5)}{=} \dfrac{\lim\limits_{x \to \infty}\left[\frac{5}{x} + 2\right]}{\lim\limits_{x \to \infty}\left[\frac{3}{x} - 1\right]}$

$\underset{(1,\overline{2},3)}{=} \dfrac{5\lim\limits_{x \to \infty} \frac{1}{x} + \lim\limits_{x \to \infty} 2}{3\lim\limits_{x \to \infty} \frac{1}{x} - \lim\limits_{x \to \infty} 1} = \dfrac{5(0) + 2}{3(0) - 1} = -2$ by (7) and Theorem 1.29.

5. $\lim\limits_{x \to -\infty} \dfrac{2x^2 - x - 1}{4x^2 + 7} = \lim\limits_{x \to -\infty} \dfrac{2 - \frac{1}{x} - \frac{1}{x^2}}{4 + \frac{7}{x^2}}$

$\underset{(5,\overline{1},\overline{2},3)}{=} \dfrac{\lim\limits_{x \to -\infty} 2 - \lim\limits_{x \to -\infty} \frac{1}{x} - \lim\limits_{x \to -\infty} \frac{1}{x^2}}{\lim\limits_{x \to -\infty} 4 + 7 \lim\limits_{x \to -\infty} \frac{1}{x^2}} = \dfrac{2 - 0 - 0}{4 + 7(0)} = \dfrac{1}{2}$ by (7) and Theorem 1.29.

7. $\lim\limits_{x \to -\infty} \dfrac{(1 - x)(2 + x)}{(1 + 2x)(2 - 3x)} = \lim\limits_{x \to -\infty} \dfrac{\left[\frac{1}{x} - 1\right]\left[\frac{2}{x} + 1\right]}{\left[\frac{1}{x} + 2\right]\left[\frac{2}{x} - 3\right]}$

$= \dfrac{\left[\lim\limits_{x \to -\infty} \frac{1}{x} - 1\right]\left[\lim\limits_{x \to -\infty} \frac{2}{x} + 1\right]}{\left[\lim\limits_{x \to -\infty} \frac{1}{x} + 2\right]\left[\lim\limits_{x \to -\infty} \frac{2}{x} - 3\right]}$ $[by\ (5, 4, 1, 2, 7)] = \dfrac{(0 - 1)(0 + 1)}{(0 + 2)(0 - 3)} = \dfrac{1}{6}.$

9. $\lim\limits_{x \to \infty} \dfrac{1}{3 + \sqrt{x}} = \lim\limits_{x \to \infty} \dfrac{1/\sqrt{x}}{(3/\sqrt{x}) + 1} \underset{(5,\overline{1},3)}{=} \dfrac{\lim\limits_{x \to \infty}(1/\sqrt{x})}{3\lim\limits_{x \to \infty}(1/\sqrt{x}) + \lim\limits_{x \to \infty} 1} = \dfrac{0}{3(0) + 1} = 0$

(by Theorem 1.29 with $r = \frac{1}{2}$).

[OR: Note that $0 < \dfrac{1}{3 + \sqrt{x}} < \dfrac{1}{\sqrt{x}}$ and use the Squeeze Theorem.]

11. $\lim\limits_{r \to \infty} \dfrac{r^4 - r^2 + 1}{r^5 + r^3 - r} = \lim\limits_{r \to \infty} \dfrac{\frac{1}{r} - \frac{1}{r^3} + \frac{1}{r^5}}{1 + \frac{1}{r^2} - \frac{1}{r^4}}$

$= \dfrac{\lim\limits_{r \to \infty} \frac{1}{r} - \lim\limits_{r \to \infty} \frac{1}{r^3} + \lim\limits_{r \to \infty} \frac{1}{r^5}}{\lim\limits_{r \to \infty} 1 + \lim\limits_{r \to \infty} \frac{1}{r^2} - \lim\limits_{r \to \infty} \frac{1}{r^4}} = \dfrac{0 - 0 + 0}{1 + 0 - 0} = 0.$

13. $\lim\limits_{x \to \infty} \dfrac{\sqrt{1 + 4x^2}}{4 + x} = \lim\limits_{x \to \infty} \dfrac{\sqrt{(1/x^2) + 4}}{(4/x) + 1} = \dfrac{\sqrt{0 + 4}}{0 + 1} = 2.$

15. $\lim\limits_{x \to \infty} \dfrac{\sqrt[3]{x^2 + 8}}{x + 2} = \lim\limits_{x \to \infty} \dfrac{\sqrt[3]{(1/x) + (8/x^3)}}{1 + 2/x} = \dfrac{\sqrt[3]{0 + 0}}{1 + 0} = 0.$

17. $\displaystyle\lim_{x\to\infty}\frac{1-\sqrt{x}}{1+\sqrt{x}}=\lim_{x\to\infty}\frac{(1/\sqrt{x})-1}{(1/\sqrt{x})+1}=\frac{0-1}{0+1}=-1.$

19. $\displaystyle\lim_{x\to\infty}(\sqrt{x^2+1}-\sqrt{x^2-1})=\lim_{x\to\infty}(\sqrt{x^2+1}-\sqrt{x^2-1})\frac{\sqrt{x^2+1}+\sqrt{x^2-1}}{\sqrt{x^2+1}+\sqrt{x^2-1}}$

$\displaystyle=\lim_{x\to\infty}\frac{(x^2+1)-(x^2-1)}{\sqrt{x^2+1}+\sqrt{x^2-1}}=\lim_{x\to\infty}\frac{2}{\sqrt{x^2+1}+\sqrt{x^2-1}}$

$\displaystyle=\lim_{x\to\infty}\frac{2/x}{\sqrt{1+(1/x^2)}+\sqrt{1-(1/x^2)}}=\frac{0}{\sqrt{1+0}+\sqrt{1-0}}=0.$

21. $\displaystyle\lim_{x\to\infty}(\sqrt{1+x}-\sqrt{x})=\lim_{x\to\infty}(\sqrt{1+x}-\sqrt{x})\frac{\sqrt{1+x}+\sqrt{x}}{\sqrt{1+x}+\sqrt{x}}=\lim_{x\to\infty}\frac{(1+x)-x}{\sqrt{1+x}+\sqrt{x}}$

$\displaystyle=\lim_{x\to\infty}\frac{1}{\sqrt{1+x}+\sqrt{x}}=\lim_{x\to\infty}\frac{1/\sqrt{x}}{\sqrt{(1/x)+1}+1}=\frac{0}{\sqrt{0+1}+1}=0.$

23. $\displaystyle\lim_{x\to-\infty}(\sqrt{x^2+x+1}+x)=\lim_{x\to-\infty}(\sqrt{x^2+x+1}+x)\frac{(\sqrt{x^2+x+1}-x)}{(\sqrt{x^2+x+1}-x)}$

$\displaystyle=\lim_{x\to-\infty}\frac{x+1}{(\sqrt{x^2+x+1}-x)}=\lim_{x\to-\infty}\frac{1+(1/x)}{-\sqrt{1+(1/x)+(1/x^2)}-1}=\frac{1+0}{-\sqrt{1+0+0}-1}=-\frac{1}{2}.$

25. $\displaystyle\lim_{x\to\pm\infty}\frac{1+2x}{2+x}=\lim_{x\to\pm\infty}\frac{(1/x)+2}{(2/x)+1}=\frac{0+2}{0+1}=2,$ so $y=2$ is the horizontal asymptote.

27. $\displaystyle\lim_{x\to\infty}\frac{x}{\sqrt[4]{x^4+1}}=\lim_{x\to\infty}\frac{1}{\sqrt[4]{1+(1/x^4)}}=\frac{1}{\sqrt[4]{1+0}}=1$ and $\displaystyle\lim_{x\to-\infty}\frac{x}{\sqrt[4]{x^4+1}}$

$\displaystyle=\lim_{x\to-\infty}\frac{1}{-\sqrt[4]{1+(1/x^4)}}=\frac{1}{-\sqrt[4]{1+0}}=-1,$ so $y=\pm1$ are horizontal asymptotes.

29. $\displaystyle\lim_{x\to\infty}\frac{4x-1}{x}=\lim_{x\to\infty}(4-\tfrac{1}{x})=4,$ and $\displaystyle\lim_{x\to\infty}\frac{4x^2+3x}{x^2}=\lim_{x\to\infty}(4+\tfrac{3}{x})=4.$

Therefore by the Squeeze Theorem, $\displaystyle\lim_{x\to\infty}f(x)=4.$

31. If $f(x)=x^2/2^x,$ then a calculator gives $f(0)=0,$ $f(1)=0.5,$ $f(2)=1,$ $f(3)=1.125,$

$f(4)=1,$ $f(5)=0.78125,$ $f(6)=0.5625,$ $f(7)=0.3828125,$ $f(8)=0.25,$

$f(9)=0.158203125,$ $f(10)=0.09765625,$ $f(20)\approx0.00038147,$ $f(50)\approx2.2204\times10^{-12},$

$f(100)\approx7.8886\times10^{-27}.$ It appears that $\displaystyle\lim_{x\to\infty}\frac{x^2}{2^x}=0.$

33. If $\epsilon>0$ is given, then $1/x^2<\epsilon\Leftrightarrow x^2>1/\epsilon\Leftrightarrow x>1/\sqrt{\epsilon}.$ Let $N=1/\sqrt{\epsilon}.$

Then $x>N\Rightarrow x>1/\sqrt{\epsilon}\Rightarrow\left|1/x^2-0\right|=1/x^2<\epsilon$ so $\displaystyle\lim_{x\to\infty}\frac{1}{x^2}=0.$

35. If $\epsilon > 0$ is given, then $1/\sqrt{x} < \epsilon \Leftrightarrow \sqrt{x} > 1/\epsilon \Leftrightarrow x > 1/\epsilon^2$. Let $N = 1/\epsilon^2$.

Then $x > N \Rightarrow x > 1/\epsilon^2 \Rightarrow |1/\sqrt{x} - 0| = 1/\sqrt{x} < \epsilon$ so $\lim\limits_{x \to \infty} \frac{1}{\sqrt{x}} = 0$.

37. Suppose that $\lim\limits_{x \to \infty} f(x) = L$ and let $\epsilon > 0$ be given. Then there exists $N > 0$ such that $x > N \Rightarrow |f(x) - L| < \epsilon$. Let $\delta = 1/N$. Then $0 < t < \delta \Rightarrow t < 1/N \Rightarrow 1/t > N$

$\Rightarrow |f(1/t) - L| < \epsilon$. So $\lim\limits_{t \to 0^+} f(1/t) = L = \lim\limits_{x \to \infty} f(x)$. Now suppose that $\lim\limits_{x \to -\infty} f(x) = L$

and let $\epsilon > 0$ be given. Then there exists $N < 0$ such that $x < N \Rightarrow |f(x) - L| < \epsilon$.

Let $\delta = -1/N$. Then $-\delta < t < 0 \Rightarrow t > 1/N \Rightarrow 1/t < N \Rightarrow |f(1/t) - L| < \epsilon$.

So $\lim\limits_{x \to 0^-} f(1/t) = L = \lim\limits_{x \to -\infty} f(x)$.

EXERCISES 1.7

1. (a) $\lim\limits_{x \to 3} f(x) = \infty$ (b) $\lim\limits_{x \to 7} f(x) = -\infty$ (c) $\lim\limits_{x \to -4} f(x) = -\infty$ (d) $\lim\limits_{x \to -9^-} f(x) = \infty$

(e) $\lim\limits_{x \to -9^+} f(x) = -\infty$ (f) $\lim\limits_{x \to \infty} f(x) = 2$ (g) $\lim\limits_{x \to -\infty} f(x) = -2$

(h) Vertical asymptotes: $x = -9$, $x = -4$, $x = 3$, $x = 7$.

(i) Horizontal asymptotes: $y = -2$, $y = 2$.

3. $\lim\limits_{x \to -1} \dfrac{-2}{(x+1)^6} = -\infty$ since $(x+1)^6 \to 0$ as $x \to -1$ and $\dfrac{-2}{(x+1)^6} < 0$.

5. $\lim\limits_{x \to 5^+} \dfrac{6}{x-5} = \infty$ since $x - 5 \to 0$ as $x \to 5^+$ and $\dfrac{6}{x-5} > 0$ for $x > 5$.

7. $\lim\limits_{x \to -1^-} \dfrac{x^2}{x+1} = -\infty$ since $x + 1 \to 0$ as $x \to -1^-$ and $\dfrac{x^2}{x+1} < 0$ for $x < -1$.

9. $\lim\limits_{t \to 3^-} \dfrac{t+3}{t^2 - 9} = \lim\limits_{t \to 3^-} \dfrac{1}{t-3} = -\infty$ since $t - 3 \to 0$ as $t \to 3^-$ and $\dfrac{1}{t-3} < 0$ for $t < 3$.

11. $\lim\limits_{x \to -3^-} \dfrac{x^2}{x^2 - 9} = \infty$ since $x^2 - 9 \to 0$ as $x \to -3^-$ and $\dfrac{x^2}{x^2 - 9} > 0$ for $x < -3$.

13. $\lim\limits_{x \to -2^+} \dfrac{x-1}{x^2(x+2)} = -\infty$ since $x^2(x+2) \to 0$ as $x \to -2^+$ and $\dfrac{x-1}{x^2(x+2)} < 0$

if $-2 < x < 0$.

15. $\lim\limits_{x \to -3^+} \dfrac{x^3}{x^2 + 5x + 6} = \lim\limits_{x \to -3^+} \dfrac{x^3}{(x+2)(x+3)} = \infty$ since $(x+2)(x+3) \to 0$ as $x \to -3^+$

and, for $-3 < x < -2$, $x^3 < 0$, $x + 3 > 0$, $x + 2 < 0 \Rightarrow \dfrac{x^3}{(x+2)(x+3)} > 0$.

17. $\lim\limits_{x \to 2^+} \dfrac{-3}{\sqrt[3]{x-2}} = -\infty$ since $\sqrt[3]{x-2} \to 0$ as $x \to 2^+$ and $\dfrac{-3}{\sqrt[3]{x-2}} < 0$ for $x > 2$.

19. $\sqrt[3]{x}$ is large negative when x is large negative, so $\lim\limits_{x\to-\infty}\sqrt[3]{x}=-\infty$.

21. $\lim\limits_{x\to\infty}(x+\sqrt{x})=\infty$ since $x\to\infty$ and $\sqrt{x}\to\infty$.

23. $\lim\limits_{x\to\infty}(x^2-x^4)=\lim\limits_{x\to\infty}x^2(1-x^2)=-\infty$ since $x^2\to\infty$ and $1-x^2\to-\infty$.

25. $\lim\limits_{x\to\infty}\dfrac{x^3-1}{x^4+1}=\lim\limits_{x\to\infty}\dfrac{(1/x)-(1/x^4)}{1+(1/x^4)}=\dfrac{0-0}{1+0}=0$.

27. $\lim\limits_{x\to\infty}\dfrac{x}{\sqrt{x-1}}=\lim\limits_{x\to\infty}\dfrac{x/\sqrt{x}}{\sqrt{x-1/\sqrt{x}}}=\lim\limits_{x\to\infty}\dfrac{\sqrt{x}}{\sqrt{1-1/x}}=\infty$ since $\sqrt{x}\to\infty$ and

$\sqrt{1-1/x}\to1$. [OR: Divide numerator and denominator by x instead of $\sqrt{x}$.]

29. $\lim\limits_{x\to-\infty}\dfrac{x^8+3x^4+2}{x^5+x^3}=\lim\limits_{x\to-\infty}\dfrac{1+(3/x^4)+(2/x^8)}{(1/x^3)+(1/x^5)}=-\infty$ since denominator $\to0^-$.

[OR: Divide numerator and denominator by x^5 instead of x^8.]

31.   Since $x^2-1\to0$ and $y<0$ for $-1<x<1$ and $y>0$ for $x<-1$ and $x>1$, we

have $\lim\limits_{x\to1^-}\dfrac{x^2+4}{x^2-1}=-\infty,\quad\lim\limits_{x\to1^+}\dfrac{x^2+4}{x^2-1}=\infty,\quad\lim\limits_{x\to-1^-}\dfrac{x^2+4}{x^2-1}=\infty$, and

$\lim\limits_{x\to-1^+}\dfrac{x^2+4}{x^2-1}=-\infty$, so $x=1$ and $x=-1$ are vertical asymptotes.   Also

$\lim\limits_{x\to\pm\infty}\dfrac{x^2+4}{x^2-1}=\lim\limits_{x\to\pm\infty}\dfrac{1+4/x^2}{1-1/x^2}=\dfrac{1+0}{1-0}=1$, so $y=1$ is a horizontal asymptote.

33.   Since $y=\dfrac{x^3+1}{x^3+x}=\dfrac{x^3+1}{x(x^2+1)}>0$ for $x>0$ and $y<0$ for $-1<x<0$,

$\lim\limits_{x\to0^+}\dfrac{x^3+1}{x^3+x}=\infty$ and $\lim\limits_{x\to0^-}\dfrac{x^3+1}{x^3+x}=-\infty$, so $x=0$ is a vertical asymptote.

$\lim\limits_{x\to\pm\infty}\dfrac{x^3+1}{x^3+x}=\lim\limits_{x\to\pm\infty}\dfrac{1+1/x^3}{1+1/x^2}=1$, so $y=1$ is a horizontal asymptote.

35. $y=f(x)=x^2(x-2)(1-x)$. The y–intercept is $f(0)=0$, and the x–intercepts occur

when $y=0\Rightarrow x=0,1,2$.   Notice (as in Example 6) that since x^2 is always positive the

graph does not cross the x–axis at 0, but does cross the x–axis at 1 and 2.

$\lim\limits_{x\to\infty}x^2(x-2)(1-x)=-\infty$ since the first two factors are large positive and the third

large negative when x is large positive. $\lim\limits_{x\to-\infty}x^2(x-2)(1-x)=-\infty$ because the first

and third factors are large positive and the second large negative as $x\to-\infty$.

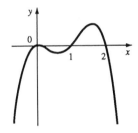

37. $y = f(x) = (x+4)^5(x-3)^4$. The y–intercept is $f(0) = 4^5(-3)^4 = 82944$. The x–intercepts occur when $y = 0 \Rightarrow x = -4, 3$. Notice (*as in Example 6*) that the graph does not cross the x–axis at 3 because $(x-3)^4$ is always positive, but does cross the x–axis at -4. $\lim\limits_{x \to \infty}(x+4)^5(x-3)^4 = \infty$ since both factors are large positive when x is large positive. $\lim\limits_{x \to -\infty}(x+4)^5(x-3)^4 = -\infty$ since the first factor is large negative and the second factor is large positive when x is large negative.

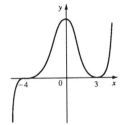

39. $\dfrac{1}{(x+3)^4} > 10000 \Leftrightarrow (x+3)^4 < \dfrac{1}{10000} \Leftrightarrow |x-(-3)| = |x+3| < \dfrac{1}{10}$.

41. Let $N < 0$ be given. Then, for $x < -1$, we have $\dfrac{5}{(x+1)^3} < N \Leftrightarrow \dfrac{5}{N} < (x+1)^3$ $\Leftrightarrow \sqrt[3]{5/N} < x+1$. Let $\delta = -\sqrt[3]{5/N}$. Then

$-1 - \delta < x < -1 \Rightarrow \sqrt[3]{5/N} < x+1 < 0 \Rightarrow \dfrac{5}{(x+1)^3} < N$, so $\lim\limits_{x \to -1^-} \dfrac{5}{(x+1)^3} = -\infty$.

43. Let M be given. Since $\lim\limits_{x \to a} f(x) = \infty$, there exists $\delta_1 > 0$ such that

$0 < |x-a| < \delta_1 \Rightarrow f(x) > M+1-c$. Since $\lim\limits_{x \to a} g(x) = c$, there exists $\delta_2 > 0$ such that $0 < |x-a| < \delta_2 \Rightarrow |g(x)-c| < 1 \Rightarrow g(x) > c-1$. Let δ be the smaller of δ_1 and

δ_2. Then $0 < |x-a| < \delta \Rightarrow f(x) + g(x) > (M+1-c) + (c-1) = M$.

Thus $\lim\limits_{x \to a} [f(x) + g(x)] = \infty$.

EXERCISES 1.8

1.   (a)   (i) $m = \lim\limits_{x \to -3} \dfrac{x^2 + 2x - 3}{x - (-3)} = \lim\limits_{x \to -3} \dfrac{(x+3)(x-1)}{x+3} = \lim\limits_{x \to -3} (x - 1) = -4.$

       (ii) $m = \lim\limits_{h \to 0} \dfrac{(-3+h)^2 + 2(-3+h) - 3}{h} = \lim\limits_{h \to 0} \dfrac{9 - 6h + h^2 - 6 + 2h - 3}{h}$

           $= \lim\limits_{h \to 0} \dfrac{h(h-4)}{h} = \lim\limits_{h \to 0} (h - 4) = -4.$

   (b)   The equation of the tangent line is $y - 3 = -4(x + 3)$ or $y = -4x - 9.$

   (c)

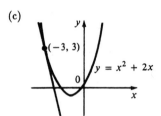

3.   Using (1.39), $m = \lim\limits_{x \to -2} \dfrac{1 - 2x - 3x^2 + 7}{x + 2} = \lim\limits_{x \to -2} \dfrac{-3x^2 - 2x + 8}{x + 2}$

   $= \lim\limits_{x \to -2} \dfrac{(-3x+4)(x+2)}{x+2} = \lim\limits_{x \to -2} (-3x + 4) = 10.$ Thus the equation of the tangent

   is $y + 7 = 10(x + 2)$ or $y = 10x + 13.$

5.   Using (1.39), $m = \lim\limits_{x \to -2} \dfrac{\frac{1}{x^2} - \frac{1}{4}}{x + 2} = \lim\limits_{x \to -2} \dfrac{4 - x^2}{4x^2(x+2)} = \lim\limits_{x \to -2} \dfrac{(2-x)(2+x)}{4x^2(x+2)}$

   $= \lim\limits_{x \to -2} \dfrac{2-x}{4x^2} = \frac{1}{4}.$ Thus the equation of the tangent is $y - \frac{1}{4} = \frac{1}{4}(x + 2)$ or $x - 4y + 3 = 0.$

7.   (a) $m = \lim\limits_{x \to a} \dfrac{\frac{2}{x+3} - \frac{2}{a+3}}{x - a} = \lim\limits_{x \to a} \dfrac{2(a-x)}{(x-a)(x+3)(a+3)} = \lim\limits_{x \to a} \dfrac{-2}{(x+3)(a+3)}$

      $= \dfrac{-2}{(a+3)^2}.$

   (b)   (i) $a = -1 \Rightarrow m = \dfrac{-2}{(-1+3)^2} = -\frac{1}{2}.$ (ii) $a = 0 \Rightarrow m = \dfrac{-2}{(0+3)^2} = -\frac{2}{9}.$

       (iii) $a = 1 \Rightarrow m = \dfrac{-2}{(1+3)^2} = -\frac{1}{8}.$

9.   Let $s(t) = 40t - 16t^2.$ $v(2) = \lim\limits_{t \to 2} \dfrac{s(t) - s(2)}{t - 2} = \lim\limits_{t \to 2} \dfrac{40t - 16t^2 - 16}{t - 2}$

   $= \lim\limits_{t \to 2} \dfrac{8(t-2)(-2t+1)}{t - 2} = \lim\limits_{t \to 2} 8(-2t + 1) = -24.$ Thus, the instantaneous velocity when $t = 2$ is -24 ft/s.

11. $v(a) = \lim\limits_{h \to 0} \dfrac{s(a+h) - s(a)}{h} = \lim\limits_{h \to 0} \dfrac{4(a+h)^3 + 6(a+h) + 2 - (4a^3 + 6a + 2)}{h}$

$= \lim\limits_{h \to 0} \dfrac{4a^3 + 12a^2h + 12ah^2 + 4h^3 + 6a + 6h + 2 - 4a^3 - 6a - 2}{h}$

$= \lim\limits_{h \to 0} \dfrac{12a^2h + 12ah^2 + 4h^3 + 6h}{h} = \lim\limits_{h \to 0} (12a^2 + 12ah + 4h^2 + 6) = 12a^2 + 6.$

So $v(1) = 12(1)^2 + 6 = 18$ m/s, $v(2) = 12(2)^2 + 6 = 54$ m/s, and
$v(3) = 12(3)^2 + 6 = 114$ m/s.

13. (a) (i) $[8, 11]$: $\dfrac{7.9 - 11.5}{3} = -1.2°/h.$ (ii) $[8, 10]$: $\dfrac{9.0 - 11.5}{2} = -1.25°/h.$

(iii) $[8, 9]$: $\dfrac{10.2 - 11.5}{1} = -1.3°/h.$

(b) The instantaneous rate of change is approximately $-1.6°/h$ at 8 P.M.

15. (a) (i) $\dfrac{\Delta C}{\Delta x} = \dfrac{C(105) - C(100)}{5} = \dfrac{6601.25 - 6500}{5} = \$20.25/\text{unit}.$

(ii) $\dfrac{\Delta C}{\Delta x} = \dfrac{C(101) - C(100)}{1} = \dfrac{6520.05 - 6500}{1} = \$20.05/\text{unit}.$

(b) $\dfrac{C(100 + h) - C(100)}{h} = \dfrac{5000 + 10(100 + h) + 0.05(100 + h)^2 - 6500}{h} = 20 + 0.05h,$

$h \ne 0$. So as h approaches 0, the rate of change of C approaches $\$20/\text{unit}$.

REVIEW EXERCISES FOR CHAPTER 1

1. False, since $\lim\limits_{x \to 4} \dfrac{2x}{x - 4}$ and $\lim\limits_{x \to 4} \dfrac{8}{x - 4}$ do not exist.

3. True, by (1.6) part 5 since $\lim\limits_{x \to 1} (x^2 + 2x - 4) = -1 \ne 0$.

5. False. For example, let $f(x) = \begin{cases} x^2 + 1 & x \ne 0 \\ 2 & x = 0 \end{cases}$

Then $f(x) > 1$ for all x, but $\lim\limits_{x \to 0} f(x) = \lim\limits_{x \to 0} (x^2 + 1) = 1$.

7. True by the definition of a limit with $\epsilon = 1$.

9. True by (1.23) with $a = 2$, $b = 5$, and $g(x) = 4x^2 - 11$.

11. $\lim\limits_{x \to 4} \sqrt{x + \sqrt{x}} = \sqrt{4 + \sqrt{4}} = \sqrt{6}$ since the function is continuous.

13. $\lim\limits_{t \to -1} \dfrac{t+1}{t^3 - t} = \lim\limits_{t \to -1} \dfrac{t+1}{t(t+1)(t-1)} = \lim\limits_{t \to -1} \dfrac{1}{t(t-1)} = \dfrac{1}{(-1)(-2)} = \dfrac{1}{2}.$

15. $\lim\limits_{h \to 0} \dfrac{(1+h)^2 - 1}{h} = \lim\limits_{h \to 0} \dfrac{1 + 2h + h^2 - 1}{h} = \lim\limits_{h \to 0} \dfrac{2h + h^2}{h} = \lim\limits_{h \to 0} (2 + h) = 2.$

17. $\lim\limits_{x \to -1} \dfrac{x^2 - x - 2}{x^2 + 3x - 2} = \dfrac{(-1)^2 - (-1) - 2}{(-1)^2 + 3(-1) - 2} = \dfrac{0}{-4} = 0.$

19. $\lim\limits_{s \to 16} \dfrac{4 - \sqrt{s}}{s - 16} = \lim\limits_{s \to 16} \dfrac{4 - \sqrt{s}}{(\sqrt{s} + 4)(\sqrt{s} - 4)} = \lim\limits_{s \to 16} \dfrac{-1}{\sqrt{s} + 4} = \dfrac{-1}{\sqrt{16} + 4} = -\dfrac{1}{8}.$

21. $\lim\limits_{x \to 8^-} \dfrac{|x - 8|}{x - 8} = \lim\limits_{x \to 8^-} \dfrac{-(x - 8)}{x - 8} = \lim\limits_{x \to 8^-} (-1) = -1.$

23. $\lim\limits_{x \to 0} \dfrac{1 - \sqrt{1 - x^2}}{x} \cdot \dfrac{1 + \sqrt{1 - x^2}}{1 + \sqrt{1 - x^2}} = \lim\limits_{x \to 0} \dfrac{1 - (1 - x^2)}{x(1 + \sqrt{1 - x^2})}$

$= \lim\limits_{x \to 0} \dfrac{x^2}{x(1 + \sqrt{1 - x^2})} = \lim\limits_{x \to 0} \dfrac{x}{1 + \sqrt{1 - x^2}} = 0.$

25. $\lim\limits_{t \to 6} \dfrac{17}{(t - 6)^2} = \infty,$ since $(t - 6)^2 \to 0$ and $\dfrac{17}{(t - 6)^2} > 0.$

27. $\lim\limits_{x \to 2^-} \dfrac{3}{\sqrt{2 - x}} = \infty,$ since $\sqrt{2 - x} \to 0$ and $\dfrac{3}{\sqrt{2 - x}} > 0.$

29. $\lim\limits_{x \to \infty} \dfrac{1 + x}{1 - x^2} = \lim\limits_{x \to \infty} \dfrac{1}{1 - x} = \lim\limits_{x \to \infty} \dfrac{1/x}{(1/x) - 1} = \dfrac{0}{0 - 1} = 0.$

31. $\lim\limits_{x \to \infty} \dfrac{\sqrt{x^2 - 9}}{2x - 6} = \lim\limits_{x \to \infty} \dfrac{\sqrt{1 - 9/x^2}}{2 - 6/x} = \dfrac{\sqrt{1 - 0}}{2 - 0} = \dfrac{1}{2}.$

33. $\lim\limits_{x \to -\infty} \dfrac{4x^5 + x^2 + 3}{2x^5 + x^3 + 1} = \lim\limits_{x \to -\infty} \dfrac{4 + 1/x^3 + 3/x^5}{2 + 1/x^2 + 1/x^5} = \dfrac{4 + 0 + 0}{2 + 0 + 0} = 2.$

35. $\lim\limits_{x \to \infty} \left[\sqrt[3]{x} - \dfrac{x}{3}\right] = \lim\limits_{x \to \infty} \sqrt[3]{x}(1 - \tfrac{1}{3}x^{2/3}) = -\infty,$ since $\sqrt[3]{x} \to \infty$ and $1 - \tfrac{1}{3}x^{2/3} \to -\infty.$

37. $\lim\limits_{x \to -2^+} \dfrac{x}{x^2 - 4} = \infty,$ since $x^2 - 4 \to 0$ and $\dfrac{x}{x^2 - 4} > 0$ for $-2 < x < 0.$

39. Given $\epsilon > 0$, we need $\delta > 0$ so that if $|x - 5| < \delta$ then $|(7x - 27) - 8| < \epsilon$

$\Leftrightarrow |7x - 35| < \epsilon \Leftrightarrow |x - 5| < \epsilon/7.$ So take $\delta = \epsilon/7.$ Then $|x - 5| < \delta$

$\Rightarrow |(7x - 27) - 8| < \epsilon.$ Thus $\lim\limits_{x \to 5} (7x - 27) = 8$ by the definition of a limit.

41. Given $\epsilon > 0$, we need $\delta > 0$ so that if $|x - 2| < \delta$ then $|x^2 - 3x - (-2)| < \epsilon.$

First, note that if $|x - 2| < 1$, then $-1 < x - 2 < 1$, so $0 < x - 1 < 2 \Rightarrow |x - 1| < 2.$

Now let $\delta = \min\{\epsilon/2, 1\}$, then $|x - 2| < \delta \Rightarrow |x^2 - 3x - (-2)| = |(x - 2)(x - 1)|$

$= |x - 2||x - 1| < (\epsilon/2)(2) = \epsilon.$ Thus $\lim\limits_{x \to 2} (x^2 - 3x) = -2$ by the definition of a limit.

43. (a) $f(x) = \sqrt{-x}$ if $x < 0$, $f(x) = 3 - x$ if $0 \le x < 3$, $f(x) = (x-3)^2$ if $x > 3$. Therefore:

(i) $\lim\limits_{x \to 0^+} f(x) = \lim\limits_{x \to 0^+} (3-x) = 3$. (ii) $\lim\limits_{x \to 0^-} f(x) = \lim\limits_{x \to 0^-} \sqrt{-x} = 0$.

(iii) Because of (i) and (ii), $\lim\limits_{x \to 0} f(x)$ does not exist.

(iv) $\lim\limits_{x \to 3^-} f(x) = \lim\limits_{x \to 3^-} (3-x) = 0$. (v) $\lim\limits_{x \to 3^+} f(x) = \lim\limits_{x \to 3^+} (x-3)^2 = 0$.

(vi) Because of (iv) and (v), $\lim\limits_{x \to 3} f(x) = 0$.

(b) f is discontinuous at 0 since $\lim\limits_{x \to 0} f(x)$ does not exist. f is discontinuous at 3 since $f(3)$ does not exist.

(c)

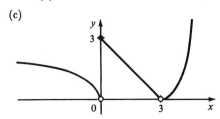

45. $f(x) = \dfrac{x+1}{x^2 + x + 1}$ is rational so it is continuous on its domain which is R.

(Note that $x^2 + x + 1 = 0$ has no real roots.)

47. $f(x) = 2x^3 + x^2 + 2$ is a polynomial, so it is continuous on $[-2, -1]$ and

$f(-2) = -10 < 0 < 1 = f(-1)$. So by the Intermediate Value Theorem there is a

number c in $(-2, -1)$ such that $f(c) = 0$, i.e., the equation $2x^3 + x^2 + 2 = 0$ has a root

in $(-2, -1)$.

49. (a) The slope of the tangent line at $(2,1)$ is $\lim\limits_{x \to 2} \dfrac{f(x) - f(2)}{x - 2} = \lim\limits_{x \to 2} \dfrac{9 - 2x^2 - 1}{x - 2}$

$= \lim\limits_{x \to 2} \dfrac{8 - 2x^2}{x - 2} = \lim\limits_{x \to 2} \dfrac{-2(x^2 - 4)}{x - 2} = \lim\limits_{x \to 2} \dfrac{-2(x-2)(x+2)}{x-2}$

$= \lim\limits_{x \to 2} -2(x+2) = -8$.

(b) The equation of this tangent line is $y - 1 = -8(x-2)$ or $8x + y = 17$.

51. (a) $s = 1 + 2t + t^2/4$. The average velocity over the time interval $[1, 1+h]$ is

$\dfrac{s(1+h) - s(1)}{h} = \dfrac{1 + 2(1+h) + (1+h)^2/4 - 13/4}{h} = \dfrac{10h + h^2}{4h} = \dfrac{10+h}{4}$. So for the

following intervals the average velocities are:

(i) $[1, 3]$: $(10 + 2)/4 = 3$ m/s.

(ii) $[1, 2]$: $(10 + 1)/4 = 2.75$ m/s.

(iii) $[1, 1.5]$: $(10 + 0.5)/4 = 2.625$ m/s.

(iv) $[1, 1.1]$: $(10 + 0.1)/4 = 2.525$ m/s.

(b) When $t = 1$ the velocity is $\lim\limits_{h \to 0} \dfrac{s(1 + h) - s(1)}{h} = \lim\limits_{h \to 0} \dfrac{10 + h}{4} = 2.5$ m/s.

53. $|f(x)| \le g(x) \Leftrightarrow -g(x) \le f(x) \le g(x)$ and $\lim\limits_{x \to a} g(x) = 0 = \lim\limits_{x \to a} -g(x)$.

Thus, by the Squeeze Theorem $\lim\limits_{x \to a} f(x) = 0$.

CHAPTER TWO

EXERCISES 2.1

1. $f'(2) = \lim\limits_{h \to 0} \dfrac{f(2+h) - f(2)}{h} = \lim\limits_{h \to 0} \dfrac{3(2+h)^2 - 5(2+h) - (3(2)^2 - 5(2))}{h}$

$= \lim\limits_{h \to 0} \dfrac{12 + 12h + 3h^2 - 10 - 5h - 12 + 10}{h} = \lim\limits_{h \to 0} \dfrac{3h^2 + 7h}{h} = \lim\limits_{h \to 0} (3h + 7) = 7$

$=$ slope of tangent at $x = 2$. So the equation of the tangent line at $(2, 2)$ is $y - 2 = 7(x - 2)$ or

$7x - y = 12$.

3. $F'(a) = \lim\limits_{h \to 0} \dfrac{F(a+h) - F(a)}{h} = \lim\limits_{h \to 0} \dfrac{\sqrt{a+h+7} - \sqrt{a+7}}{h}$

$= \lim\limits_{h \to 0} \dfrac{\sqrt{a+h+7} - \sqrt{a+7}}{h} \dfrac{\sqrt{a+h+7} + \sqrt{a+7}}{\sqrt{a+h+7} + \sqrt{a+7}} = \lim\limits_{h \to 0} \dfrac{a+h+7-a-7}{h(\sqrt{a+h+7} + \sqrt{a+7})}$

$= \lim\limits_{h \to 0} \dfrac{h}{h(\sqrt{a+h+7} + \sqrt{a+7})} = \dfrac{1}{2\sqrt{a+7}}$. Thus, $F'(2) = \dfrac{1}{2\sqrt{9}} = \dfrac{1}{6}$ and the equation of the

tangent line is $y - 3 = \frac{1}{6}(x - 2)$ or $x - 6y + 16 = 0$.

5. $v(2) = f'(2) = \lim\limits_{h \to 0} \dfrac{f(2+h) - f(2)}{h} = \lim\limits_{h \to 0} \dfrac{(2+h)^2 - 6(2+h) - 5 - (2^2 - 6(2) - 5)}{h}$

$= \lim\limits_{h \to 0} \dfrac{4 + 4h + h^2 - 12 - 6h - 5 - 4 + 12 + 5}{h} = \lim\limits_{h \to 0} \dfrac{h^2 - 2h}{h} = \lim\limits_{h \to 0} (h - 2) = -2 \text{ m/s}.$

7. $f'(a) = \lim\limits_{h \to 0} \dfrac{f(a+h) - f(a)}{h} = \lim\limits_{h \to 0} \dfrac{1 + (a+h) - 2(a+h)^2 - (1 + a - 2a^2)}{h}$

$= \lim\limits_{h \to 0} \dfrac{h - 4ah - 2h^2}{h} = \lim\limits_{h \to 0} (1 - 4a - 2h) = 1 - 4a.$

9. $f'(a) = \lim\limits_{h \to 0} \dfrac{f(a+h) - f(a)}{h} = \lim\limits_{h \to 0} \dfrac{\dfrac{a+h}{2(a+h)-1} - \dfrac{a}{2a-1}}{h}$

$= \lim\limits_{h \to 0} \dfrac{(a+h)(2a-1) - a(2a+2h-1)}{h(2a+2h-1)(2a-1)} = \lim\limits_{h \to 0} \dfrac{-h}{h(2a+2h-1)(2a-1)}$

$= \lim\limits_{h \to 0} \dfrac{-1}{(2a+2h-1)(2a-1)} = -\dfrac{1}{(2a-1)^2}.$

11. $f'(a) = \lim\limits_{h \to 0} \dfrac{f(a+h) - f(a)}{h} = \lim\limits_{h \to 0} \dfrac{\dfrac{2}{\sqrt{3-(a+h)}} - \dfrac{2}{\sqrt{3-a}}}{h} = \lim\limits_{h \to 0} \dfrac{2(\sqrt{3-a} - \sqrt{3-a-h})}{h\sqrt{3-a-h}\sqrt{3-a}}$

$= \lim\limits_{h \to 0} \dfrac{2(\sqrt{3-a} - \sqrt{3-a-h})}{h\sqrt{3-a-h}\sqrt{3-a}} \cdot \dfrac{(\sqrt{3-a} + \sqrt{3-a-h})}{(\sqrt{3-a} + \sqrt{3-a-h})}$

$$= \lim_{h \to 0} \frac{2\left(3 - a - (3 - a - h)\right)}{h\sqrt{3 - a - h}\sqrt{3 - a}\left(\sqrt{3 - a} + \sqrt{3 - a - h}\right)} = \lim_{h \to 0} \frac{2}{\sqrt{3 - a - h}\sqrt{3 - a}\left(\sqrt{3 - a} + \sqrt{3 - a - h}\right)}$$

$$= \frac{2}{\sqrt{3 - a}\sqrt{3 - a}\left(2\sqrt{3 - a}\right)} = \frac{1}{(3 - a)^{3/2}}.$$

13.   (a) $f'(a) = \lim\limits_{x \to a} \dfrac{f(x) - f(a)}{x - a} = \lim\limits_{x \to a} \dfrac{x^{1/3} - a^{1/3}}{x - a} = \lim\limits_{x \to a} \dfrac{x^{1/3} - a^{1/3}}{(x^{1/3})^3 - (a^{1/3})^3}$

$$= \lim_{x \to a} \frac{x^{1/3} - a^{1/3}}{(x^{1/3} - a^{1/3})\left((x^{1/3})^2 + x^{1/3} a^{1/3} + (a^{1/3})^2\right)} = \lim_{x \to a} \frac{1}{x^{2/3} + x^{1/3} a^{1/3} + a^{2/3}}$$

$$= \frac{1}{a^{2/3} + a^{1/3} a^{1/3} + a^{2/3}} = \frac{1}{3a^{2/3}}.$$

(b) $f'(a) = \lim\limits_{h \to 0} \dfrac{f(a + h) - f(a)}{h} = \lim\limits_{h \to 0} \dfrac{(a + h)^{1/3} - a^{1/3}}{(a + h) - a}$

$$= \lim_{h \to 0} \frac{(a + h)^{1/3} - a^{1/3}}{\left((a + h)^{1/3} - a^{1/3}\right)\left(\left((a + h)^{1/3}\right)^2 + (a + h)^{1/3} a^{1/3} + \left(a^{1/3}\right)^2\right)}$$

$$= \lim_{h \to 0} \frac{1}{(a + h)^{2/3} + (a + h)^{1/3} a^{1/3} + a^{2/3}} = \frac{1}{a^{2/3} + a^{1/3} a^{1/3} + a^{2/3}} = \frac{1}{3a^{2/3}}.$$

15. $\lim\limits_{h \to 0} \dfrac{\sqrt{1 + h} - 1}{h} = f'(1)$ where $f(x) = \sqrt{x}$. (*Or* $f'(0)$ *where* $f(x) = \sqrt{1 + x}$. *The answers to*

Exercises 15–20 are not unique.)

17. $\lim\limits_{x \to 1} \dfrac{x^9 - 1}{x - 1} = f'(1)$ where $f(x) = x^9$. (See equation 2.3.)

19. $\lim\limits_{t \to 0} \dfrac{\sin\left(\frac{\pi}{2} + t\right) - 1}{t} = f'(\frac{\pi}{2})$ where $f(x) = \sin x$.

21. $f'(x) = \lim\limits_{h \to 0} \dfrac{f(x + h) - f(x)}{h} = \lim\limits_{h \to 0} \dfrac{5(x + h) + 3 - (5x + 3)}{h} = \lim\limits_{h \to 0} \dfrac{5h}{h} = \lim\limits_{h \to 0} 5 = 5.$

Domain of f = domain of f' = R.

23. $f'(x) = \lim\limits_{h \to 0} \dfrac{f(x + h) - f(x)}{h} = \lim\limits_{h \to 0} \dfrac{(x + h)^3 - (x + h)^2 + 2(x + h) - (x^3 - x^2 + 2x)}{h}$

$$= \lim_{h \to 0} \frac{3x^2 h + 3xh^2 + h^3 - 2xh - h^2 + 2h}{h} = \lim_{h \to 0} (3x^2 + 3xh + h^2 - 2x - h + 2)$$

$= 3x^2 - 2x + 2$. Domain of f = domain of f' = R.

25. $f'(x) = \lim\limits_{h \to 0} \dfrac{f(x + h) - f(x)}{h} = \lim\limits_{h \to 0} \dfrac{x + h - \dfrac{2}{x + h} - \left(x - \dfrac{2}{x}\right)}{h} = \lim\limits_{h \to 0} \dfrac{h + \dfrac{2(x + h) - 2x}{x(x + h)}}{h}$

$$= \lim_{h \to 0} \left(1 + \frac{2}{x(x + h)}\right) = 1 + \frac{2}{x^2}.$$ Domain of f = domain of f' = $\{x \mid x \neq 0\}$.

27. $g'(x) = \lim\limits_{h \to 0} \dfrac{g(x+h) - g(x)}{h} = \lim\limits_{h \to 0} \dfrac{\sqrt{1 + 2(x+h)} - \sqrt{1 + 2x}}{h} \left(\dfrac{\sqrt{1 + 2(x+h)} + \sqrt{1 + 2x}}{\sqrt{1 + 2(x+h)} + \sqrt{1 + 2x}} \right)$

$= \lim\limits_{h \to 0} \dfrac{1 + 2x + 2h - (1 + 2x)}{h \left(\sqrt{1 + 2(x+h)} + \sqrt{1 + 2x} \right)} = \lim\limits_{h \to 0} \dfrac{2}{\sqrt{1 + 2(x+h)} + \sqrt{1 + 2x}} = \dfrac{1}{\sqrt{1 + 2x}}$. Domain of

$g = [-\tfrac{1}{2}, \infty)$, domain of $g' = (-\tfrac{1}{2}, \infty)$.

29. $G'(x) = \lim\limits_{h \to 0} \dfrac{G(x+h) - G(x)}{h} = \lim\limits_{h \to 0} \dfrac{\dfrac{4 - 3(x+h)}{2 + (x+h)} - \dfrac{4 - 3x}{2 + x}}{h}$

$= \lim\limits_{h \to 0} \dfrac{(4 - 3x - 3h)(2 + x) - (4 - 3x)(2 + x + h)}{h(2 + x + h)(2 + x)} = \lim\limits_{h \to 0} \dfrac{-10h}{h(2 + x + h)(2 + x)}$

$= \lim\limits_{h \to 0} \dfrac{-10}{(2 + x + h)(2 + x)} = \dfrac{-10}{(2 + x)^2}$. Domain of G = domain of $G' = \{x \mid x \neq -2\}$.

31. $f'(x) = \lim\limits_{h \to 0} \dfrac{f(x+h) - f(x)}{h} = \lim\limits_{h \to 0} \dfrac{(x+h)^4 - x^4}{h} = \lim\limits_{h \to 0} \dfrac{4x^3h + 6x^2h^2 + 4xh^3 + h^4}{h}$

$= \lim\limits_{h \to 0} (4x^3 + 6x^2h + 4xh^2 + h^3) = 4x^3$. Domain of f = domain of $f' = \mathbb{R}$.

33. $f(x) = x \Rightarrow f'(x) = \lim\limits_{h \to 0} \dfrac{x + h - x}{h} = \lim\limits_{h \to 0} 1 = 1$.

$f(x) = x^2 \Rightarrow f'(x) = \lim\limits_{h \to 0} \dfrac{(x+h)^2 - x^2}{h} = \lim\limits_{h \to 0} \dfrac{2xh + h^2}{h} = \lim\limits_{h \to 0} (2x + h) = 2x$.

$f(x) = x^3 \Rightarrow f'(x) = \lim\limits_{h \to 0} \dfrac{(x+h)^3 - x^3}{h} = \lim\limits_{h \to 0} \dfrac{3x^2h + 3xh^2 + h^3}{h} = \lim\limits_{h \to 0} (3x^2 + 3xh + h^2)$

$= 3x^2$. $f(x) = x^4 \Rightarrow f'(x) = 4x^3$ from Exercise 31.

Guess: The derivative of $f(x) = x^n$ is $f'(x) = nx^{n-1}$. Test for $n = 5$:

$f(x) = x^5 \Rightarrow f'(x) = \lim\limits_{h \to 0} \dfrac{(x+h)^5 - x^5}{h} = \lim\limits_{h \to 0} \dfrac{5x^4h + 10x^3h^2 + 10x^2h^3 + 5xh^4 + h^5}{h}$

$= \lim\limits_{h \to 0} (5x^4 + 10x^3h + 10x^2h^2 + 5xh^3 + h^4) = 5x^4$.

35.

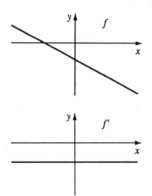

37.

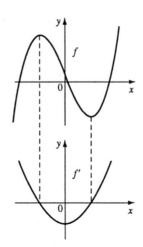

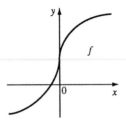

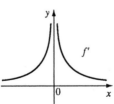

39.

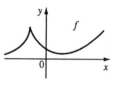

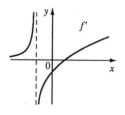

41.

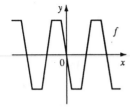

43.

45.

$$f(x) = |x - 6| = \begin{cases} x - 6 & \text{if } x \geq 6 \\ 6 - x & \text{if } x < 6 \end{cases} \qquad \lim_{x \to 6+} \frac{f(x) - f(6)}{x - 6} = \lim_{x \to 6+} \frac{|x - 6| - 0}{x - 6}$$

$$= \lim_{x \to 6+} \frac{x - 6}{x - 6} = \lim_{x \to 6+} 1 = 1. \text{ But } \lim_{x \to 6-} \frac{f(x) - f(6)}{x - 6} = \lim_{x \to 6-} \frac{|x - 6| - 0}{x - 6}$$

$$= \lim_{x \to 6-} \frac{6 - x}{x - 6} = \lim_{x \to 6-} (-1) = -1. \text{ So } f'(6) = \lim_{x \to 6} \frac{f(x) - f(6)}{x - 6} \text{ does not exist.}$$

However $f'(x) = \begin{cases} 1 & \text{if } x > 6 \\ -1 & \text{if } x < 6 \end{cases}$

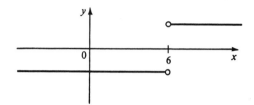

47. (a)
$$f(x) = x|x| = \begin{cases} x^2 & \text{if } x \geq 0 \\ -x^2 & \text{if } x < 0 \end{cases}$$

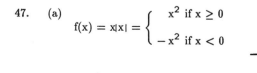

(b) Since $f(x) = x^2$ for $x \geq 0$, we have $f'(x) = 2x$ for $x > 0$. Since $f(x) = -x^2$ for $x < 0$, we have $f'(x) = -2x$ for $x < 0$. At $x = 0$, we have $f'(0) = \lim\limits_{x \to 0} \dfrac{f(x) - f(0)}{x - 0}$

$= \lim\limits_{x \to 0} \dfrac{x|x|}{x} = \lim\limits_{x \to 0} |x| = 0$ (*Example* 1.3.8). So f is differentiable at 0. Thus f is

differentiable for all x.

(c)
From part (b) we have $f'(x) = \begin{cases} 2x & \text{if } x \geq 0 \\ -2x & \text{if } x < 0 \end{cases} = 2|x|.$

49. (a) $f'_-(0.6) = \lim\limits_{h \to 0^-} \dfrac{f(0.6 + h) - f(0.6)}{h} = \lim\limits_{h \to 0^-} \dfrac{|5(0.6 + h) - 3| - |3 - 3|}{h}$

$= \lim\limits_{h \to 0^-} \dfrac{|3 + 5h - 3|}{h} = \lim\limits_{h \to 0^-} \dfrac{-5h}{h} = -5.$

$f'_+(0.6) = \lim\limits_{h \to 0^+} \dfrac{f(0.6 + h) - f(0.6)}{h} = \lim\limits_{h \to 0^+} \dfrac{|5(0.6 + h) - 3| - |3 - 3|}{h}$

$= \lim\limits_{h \to 0^+} \dfrac{5h}{h} = 5.$

(b) Since $f'_-(0.6) \neq f'_+(0.6)$, $f'(0.6)$ does not exist.

51. Since $f(x) = x \sin(1/x)$ when $x \neq 0$ and $f(0) = 0$, we have $f'(0) = \lim\limits_{h \to 0} \dfrac{f(0 + h) - f(0)}{h}$

$= \lim\limits_{h \to 0} \dfrac{h \sin(1/h) - 0}{h} = \lim\limits_{h \to 0} \sin\frac{1}{h}$. This limit does not exist since $\sin\frac{1}{h}$ takes the

values -1 and 1 on any interval containing 0. (*Compare with Example* 1.2.4.)

53. (a) If f is even, then $f'(-x) = \lim\limits_{h \to 0} \dfrac{f(-x + h) - f(-x)}{h} = \lim\limits_{h \to 0} \dfrac{-f(x - h) + f(x)}{h}$

$$= -\lim_{h \to 0} \frac{f(x-h) - f(x)}{-h} \ [Let \ \Delta x = -h.] \ = -\lim_{\Delta x \to 0} \frac{f(x + \Delta x) - f(x)}{\Delta x} = -f'(x).$$

Therefore f' is odd.

(b) If f is odd, then $f'(-x) = \lim_{h \to 0} \dfrac{f(-x+h) - f(-x)}{h} = \lim_{h \to 0} \dfrac{-f(x-h) + f(x)}{h}$

$$= \lim_{h \to 0} \frac{f(x-h) - f(x)}{-h} \ [Let \ \Delta x = -h] \ = \lim_{\Delta x \to 0} \frac{f(x + \Delta x) - f(x)}{\Delta x} = f'(x). \ \text{Therefore}$$

f' is even.

EXERCISES 2.2

1. $f(x) = x^2 - 10x + 100 \Rightarrow f'(x) = 2x - 10.$

3. $V(r) = \frac{4}{3}\pi r^3 \Rightarrow V'(r) = \frac{4}{3}\pi (3r^2) = 4\pi r^2.$

5. $F(x) = (16x)^3 = 4096x^3 \Rightarrow F'(x) = 4096(3x^2) = 12288x^2.$

7. $Y(t) = 6t^{-9} \Rightarrow Y'(t) = 6(-9)t^{-10} = -54t^{-10}.$

9. $g(x) = x^2 + 1/x^2 = x^2 + x^{-2} \Rightarrow g'(x) = 2x + (-2)x^{-3} = 2x - 2/x^3.$

11. $h(x) = \frac{x+2}{x-1} \Rightarrow h'(x) = \dfrac{(x-1)D(x+2) - (x+2)D(x-1)}{(x-1)^2} = \dfrac{x-1-(x+2)}{(x-1)^2} = \dfrac{-3}{(x-1)^2}$

13. $G(s) = (s^2 + s + 1)(s^2 + 2) \Rightarrow G'(s) = (2s+1)(s^2 + 2) + (s^2 + s + 1)(2s)$

$$= 4s^3 + 3s^2 + 6s + 2.$$

15. $y = (x^2 + 4x + 3)/\sqrt{x} = x^{3/2} + 4x^{1/2} + 3x^{-1/2}$

$\Rightarrow y' = \frac{3}{2}x^{1/2} + 4(\frac{1}{2})x^{-1/2} + 3(-\frac{1}{2})x^{-3/2} = \frac{3}{2}\sqrt{x} + 2/\sqrt{x} - 3/(2x\sqrt{x}).$

[Another method: Use the Quotient Rule.]

17. $y = \sqrt{5x} = \sqrt{5}x^{1/2} \Rightarrow y' = \sqrt{5}(\frac{1}{2})x^{-1/2} = \sqrt{5}/(2\sqrt{x}).$

19. $y = \dfrac{1}{x^4 + x^2 + 1} \Rightarrow y' = \dfrac{(x^4 + x^2 + 1)(0) - 1(4x^3 + 2x)}{(x^4 + x^2 + 1)^2} = -\dfrac{4x^3 + 2x}{(x^4 + x^2 + 1)^2}.$

21. $y = ax^2 + bx + c \Rightarrow y' = 2ax + b.$

23. $y = \dfrac{3t - 7}{t^2 + 5t - 4} \Rightarrow y' = \dfrac{(t^2 + 5t - 4)(3) - (3t - 7)(2t + 5)}{(t^2 + 5t - 4)^2} = \dfrac{-3t^2 + 14t + 23}{(t^2 + 5t - 4)^2}.$

25. $y = x + \sqrt[5]{x^2} = x + x^{2/5} \Rightarrow y' = 1 + \frac{2}{5}x^{-3/5} = 1 + 2/(5\sqrt[5]{x^3}).$

27. $u = x^{\sqrt{2}} \Rightarrow u' = \sqrt{2}x^{\sqrt{2} - 1}.$

29. $v = x\sqrt{x} + 1/(x^2\sqrt{x}) = x^{3/2} + x^{-5/2} \Rightarrow v' = \frac{3}{2}x^{1/2} - \frac{5}{2}x^{-7/2} = \frac{3}{2}\sqrt{x} - 5/(2x^3\sqrt{x}).$

31. $f(x) = \dfrac{x}{x + c/x} \Rightarrow f'(x) = \dfrac{(x + c/x)(1) - x(1 - c/x^2)}{(x + c/x)^2} = \dfrac{2cx}{(x^2 + c)^2}.$

33. $f(x) = \dfrac{x^5}{x^3 - 2} \Rightarrow f'(x) = \dfrac{(x^3 - 2)(5x^4) - x^5(3x^2)}{(x^3 - 2)^2} = \dfrac{2x^4(x^3 - 5)}{(x^3 - 2)^2}.$

35. $s = \dfrac{2 - 1/t}{t + 1} \Rightarrow s' = \dfrac{(t + 1)(1/t^2) - (2 - 1/t)(1)}{(t + 1)^2} = \dfrac{1 + 2t - 2t^2}{t^2(t + 1)^2}.$

37. $P(x) = a_n x^n + a_{n-1} x^{n-1} + \cdots + a_2 x^2 + a_1 x + a_0$

$\Rightarrow P'(x) = na_n x^{n-1} + (n - 1)a_{n-1} x^{n-2} + \cdots + 2a_2 x + a_1.$

39. $y = f(x) = x + 4/x \Rightarrow f'(x) = 1 - 4/x^2.$ So the slope of the tangent line at $(2, 4)$ is $f'(2) = 0$ and the equation is $y - 4 = 0$ or $y = 4$.

41. $y = f(x) = \dfrac{1}{x^2 + 1} \Rightarrow f'(x) = \dfrac{(x^2 + 1)(0) - 1(2x)}{(x^2 + 1)^2} = \dfrac{-2x}{(x^2 + 1)^2}.$ So the slope of the

tangent line at $\left(-1, \frac{1}{2}\right)$ is $f'(-1) = \frac{1}{2}$ and the equation is $y - \frac{1}{2} = \frac{1}{2}(x + 1)$ or $x - 2y + 2 = 0$.

43. $y = x\sqrt{x} = x^{3/2} \Rightarrow y' = \frac{3}{2}\sqrt{x}$ so the tangent line is parallel to $3x - y + 6 = 0$ when $\frac{3}{2}\sqrt{x} = 3$

$\Leftrightarrow \sqrt{x} = 2 \Leftrightarrow x = 4.$ So the point is $(4, 8)$.

45. $y = x^3 - x^2 - x + 1$ has a horizontal tangent when $y' = 3x^2 - 2x - 1 = 0$

$\Leftrightarrow (3x + 1)(x - 1) = 0 \Leftrightarrow x = 1$ or $-\frac{1}{3}.$ Therefore the points are $(1, 0)$ and $\left(-\frac{1}{3}, \frac{32}{27}\right)$.

47. If $y = f(x) = \dfrac{x}{x + 1}$ then $f'(x) = \dfrac{(x + 1)(1) - x(1)}{(x + 1)^2} = \dfrac{1}{(x + 1)^2}.$ When $x = a$, the equation of

the tangent line is $y - \dfrac{a}{a + 1} = \dfrac{1}{(a + 1)^2}(x - a).$ This line passes through $(1, 2)$ when

$2 - \dfrac{a}{a + 1} = \dfrac{1}{(a + 1)^2}(1 - a) \Leftrightarrow 2(a + 1)^2 = a(a + 1) + (1 - a) = a^2 + 1$

$\Leftrightarrow a^2 + 4a + 1 = 0.$ The quadratic formula gives the roots of this equation as $-2 \pm \sqrt{3}$, so there are two such tangent lines, which touch the curve at $\left(-2 + \sqrt{3}, (1 - \sqrt{3})/2\right)$ and $\left(-2 - \sqrt{3}, (1 + \sqrt{3})/2\right)$.

49. $y = 6x^3 + 5x - 3 \Rightarrow m = y' = 18x^2 + 5$, but $x^2 \geq 0$ for all x so $m \geq 5$ for all x.

51. $y = f(x) = 1 - x^2 \Rightarrow f'(x) = -2x$, so the tangent line at $(2, -3)$ has slope $f'(2) = -4$. The

normal line has slope $-1/(-4) = \frac{1}{4}$ and equation $y + 3 = \frac{1}{4}(x - 2)$ or $x - 4y = 14$.

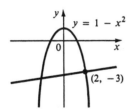

53. $y = f(x) = \sqrt[3]{x} = x^{1/3} \Rightarrow f'(x) = \frac{1}{3}x^{-2/3}$, so the tangent line at $(-8, -2)$ has slope $f'(-8) = 1/12$. The normal line has slope $-1/(1/12) = -12$ and equation $y + 2 = -12(x + 8)$ or $12x + y + 98 = 0$.

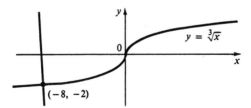

55. If the normal line has slope 16, then the tangent has slope $-\frac{1}{16}$, so $y' = 4x^3 = -\frac{1}{16} \Rightarrow$ $x^3 = -\frac{1}{64} \Rightarrow x = -\frac{1}{4}$. The point is $(-\frac{1}{4}, \frac{1}{256})$.

57. (a) $(fg)'(5) = f'(5)g(5) + f(5)g'(5) = 6(-3) + 1(2) = -16$.

(b) $\left(\frac{f}{g}\right)'(5) = \dfrac{f'(5)g(5) - f(5)g'(5)}{(g(5))^2} = \dfrac{6(-3) - 1(2)}{(-3)^2} = -\dfrac{20}{9}$.

(c) $\left(\frac{g}{f}\right)'(5) = \dfrac{g'(5)f(5) - g(5)f'(5)}{(f(5))^2} = \dfrac{2(1) - (-3)(6)}{1^2} = 20$.

59. (a) $(fgh)' = ((fg)h)' = (fg)'h + (fg)h' = (f'g + fg')h + (fg)h' = f'gh + fg'h + fgh'$.

(b) Putting $f = g = h$ in part (a), we have $\frac{d}{dx}\left(f(x)\right)^3 = (fff)' = f'ff + ff'f + fff' = 3fff'$ $= 3\left(f(x)\right)^2 f'(x)$.

61. $y = \sqrt{x}(x^4 + x + 1)(2x - 3)$. Using Exercise 59(a), we have
$y' = \left[1/(2\sqrt{x})\right](x^4 + x + 1)(2x - 3) + \sqrt{x}(4x^3 + 1)(2x - 3) + \sqrt{x}(x^4 + x + 1)(2)$
$= (x^4 + x + 1)(2x - 3)/(2\sqrt{x}) + \sqrt{x}\left[(4x^3 + 1)(2x - 3) + 2(x^4 + x + 1)\right]$.

63. $f(x) = 2 - x$ if $x \le 1$ and $f(x) = x^2 - 2x + 2$ if $x > 1$. Now we compute the right and left hand derivatives defined in Exercise 47 in Section 2.1. $f'_-(1) = \lim\limits_{h \to 0^-} \dfrac{f(1 + h) - f(1)}{h}$

$= \lim\limits_{h \to 0^-} \dfrac{2 - (1 + h) - 1}{h} = \lim\limits_{h \to 0^-} \dfrac{-h}{h} = \lim\limits_{h \to 0^-} -1 = -1$ and

$f'_+(1) = \lim\limits_{h \to 0^+} \dfrac{f(1 + h) - f(1)}{h} = \lim\limits_{h \to 0^+} \dfrac{(1 + h)^2 - 2(1 + h) + 2 - 1}{h} = \lim\limits_{h \to 0^+} \dfrac{h^2}{h}$

$= \lim\limits_{h \to 0^+} h = 0$. Thus $f'(1)$ does not exist since $f'_-(1) \ne f'_+(1)$, so f is not differentiable at 1. But $f'(x) = -1$ for $x < 1$ and $f'(x) = 2x - 2$ if $x > 1$.

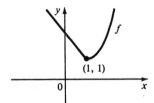 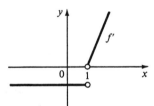

65. (a) Note that $x^2 - 9 < 0$ for $x^2 < 9 \Leftrightarrow |x| < 3 \Leftrightarrow -3 < x < 3$. So

$$f(x) = \begin{cases} x^2 - 9 & \text{if } x \le -3 \\ -x^2 + 9 & \text{if } -3 < x < 3 \\ x^2 - 9 & \text{if } x \ge 3 \end{cases} \Rightarrow f'(x) = \begin{cases} 2x & \text{if } x < -3 \\ -2x & \text{if } -3 < x < 3 \\ 2x & \text{if } x > 3 \end{cases}$$

To show that $f'(3)$ does not exist we investigate $\lim\limits_{h \to 0} \dfrac{f(3+h) - f(3)}{h}$ by computing the left

and right hand derivatives defined in Exercise 47 in Section 2.1.

$$f'_+(3) \lim_{h \to 0^+} \frac{f(3+h) - f(3)}{h} = \lim_{h \to 0^+} \frac{((3+h)^2 + 9) - 0}{h} = \lim_{h \to 0^+} \frac{6h + h^2}{h}$$

$$= \lim_{h \to 0^+} (6 + h) = 6. \quad f'_-(3) = \lim_{h \to 0^-} \frac{f(3+h) - f(3)}{h} = \lim_{h \to 0^-} \frac{(-(3+h)^2 + 9) - 0}{h}$$

$$= \lim_{h \to 0^-} (-6 + h) = -6. \quad \text{Since the right and left limits are different,}$$

$$\lim_{h \to 0} \frac{f(3+h) - f(3)}{h}$$

does not exist, that is, $f'(3)$ does not exist. Similarly, $f'(-3)$ does not exist. Therefore f is

not differentiable at 3 or -3.

(b)

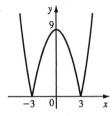 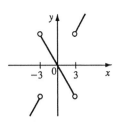

67. $F = f/g \Rightarrow f = Fg \Rightarrow f' = F'g + Fg' \Rightarrow F' = \dfrac{f' - Fg'}{g} = \dfrac{f' - (f/g)g'}{g} = \dfrac{f'g - fg'}{g^2}.$

EXERCISES 2.3

1. (a) $v(t) = f'(t) = 2t - 6.$

   (b) $v(2) = 2(2) - 6 = -2$ ft/s.

   (c) It is at rest when $v(t) = 2t - 6 = 0 \Leftrightarrow t = 3.$

   (d) It moves in the positive direction when $2t - 6 > 0 \Leftrightarrow t > 3.$

   (e) Distance in positive direction $= |f(4) - f(3)| = |1 - 0| = 1$ ft.

   Distance in negative direction $= |f(3) - f(0)| = |0 - 9| = 9$ ft.

   Total distance traveled $= 1 + 9 = 10$ ft.

   (f)

   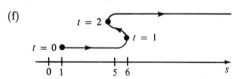

3. (a) $v(t) = f'(t) = 6t^2 - 18t + 12.$

   (b) $v(2) = 6(2)^2 - 18(2) + 12 = 0$ ft/s.

   (c) It is at rest when $v(t) = 6t^2 - 18t + 12 = 6(t-1)(t-2) = 0 \Leftrightarrow t = 1$ or 2.

   (d) It moves in the positive direction when $6(t-1)(t-2) > 0 \Leftrightarrow 0 \le t < 1$ or $t > 2$.

   (e) Distance in positive direction $= |f(4) - f(2)| + |f(1) - f(0)| = |33 - 5| + |6 - 1| = 33$ ft.

   Distance in negative direction $= |f(2) - f(1)| = |5 - 6| = 1$ ft. Total distance

   traveled $= 33 + 1 = 34$ ft.

   (f)

5. (a) $v(t) = s'(t) = \dfrac{(t^2 + 1)(1) - t(2t)}{(t^2 + 1)^2} = \dfrac{1 - t^2}{(t^2 + 1)^2}.$

   (b) $v(2) = \dfrac{1 - (2)^2}{(2^2 + 1)^2} = -\dfrac{3}{25}$ ft/s.

   (c) It is at rest when $v = 0 \Leftrightarrow 1 - t^2 = 0 \Leftrightarrow t = 1.$

   (d) It moves in the positive direction when $v > 0 \Leftrightarrow 1 - t^2 > 0 \Leftrightarrow t^2 < 1$

   $\Leftrightarrow 0 \le t < 1.$

   (e) Distance in positive direction $= |s(1) - s(0)| = |\frac{1}{2} - 0| = \frac{1}{2}$ ft.

   Distance in negative direction $= |s(4) - s(1)| = |\frac{4}{17} - \frac{1}{2}| = \frac{9}{34}$ ft.

Total distance traveled $= \frac{1}{2} + \frac{9}{34} = \frac{13}{17}$ ft.

(f)

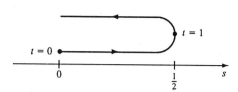

7. $s(t) = t^3 - 4.5t^2 - 7t \Rightarrow v(t) = s'(t) = 3t^2 - 9t - 7 = 5 \Leftrightarrow 3t^2 - 9t - 12 = 0$

    $\Leftrightarrow 3(t-4)(t+1) = 0 \Leftrightarrow t = 4$ or -1. Since $t \geq 0$ the particle reaches a velocity of 5 m/s at

    $t = 4$.

9. (a) $V(x) = x^3 \Rightarrow$ the average rate of change is:

    (i) $\dfrac{V(6) - V(5)}{6 - 5} = 6^3 - 5^3 = 216 - 125 = 91.$

    (ii) $\dfrac{V(5.1) - V(5)}{5.1 - 5} = \dfrac{(5.1)^3 - 5^3}{0.1} = 76.51.$

    (iii) $\dfrac{V(5.01) - V(5)}{5.01 - 5} = \dfrac{(5.01)^3 - 5^3}{0.01} = 75.1501.$

    (b) $V'(x) = 3x^2$, $V'(5) = 75.$

    (c) The surface area is $S(x) = 6x^2$, so $V'(x) = 3x^2 = \frac{1}{2}(6x^2) = \frac{1}{2}S(x).$

11. After t seconds the radius is $r = 60t$, so the area is $A(t) = \pi(60t)^2 = 3600\pi t^2$

    $\Rightarrow A'(t) = 7200\pi t \Rightarrow$

    (a) $A'(1) = 7200\pi$ cm^2/s.   (b) $A'(3) = 21600\pi$ cm^2/s.   (c) $A'(5) = 36000\pi$ cm^2/s.

13. $S(r) = 4\pi r^2 \Rightarrow S'(r) = 8\pi r.$ (a) $S'(1) = 8\pi$ ft^2/ft.

    (b) $S'(2) = 16\pi$ ft^2/ft.   (c) $S'(3) = 24\pi$ ft^2/ft.

15. $f(x) = 3x^2$, so the linear density at x is $\rho(x) = f'(x) = 6x.$

    (a) $\rho(1) = 6$ kg/m.   (b) $\rho(2) = 12$ kg/m.   (c) $\rho(3) = 18$ kg/m.

17. $Q(t) = t^3 - 2t^2 + 6t + 2$, so the current is $Q'(t) = 3t^2 - 4t + 6.$

    (a) $Q'(0.5) = 3(0.5)^2 - 4(0.5) + 6 = 4.75$ A.

    (b) $Q'(1) = 3(1)^2 - 4(1) + 6 = 5$ A.

19. (a) $PV = C \Rightarrow V = \dfrac{C}{P} \Rightarrow \dfrac{dV}{dP} = -\dfrac{C}{P^2}.$

    (b) $\beta = -\dfrac{1}{V}\dfrac{dV}{dP} = -\dfrac{1}{V}\left(-\dfrac{C}{P^2}\right) = \dfrac{C}{(PV)P} = \dfrac{C}{CP} = \dfrac{1}{P}.$

63

21.   (a)   rate of reaction $= \dfrac{d[C]}{dt} = \dfrac{a^2k\,(akt+1) - (a^2kt)\,(ak)}{(akt+1)^2} = \dfrac{a^2k[akt+1-akt]}{(akt+1)^2}$

$= \dfrac{a^2k}{(akt+1)^2}.$

(b) $a - x = a - \dfrac{a^2kt}{akt+1} = \dfrac{a^2kt + a - a^2kt}{akt+1} = \dfrac{a}{akt+1}.$

So $k(a-x)^2 = k\left(\dfrac{a}{akt+1}\right)^2 = \dfrac{a^2k}{(akt+1)^2} = \dfrac{dx}{dt}.$

23. $m(t) = 5 - 0.02t^2 \Rightarrow m'(t) = -0.04t \Rightarrow m'(1) = -0.04.$

25. $v(r) = \dfrac{P}{4\eta\ell}\left(R^2 - r^2\right) \Rightarrow v'(r) = \dfrac{P}{4\eta\ell}(-2r) = -\dfrac{Pr}{2\eta\ell}.$ When $\ell = 3$, $P = 3000$ and $\eta = 0.027$,

we have $v'(0.005) = -\dfrac{3000\,(0.005)}{2\,(0.027)\,(3)} \approx -92.6$ cm/s/cm.

27. $C(x) = 420 + 1.5x + 0.002x^2 \Rightarrow C'(x) = 1.5 + 0.004x$

$\Rightarrow C'(100) = 1.5 + (0.004)\,(100) = \$1.90/\text{item}.$

$C(101) - C(100) = (420 + 151.5 + 20.402) - (420 + 150 + 20) = \$1.902/\text{item}.$

29. $C(x) = 2000 + 3x + 0.01x^2 + 0.0002x^3 \Rightarrow C'(x) = 3 + 0.02x + 0.0006x^2$

$\Rightarrow C'(100) = 3 + 0.02\,(100) + 0.0006\,(10000) = 3 + 2 + 6 = \$11/\text{item}.$

$C(101) - C(100) = (2000 + 303 + 102.1 + 206.0602) - (2000 + 300 + 100 + 200)$

$= 11.0702 \approx \$11.07/\text{item}.$

EXERCISES 2.4

1. $\displaystyle\lim_{x\to0} (x^2 + \cos x) = \lim_{x\to0} x^2 + \lim_{x\to0} \cos x = 0^2 + \cos 0 = 0 + 1 = 1.$

3. $\displaystyle\lim_{x\to\pi/3} (\sin x - \cos x) = \sin\tfrac{\pi}{3} - \cos\tfrac{\pi}{3} = \tfrac{\sqrt{3}}{2} - \tfrac{1}{2}.$

5. $\displaystyle\lim_{x\to\pi/4} \dfrac{\sin x}{3x} = \dfrac{\sin(\pi/4)}{3\pi/4} = \dfrac{1/\sqrt{2}}{3\pi/4} = \dfrac{2\sqrt{2}}{3\pi}.$

7. $\displaystyle\lim_{t\to-3\pi} t^3 \sin^4 t = \left(\lim_{t\to-3\pi} t\right)^3 \left(\lim_{t\to-3\pi} \sin t\right)^4 = (-3\pi)^3 (0)^4 = 0.$

9. $\displaystyle\lim_{t\to0} \dfrac{\sin 5t}{t} = \lim_{t\to0} \dfrac{5\sin 5t}{5t} = 5\lim_{t\to0} \dfrac{\sin 5t}{5t} = 5\cdot 1 = 5.$

11. $\displaystyle\lim_{\theta\to0} \dfrac{\sin(\cos\theta)}{\sec\theta} = \dfrac{\sin\left(\lim_{\theta\to0} \cos\theta\right)}{\lim_{\theta\to0} \sec\theta} = \dfrac{\sin 1}{1} = \sin 1.$

13. $\displaystyle\lim_{x\to\pi/4} \dfrac{\tan x}{4x} = \dfrac{\tan(\pi/4)}{4\,(\pi/4)} = \tfrac{1}{\pi}.$

15. $\lim\limits_{\theta\to 0}\dfrac{\sin^2\theta}{\theta}=\lim\limits_{\theta\to 0}\left(\dfrac{\sin\theta}{\theta}\right)\sin\theta=\lim\limits_{\theta\to 0}\dfrac{\sin\theta}{\theta}\lim\limits_{\theta\to 0}\sin\theta=1\cdot 0=0.$

17. $\lim\limits_{x\to 0}\dfrac{\tan 3x}{3\tan 2x}=\lim\limits_{x\to 0}\dfrac{\frac{\tan 3x}{3x}}{2\frac{\tan 2x}{2x}}=\dfrac{1}{2}\dfrac{\lim\limits_{x\to 0}\frac{\sin 3x}{3x}\cdot\frac{1}{\cos 3x}}{\lim\limits_{x\to 0}\frac{\sin 2x}{2x}\cdot\frac{1}{\cos 2x}}=\dfrac{1}{2}\dfrac{1\cdot 1}{1\cdot 1}=\dfrac{1}{2}.$

19. $\lim\limits_{t\to 0}\dfrac{\sin^2 3t}{t^2}=\lim\limits_{t\to 0}9\left(\dfrac{\sin 3t}{3t}\right)^2=9\left(\lim\limits_{t\to 0}\dfrac{\sin 3t}{3t}\right)^2=9(1)^2=9.$

21. $\dfrac{d}{dx}(\csc x)=\dfrac{d}{dx}\left(\dfrac{1}{\sin x}\right)=\dfrac{(\sin x)(0)-1(\cos x)}{\sin^2 x}=\dfrac{-\cos x}{\sin^2 x}=-\dfrac{1}{\sin x}\cdot\dfrac{\cos x}{\sin x}$

$=-\csc x\cot x.$

23. $\dfrac{d}{dx}(\cot x)=\dfrac{d}{dx}\left(\dfrac{\cos x}{\sin x}\right)=\dfrac{(\sin x)(-\sin x)-(\cos x)(\cos x)}{\sin^2 x}=-\dfrac{\sin^2 x+\cos^2 x}{\sin^2 x}$

$=-\dfrac{1}{\sin^2 x}=-\csc^2 x.$

25. $y=\sin x+\cos x\Rightarrow dy/dx=\cos x-\sin x.$

27. $y=\csc x\cot x\Rightarrow dy/dx=(-\csc x\cot x)\cot x+\csc x(-\csc^2 x)=-\csc x(\cot^2 x+\csc^2 x).$

29. $y=\dfrac{\tan x}{x}\Rightarrow\dfrac{dy}{dx}=\dfrac{x\sec^2 x-\tan x}{x^2}.$

31. $y=\dfrac{x}{\sin x+\cos x}\Rightarrow\dfrac{dy}{dx}=\dfrac{(\sin x+\cos x)-x(\cos x-\sin x)}{(\sin x+\cos x)^2}$

$=\dfrac{(1+x)\sin x+(1-x)\cos x}{\sin^2 x+\cos^2 x+2\sin x\cos x}=\dfrac{(1+x)\sin x+(1-x)\cos x}{1+\sin 2x}.$

33. $y=x^{-3}\sin x\tan x\Rightarrow dy/dx=-3x^{-4}\sin x\tan x+x^{-3}\cos x\tan x+x^{-3}\sin x\sec^2 x$

$=x^{-4}\sin x(-3\tan x+x+x\sec^2 x).$

35. $y=\dfrac{x^2\tan x}{\sec x}\Rightarrow\dfrac{dy}{dx}=\dfrac{\sec x(2x\tan x+x^2\sec^2 x)-x^2\tan x\sec x\tan x}{\sec^2 x}$

$=\dfrac{2x\tan x+x^2(\sec^2 x-\tan^2 x)}{\sec x}=\dfrac{2x\tan x+x^2}{\sec x}.$

[Another method: Write $y=x^2\sin x$. Then $y'=2x\sin x+x^2\cos x$.]

37. $y=\tan x\Rightarrow y'=\sec^2 x\Rightarrow$ The slope of the tangent line at $(\frac{\pi}{4},1)$ is $\sec^2\left(\frac{\pi}{4}\right)=2$ and the
equation is $y-1=2\left(x-\frac{\pi}{4}\right)$ or $4x-2y=\pi-2.$

39. $y=x+2\sin x$ has a horizontal tangent when $y'=1+2\cos x=0\Leftrightarrow\cos x=-\frac{1}{2}$

$\Leftrightarrow x=(2n+1)\pi\pm\frac{\pi}{3}$, n an integer.

41. $\lim\limits_{\theta\to 0}\dfrac{\cos\theta-1}{\theta}=\lim\limits_{\theta\to 0}\dfrac{1-2\sin^2(\theta/2)-1}{\theta}=\lim\limits_{\theta\to 0}\dfrac{-\sin^2(\theta/2)}{(\theta/2)}$

$=-\lim\limits_{\theta\to 0}\dfrac{\sin(\theta/2)}{(\theta/2)}\lim\limits_{\theta\to 0}\sin(\theta/2)=-1\cdot 0=0.$

43. $\lim\limits_{x \to 0} \dfrac{\cot 2x}{\csc x} = \lim\limits_{x \to 0} \dfrac{\cos 2x \sin x}{\sin 2x} = \lim\limits_{x \to 0} \cos 2x \left(\dfrac{\frac{\sin x}{x}}{\frac{\sin 2x}{x}} \right)$

$= \lim\limits_{x \to 0} \cos 2x \dfrac{\lim\limits_{x \to 0} \frac{\sin x}{x}}{2 \lim\limits_{x \to 0} \frac{\sin 2x}{2x}} = 1 \cdot \dfrac{1}{2 \cdot 1} = \dfrac{1}{2}.$

45. $\lim\limits_{x \to \pi} \dfrac{\tan x}{\sin 2x} = \lim\limits_{x \to \pi} \dfrac{\sin x}{\cos x \, (2 \sin x \, \cos x)} = \lim\limits_{x \to \pi} \dfrac{1}{2 \cos^2 x} = \dfrac{1}{2(-1)^2} = \dfrac{1}{2}.$

47. Divide numerator and denominator by θ. (*$\sin \theta$ also works.*)

$\lim\limits_{\theta \to 0} \dfrac{\sin \theta}{\theta + \tan \theta} = \lim\limits_{\theta \to 0} \dfrac{\frac{\sin \theta}{\theta}}{1 + \frac{\sin \theta}{\theta} \cdot \frac{1}{\cos \theta}} = \dfrac{\lim\limits_{\theta \to 0} \frac{\sin \theta}{\theta}}{1 + \lim\limits_{\theta \to 0} \frac{\sin \theta}{\theta} \lim\limits_{\theta \to 0} \frac{1}{\cos \theta}}$

$= \dfrac{1}{1 + 1 \cdot 1} = \dfrac{1}{2}.$

49. $\lim\limits_{x \to 0^+} \sqrt{x} \csc \sqrt{x} = \lim\limits_{x \to 0^+} \dfrac{\sqrt{x}}{\sin \sqrt{x}} = \left(\lim\limits_{x \to 0^+} \dfrac{\sin \sqrt{x}}{\sqrt{x}} \right)^{-1} = 1^{-1} = 1.$

51. (a) $\dfrac{d}{dx} \tan x = \dfrac{d}{dx} \dfrac{\sin x}{\cos x} \Rightarrow \sec^2 x = \dfrac{\cos x \cos x - \sin x \, (-\sin x)}{\cos^2 x} = \dfrac{\cos^2 x + \sin^2 x}{\cos^2 x}.$ So

$\sec^2 x = \dfrac{1}{\cos^2 x}.$

(b) $\dfrac{d}{dx} \sec x = \dfrac{d}{dx} \dfrac{1}{\cos x} \Rightarrow \sec x \tan x = \dfrac{(\cos x)(0) - 1(-\sin x)}{\cos^2 x}.$ So $\sec x \tan x = \dfrac{\sin x}{\cos^2 x}.$

(c) $\dfrac{d}{dx} (\sin x + \cos x) = \dfrac{d}{dx} \dfrac{1 + \cot x}{\csc x} \Rightarrow$

$\cos x - \sin x = \dfrac{\csc x \, (-\csc^2 x) - (1 + \cot x)(-\csc x \cot x)}{\csc^2 x} = \dfrac{-\csc^2 x + \cot^2 x + \cot x}{\csc x}.$

So $\cos x - \sin x = \dfrac{\cot x - 1}{\csc x}.$

EXERCISES 2.5

1. $y = u^2, \; u = x^2 + 2x + 3.$

(a) $\dfrac{dy}{dx} = \dfrac{dy}{du} \dfrac{du}{dx} = 2u \, (2x + 2) = 4u \, (x + 1).$ When $x = 1$, $u = 1^2 + 2\,(1) + 3 = 6$, so

$\left. \dfrac{dy}{dx} \right|_{x=1} = 4\,(6)(1 + 1) = 48.$

(b) $y = u^2 = (x^2 + 2x + 3)^2 = x^4 + 4x^2 + 9 + 4x^3 + 6x^2 + 12x$

$= x^4 + 4x^3 + 10x^2 + 12x + 9$ so $\dfrac{dy}{dx} = 4x^3 + 12x^2 + 20x + 12$ and

$\dfrac{dy}{dx}\Big|_{x=1} = 4(1)^3 + 12(1)^2 + 20(1) + 12 = 48.$

3. $y = u^3, u = x + \frac{1}{x}.$

(a) $\dfrac{dy}{dx} = \dfrac{dy}{du}\dfrac{du}{dx} = 3u^2(1 - 1/x^2).$ When $x = 1$, $u = 1 + \frac{1}{1} = 2$, so

$\dfrac{dy}{dx}\Big|_{x=1} = 3(2)^2(1 - 1/1^2) = 0.$

(b) $y = u^3 = \left(x + \frac{1}{x}\right)^3 = x^3 + 3x^2\left(\frac{1}{x}\right) + 3x\left(\frac{1}{x}\right)^2 + \left(\frac{1}{x}\right)^3 = x^3 + 3x + 3x^{-1} + x^{-3}$, so

$\dfrac{dy}{dx} = 3x^2 + 3 - 3x^{-2} - 3x^{-4}$ and $\dfrac{dy}{dx}\Big|_{x=1} = 3(1)^2 + 3 - 3(1)^{-2} - 3(1)^{-4} = 0.$

5. $F(x) = (x^2 + 4x + 6)^5 \Rightarrow F'(x) = 5(x^2 + 4x + 6)^4 \dfrac{d}{dx}(x^2 + 4x + 6)$

$= 5(x^2 + 4x + 6)^4(2x + 4) = 10(x^2 + 4x + 6)^4(x + 2).$

7. $G(x) = (3x - 2)^{10}(5x^2 - x + 1)^{12}$

$\Rightarrow G'(x) = 10(3x - 2)^9(3)(5x^2 - x + 1)^{12} + (3x - 2)^{10}(12)(5x^2 - x + 1)^{11}(10x - 1)$

$= 30(3x - 2)^9(5x^2 - x + 1)^{12} + 12(3x - 2)^{10}(5x^2 - x + 1)^{11}(10x - 1).$

[This can be simplified to $6(3x - 2)^9(5x^2 - x + 1)^{11}(85x^2 - 51x + 9).$]

9. $f(t) = (2t^2 - 6t + 1)^{-8} \Rightarrow f'(t) = -8(2t^2 - 6t + 1)^{-9}(4t - 6)$

$= -16(2t^2 - 6t + 1)^{-9}(2t - 3).$

11. $g(x) = \sqrt{x^2 - 7x} = (x^2 - 7x)^{1/2} \Rightarrow g'(x) = \frac{1}{2}(x^2 - 7x)^{-1/2}(2x - 7) = \dfrac{2x - 7}{2\sqrt{x^2 - 7x}}.$

13. $h(t) = (t - 1/t)^{3/2} \Rightarrow h'(t) = \frac{3}{2}(t - 1/t)^{1/2}(1 + 1/t^2).$

15. $F(y) = \left(\dfrac{y - 6}{y + 7}\right)^3 \Rightarrow F'(y) = 3\left(\dfrac{y - 6}{y + 7}\right)^2 \dfrac{(y + 7)(1) - (y - 6)(1)}{(y + 7)^2} = 3\left(\dfrac{y - 6}{y + 7}\right)^2 \dfrac{13}{(y + 7)^2}$

$= \dfrac{39(y - 6)^2}{(y + 7)^4}.$

17. $f(z) = (2z - 1)^{-1/5} \Rightarrow f'(z) = -\frac{1}{5}(2z - 1)^{-6/5}(2) = -\frac{2}{5}(2z - 1)^{-6/5}.$

19. $y = (2x - 5)^4(8x^2 - 5)^{-3}$

$\Rightarrow y' = 4(2x - 5)^3(2)(8x^2 - 5)^{-3} + (2x - 5)^4(-3)(8x^2 - 5)^{-4}(16x)$

$= 8(2x - 5)^3(8x^2 - 5)^{-3} - 48x(2x - 5)^4(8x^2 - 5)^{-4}.$

[This simplifies to $8(2x - 5)^3(8x^2 - 5)^{-4}(-4x^2 + 30x - 5).$]

21. $y = \tan 3x \Rightarrow y' = \sec^2 3x \frac{d}{dx}(3x) = 3 \sec^2 3x.$

23. $y = \cos(x^3) \Rightarrow y' = -\sin(x^3)(3x^2) = -3x^2 \sin(x^3).$

25. $y = (1 + \cos^2 x)^6 \Rightarrow y' = 6(1 + \cos^2 x)^5 \, 2 \cos x (-\sin x) = -12 \cos x \sin x \, (1 + \cos^2 x)^5.$

27. $y = \cos(\tan x) \Rightarrow y' = -\sin(\tan x) \sec^2 x.$

29. $y = \sec^2 2x - \tan^2 2x \Rightarrow y' = 2 \sec 2x (\sec 2x \tan 2x)(2) - 2 \tan 2x \sec^2(2x)(2) = 0.$
    [Easier method: $y = \sec^2 2x - \tan^2 2x = 1 \Rightarrow y' = 0.$]

31. $y = \csc(x/3) \Rightarrow y' = -(1/3) \csc(x/3) \cot(x/3).$

33. $y = \sin^3 x + \cos^3 x \Rightarrow y' = 3 \sin^2 x \cos x + 3 \cos^2 x \, (-\sin x)$
    $= 3 \sin x \cos x \, (\sin x - \cos x).$

35. $y = \sin(1/x) \Rightarrow y' = \cos(1/x)(-1/x^2) = -\cos(1/x)/x^2.$

37. $y = \dfrac{1 + \sin 2x}{1 - \sin 2x} \Rightarrow y' = \dfrac{(1 - \sin 2x)(2 \cos 2x) - (1 + \sin 2x)(-2 \cos 2x)}{(1 - \sin 2x)^2} = \dfrac{4 \cos 2x}{(1 - \sin 2x)^2}.$

39. $y = \tan^2(x^3) \Rightarrow y' = 2 \tan(x^3) \sec^2(x^3)(3x^2) = 6x^2 \tan(x^3) \sec^2(x^3).$

41. $y = \cos^2 \left(\dfrac{1 - \sqrt{x}}{1 + \sqrt{x}} \right)$

    $\Rightarrow y' = 2 \cos \left(\dfrac{1 - \sqrt{x}}{1 + \sqrt{x}} \right)(-1) \sin \left(\dfrac{1 - \sqrt{x}}{1 + \sqrt{x}} \right) \dfrac{(1 + \sqrt{x})[-1/(2\sqrt{x})] - (1 - \sqrt{x})[1/(2\sqrt{x})]}{(1 + \sqrt{x})^2}$

    $= 2 \cos \left(\dfrac{1 - \sqrt{x}}{1 + \sqrt{x}} \right) \sin \left(\dfrac{1 - \sqrt{x}}{1 + \sqrt{x}} \right) \Big/ \left[\sqrt{x} \, (1 + \sqrt{x})^2 \right].$

43. $y = \cos^2(\cos x) + \sin^2(\cos x) = 1 \Rightarrow y' = 0.$

45. $y = \sqrt{x + \sqrt{x}} \Rightarrow y' = \frac{1}{2}(x + \sqrt{x})^{-1/2}(1 + \frac{1}{2}x^{-1/2}) = \dfrac{1}{2\sqrt{x + \sqrt{x}}} \left(1 + \dfrac{1}{2\sqrt{x}} \right).$

47. $f(x) = \left(x^3 + (2x - 1)^3 \right)^3 \Rightarrow f'(x) = 3 \left(x^3 + (2x - 1)^3 \right)^2 \left(3x^2 + 3(2x - 1)^2(2) \right)$
    $= 9 \left(x^3 + (2x - 1)^3 \right)^2 (9x^2 - 8x + 2).$

49. $p(t) = \left((1 + 2/t)^{-1} + 3t \right)^{-2}$
    $\Rightarrow p'(t) = -2 \left((1 + 2/t)^{-1} + 3t \right)^{-3} \left(-(1 + 2/t)^{-2}(-2/t^2) + 3 \right)$
    $= -2 \left((1 + 2/t)^{-1} + 3t \right)^{-3} \left(2(t + 2)^{-2} + 3 \right).$

51. $y = \sin(\tan \sqrt{\sin x}) \Rightarrow y' = \cos(\tan \sqrt{\sin x})(\sec^2 \sqrt{\sin x})[1/(2\sqrt{\sin x})](\cos x).$

68

53. $y = f(x) = (x^3 - x^2 + x - 1)^{10} \Rightarrow f'(x) = 10(x^3 - x^2 + x - 1)^9 (3x^2 - 2x + 1)$. The slope of the tangent at $(1,0)$ is $f'(1) = 0$ and its equation is $y - 0 = 0(x - 1)$ or $y = 0$.

55. $y = f(x) = 8/\sqrt{4 + 3x} \Rightarrow f'(x) = 8(-\frac{1}{2})(4 + 3x)^{-3/2}(3) = -12(4 + 3x)^{-3/2}$. The slope of the tangent at $(4,2)$ is $f'(4) = -3/16$ and its equation is $y - 2 = (-3/16)(x - 4)$ or $3x + 16y = 44$.

57. $y = f(x) = \cot^2 x \Rightarrow y' = 2\cot x (-\csc^2 x) = -2\cot x \csc^2 x$. The slope of the tangent at $(\pi/4, 1)$ is $f'(\pi/4) = -2(1)(\sqrt{2})^2 = -4$ and its equation is $y - 1 = -4(x - \pi/4)$ or $4x + y = \pi + 1$.

59. For the tangent line to be horizontal $f'(x) = 0$. $f(x) = 2\sin x + \sin^2 x$
$\Rightarrow f'(x) = 2\cos x + 2\sin x \cos x = 0 \Leftrightarrow 2\cos x(1 + \sin x) = 0 \Leftrightarrow \cos x = 0$ or $\sin x = -1$, so $x = (n + \frac{1}{2})\pi$ or $(2n + \frac{3}{2})\pi$ where n is any integer. So the points on the curve with a horizontal tangent are $((2n + \frac{1}{2})\pi, 3)$ and $((2n + \frac{3}{2})\pi, -1)$ where n is any integer.

61. $F(x) = f(g(x)) \Rightarrow F'(x) = f'(g(x))g'(x)$, so $F'(3) = f'(g(3))g'(3) = f'(6)g'(3) = 7 \cdot 4 = 28$

63. $f(x) = x^2 \sec^2 3x \Rightarrow f'(x) = 2x\sec^2 3x + x^2 (2\sec 3x)(\sec 3x \tan 3x)(3)$
$= 2x\sec^2 3x (1 + 3x\tan 3x)$. Domain of f = domain of $f' = \{x \mid \cos 3x \neq 0\}$
$= \{x \mid x \neq (2n - 1)\pi/6, \text{ n an integer}\}$.

65. $f(x) = \sqrt{\cos\sqrt{x}} \Rightarrow f'(x) = \frac{1}{2}(\cos\sqrt{x})^{-1/2}(-\sin\sqrt{x})(\frac{1}{2})x^{-1/2} = -\sin\sqrt{x}/(4\sqrt{x}\sqrt{\cos\sqrt{x}})$.

Domain of $f = \{x \mid x \geq 0 \text{ and } \cos\sqrt{x} \geq 0\}$

$= \{x \mid 0 \leq x \leq \pi^2/4 \text{ or } ((4n - 1)\pi/2)^2 \leq x \leq ((4n + 1)\pi/2)^2 \text{ for some } n = 1, 2, \dots\}$.

Domain of $f' = \{x \mid x > 0 \text{ and } \cos\sqrt{x} > 0\}$

$= \{x \mid 0 < x < \pi^2/4 \text{ or } ((4n - 1)\pi/2)^2 < x < ((4n + 1)\pi/2)^2 \text{ for some } n = 1, 2, \dots\}$.

67. $s(t) = 10 + \frac{1}{4}\sin(10\pi t) \Rightarrow$ the velocity after t seconds is
$v(t) = s'(t) = \frac{1}{4}\cos(10\pi t)(10\pi) = \frac{5\pi}{2}\cos(10\pi t)$ cm/s.

69. (a) $\frac{dB}{dt} = \left(0.35\cos(2\pi t/5.4)\right)(2\pi/5.4) = \frac{7}{54}\pi\cos(2\pi t/5.4)$.

(b) At $t = 1$, $\frac{dB}{dt} = \frac{7}{54}\pi\cos\frac{2\pi}{5.4} \approx 0.1613$.

71. (a) Since h is differentiable on $[0, \infty)$ and $\sqrt{x}$ is differentiable on $(0, \infty)$, it follows that $G(x) = h(\sqrt{x})$ is differentiable on $(0, \infty)$.

(b) By the Chain Rule, $G'(x) = h'(\sqrt{x})(d/dx)\sqrt{x} = h'(\sqrt{x})/(2\sqrt{x})$.

73. (a) $F(x) = f(\cos x) \Rightarrow F'(x) = f'(\cos x)(d/dx)(\cos x) = -\sin x\, f'(\cos x)$.

(b) $G(x) = \cos(f(x)) \Rightarrow G'(x) = -\sin(f(x))f'(x)$.

75. (a) If f is even, then $f(x) = f(-x)$. Using the Chain Rule to differentiate this equation, we get
$f'(x) = f'(-x)(d/dx)(-x) = -f'(-x)$. Thus $f'(-x) = -f'(x)$, so f' is odd.

(b) If f is odd, then $f(x) = -f(-x)$. Differentiating this equation, we get
$f'(x) = -f'(-x)(-1) = f'(-x)$, so f' is even.

77. $\frac{d}{dx}(\sin^n x \cos nx) = n \sin^{n-1} x \cos x \cos nx + \sin^n x (-n \sin nx)$
$= n \sin^{n-1} x (\cos nx \cos x - \sin nx \sin x) = n \sin^{n-1} x \cos(n+1)x.$

79. $f(x) = |x| = \sqrt{x^2} \Rightarrow f'(x) = \frac{1}{2}(x^2)^{-1/2}(2x) = x/\sqrt{x^2} = x/|x|.$

81. Using Exercise 79, we have $h(x) = x|2x - 1| \Rightarrow h'(x) = |2x - 1| + x\frac{2x - 1}{|2x - 1|}(2)$
$= |2x - 1| + \frac{2x(2x - 1)}{|2x - 1|}.$

83. Since $\theta° = (\pi/180)\theta$ rad, we have $\frac{d}{d\theta}(\sin \theta°) = \frac{d}{d\theta}(\sin \frac{\pi}{180}\theta) = \frac{\pi}{180}\cos \frac{\pi}{180}\theta = \frac{\pi}{180}\cos \theta°.$

EXERCISES 2.6

1. (a) $x^2 + 3x + xy = 5 \Rightarrow 2x + 3 + y + xy' = 0 \Rightarrow y' = -(2x + y + 3)/x.$

(b) $x^2 + 3x + xy = 5 \Rightarrow y = \frac{5 - x^2 - 3x}{x} = \frac{5}{x} - x - 3 \Rightarrow y' = -\frac{5}{x^2} - 1.$

(c) $y' = -\frac{2x + y + 3}{x} = \frac{-2x - 3 - (-3 - x + 5/x)}{x} = -1 - \frac{5}{x^2}.$

3. (a) $\frac{1}{x} + \frac{1}{y} = 3 \Rightarrow -\frac{1}{x^2} - \frac{1}{y^2}y' = 0 \Rightarrow y' = -\frac{y^2}{x^2}.$

(b) $\frac{1}{y} = 3 - \frac{1}{x} = \frac{3x - 1}{x} \Rightarrow y = \frac{x}{3x - 1} \Rightarrow y' = \frac{(3x - 1) - (x)(3)}{(3x - 1)^2} = -\frac{1}{(3x - 1)^2}.$

(c) $y' = -\frac{y^2}{x^2} = -\frac{x^2/(3x - 1)^2}{x^2} = -\frac{1}{(3x - 1)^2}.$

5. (a) $2y^2 + xy = x^2 + 3 \Rightarrow 4yy' + y + xy' = 2x \Rightarrow y' = \frac{2x - y}{x + 4y}.$

(b) Use the quadratic formula: $2y^2 + xy - (x^2 + 3) = 0$

$\Rightarrow y = \frac{1}{4}\left(-x \pm \sqrt{x^2 + 8(x^2 + 3)}\right) = \frac{1}{4}\left(-x \pm \sqrt{9x^2 + 24}\right) \Rightarrow y' = \frac{1}{4}\left(-1 \pm \frac{9x}{\sqrt{9x^2 + 24}}\right).$

(c) $y' = \frac{2x - y}{x + 4y} = \frac{2x - (-x \pm \sqrt{9x^2 + 24})/4}{x + (-x \pm \sqrt{9x^2 + 24})} = \frac{1}{4}\left(-1 \pm \frac{9x}{\sqrt{9x^2 + 24}}\right).$

7. $x^2 - xy + y^3 = 8 \Rightarrow 2x - y - xy' + 3y^2y' = 0 \Rightarrow y' = (y - 2x)/(3y^2 - x).$

9. $2y^2 + \sqrt[3]{xy} = 3x^2 + 17 \Rightarrow 4yy' + \frac{1}{3}x^{-2/3}y^{1/3} + \frac{1}{3}x^{1/3}y^{-2/3}y' = 6x$
$\Rightarrow y' = \frac{6x - \frac{1}{3}x^{-2/3}y^{1/3}}{4y + \frac{1}{3}x^{1/3}y^{-2/3}} = \frac{18x - x^{-2/3}y^{1/3}}{12y + x^{1/3}y^{-2/3}}.$

11. $x^4 + y^4 = 16 \Rightarrow 4x^3 + 4y^3 y' = 0 \Rightarrow y' = -x^3/y^3.$

13. $2xy = (x^2 + y^2)^{3/2} \Rightarrow 2y + 2xy' = \frac{3}{2}(x^2 + y^2)^{1/2}(2x + 2yy')$

$\Rightarrow y' = \dfrac{3x(x^2 + y^2)^{1/2} - 2y}{2x - 3y(x^2 + y^2)^{1/2}}.$

15. $\frac{y}{x-y} = x^2 + 1 \Rightarrow 2x = \dfrac{(x-y)y' - y(1-y')}{(x-y)^2} = \dfrac{xy' - y}{(x-y)^2} \Rightarrow y' = \frac{y}{x} + 2(x-y)^2.$

[Another method: Write the equation as $y = (x-y)(x^2 + 1) = x^3 + x - yx^2 - y$. This gives $y' = (3x^2 + 1 - 2xy)/(x^2 + 2)$.]

17. $\cos(x-y) = y \sin x \Rightarrow -\sin(x-y)(1-y') = y' \sin x + y \cos x$

$\Rightarrow y' = \dfrac{\sin(x-y) + y \cos x}{\sin(x-y) - \sin x}.$

19. $xy = \cot(xy) \Rightarrow y + xy' = -\csc^2(xy)(y + xy') \Rightarrow (y + xy')(1 + \csc^2(xy)) = 0$

$\Rightarrow y + xy' = 0 \Rightarrow y' = -y/x.$

21. $y^4 + x^2 y^2 + yx^4 = y + 1 \Rightarrow 4y^3 + 2x\dfrac{dx}{dy} y^2 + 2x^2 y + x^4 + 4yx^3 \dfrac{dx}{dy} = 1$

$\Rightarrow \dfrac{dx}{dy} = \dfrac{1 - 4y^3 - 2x^2 y - x^4}{2xy^2 + 4yx^3}.$

23. $x\big(f(x)\big)^3 + xf(x) = 6 \Rightarrow \big(f(x)\big)^3 + 3x\big(f(x)\big)^2 f'(x) + f(x) + xf'(x) = 0$

$\Rightarrow f'(x) = -\dfrac{\big(f(x)\big)^3 + f(x)}{3x\big(f(x)\big)^2 + x} \Rightarrow f'(3) = -\dfrac{(1)^3 + 1}{3(3)(1)^2 + 3} = -\frac{1}{6}.$

25. $\frac{x^2}{16} - \frac{y^2}{9} = 1 \Rightarrow \frac{x}{8} - \frac{2yy'}{9} = 0 \Rightarrow y' = \frac{9x}{16y}.$ When $x = -5$ and $y = \frac{9}{4}$ we have

$y' = \dfrac{9(-5)}{16(9/4)} = -\frac{5}{4}$ so the equation of the tangent is $y - (\frac{9}{4}) = (-\frac{5}{4})(x + 5)$ or

$5x + 4y + 16 = 0.$

27. $y^2 = x^3(2 - x) = 2x^3 - x^4 \Rightarrow 2yy' = 6x^2 - 4x^3 \Rightarrow y' = (3x^2 - 2x^3)/y.$ When $x = y = 1$,

$y' = (3(1)^2 - 2(1)^3)/1 = 1$, so the equation of the tangent line is $y - 1 = 1(x - 1)$ or $y = x.$

29. $2(x^2 + y^2)^2 = 25(x^2 - y^2) \Rightarrow 4(x^2 + y^2)(2x + 2yy') = 25(2x - 2yy') \Rightarrow$

$y' = \dfrac{25x - 4x(x^2 + y^2)}{25y + 4y(x^2 + y^2)}.$ When $x = 3$ and $y = 1$, $y' = \frac{75 - 120}{25 + 40} = -\frac{9}{13}$ so the equation

of the tangent is $y - 1 = (-\frac{9}{13})(x - 3)$ or $9x + 13y = 40.$

31. From question 29, a tangent to the lemniscate will be horizontal $\Leftrightarrow y' = 0$

$\Rightarrow 25x - 4x(x^2 + y^2) = 0 \Rightarrow x^2 + y^2 = 25/4.$ (Note that $x = 0 \Rightarrow y = 0$ and there is no horizontal tangent at the origin.) Putting this in the equation of the lemniscate, we get $x^2 - y^2 = 25/8$. Solving these two equations we have $x^2 = 75/16$ and $y^2 = 25/16$, so the points are $(\pm 5\sqrt{3}/4, \pm 5/4)$.

33. $\dfrac{x^2}{a^2} - \dfrac{y^2}{b^2} = 1 \Rightarrow \dfrac{2x}{a^2} - \dfrac{2yy'}{b^2} = 0 \Rightarrow y' = \dfrac{b^2 x}{a^2 y} \Rightarrow$ the equation of the tangent at (x_0, y_0) is

$y - y_0 = \dfrac{b^2 x_0}{a^2 y_0}(x - x_0)$. Multiplying both sides by $\dfrac{y_0}{b^2}$ gives $\dfrac{y_0 y}{b^2} - \dfrac{y_0^2}{b^2} = \dfrac{x_0 x}{a^2} - \dfrac{x_0^2}{a^2}$. Since

(x_0, y_0) lies on the hyperbola, we have $\dfrac{x_0 x}{a^2} - \dfrac{y_0 y}{b^2} = \dfrac{x_0^2}{a^2} - \dfrac{y_0^2}{b^2} = 1$.

35. If the circle has radius r, its equation is $x^2 + y^2 = r^2 \Rightarrow 2x + 2yy' = 0 \Rightarrow y' = -x/y$, so the slope of the tangent line at $P(x_0, y_0)$ is $-x_0/y_0$. The slope of OP is $y_0/x_0 = -1/(-x_0/y_0)$, so the tangent line is perpendicular to OP.

37. $2x^2 + y^2 = 3$ and $x = y^2$ intersect when $2x^2 + x - 3 = (2x + 3)(x - 1) = 0 \Leftrightarrow x = -3/2$ or 1, but $-3/2$ is extraneous. $2x^2 + y^2 = 3 \Rightarrow 4x + 2yy' = 0 \Rightarrow y' = -2x/y$ and $x = y^2 \Rightarrow 1 = 2yy' \Rightarrow y' = 1/(2y)$. At $(1, 1)$ the slopes are $m_1 = -2$ and $m_2 = 1/2$, so the curves are orthogonal there. By symmetry they are also orthogonal at $(1, -1)$.

39. $x^2 + y^2 = r^2$ is a circle with center O and $ax + by = 0$ is a line through O. By Exercise 35, the curves are orthogonal.

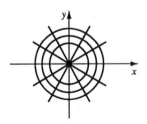

41. $y = cx^2 \Rightarrow y' = 2cx$ and $x^2 + 2y^2 = k \Rightarrow 2x + 4yy' = 0 \Rightarrow y' = -\dfrac{x}{2y} = -\dfrac{x}{2cx^2} = -\dfrac{1}{2cx}$

so the curves are orthogonal.

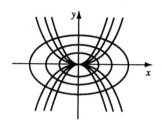

43. $x^2y^2 + xy = 2 \Rightarrow 2xy^2 + 2x^2yy' + y + xy' = 0 \Leftrightarrow y'(2x^2y + x) = -2xy^2 - y$

$\Leftrightarrow y' = -\dfrac{2xy^2 + y}{2x^2y + x}$. So $-\dfrac{2xy^2 + y}{2x^2y + x} = -1 \Leftrightarrow 2xy^2 + y = 2x^2y + x \Leftrightarrow$

$y(2xy + 1) = x(2xy + 1) \Leftrightarrow (2xy + 1)(y - x) = 0 \Leftrightarrow y = x$ or $xy = -\frac{1}{2}$. But

$xy = -\frac{1}{2} \Rightarrow x^2y^2 + xy = \frac{1}{4} - \frac{1}{2} \neq 2$ so we must have $x = y$, then $x^2y^2 + xy = 2 \Rightarrow$

$x^4 + x^2 = 2 \Leftrightarrow x^4 + x^2 - 2 = 0 \Leftrightarrow (x^2 + 2)(x^2 - 1) = 0$. So $x^2 = -2$, which is impossible,

or $x^2 = 1 \Leftrightarrow x = \pm 1$. So the points on the curve where the tangent line has a slope of -1 are

$(-1, -1)$ and $(1, 1)$.

EXERCISES 2.7

1. $f(x) = x^4 - 3x^3 + 16x \Rightarrow f'(x) = 4x^3 - 9x^2 + 16 \Rightarrow f''(x) = 12x^2 - 18x$.

3. $h(x) = \sqrt{x^2 + 1} \Rightarrow h'(x) = \frac{1}{2}(x^2 + 1)^{-1/2}(2x) = \dfrac{x}{\sqrt{x^2 + 1}}$

$\Rightarrow h''(x) = \dfrac{\sqrt{x^2 + 1} - x(x/\sqrt{x^2 + 1})}{x^2 + 1} = \dfrac{x^2 + 1 - x^2}{(x^2 + 1)^{3/2}} = \dfrac{1}{(x^2 + 1)^{3/2}}$.

5. $F(s) = (3s + 5)^8 \Rightarrow F'(s) = 8(3s + 5)^7(3) = 24(3s + 5)^7 \Rightarrow F''(s) = 168(3s + 5)^6(3)$

$= 504(3s + 5)^6$.

7. $y = \frac{x}{1-x} \Rightarrow y' = \frac{1(1-x)-x(-1)}{(1-x)^2} = \frac{1}{(1-x)^2} \Rightarrow y'' = -2(1-x)^{-3}(-1) = \frac{2}{(1-x)^3}.$

9. $y = (1-x^2)^{3/4} \Rightarrow y' = \frac{3}{4}(1-x^2)^{-1/4}(-2x) = -\frac{3}{2}x(1-x^2)^{-1/4} \Rightarrow$

   $y'' = -\frac{3}{2}(1-x^2)^{-1/4} - \frac{3}{2}x(-\frac{1}{4})(1-x^2)^{-5/4}(-2x) = -\frac{3}{2}(1-x^2)^{-1/4} - \frac{3}{4}x^2(1-x^2)^{-5/4} =$

   $\frac{3}{4}(1-x^2)^{-5/4}(x^2-2).$

11. $H(t) = \tan^3(2t-1)$

   $\Rightarrow H'(t) = 3\tan^2(2t-1)\sec^2(2t-1)(2) = 6\tan^2(2t-1)\sec^2(2t-1)$

   $\Rightarrow H''(t) = 12\tan(2t-1)\sec^2(2t-1)(2)\sec^2(2t-1)$

   $\qquad + 6\tan^2(2t-1)2\sec(2t-1)\sec(2t-1)\tan(2t-1)(2).$

   $= 24\tan(2t-1)\sec^4(2t-1) + 24\tan^3(2t-1)\sec^2(2t-1).$

13. $F(r) = \sec\sqrt{r} \Rightarrow F'(r) = (\sec\sqrt{r}\,\tan\sqrt{r})/(2\sqrt{r}) \Rightarrow F''(r) =$

   $$\frac{2\sqrt{r}\left(\{(\sec\sqrt{r}\,\tan\sqrt{r})/2\sqrt{r}\}\tan\sqrt{r} + \sec\sqrt{r}\,(\sec^2\sqrt{r})/2\sqrt{r}\right) - \left(\sec\sqrt{r}\,\tan\sqrt{r}\right)/\sqrt{r}}{4r}$$

   $= (\sec\sqrt{r}\,\tan^2\sqrt{r} + \sec^3\sqrt{r} - r^{-1/2}\sec\sqrt{r}\,\tan\sqrt{r})/(4r).$

15. $y = ax^2 + bx + c \Rightarrow y' = 2ax + b \Rightarrow y'' = 2a \Rightarrow y''' = 0.$

17. $y = \sqrt{5t-1} \Rightarrow y' = \frac{1}{2}(5t-1)^{-1/2}(5) = \frac{5}{2}(5t-1)^{-1/2}$

   $\Rightarrow y'' = -\frac{5}{4}(5t-1)^{-3/2}(5) = -\frac{25}{4}(5t-1)^{-3/2}$

   $\Rightarrow y''' = \frac{75}{8}(5t-1)^{-5/2}(5) = \frac{375}{8}(5t-1)^{-5/2}.$

19. $f(x) = (2-3x)^{-1/2} \Rightarrow f(0) = 2^{-1/2} = 1/\sqrt{2}.$

   $f'(x) = -\frac{1}{2}(2-3x)^{-3/2}(-3) = \frac{3}{2}(2-3x)^{-3/2} \Rightarrow f'(0) = \frac{3}{2}(2)^{-3/2} = 3/(4\sqrt{2}).$

   $f''(x) = -\frac{9}{4}(2-3x)^{-5/2}(-3) = \frac{27}{4}(2-3x)^{-5/2} \Rightarrow f''(0) = \frac{27}{4}(2)^{-5/2} = 27/(16\sqrt{2}).$

   $f'''(x) = \frac{405}{8}(2-3x)^{-7/2} \Rightarrow f'''(0) = \frac{405}{8}(2)^{-7/2} = 405/(64\sqrt{2}).$

21. $f(\theta) = \cot\theta \Rightarrow f'(\theta) = -\csc^2\theta \Rightarrow f''(\theta) = -2\csc\theta\,(-\csc\theta\cot\theta) = 2\csc^2\theta\cot\theta$

   $\Rightarrow f'''(\theta) = 2(-2\csc^2\theta\cot\theta)\cot\theta + 2\csc^2\theta(-\csc^2\theta) = -2\csc^2\theta(2\cot^2\theta + \csc^2\theta)$

   $\Rightarrow f'''(\pi/6) = -2(2)^2\left(2(\sqrt{3})^2 + (2)^2\right) = -80.$

23. $x^3 + y^3 = 1 \Rightarrow 3x^2 + 3y^2y' = 0 \Rightarrow y' = -\frac{x^2}{y^2} \Rightarrow y'' = -\frac{2xy^2 - 2x^2yy'}{y^4}$

   $= -\frac{2xy^2 - 2x^2y(-x^2/y^2)}{y^4} = -\frac{2xy^3 + 2x^4}{y^5} = -\frac{2x(y^3 + x^3)}{y^5} = -\frac{2x}{y^5}$ since x and y

   must satisfy the original equation of $x^3 + y^3 = 1.$

25. $x^2 + 6xy + y^2 = 8 \Rightarrow 2x + 6y + 6xy' + 2yy' = 0 \Rightarrow y' = -\dfrac{x + 3y}{3x + y} \Rightarrow$

$y'' = -\dfrac{(1 + 3y')(3x + y) - (x + 3y)(3 + y')}{(3x + y)^2} = \dfrac{8(y - xy')}{(3x + y)^2} = \dfrac{8(y - x(-x - 3y)/(3x + y))}{(3x + y)^2}$

$= \dfrac{8(y(3x + y) + x(x + 3y))}{(3x + y)^3} = \dfrac{8(x^2 + 6xy + y^2)}{(3x + y)^3} = \dfrac{64}{(3x + y)^3}$ since x and y must satisfy

the original equation of $x^2 + 6xy + y^2 = 8$.

27. $f(x) = x - x^2 + x^3 - x^4 + x^5 - x^6 \Rightarrow f'(x) = 1 - 2x + 3x^2 - 4x^3 + 5x^4 - 6x^5$

$\Rightarrow f''(x) = -2 + 6x - 12x^2 + 20x^3 - 30x^4 \Rightarrow f'''(x) = 6 - 24x + 60x^2 - 120x^3$

$\Rightarrow f^{(4)}(x) = -24 + 120x - 360x^2 \Rightarrow f^{(5)}(x) = 120 - 720x \Rightarrow f^{(6)}(x) = -720$

$\Rightarrow f^{(n)}(x) = 0$ for $7 \le n \le 73$.

29. $f(x) = x^n \Rightarrow f'(x) = nx^{n-1} \Rightarrow f''(x) = n(n-1)x^{n-2} \Rightarrow \cdots \Rightarrow$

$f^{(n)}(x) = n(n-1)(n-2)\cdots 2 \cdot 1 x^{n-n} = n!$.

31. $f(x) = 1/(3x^3) = \frac{1}{3}x^{-3} \Rightarrow f'(x) = \frac{1}{3}(-3)x^{-4} \Rightarrow f''(x) = \frac{1}{3}(-3)(-4)x^{-5}$

$\Rightarrow f'''(x) = \frac{1}{3}(-3)(-4)(-5)x^{-6} \Rightarrow \cdots \Rightarrow f^{(n)}(x) = \frac{1}{3}(-3)(-4)\cdots(-(n+2))x^{-(n+3)}$

$= (-1)^n 3 \cdot 4 \cdot 5 \cdots (n+2)/\left(3x^{n+3}\right) = (-1)^n(n+2)!/\left(6x^{n+3}\right)$.

33. In general, $Df(2x) = 2f'(2x)$, $D^2f(2x) = 4f''(2x)$, $\cdots$, $D^nf(2x) = 2^nf^{(n)}(2x)$. Since $f(x) = \cos x$ and $50 = 4(12) + 2$, we have $f^{(50)}(x) = f^{(2)}(x) = -\cos x$, so $D^{50}\cos 2x = -2^{50}\cos 2x$.

35. (a) $s = t^3 - 3t \Rightarrow v(t) = s'(t) = 3t^2 - 3 \Rightarrow a(t) = v'(t) = 6t$.

 (b) $a(1) = 6(1) = 6$ m/s^2.

 (c) $v(t) = 3t^2 - 3 = 0$ when $t^2 = 1$, i.e., $t = 1$ and $a(1) = 6$ m/s^2.

37. (a) $s = At^2 + Bt + C \Rightarrow v(t) = s'(t) = 2At + B \Rightarrow a(t) = v'(t) = 2A$.

 (b) $a(1) = 2A$ m/s^2.

 (c) $2A$ m/s^2 (since $a(t)$ is constant).

39. (a) $s(t) = t^4 - 4t^3 + 2 \Rightarrow v(t) = s'(t) = 4t^3 - 12t^2 \Rightarrow a(t) = v'(t) = 12t^2 - 24t$

 $= 12t(t - 2) = 0$ when $t = 0$ or 2.

 (b) $s(0) = 2$ m, $v(0) = 0$ m/s, $s(2) = -14$ m, $v(2) = -16$ m/s.

41. (a) $y(t) = A\sin \omega t \Rightarrow v(t) = y'(t) = A\omega \cos \omega t \Rightarrow a(t) = v'(t) = -A\omega^2 \sin \omega t$.

 (b) $a(t) = -A\omega^2 \sin \omega t = -\omega^2 y(t)$.

 (c) $|v(t)| = A\omega|\cos \omega t|$ is a maximum when $\cos \omega t = \pm 1 \Leftrightarrow \sin \omega t = 0 \Leftrightarrow$

 $a(t) = -A\omega^2 \sin^2 \omega t = 0$.

43. Let $P(x) = ax^2 + bx + c$. Then $P'(x) = 2ax + b$ and $P''(x) = 2a$. $P''(2) = 2 \Rightarrow 2a = 2$

$\Rightarrow a = 1.$ $P'(2) = 3 \Rightarrow 4a + b = 4 + b = 3 \Rightarrow b = -1.$ $P(2) = 5 \Rightarrow 2^2 - 2 + c = 5$
$\Rightarrow c = 3.$ So $P(x) = x^2 - x + 3.$

45. $P(x) = c_n x^n + c_{n-1} x^{n-1} + \cdots + c_1 x + c_0 \Rightarrow P'(x) = n c_n x^{n-1} + (n-1) c_{n-1} x^{n-2} + \cdots \Rightarrow$
$P''(x) = n(n-1) c_n x^{n-2} + \cdots \Rightarrow P^{(n)}(x) = n(n-1)(n-2)\cdots(1) c_n x^{n-n} = n! c_n$ which is a
constant. Therefore $P^{(m)}(x) = 0$ for $m > n.$

47. $f(x) = x g(x^2) \Rightarrow f'(x) = g(x^2) + x g'(x^2) 2x = g(x^2) + 2x^2 g'(x^2)$
$\Rightarrow f''(x) = 2x g'(x^2) + 4x g'(x^2) + 4x^3 g''(x^2) = 6x g'(x^2) + 4x^3 g''(x^2).$

49. $f(x) = g(\sqrt{x}) \Rightarrow f'(x) = \dfrac{g'(\sqrt{x})}{2\sqrt{x}} \Rightarrow f''(x) = \dfrac{\dfrac{g''(\sqrt{x})}{2\sqrt{x}} \cdot 2\sqrt{x} - g'(\sqrt{x})/\sqrt{x}}{4x} = \dfrac{g''(\sqrt{x}) - \dfrac{g'(\sqrt{x})}{\sqrt{x}}}{4x}.$

51. The Chain Rule says $\dfrac{dy}{dx} = \dfrac{dy}{du}\dfrac{du}{dx}$ and so $\dfrac{d^2y}{dx^2} = \dfrac{d}{dx}\left(\dfrac{dy}{dx}\right) = \dfrac{d}{dx}\left(\dfrac{dy}{du}\dfrac{du}{dx}\right)$

$= \left(\dfrac{d}{dx}\left(\dfrac{dy}{du}\right)\right)\dfrac{du}{dx} + \dfrac{dy}{du}\dfrac{d}{dx}\left(\dfrac{du}{dx}\right)$ *(Product Rule)*

$= \left(\dfrac{d}{du}\left(\dfrac{dy}{du}\right)\dfrac{du}{dx}\right)\dfrac{du}{dx} + \dfrac{dy}{du}\dfrac{d^2u}{dx^2} = \dfrac{d^2y}{du^2}\left(\dfrac{du}{dx}\right)^2 + \dfrac{dy}{du}\dfrac{d^2u}{dx^2}.$

EXERCISES 2.8

1. $V = x^3 \Rightarrow \dfrac{dV}{dt} = 3x^2 \dfrac{dx}{dt}.$

3. $xy = 1 \Rightarrow x\dfrac{dy}{dt} + y\dfrac{dx}{dt} = 0.$ If $\dfrac{dx}{dt} = 4$ and $x = 2,$ then $y = \dfrac{1}{2},$ so $\dfrac{dy}{dt} = -\dfrac{y}{x}\dfrac{dx}{dt}$
$= -\dfrac{1/2}{2}(4) = -1.$

5. If the radius is r and the diameter $x,$ then $V = \dfrac{4}{3}\pi r^3 = \dfrac{\pi}{6}x^3 \Rightarrow -1 = \dfrac{dV}{dt} = \dfrac{\pi}{2}x^2\dfrac{dx}{dt}$
$\Rightarrow \dfrac{dx}{dt} = -\dfrac{2}{\pi x^2}.$ When $x = 10,$ $\dfrac{dx}{dt} = -\dfrac{2}{\pi(100)} = -\dfrac{1}{50\pi}.$ So the rate of decrease is
$1/50\pi$ cm/min.

7.

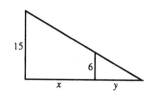

We are given that $dx/dt = 5$ ft/s. By similar triangles,

$\frac{15}{6} = \frac{x+y}{y} \Rightarrow y = \frac{2}{3}x$.

(a) The shadow moves at a rate of $\frac{d}{dt}(x+y) = \frac{d}{dt}(x+\frac{2}{3}x)$

$= \frac{5}{3}\frac{dx}{dt} = \frac{5}{3}(5) = \frac{25}{3}$ ft/s.

(b) The shadow lengthens at a rate of $\frac{dy}{dt} = \frac{d}{dt}(\frac{2}{3}x)$

$= \frac{2}{3}\frac{dx}{dt} = \frac{2}{3}(5) = \frac{10}{3}$ ft/s.

9.

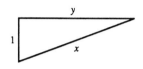

We are given that $dx/dt = 500$ mi/h. By the

Pythagorean Theorem, $y^2 = x^2 + 1$, so $2y\frac{dy}{dt} = 2x\frac{dx}{dt}$

$\Rightarrow \frac{dy}{dt} = \frac{x}{y}\frac{dx}{dt} = 500\frac{x}{y}$. When $y = 2$, $x = \sqrt{3}$, so

$\frac{dy}{dt} = 500(\sqrt{3}/2) = 250\sqrt{3}$ mi/h.

11.

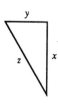

We are given that $\frac{dx}{dt} = 60$ mi/h and $\frac{dy}{dt} = 25$ mi/h.

$z^2 = x^2 + y^2 \Rightarrow 2z\frac{dz}{dt} = 2x\frac{dx}{dt} + 2y\frac{dy}{dt}$. After 2 hours,

$x = 120$ and $y = 50 \Rightarrow z = 130$, so $\frac{dz}{dt} = \frac{1}{z}\left(x\frac{dx}{dt} + y\frac{dy}{dt}\right)$

$= \frac{120(60) + 50(25)}{130} = 65$ mi/h.

13.

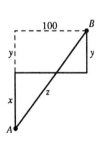

We are given that $\frac{dx}{dt} = 35$ km/h and $\frac{dy}{dt} = 25$ km/h.

$z^2 = (x+y)^2 + 100^2 \Rightarrow 2z\frac{dz}{dt} = 2(x+y)\left(\frac{dx}{dt} + \frac{dy}{dt}\right)$. At 4:00

P.M., $x = 140$ and $y = 100 \Rightarrow z = 260$, so

$\frac{dz}{dt} = \frac{x+y}{z}\left(\frac{dx}{dt} + \frac{dy}{dt}\right) = \frac{140+100}{260}(35+25) = \frac{720}{13}$

≈ 55.4 km/h.

15. $A = \frac{bh}{2}$, where b is the base and h is the altitude. We are given that $\frac{dh}{dt} = 1$ and $\frac{dA}{dt} = 2$. So

$2 = \frac{dA}{dt} = \frac{1}{2}b\frac{dh}{dt} + \frac{1}{2}h\frac{db}{dt} = \frac{1}{2}b + \frac{1}{2}h\frac{db}{dt} \Rightarrow \frac{db}{dt} = \frac{4-b}{h}$. When $h = 10$ and $A = 100$, we have

$b = 20$, so $\frac{db}{dt} = \frac{4-20}{10} = -1.6$ cm/min.

17.

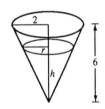

If C = the rate at which water is pumped in, then

$\frac{dV}{dt} = C - 10000$, where $V = \frac{1}{3}\pi r^2 h$ is the volume at time t.

By similar triangles, $\frac{r}{2} = \frac{h}{6} \Rightarrow r = \frac{h}{3} \Rightarrow V = \frac{1}{3}\pi\left(\frac{h}{3}\right)^2 h$

$= \frac{\pi}{27}h^3 \Rightarrow \frac{dV}{dt} = \frac{\pi}{9}h^2\frac{dh}{dt}$. When h = 200, $\frac{dh}{dt} = 20$, so

$C - 10000 = \frac{\pi}{9}(200)^2(20) \Rightarrow C = 10000 + \frac{800000}{9}\pi$

$\approx 2.89 \times 10^5$ cm^3/min.

19.

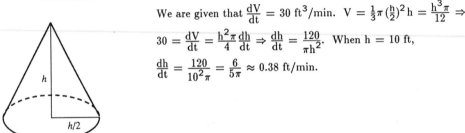

$V = \frac{1}{2}(0.3 + (0.3 + 2a))h(10)$, where $\frac{a}{h} = \frac{0.25}{0.5} = \frac{1}{2}$ so 2a = h

$\Rightarrow V = 5(0.6 + h)h = 3h + 5h^2 \Rightarrow 0.2 = \frac{dV}{dt} = (3 + 10h)\frac{dh}{dt}$

$\Rightarrow \frac{dh}{dt} = \frac{0.2}{3 + 10h}$. When h = 0.3, $\frac{dh}{dt} = \frac{0.2}{3 + 10(0.3)}$

$= \frac{0.2}{6}$ m/min $= \frac{10}{3}$ cm/min.

21.

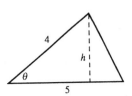

We are given that $\frac{dV}{dt} = 30$ ft^3/min. $V = \frac{1}{3}\pi\left(\frac{h}{2}\right)^2 h = \frac{h^3\pi}{12} \Rightarrow$

$30 = \frac{dV}{dt} = \frac{h^2\pi}{4}\frac{dh}{dt} \Rightarrow \frac{dh}{dt} = \frac{120}{\pi h^2}$. When h = 10 ft,

$\frac{dh}{dt} = \frac{120}{10^2\pi} = \frac{6}{5\pi} \approx 0.38$ ft/min.

23.

$A = \frac{bh}{2}$, but b = 5 m and h = $4\sin\theta$ so A = $10\sin\theta$. We are

given $\frac{d\theta}{dt} = 0.06$ rad/s. $\frac{dA}{dt} = 10\cos\theta\frac{d\theta}{dt} = 0.6\cos\theta$. When

$\theta = \frac{\pi}{3}$, $\frac{dA}{dt} = 10(0.06)(\cos\frac{\pi}{3}) = (0.6)(\frac{1}{2}) = 0.3$ m^2/s.

25. $PV = C \Rightarrow P\frac{dV}{dt} + V\frac{dP}{dt} = 0 \Rightarrow \frac{dV}{dt} = -\frac{V}{P}\frac{dP}{dt}$. When V = 600, P = 150 and $\frac{dP}{dt} = 20$, we

have $\frac{dV}{dt} = -\frac{600}{150}(20) = -80$, so the volume is decreasing at a rate of 80 cm^3/min.

27.

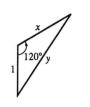

We are given that $\frac{dx}{dt} = 30$ km/h. By the Law of Cosines,

$$y^2 = x^2 + 1 - 2x\cos 120° = x^2 + 1 - 2x\left(-\tfrac{1}{2}\right) = x^2 + x + 1,$$

so $2y\frac{dy}{dt} = 2x\frac{dx}{dt} + \frac{dx}{dt} \Rightarrow \frac{dy}{dt} = \frac{2x+1}{2y}\frac{dx}{dt}$. After 1 minute,

$x = \frac{300}{60} = 5 \Rightarrow y = \sqrt{31} \Rightarrow$

$\frac{dy}{dt} = \frac{2(5)+1}{2\sqrt{31}}(300) = \frac{1650}{\sqrt{31}} \approx 296$ km/h.

29.

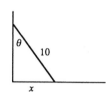

We are given that $\frac{dx}{dt} = 2$ ft/s. $x = 10\sin\theta$

$\Rightarrow \frac{dx}{dt} = 10\cos\theta\,\frac{d\theta}{dt}$. When $\theta = \pi/4$, $\frac{d\theta}{dt} = \frac{2}{10(1/\sqrt{2})}$

$= \frac{\sqrt{2}}{5}$ rad/s.

EXERCISES 2.9

1. $y = x^5 \Rightarrow dy = 5x^4 dx$.

3. $y = \sqrt{x^4 + x^2 + 1} \Rightarrow dy = \tfrac{1}{2}(x^4 + x^2 + 1)^{-1/2}(4x^3 + 2x)dx = \dfrac{2x^3 + x}{\sqrt{x^4 + x^2 + 1}}dx$.

5. $y = \frac{x-2}{2x+3} \Rightarrow dy = \dfrac{(2x+3)-(x-2)(2)}{(2x+3)^2}dx = \dfrac{7}{(2x+3)^2}dx$.

7. $y = \sin 2x \Rightarrow dy = 2\cos 2x\,dx$.

9. (a) $y = 1 - x^2 \Rightarrow dy = -2x\,dx$.

    (b) When $x = 5$ and $dx = \tfrac{1}{2}$, $dy = -2(5)\left(\tfrac{1}{2}\right) = -5$.

11. (a) $y = (x^2 + 5)^3 \Rightarrow dy = 3(x^2 + 5)^2\,2x\,dx = 6x(x^2 + 5)^2\,dx$.

    (b) When $x = 1$ and $dx = 0.05$, $dy = 6(1)(1^2 + 5)^2(0.05) = 10.8$.

13. (a) $y = (3x+2)^{-1/3} \Rightarrow dy = -\tfrac{1}{3}(3x+2)^{-4/3}(3) = -(3x+2)^{-4/3}\,dx$.

    (b) When $x = 2$ and $dx = -0.04$, $dy = -8^{-4/3}(-0.04) = 0.0025$.

15. (a) $y = \cos x \Rightarrow dy = -\sin x\,dx$.

    (b) When $x = \pi/6$ and $dx = 0.05$, $dy = -\tfrac{1}{2}(0.05) = -0.025$.

17. $y = x^2$, $x = 1$, $\Delta x = 0.5 \Rightarrow \Delta y = (1.5)^2 - 1^2 = 1.25$. $dy = 2x\,dx = 2(1)(0.5) = 1$.

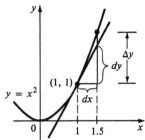

19. $y = 6 - x^2$, $x = -2$, $\Delta x = 0.4 \Rightarrow \Delta y = \left(6 - (-1.6)^2\right) - \left(6 - (-2)^2\right) = 1.44$.

$dy = -2x\,dx = -2(-2)(0.4) = 1.6$.

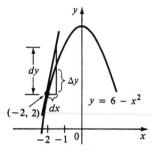

21. $y = f(x) = 2x^3 + 3x - 4$, $x = 3 \Rightarrow dy = (6x^2 + 3)\,dx = 57\,dx$.

$\Delta x = 1 \Rightarrow \Delta y = f(4) - f(3) = 136 - 59 = 77$, $dy = 57(1) = 57$, $\Delta y - dy = 77 - 57 = 20$.

$\Delta x = 0.5 \Rightarrow \Delta y = f(3.5) - f(3) = 92.25 - 59 = 33.25$, $dy = 57(0.5) = 28.5$,

$\Delta y - dy = 33.25 - 28.5 = 4.75$.

$\Delta x = 0.1 \Rightarrow \Delta y = f(3.1) - f(3) = 64.882 - 59 = 5.882$, $dy = 57(0.1) = 5.7$,

$\Delta y - dy = 5.882 - 5.7 = 0.182$.

$\Delta x = 0.01 \Rightarrow \Delta y = f(3.01) - f(3) = 59.571802 - 59 = 0.571802$, $dy = 57(0.01) = 0.57$,

$\Delta y - dy = 0.571802 - 0.57 = 0.001802$.

23. $y = f(x) = \sqrt{x} \Rightarrow dy = (1/2\sqrt{x})\,dx$. When $x = 36$ and $dx = 0.1$,

$dy = [1/(2\sqrt{36})](0.1) = 1/120$, so $\sqrt{36.1} = f(36.1) \approx f(36) + dy = \sqrt{36} + 1/120 \approx 6.0083$.

25. $y = \sqrt[3]{x} \Rightarrow dy = \frac{1}{3}x^{-2/3}\,dx$. When $x = 216$ and $dx = 2$, $dy = \frac{1}{3}(216)^{-2/3}(2) = 1/54$, so

$\sqrt[3]{218} = f(218) \approx f(216) + dy = 6 + 1/54 \approx 6.0185$.

27. $y = f(x) = 1/x \Rightarrow dy = (-1/x^2)\,dx$. When $x = 10$ and $dx = 0.1$,

$dy = (-1/100)(0.1) = -0.001$, so $1/10.1 = f(10.1) \approx f(10) + dy = 0.1 - 0.001 = 0.099$

29. $y = f(x) = \sin x \Rightarrow dy = \cos x\,dx$. When $x = \pi/3$ and $dx = -\pi/180$,

$dy = \cos(\pi/3)(-\pi/180) = -\pi/360$, so $\sin 59° = f(59\pi/180) \approx f(\pi/3) + dy$

$= \sqrt{3}/2 - \pi/360 \approx 0.857$.

31. $y = f(x) = \tan x \Rightarrow dy = \sec^2 x \, dx$. When $x = \pi/4$ and $dx = \pi/90$,

    $dy = \sec^2(\pi/4)(\pi/90) = 2\pi/90 = \pi/45$, so $\tan 47° = f(47\pi/180) \approx f(\pi/4) + dy$

    $= 1 + \pi/45 \approx 1.07$.

33. $L(x) = f(1) + f'(1)(x-1)$. $f(x) = x^3 \Rightarrow f'(x) = 3x^2$ so $f(1) = 1$ and $f'(1) = 3$. So

    $L(x) = 1 + 3(x-1) = 3x - 2$.

35. $f(x) = 1/x \Rightarrow f'(x) = -1/x^2$. So $f(4) = \frac{1}{4}$ and $f'(4) = -\frac{1}{16}$. So

    $L(x) = f(4) + f'(4)(x-4) = \frac{1}{4} + (-\frac{1}{16})(x-4) = \frac{1}{2} - \frac{x}{16}$.

37. $f(x) = \sqrt{1+x} \Rightarrow f'(x) = \dfrac{1}{2\sqrt{1+x}}$ so $f(0) = 1$ and $f'(0) = \frac{1}{2}$. So

    $f(x) \approx f(0) + f'(0)(x-0) = 1 + \frac{1}{2}(x-0) = 1 + \frac{1}{2}x$.

39. $f(x) = \dfrac{1}{(1+2x)^4} \Rightarrow f'(x) = \dfrac{-8}{(1+2x)^5}$ so $f(0) = 1$ and $f'(0) = -8$. So

    $f(x) \approx f(0) + f'(0)(x-0) = 1 + (-8)(x-0) = 1 - 8x$.

41. $f(x) = \sqrt{1-x} \Rightarrow f'(x) = \dfrac{-1}{2\sqrt{1-x}}$ so $f(0) = 1$ and $f'(0) = -\frac{1}{2}$. So

    $\sqrt{1-x} = f(x) \approx f(0) + f'(0)(x-0) = 1 + (-\frac{1}{2})(x-0) = 1 - \frac{1}{2}x$. So

    $\sqrt{0.9} = \sqrt{1-0.1} \approx 1 - \frac{1}{2}(0.1) = 0.95$ and $\sqrt{0.99} = \sqrt{1-0.01} \approx 1 - \frac{1}{2}(0.01) = 0.995$.

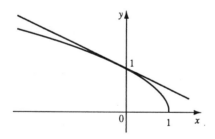

43. (a) If x is the edge length, then $V = x^3 \Rightarrow dV = 3x^2 \, dx$. When $x = 30$ and $dx = 0.1$,

    $dV = 3(30)^2(0.1) = 270$, so the maximum error ≈ 270 cm^3.

    (b) $S = 6x^2 \Rightarrow dS = 12x \, dx$. When $x = 30$ and $dx = 0.1$, $dS = 12(30)(0.1) = 36$, so the

    maximum error ≈ 36 cm^2.

45. (a) For a sphere of radius r, the circumference is $C = 2\pi r$ and the surface area is $S = 4\pi r^2$, so

    $r = C/(2\pi) \Rightarrow S = 4\pi(C/2\pi)^2 = C^2/\pi \Rightarrow dS = (2/\pi)C \, dC$. When $C = 84$ and

    $dC = 0.5$, $dS = (2/\pi)(84)(0.5) = 84/\pi$, so the maximum error $\approx 84/\pi \approx 27$ cm^2.

    (b) Relative error $\approx \dfrac{dS}{S} = \dfrac{84/\pi}{84^2/\pi} = \dfrac{1}{84} \approx 0.012$.

47. (a) $V = \pi r^2 h \Rightarrow \Delta V \approx dV = 2\pi rh\, dr = 2\pi rh\, \Delta r.$

(b) $\Delta V = \pi(r + \Delta r)^2 h - \pi r^2 h$, so the error is $\Delta V - dv = \pi(r + \Delta r)^2 h - \pi r^2 h - 2\pi rh\, \Delta r$

$= \pi(\Delta r)^2 h.$

49. (a) $dc = \frac{dc}{dx}\, dx = 0\, dx = 0.$ (b) $d(cu) = \frac{d}{dx}(cu)\, dx = c\frac{du}{dx}\, dx = c\, du.$

(c) $d(u + v) = \frac{d}{dx}(u + v)\, dx = \left(\frac{du}{dx} + \frac{dv}{dx}\right)dx = \frac{du}{dx}\, dx + \frac{dv}{dx}\, dx = du + dv.$

(d) $d(uv) = \frac{d}{dx}(uv)\, dx = \left(u\frac{dv}{dx} + v\frac{du}{dx}\right)dx = u\frac{dv}{dx}\, dx + v\frac{du}{dx}\, dx = u\, dv + v\, du.$

(e) $d\left(\frac{u}{v}\right) = \frac{d}{dx}\left(\frac{u}{v}\right)dx = \dfrac{v\frac{du}{dx} - u\frac{dv}{dx}}{v^2}\, dx = \dfrac{v\frac{du}{dx}\, dx - u\frac{dv}{dx}\, dx}{v^2} = \dfrac{v\, du - u\, dv}{v^2}.$

(f) $d(x^n) = \frac{d}{dx}(x^n)\, dx = nx^{n-1}\, dx.$

EXERCISES 2.10

1. $f(x) = x^3 + x + 1 \Rightarrow f'(x) = 3x^2 + 1$, so $x_{n+1} = x_n - \dfrac{x_n^3 + x_n + 1}{3x_n^2 + 1}$. $x_1 = -1 \Rightarrow$

$x_2 = -1 - \dfrac{-1 - 1 + 1}{3 \cdot 1 + 1} = -0.75 \Rightarrow x_3 = -0.75 - \dfrac{(-0.75)^3 - 0.75 + 1}{3(-0.75)^2 + 1} \approx -0.6860.$

3. $f(x) = x^5 - 10 \Rightarrow f'(x) = 5x^4$, so $x_{n+1} = x_n - \dfrac{x_n^5 - 10}{5x_n^4}$. $x_1 = 1.5$

$\Rightarrow x_2 = 1.5 - \dfrac{(1.5)^5 - 10}{5(1.5)^4} \approx 1.5951 \Rightarrow x_3 = 1.5951 - \dfrac{f(1.5951)}{f'(1.5951)} \approx 1.5850.$

5. Finding $\sqrt[4]{22}$ is equivalent to finding the positive root of $x^4 - 22 = 0$ so we take

$f(x) = x^4 - 22 \Rightarrow f'(x) = 4x^3$ and $x_{n+1} = x_n - \dfrac{x_n^4 - 22}{4x_n^3}$. Taking $x_1 = 2$, we get

$x_2 \approx 2.1875$, $x_3 \approx 2.166059$, $x_4 \approx 2.165737$ and $x_5 \approx 2.165737$. Thus $\sqrt[4]{22} \approx 2.165737$ to 6 decimal places.

7. $f(x) = x^3 - 2x - 1 \Rightarrow f'(x) = 3x^2 - 2$, so $x_{n+1} = x_n - \dfrac{x_n^3 - 2x_n - 1}{3x_n^2 - 2}$. Taking $x_1 = 1.5$,

we get $x_2 \approx 1.631579$, $x_3 \approx 1.618184$, $x_4 \approx 1.618034$ and $x_5 \approx 1.618034$. So the root is 1.618034 to 6 decimal places.

9. $f(x) = x^4 + x^3 - 22x^2 - 2x + 41 \Rightarrow f'(x) = 4x^3 + 3x^2 - 44x - 2$, so

$x_{n+1} = x_n - \dfrac{x_n^4 + x_n^3 - 22x_n^2 - 2x_n + 41}{4x_n^3 + 3x_n^2 - 44x_n - 2}$. Taking $x_1 = 4$, we get $x_2 \approx 3.992063$,

$x_3 \approx 3.992020$ and $x_4 \approx 3.992020$. So the root in the interval $[3, 4]$ is 3.992020 to 6 decimal places.

11.

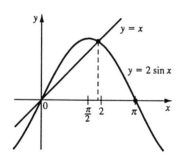

From the graph it appears that there is a root near 2, so we take $x_1 = 2$. Write the equation as $f(x) = 2\sin x - x = 0$. Then $f'(x) = 2\cos x - 1$, so $x_{n+1} = x_n - \dfrac{2\sin x_n - x_n}{2\cos x_n - 1} \Rightarrow x_1 = 2$, $x_2 \approx 1.900996$, $x_3 \approx 1.895512$, $x_4 \approx 1.895494$ and $x_5 \approx 1.895494$. So the root is 1.895494 to 6 decimal places.

13. $f(x) = x^3 - 4x + 1 \Rightarrow f'(x) = 3x^2 - 4$, so $x_{n+1} = x_n - \dfrac{x_n^3 - 4x_n + 1}{3x_n^2 - 4}$. Observe that $f(-3) = -14$, $f(-2) = 1$, $f(0) = 1$, $f(1) = -2$ and $f(2) = 1$ so there are roots in $[-3, -2]$, $[0, 1]$ and $[1, 2]$.

$[-3, -2]$	$[0, 1]$	$[1, 2]$
$x_1 = -2$	$x_1 = 0$	$x_1 = 2$
$x_2 = -2.125$	$x_2 = 0.25$	$x_2 = 1.875$
$x_3 \approx -2.114975$	$x_3 \approx 0.254098$	$x_3 \approx 1.860979$
$x_4 \approx -2.114908$	$x_4 \approx 0.254102$	$x_4 \approx 1.860806$
$x_5 \approx -2.114908$	$x_5 \approx 0.254102$	$x_5 \approx 1.860806$

To 6 decimal places, the roots are -2.114908, 0.254102 and 1.860806.

15. $f(x) = x^4 + x^2 - x - 1 \Rightarrow f'(x) = 4x^3 + 2x - 1$, so $x_{n+1} = x_n - \dfrac{x_n^4 + x_n^2 - x_n - 1}{4x_n^3 + 2x_n - 1}$. Note that $f(1) = 0$, so $x = 1$ is a root. Also $f(-1) = 2$ and $f(0) = -1$, so there is a root in $[-1, 0]$. A sketch shows that these are the only roots. Taking $x_1 = -0.5$, we have $x_2 = -0.575$, $x_3 \approx -0.569867$, $x_4 \approx -0.569840$ and $x_5 \approx -0.569840$. The roots are 1 and -0.569840, to 6 decimal places.

17.

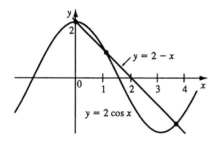

Obviously $x = 0$ is a root. From the sketch, there appear to be roots near 1 and 3.5. Write the equation as $f(x) = 2\cos x + x - 2 = 0$. Then $f'(x) = -2\sin x + 1$, so $x_{n+1} = x_n - \dfrac{2\cos x_n + x_n - 2}{1 - 2\sin x_n}$. Taking $x_1 = 1$, we get $x_2 \approx 1.118026$, $x_3 \approx 1.109188$, $x_4 \approx 1.109144$ and $x_5 \approx 1.109144$. Taking $x_1 = 3.5$, we get $x_2 \approx 3.719159$, $x_3 \approx 3.698331$,

$x_4 \approx 3.698154$ and $x_5 \approx 3.698154$. To 6 decimal places the roots are 0, 1.109144 and 3.698154.

19. (a) $f(x) = x^2 - a \Rightarrow f'(x) = 2x$, so Newton's Method gives $x_{n+1} = x_n - \dfrac{x_n^2 - a}{2x_n}$

   $= x_n - \frac{1}{2}x_n + \dfrac{a}{2x_n} = \frac{1}{2}\left(x_n + \dfrac{a}{x_n}\right)$.

   (b) Using (a) with $x_1 = 30$, we get $x_2 \approx 31.666667$, $x_3 \approx 31.622807$, $x_4 \approx 31.622777$ and $x_5 \approx 31.622777$. So $\sqrt{1000} \approx 31.622777$.

21. If we attempt to compute x_2 we get $x_2 = x_1 - \dfrac{f(x_1)}{f'(x_1)}$, but $f(x) = x^3 - 3x + 6$

   $\Rightarrow f'(x_1) = 3x_1^2 - 3 = 3(1)^2 - 3 = 0$. For Newton's Method to work $f'(x_n) \neq 0$ (no horizontal tangents).

23. For $f(x) = x^{1/3}$, $f'(x) = \frac{1}{3}x^{-2/3}$ and $x_{n+1} = x_n - \dfrac{f(x_n)}{f'(x_n)} = x_n - \dfrac{x_n^{1/3}}{\frac{1}{3}x_n^{-2/3}}$

   $= x_n - 3x_n = -2x_n$. Therefore each successive approximation becomes twice as large as the previous one in absolute value, so the sequence of approximations fails to converge to the root, which is 0.

REVIEW EXERCISES FOR CHAPTER 2

1. False, see the warning after Theorem 2.8.

3. False. See the discussion before the Product Rule (2.13).

5. True, by the Chain Rule.

7. False. $f(x) = |x^2 + x| = x^2 + x$ for $x \geq 0$ or $x \leq -1$ and $|x^2 + x| = -(x^2 + x)$ for $-1 < x < 0$. So $f'(x) = 2x + 1$ for $x > 0$ or $x < -1$ and $f'(x) = -(2x + 1)$ for $-1 < x < 0$. But $|2x + 1| = 2x + 1$ for $x \geq -1/2$ and $|2x + 1| = -2x - 1$ for $x < -1/2$.

9. True. $g(x) = x^5 \Rightarrow g'(x) = 5x^4 \Rightarrow g'(2) = 5(2)^4 = 80$ and by the definition of derivative

   $\displaystyle\lim_{x \to 2} \dfrac{g(x) - g(2)}{x - 2} = g'(2) = 80$.

11. $f(x) = x^3 + 5x + 4 \Rightarrow f'(x) = \displaystyle\lim_{h \to 0} \dfrac{f(x + h) - f(x)}{h}$

   $= \displaystyle\lim_{h \to 0} \dfrac{(x + h)^3 + 5(x + h) + 4 - (x^3 + 5x + 4)}{h} = \displaystyle\lim_{h \to 0} \dfrac{3x^2 h + 3xh^2 + h^3 + 5h}{h}$.

   $= \displaystyle\lim_{h \to 0} (3x^2 + 3xh + h^2 + 5) = 3x^2 + 5$.

13. $f(x) = \sqrt{3 - 5x} \Rightarrow f'(x) = \lim\limits_{h \to 0} \dfrac{f(x + h) - f(x)}{h} = \lim\limits_{h \to 0} \dfrac{\sqrt{3 - 5(x + h)} - \sqrt{3 - 5x}}{h}$

$= \lim\limits_{h \to 0} \dfrac{\sqrt{3 - 5x - 5h} - \sqrt{3 - 5x}}{h} \left(\dfrac{\sqrt{3 - 5x - 5h} + \sqrt{3 - 5x}}{\sqrt{3 - 5x - 5h} + \sqrt{3 - 5x}} \right) = \lim\limits_{h \to 0} \dfrac{-5h}{h(\sqrt{3 - 5x - 5h} + \sqrt{3 - 5x})}$

$= \lim\limits_{h \to 0} \dfrac{-5}{\sqrt{3 - 5x - 5h} + \sqrt{3 - 5x}} = \dfrac{-5}{2\sqrt{3 - 5x}}.$

15. $y = (x + 2)^8 (x + 3)^6 \Rightarrow y' = 6(x + 3)^5 (x + 2)^8 + 8(x + 2)^7 (x + 3)^6$

$= 2(7x + 18)(x + 2)^7 (x + 3)^5.$

17. $y = \dfrac{x}{\sqrt{9 - 4x}} \Rightarrow y' = \dfrac{\sqrt{9 - 4x} - x[-4/(2\sqrt{9 - 4x})]}{9 - 4x} = \dfrac{9 - 4x + 2x}{(9 - 4x)^{3/2}} = \dfrac{9 - 2x}{(9 - 4x)^{3/2}}.$

19. $x^2 y^3 + 3y^2 = x - 4y \Rightarrow 2xy^3 + 3x^2 y^2 y' + 6yy' = 1 - 4y' \Rightarrow y' = \dfrac{1 - 2xy^3}{3x^2 y^2 + 6y + 4}.$

21. $y = \sqrt{x\sqrt{x\sqrt{x}}} = \left(x(x^{3/2})^{1/2}\right)^{1/2} = \left(x(x^{3/4})\right)^{1/2} = x^{7/8} \Rightarrow y' = \tfrac{7}{8}x^{-1/8}.$

23. $y = \dfrac{x}{8 - 3x} \Rightarrow y' = \dfrac{(8 - 3x) - x(-3)}{(8 - 3x)^2} = \dfrac{8}{(8 - 3x)^2}.$

25. $y = (x\tan x)^{1/5} \Rightarrow y' = \tfrac{1}{5}(x\tan x)^{-4/5}(\tan x + x\sec^2 x).$

27. $x^2 = y(y + 1) = y^2 + y \Rightarrow 2x = 2yy' + y' \Rightarrow y' = 2x/(2y + 1).$

29. $y = \dfrac{(x - 1)(x - 4)}{(x - 2)(x - 3)} = \dfrac{x^2 - 5x + 4}{x^2 - 5x + 6}$

$\Rightarrow y' = \dfrac{(x^2 - 5x + 6)(2x - 5) - (x^2 - 5x + 4)(2x - 5)}{(x^2 - 5x + 6)^2} = \dfrac{2(2x - 5)}{(x - 2)^2 (x - 3)^2}.$

31. $y = \tan\sqrt{1 - x} \Rightarrow y' = (\sec^2\sqrt{1 - x})[1/(2\sqrt{1 - x})](-1) = -(\sec^2\sqrt{1 - x})/(2\sqrt{1 - x}).$

33. $y = \sin\left(\tan\sqrt{1 + x^3}\right) \Rightarrow y' = \cos\left(\tan\sqrt{1 + x^3}\right)\left(\sec^2\sqrt{1 + x^3}\right)\left[3x^2/\left(2\sqrt{1 + x^3}\right)\right].$

35. $y = \cot(3x^2 + 5) \Rightarrow y' = -\csc^2(3x^2 + 5)(6x) = -6x\csc^2(3x^2 + 5).$

37. $y = (\sin mx)/x \Rightarrow y' = (mx\cos mx - \sin mx)/x^2.$

39. $y = \cos^2(\tan x) \Rightarrow y' = 2\cos(\tan x)(-\sin(\tan x))\sec^2 x = -\sin(2\tan x)\sec^2 x.$

41. $y = \sqrt{7 - x^2}(x^3 + 7)^5 \Rightarrow y' = \tfrac{1}{2}(7 - x^2)^{-1/2}(-2x)(x^3 + 7)^5 + \sqrt{7 - x^2}\left(5(x^3 + 7)^4(3x^2)\right)$

$= -x(x^3 + 7)^5/\sqrt{7 - x^2} + 15x^2(x^3 + 7)^4\sqrt{7 - x^2}.$

43. $f(x) = (2x - 1)^{-5} \Rightarrow f'(x) = -5(2x - 1)^{-6}(2) = -10(2x - 1)^{-6}$

$\Rightarrow f''(x) = 60(2x - 1)^{-7}(2) = 120(2x - 1)^{-7} \Rightarrow f''(0) = 120(-1)^{-7} = -120.$

45. $x^6 + y^6 = 1 \Rightarrow 6x^5 + 6y^5 y' = 0 \Rightarrow y' = -\dfrac{x^5}{y^5} \Rightarrow y'' = -\dfrac{5x^4 y^5 - x^5 (5y^4 y')}{y^{10}}$

$= -\dfrac{5x^4 y^5 - 5x^5 y^4 (-x^5/y^5)}{y^{10}} = -\dfrac{5x^4 y^6 + 5x^{10}}{y^{11}} = -\dfrac{5x^4 (y^6 + x^6)}{y^{11}} = -\dfrac{5x^4}{y^{11}}.$

47. $y = \dfrac{x}{x^2 - 2} \Rightarrow y' = \dfrac{(x^2 - 2) - x(2x)}{(x^2 - 2)^2} = \dfrac{-x^2 - 2}{(x^2 - 2)^2}$. When $x = 2$, $y' = \dfrac{-2^2 - 2}{(2^2 - 2)^2}$

$= -3/2$, so the equation of the tangent at $(2, 1)$ is $y - 1 = (-3/2)(x - 2)$ or

$3x + 2y - 8 = 0$.

49. $y = \tan x \Rightarrow y' = \sec^2 x$. When $x = \pi/3$, $y' = 2^2 = 4$, so the tangent at $(\pi/3, \sqrt{3})$ is

$y - \sqrt{3} = 4(x - \pi/3)$ or $y = 4x + \sqrt{3} - 4\pi/3$.

51. $y = \sin x + \cos x \Rightarrow y' = \cos x - \sin x = 0 \Leftrightarrow \cos x = \sin x$ and $0 \le x \le 2\pi \Leftrightarrow x = \pi/4$ and

$5\pi/4$, so the points are $(\pi/4, \sqrt{2})$ and $(5\pi/4, -\sqrt{2})$.

53. (a) $y = t^3 - 12t + 3 \Rightarrow v(t) = y' = 3t^2 - 12$, $a(t) = v'(t) = 6t$.

(b) $v(t) = 3(t^2 - 4) > 0$ when $t > 2$, so it moves upward when $t > 2$ and downward when

$0 \le t < 2$.

(c) Distance upward $= y(3) - y(2) = -6 - (-13) = 7$. Distance

downward $= y(0) - y(2) = 3 - (-13) = 16$. Total distance $= 7 + 16 = 23$.

55. $f(x) = (x - a)(x - b)(x - c) \Rightarrow f'(x) = (x - b)(x - c) + (x - a)(x - c) + (x - a)(x - b)$.

So $\dfrac{f'(x)}{f(x)} = \dfrac{(x - b)(x - c) + (x - a)(x - c) + (x - a)(x - b)}{(x - a)(x - b)(x - c)} = \dfrac{1}{x - a} + \dfrac{1}{x - b} + \dfrac{1}{x - c}$.

57. (a) $h'(x) = f'(x)g(x) + f(x)g'(x) \Rightarrow h'(2) = f'(2)g(2) + f(2)g'(2) = (-2)(5) + (3)(4) = 2$.

(b) $F'(x) = f'(g(x))g'(x) \Rightarrow F'(2) = f'(g(2))g'(2) = f'(5)(4) = 11 \cdot 4 = 44$.

59. $f(x) = x^2 g(x) \Rightarrow f'(x) = 2xg(x) + x^2 g'(x)$.

61. $f(x) = \big(g(x)\big)^2 \Rightarrow f'(x) = 2g(x)g'(x)$.

63. $f(x) = g(g(x)) \Rightarrow f'(x) = g'(g(x))g'(x)$.

65. $h(x) = \dfrac{f(x)g(x)}{f(x) + g(x)} \Rightarrow h'(x) = \dfrac{\Big(f'(x)g(x) + f(x)g'(x)\Big)\Big(f(x) + g(x)\Big) - f(x)g(x)\Big(f'(x) + g'(x)\Big)}{[f(x) + g(x)]^2}.$

$= \dfrac{f'(x)\big[g(x)\big]^2 + g'(x)\big[f(x)\big]^2}{[f(x) + g(x)]^2}.$

67. Using the Chain Rule repeatedly, $h(x) = f\Big(g(\sin 4x)\Big)$

$\Rightarrow h'(x) = f'\Big(g(\sin 4x)\Big) \cdot \dfrac{d}{dx}[g(\sin 4x)] = f'\Big(g(\sin 4x)\Big) \cdot g'(\sin 4x) \cdot \dfrac{d}{dx}(\sin 4x)$

$= f'\Big(g(\sin 4x)\Big)g'(\sin 4x)(\cos 4x)(4).$

69. $\rho = x(1 + \sqrt{x}) = x + x^{3/2} \Rightarrow d\rho/dx = 1 + \frac{3}{2}\sqrt{x}$, so the density when $x = 4$ is
$1 + \frac{3}{2}\sqrt{4} = 4$ kg/m.

71. If $x =$ edge length, then $V = x^3 \Rightarrow dV/dt = 3x^2\,dx/dt = 10 \Rightarrow dx/dt = 10/(3x^2)$ and
$S = 6x^2 \Rightarrow dS/dt = (12x)\,dx/dt = 12x[10/(3x^2)] = 40/x$. When $x = 30$,
$dS/dt = 40/30 = 4/3$ cm^2/min.

73.

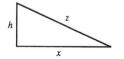

Given $dh/dt = 5$ and $dx/dt = 15$, find dz/dt.

$$z^2 = x^2 + h^2 \Rightarrow 2z\frac{dz}{dt} = 2x\frac{dx}{dt} + 2h\frac{dh}{dt}$$

$$\Rightarrow \frac{dz}{dt} = \frac{1}{z}(15x + 5h). \text{ When } t = 3,$$

$$h = 45 + 3\,(5) = 60 \text{ and } x = 15\,(3) = 45$$

$$\Rightarrow z = 75, \text{ so } \frac{dz}{dt} = \frac{1}{75}(15\,(45) + 5\,(60))$$

$$= 13 \text{ ft/s.}$$

75.

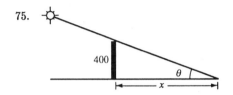

We are given $\frac{d\theta}{dt} = -0.25$ rad/h. $x = 400\cot\theta$
$\Rightarrow \frac{dx}{dt} = -400\csc^2\theta\frac{d\theta}{dt}$. When $\theta = \pi/6$,
$\frac{dx}{dt} = -400\,(2)^2\,(-0.25) = 400$ ft/h.

77. $y = x^3 - 2x^2 + 1 \Rightarrow dy = (3x^2 - 4x)\,dx$. When $x = 2$ and $dx = 0.2$,
$dy = \left(3\,(2)^2 - 4\,(2)\right)(0.2) = 0.8$.

79. $f(x) = \sqrt[3]{1 + 3x} = (1 + 3x)^{1/3} \Rightarrow f'(x) = (1 + 3x)^{-2/3}$ so $L(x) = f(0) + f'(0)(x - 0)$
$= 1^{1/3} + 1^{-2/3}x = 1 + x$. Thus $\sqrt[3]{1 + 3x} \approx 1 + x \Rightarrow \sqrt[3]{1.03} = \sqrt[3]{1 + 3(0.01)}$
$\approx 1 + (0.01) = 1.01$.

81. $f(x) = x^4 + x - 1 \Rightarrow f'(x) = 4x^3 + 1 \Rightarrow x_{n+1} = x_n - \dfrac{x_n^4 + x_n - 1}{4x_n^3 + 1}$. If $x_1 = 0.5$ then
$x_2 \approx 0.791667$, $x_3 \approx 0.729862$, $x_4 \approx 0.724528$, $x_5 \approx 0.724492$ and $x_6 \approx 0.724492$, so, to 6
decimal places, the root is 0.724492.

83. $y = x^6 + 2x^2 - 8x + 3$ has a horizontal tangent when $y' = 6x^5 + 4x - 8 = 0$. Let
$f(x) = 6x^5 + 4x - 8$. Then $f'(x) = 30x^4 + 4$, so $x_{n+1} = x_n - \dfrac{6x_n^5 + 4x_n - 8}{30x_n^4 + 4}$. A sketch
shows that the root is near 1, so we take $x_1 = 1$. Then $x_2 \approx 0.9412$, $x_3 \approx 0.9341$, $x_4 \approx 0.9340$
and $x_5 \approx 0.9340$. Thus, to 4 decimal places, the point is $(0.9340, -2.0634)$.

85. $\displaystyle\lim_{h \to 0}\frac{(2 + h)^6 - 64}{h} = \frac{d}{dx}x^6\Big|_{x=2} = 6(2)^5 = 192$.

87.

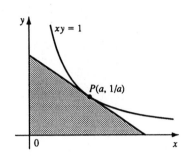

$xy = 1 \Rightarrow y = \frac{1}{x} \Rightarrow y' = -\frac{1}{x^2}$. Let any point on the hyperbola be $P(a, 1/a)$. At P the tangent line is $y - \frac{1}{a} = -\frac{1}{a^2}(x - a) \Leftrightarrow$ $y = -\frac{1}{a^2}x + \frac{2}{a}$. From this equation we find that the x- and y-intercepts of the tangent line are 2a and 2/a respectively. The area of the triangle is $A = \frac{1}{2}bh = \frac{1}{2}(2/a)(2a) = 2$.

PROBLEMS PLUS (page 192)

1. Let a be the x–coordinate of Q. Then $y = 1 - x^2 \Rightarrow$ the slope at $Q = y'(a) = -2a$. But since the triangle is equilateral, $\angle ACB = 60°$, so that the slope at $Q = \tan 120° = -\sqrt{3}$. Therefore we must have that $-2a = -\sqrt{3} \Rightarrow a = \frac{\sqrt{3}}{2}$. Therefore the point Q has coordinates $(\sqrt{3}/2, 1 - (\sqrt{3}/2)^2) = (\sqrt{3}/2, \frac{1}{4})$, and by symmetry P has coordinates $(-\sqrt{3}/2, \frac{1}{4})$.

3. $1 + x + x^2 + \cdots + x^{100} = \frac{1 - x^{101}}{1 - x}$ $(x \neq 1)$. If $x = 1$, then the sum is clearly equal to $101 > 0$. If $x \geq 0$, then we have a sum of positive terms which is clearly positive. And if $x < 0$ then $x^{101} < 0 \Rightarrow 1 - x > 0$ and $1 - x^{101} > 0 \Rightarrow \frac{1 - x^{101}}{1 - x} > 0$. Therefore
$1 + x + x^2 + \cdots + x^{100} = \frac{1 - x^{101}}{1 - x} \geq 0$ for all x.

5. Sketch the graph for each quadrant separately:

In the first quadrant, $x, y \geq 0 \Rightarrow |x| = x$ and $|y| = y$ so that the inequality becomes $x + y \leq 1$, or $y \leq 1 - x$. In the second quadrant, $x \leq 0$, $y \geq 0 \Rightarrow |x| = -x$, and $|y| = y$ and the inequality becomes $-x + y \leq 1$, or $y \leq 1 + x$. Similarly in the third and fourth quadrants the inequality becomes $y \geq -x - 1$ and $y \geq x - 1$ respectively.

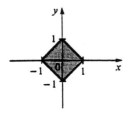

7. For $-\frac{1}{2} < x < \frac{1}{2}$ we have $2x - 1 < 0$, so $|2x - 1| = -(2x - 1)$ and $2x + 1 > 0 \Rightarrow$
$|2x + 1| = 2x + 1$. Therefore $\lim_{x \to 0} \frac{|2x - 1| - |2x + 1|}{x} = \lim_{x \to 0} \frac{-(2x - 1) - (2x + 1)}{x} = \lim_{x \to 0} \frac{-4x}{x}$
$= \lim_{x \to 0} -4 = -4.$

9. Let S_n be the statement that $\dfrac{d^n}{dx^n}(\sin^4 x + \cos^4 x) = 4^{n-1}\cos(4x + n\frac{\pi}{2})$.

S_1 is true because $\dfrac{d}{dx}(\sin^4 x + \cos^4 x) = 4\sin^3 x\cos x - 4\cos^3 x\sin x$

$= 4\sin x\cos x(\sin^2 x - \cos^2 x) = -4\sin x\cos x\cos 2x = -2\sin 2x\cos 2x = -\sin 4x$

$= \cos[\frac{\pi}{2} - (-4x)] = \cos(\frac{\pi}{2} + 4x) = 4^{n-1}\cos(4x + n\frac{\pi}{2})$ when $n = 1$.

Assume S_k is true, that is, $\dfrac{d^k}{dx^k}(\sin^4 x + \cos^4 x) = 4^{k-1}\cos(4x + k\frac{\pi}{2})$.

Then $\dfrac{d^{k+1}}{dx^{k+1}}(\sin^4 x + \cos^4 x) = \dfrac{d}{dx}\left(\dfrac{d^k}{dx^k}(\sin^4 x + \cos^4 x)\right)$

$= \dfrac{d}{dx}[4^{k-1}\cos(4x + k\frac{\pi}{2})] = -4^{k-1}\sin(4x + k\frac{\pi}{2})\cdot\dfrac{d}{dx}(4x + k\frac{\pi}{2})$

$= -4^k\sin(4x + k\frac{\pi}{2}) = 4^k\sin(-4x - k\frac{\pi}{2}) = 4^k\cos[\frac{\pi}{2} - (-4x - k\frac{\pi}{2})] = 4^k\cos[4x + (k+1)\frac{\pi}{2}]$

which shows that S_{k+1} is true. Therefore $\dfrac{d^n}{dx^n}(\sin^4 x + \cos^4 x) = 4^{n-1}\cos(4x + n\frac{\pi}{2})$ for every

positive integer n by mathematical induction.

ANOTHER PROOF: First write $\sin^4 x + \cos^4 x = (\sin^2 x + \cos^2 x)^2 - 2\sin^2 x\cos^2 x$

$= 1 - \frac{1}{2}\sin^2 2x = 1 - \frac{1}{4}(1 - \cos 4x) = \frac{3}{4} + \frac{1}{4}\cos 4x$. Then we have

$\dfrac{d^n}{dx^n}(\sin^4 x + \cos^4 x) = \dfrac{d^n}{dx^n}(\frac{3}{4} + \frac{1}{4}\cos 4x) = \frac{1}{4}\cdot 4^n\cos(4x + \frac{n\pi}{2}) = 4^{n-1}\cos(4x + \frac{n\pi}{2})$.

11. (a) $D = \{x \mid 3 - x \geq 0,\ 2 - \sqrt{3 - x} \geq 0,\ 1 - \sqrt{2 - \sqrt{3 - x}} \geq 0\}$

$= \{x \mid 3 \geq x,\ 2 \geq \sqrt{3 - x},\ 1 \geq \sqrt{2 - \sqrt{3 - x}}\}$

$= \{x \mid 3 \geq x,\ 4 \geq 3 - x,\ 1 \geq 2 - \sqrt{3 - x}\}$

$= \{x \mid x \leq 3,\ x \geq -1,\ 1 \leq \sqrt{3 - x}\}$

$= \{x \mid x \leq 3,\ x \geq -1,\ 1 \leq 3 - x\}$

$= \{x \mid x \leq 3,\ x \geq -1,\ x \leq 2\} = \{x \mid -1 \leq x \leq 2\} = [-1, 2]$

(b) $f(x) = \sqrt{1 - \sqrt{2 - \sqrt{3 - x}}}$

$\Rightarrow f'(x) = \dfrac{1}{2\sqrt{1 - \sqrt{2 - \sqrt{3 - x}}}}\dfrac{d}{dx}\left(1 - \sqrt{2 - \sqrt{3 - x}}\right)$

$= \dfrac{1}{2\sqrt{1 - \sqrt{2 - \sqrt{3 - x}}}}\cdot\dfrac{-1}{2\sqrt{2 - \sqrt{3 - x}}}\dfrac{d}{dx}\left(2 - \sqrt{3 - x}\right)$

$$= -\frac{1}{8\sqrt{1-\sqrt{2-\sqrt{3-x}}}\sqrt{2-\sqrt{3-x}}\sqrt{3-x}}$$

13. (a) For $x \geq 0$, $h(x) = \left|x^2 - 6|x| + 8\right|$

$$= \left|x^2 - 6x + 8\right| = |(x-2)(x-4)| = \begin{cases} x^2 - 6x + 8 & \text{if } 0 \leq x \leq 2 \\ -x^2 + 6x - 8 & \text{if } 2 < x < 4 \\ x^2 - 6x + 8 & \text{if } x \geq 4 \end{cases}$$

and for $x < 0$, $h(x) = \left|x^2 - 6|x| + 8\right|$

$$= \left|x^2 + 6x + 8\right| = |(x+2)(x+4)| = \begin{cases} x^2 + 6x + 8 & \text{if } x \leq -4 \\ -x^2 - 6x - 8 & \text{if } -4 < x < -2 \\ x^2 + 6x + 8 & \text{if } -2 \leq x < 0 \end{cases}$$

OR: Use the fact that h is an even function and reflect the part of the graph for $x \geq 0$ about the y-axis.

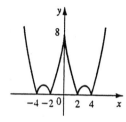

(b) To find where h is differentiable we check the points 0, 2 and 4 by computing the

left and right hand derivatives, $h'_-(0) = \lim\limits_{t \to 0^-} \dfrac{h(0+t) - h(0)}{t}$

$$= \lim_{t \to 0^-} \frac{(0+t)^2 + 6(0+t) + 8 - 8}{t} = \lim_{t \to 0^-} (t+6) = 6.$$

$$h'_+(0) = \lim_{t \to 0^+} \frac{h(0+t) - h(0)}{t} = \lim_{t \to 0^+} \frac{(0+t)^2 - 6(0+t) + 8 - 8}{t}$$

$$= \lim_{t \to 0^+} (t-6) = -6 \neq h'_-(0) \text{ so that h is not differentiable at 0. Similarly h is not}$$

differentiable at ± 2 or ± 4. This can also be seen from the graph.

15. Let the coordinates of P be (a, a^3). Then the slope at P is $3a^2$ and the equation of the tangent

at P is $y - a^3 = 3a^2(x - a)$ or $y = 3a^2 x - 2a^3$. To find Q we solve the equations $y = x^3$ and

$y = 3a^2 x - 2a^3$. This gives $x^3 - 3a^2 x + 2a^3 = 0$. Note that $(x - a)^2$ must be a factor since

$x = a$ is a double root. So we have $(x - a)^2(x + 2a) = 0$. The other root is $x = -2a$, so Q is

$(-2a, -8a^3)$. The slope at Q is $3(-2a)^2 = 12a^2 = 4(3a^2)$ which is four times the slope at P.

17. (a) Since f is differentiable at 0, f is continuous at 0 so

$$f(0) = \lim_{x \to 0} f(x) = \lim_{x \to 0} \frac{f(x)}{x} \cdot x = \lim_{x \to 0} \frac{f(x)}{x} \cdot \lim_{x \to 0} x = 4 \cdot 0 = 0$$

(b) $f'(0) = \lim_{x \to 0} \dfrac{f(x) - f(0)}{x - 0} = \lim_{x \to 0} \dfrac{f(x)}{x} = 4$ [since $f(0) = 0$ from (a)]

(c) $\displaystyle \lim_{x \to 0} \frac{g(x)}{f(x)} = \lim_{x \to 0} \frac{\frac{g(x)}{x}}{\frac{f(x)}{x}} = \frac{\lim_{x \to 0} \frac{g(x)}{x}}{\lim_{x \to 0} \frac{f(x)}{x}} = \frac{2}{4} = \frac{1}{2}.$

19. $|f(x)| \le x^2 \Rightarrow -x^2 \le f(x) \le x^2 \Rightarrow -0^2 \le f(0) \le 0^2 \Rightarrow f(0) = 0.$

Now $\left| \dfrac{f(x) - f(0)}{x - 0} \right| = \left| \dfrac{f(x)}{x} \right| = \dfrac{|f(x)|}{|x|} \le \dfrac{x^2}{|x|} = \dfrac{|x^2|}{|x|} = \left| \dfrac{x^2}{x} \right| = |x|$

$\Rightarrow -|x| \le \dfrac{f(x) - f(0)}{x - 0} \le |x|.$ But $\lim_{x \to 0} -|x| = 0 = \lim_{x \to 0} |x|$ so that by the Squeeze Theorem

$\displaystyle \lim_{x \to 0} \frac{f(x) - f(0)}{x - 0} = f'(0) = 0$ which shows that f is differentiable at 0.

21. (a) If the two lines L_1 and L_2 have slopes m_1 and m_2 and angles of inclination ϕ_1 and ϕ_2,

then $m_1 = \tan \phi_1$ and $m_2 = \tan \phi_2$. The figure shows that $\phi_2 = \phi_1 + \alpha$ and so

$\alpha = \phi_2 - \phi_1$. Therefore using the identity for $\tan(x - y)$, we have

$\tan \alpha = \tan(\phi_2 - \phi_1) = \dfrac{\tan \phi_2 - \tan \phi_1}{1 + \tan \phi_2 \tan \phi_1}$ and so $\tan \alpha = \dfrac{m_2 - m_1}{1 + m_1 m_2}.$

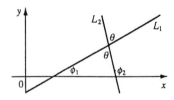

(b) (i) The parabolas intersect when $x^2 = (x - 2)^2 \Rightarrow x = 1$. If $y = x^2$, then $y' = 2x$, so

the slope of the tangent to $y = x^2$ at $(1, 1)$ is $m_1 = 2(1) = 2$. If $y = (x - 2)^2$, then

$y' = 2(x - 2)$, so the slope of the tangent to $y = (x - 2)^2$ at $(1, 1)$ is

$m_2 = 2(1 - 2) = -2$. Therefore $\tan \alpha = \frac{m_2 - m_1}{1 + m_1 m_2} = \frac{-2 - 2}{1 + 2(-2)} = \frac{4}{3}$ and so

$\alpha = \tan^{-1}\left(\frac{4}{3}\right) \approx 53°$.

(ii) $x^2 - y^2 = 3$ and $x^2 - 4x + y^2 + 3 = 0$ intersect when

$x^2 - 4x + x^2 = 0 \Leftrightarrow 2x(x - 2) = 0 \Leftrightarrow x = 0$ or 2, but 0 is extraneous. If $x^2 - y^2 = 3$

then $2x - 2yy' = 0 \Rightarrow y' = \frac{x}{y}$ and

$x - 4x + y^2 + 3 = 0 \Rightarrow 2x - 4 + 2yy' = 0 \Rightarrow y' = \frac{2 - x}{y}$. At $(2, 1)$ the slopes are

$m_1 = 2$ and $m_2 = 0$ so $\tan \alpha = \frac{0 - 2}{1 + 2 \cdot 0} = -2 \Rightarrow \alpha \approx 117°$. At $(2, -1)$ the slopes are

$m_1 = -2$ and $m_2 = 0$ so $\tan \alpha = \frac{0 - (-2)}{1 + (-2)(0)} = 2 \Rightarrow \alpha \approx 63°$.

CHAPTER 3

EXERCISES 3.1

1.

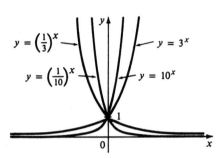

$y = \left(\frac{1}{3}\right)^x$ $y = 3^x$

$y = \left(\frac{1}{10}\right)^x$ $y = 10^x$

3. $y = 100^x$

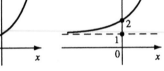

5. $y = (0.9)^x$

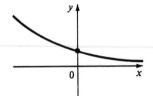

7. $y = 2^x$ $y = 2^x + 1$

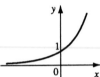

9. $y = 3^x$ $y = 3^{-x}$

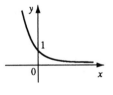

11. $y = -3^{-x}$

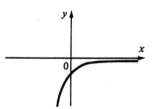

13. $y = 5^{3x} = (5^3)^x = 125^x$ $y = 5^{-3x} = 125^{-x}$

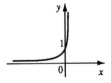

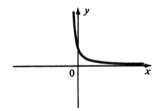

15. $y = 2^{x-1}$

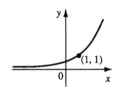

$y = -2^{x-1}$

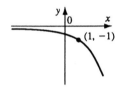

$y = 3 - 2^{x-1}$

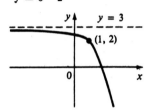

17. $\lim\limits_{x \to \infty} (1.1)^x = \infty$ by (3.3) since $1.1 > 1$

19. $\lim\limits_{x \to \infty} \pi^{-x} = \lim\limits_{x \to \infty} \left[\frac{1}{\pi}\right]^x = 0$ by (3.3) since $0 < \frac{1}{\pi} < 1$

$\left[\text{OR: } \lim\limits_{x \to \infty} \pi^{-x} = 0 \text{ since } \pi > 1 \text{ and } -x \to -\infty \text{ as } x \to \infty \right]$

21. $\lim\limits_{x \to \infty} 5^{2x+5} = \infty$ since $2x + 5 \to \infty$ as $x \to \infty$

23. $\lim\limits_{x \to \infty} \left[2^{-0.8x} + \frac{1}{x}\right] = \lim\limits_{x \to \infty} 2^{-0.8x} + \lim\limits_{x \to \infty} \frac{1}{x} = 0 + 0 = 0$ since $-0.8x \to -\infty$

25. $\lim\limits_{x \to -\infty} \left[\frac{\pi}{4}\right]^x = \infty$ since $0 < \frac{\pi}{4} < 1$

27. $\lim\limits_{x \to \pi/2^-} 2^{\tan x} = \infty$ since $\tan x \to \infty$ as $x \to \pi/2^-$

29. $\lim\limits_{x \to \infty} 3^{1/x} = 3^0 = 1$ since $\frac{1}{x} \to 0$ as $x \to \infty$

31. $\lim\limits_{x \to 0^+} 3^{1/x} = \infty$ since $\frac{1}{x} \to \infty$ as $x \to 0^+$

33. Divide numerator and denominator by 10^x: $\lim\limits_{x \to \infty} \frac{10^x}{10^x + 1} = \lim\limits_{x \to \infty} \frac{1}{1 + 10^{-x}} = \frac{1}{1+0} = 1$

35. $\lim\limits_{t \to 0^+} \pi^{\csc t} = \infty$ since $\csc t \to \infty$ as $t \to 0^+$

37. (a) $\lim\limits_{x \to 2^+} 4^{x/(2-x)} = 0$ since $\frac{x}{2-x} \to -\infty$ as $x \to 2^+$

(b) $\lim\limits_{x \to 2^-} 4^{x/(2-x)} = \infty$ since $\frac{x}{2-x} \to \infty$ as $x \to 2^-$

(c) $\lim\limits_{x \to \infty} 4^{x/(2-x)} = \frac{1}{4}$ since $\frac{x}{2-x} = \frac{1}{2/x - 1} \to -1$ as $x \to \infty$

39. $\dfrac{f(x+h) - f(x)}{h} = \dfrac{10^{x+h} - 10^x}{h} = \dfrac{10^x 10^h - 10^x}{h} = \dfrac{10^x (10^h - 1)}{h}$

EXERCISES 3.2

1.　(a)　$y = 4^x$

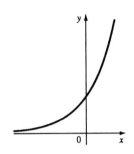

(b)

h	0.1	0.01	0.001	0.0001
$\dfrac{4^h - 1}{h}$	1.4870	1.3959	1.3873	1.3864

$\dfrac{4^h - 1}{h}$ represents the slope of the secant line through the points $(0, 4^0)$ and $(h, 4^h)$

(c)　$\displaystyle \lim_{h \to 0} \frac{4^h - 1}{h} \approx 1.39.$

(d)　It represents $f'(0)$ for $f(x) = 4^x$, which is the slope

of the tangent line to $y = 4^x$ at the point $(0, 1)$

3.　$f(x) = e^{\sqrt{x}} \Rightarrow f'(x) = e^{\sqrt{x}} / (2\sqrt{x})$

5.　$y = xe^{2x} \Rightarrow y' = e^{2x} + xe^{2x} (2) = e^{2x} (1 + 2x)$

7.　$h(t) = \sqrt{1 - e^t} \Rightarrow h'(t) = -e^t / (2\sqrt{1 - e^t})$

9.　$y = e^{x \cos x} \Rightarrow y' = e^{x \cos x} (\cos x - x \sin x)$

11.　$y = e^{-1/x} \Rightarrow y' = e^{-1/x} / x^2$

13.　$y = \tan(e^{3x - 2}) \Rightarrow y' = 3e^{3x - 2} \sec^2(e^{3x - 2})$

15.　$y = \dfrac{e^{3x}}{1 + e^x} \Rightarrow y' = \dfrac{3e^{3x}(1 + e^x) - e^{3x}(e^x)}{(1 + e^x)^2} = \dfrac{3e^{3x} + 3e^{4x} - e^{4x}}{(1 + e^x)^2} = \dfrac{3e^{3x} + 2e^{4x}}{(1 + e^x)^2}$

17.　$y = x^e \Rightarrow y' = ex^{e - 1}$

19.　$y = f(x) = e^{-x} \sin x \Rightarrow f'(x) = -e^{-x} \sin x + e^{-x} \cos x \Rightarrow f'(\pi) = $

　　$e^{-\pi} (\cos \pi - \sin \pi) = -e^{-\pi}$, so the equation of the tangent at $(\pi , 0)$

　　is $y - 0 = -e^{-\pi} (x - \pi)$ or $x + e^\pi y = \pi$.

21.　$\cos (x - y) = xe^x \Rightarrow -\sin(x - y) (1 - y') = e^x + xe^x \Rightarrow y' = 1 + \dfrac{e^x(1 + x)}{\sin (x - y)}$

23.　$y = e^{2x} + e^{-3x} \Rightarrow y' = 2e^{2x} - 3e^{-3x} \Rightarrow y'' = 4e^{2x} + 9e^{-3x}$, so

　　$y'' + y' - 6y = (4e^{2x} + 9e^{-3x}) + (2e^{2x} - 3e^{-3x}) - 6(e^{2x} + e^{-3x}) = 0$

25.　$y = e^{rx} \Rightarrow y' = re^{rx} \Rightarrow y'' = r^2 e^{rx}$, so $y'' + 5y' - 6y = r^2 e^{rx} + 5re^{rx} - 6e^{rx}$

　　$= e^{rx}(r^2 + 5r - 6) = e^{rx}(r + 6)(r - 1) = 0 \Rightarrow (r + 6)(r - 1) = 0 \Rightarrow r = 1, -6.$

27.　$f(x) = e^{-2x} \Rightarrow f'(x) = -2e^{-2x} \Rightarrow f''(x) = (-2)^2 e^{-2x} \Rightarrow$

　　$f'''(x) = (-2)^3 e^{-2x} \Rightarrow \cdots \Rightarrow f^{(8)}(x) = (-2)^8 e^{-2x} = 256 e^{-2x}$

29.　(a)　$v(t) = ce^{-kt} \Rightarrow a(t) = v'(t) = -kce^{-kt} = -k v(t)$

　　(b)　$v(0) = ce^0 = c$, so c is the initial velocity.

(c) $v(t) = c\,e^{-kt} = \frac{c}{2} \Rightarrow e^{-kt} = \frac{1}{2} \Rightarrow -kt = \ln\frac{1}{2} = -\ln 2 \Rightarrow t = (\ln 2)/k$

31. $y = e^{x}$ $y = e^{-x}$

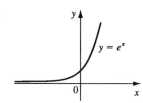
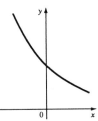

33.   Divide numerator and denominator by e^{3x} : $\displaystyle\lim_{x\to\infty} \frac{e^{3x} - e^{-3x}}{e^{3x} + e^{-3x}} = \lim_{x\to\infty} \frac{1 - e^{-6x}}{1 + e^{-6x}} = \frac{1 - 0}{1 + 0} = 1$

35. $\displaystyle\lim_{x\to 1^{-}} e^{2/(x-1)} = 0$ since $\dfrac{2}{x-1} \to -\infty$ as $x \to 1^{-}$

37. $\displaystyle\lim_{x\to\pi/2^{-}} \frac{2}{1 + e^{\tan x}} = 0$ since $\tan x \to \infty \Rightarrow e^{\tan x} \to \infty$

39. $f(x) = e^{x} + x$ is continuous on $\mathbb{R}$ and $f(-1) = e^{-1} - 1 < 0 < 1 = f(0)$, so

        by the Intermediate Value Theorem $e^{x} + x = 0$ has a root in $(-1, 0)$.

EXERCISES 3.3

1.   Not one–to–one   3.   One–to–one   5.   Not one–to–one

7. $x_1 \neq x_2 \Rightarrow 7x_1 \neq 7x_2 \Rightarrow 7x_1 - 3 \neq 7x_2 - 3 \Rightarrow f(x_1) \neq f(x_2)$, so f is 1–1.

9. $x_1 \neq x_2 \Rightarrow \sqrt{x_1} \neq \sqrt{x_2} \Rightarrow g(x_1) \neq g(x_2)$, so g is 1–1.

11. $h(x) = x^4 + 5 \Rightarrow h(1) = 6 = h(-1)$, so h is not 1–1.

        $h(x_1) \neq h(x_2)$, so h is one–to–one.

13. $x_1 \neq x_2 \Rightarrow 4x_1 \neq 4x_2 \Rightarrow 4x_1 + 7 \neq 4x_2 + 7 \Rightarrow f(x_1) \neq f(x_2)$, so f is 1–1.

        $y = 4x + 7 \Rightarrow 4x = y - 7 \Rightarrow x = (y - 7)/4.$

        Interchange x and y : $y = (x - 7)/4$. So $f^{-1}(x) = (x - 7)/4$.

15. $f(x) = \dfrac{1 + 3x}{5 - 2x}$.   If $f(x_1) = f(x_2)$, then $\dfrac{1 + 3x_1}{5 - 2x_1} = \dfrac{1 + 3x_2}{5 - 2x_2} \Rightarrow$

        $5 + 15x_1 - 2x_2 - 6x_1x_2 = 5 - 2x_1 + 15x_2 - 6x_1x_2 \Rightarrow 17x_1 = 17x_2 \Rightarrow x_1 = x_2$,

        so f is one–to–one. $y = \dfrac{1 + 3x}{5 - 2x} \Rightarrow 5y - 2xy = 1 + 3x \Rightarrow x(3 + 2y) = 5y - 1 \Rightarrow$

$x = \dfrac{5y-1}{2y+3}$. Interchange x and y : $y = \dfrac{5x-1}{2x+3}$ So $f^{-1}(x) = \dfrac{5x-1}{2x+3}$.

17. $x_1 \neq x_2 \Rightarrow 5x_1 \neq 5x_2 \Rightarrow 2+5x_1 \neq 2+5x_2 \Rightarrow \sqrt{2+5x_1} \neq \sqrt{2+5x_2} \Rightarrow$

$f(x_1) \neq f(x_2)$, so f is 1-1. $y = \sqrt{2+5x} \Rightarrow y^2 = 2+5x$ and $y \geq 0 \Rightarrow$

$5x = y^2 - 2 \Rightarrow x = (y^2-2)/5$, $y \geq 0$. Interchange x and y :

$y = (x^2-2)/5$, $x \geq 0$. So $f^{-1}(x) = (x^2-2)/5$, $x \geq 0$.

19. $f(x) = 2x+1$ (a) $x_1 \neq x_2 \Rightarrow 2x_1 \neq 2x_2 \Rightarrow 2x_1+1 \neq 2x_2+1 \Rightarrow$

$f(x_1) \neq f(x_2)$, so f is 1-1. (b) $f(1) = 3 \Rightarrow g(3) = 1$. Also $f'(x) = 2$,

so $g'(3) = 1/f'(x) = 1/2$ (c) $y = 2x+1 \Rightarrow x = (y-1)/2$.

Interchanging x and y gives : $y = (x-1)/2$, so $f^{-1}(x) = (x-1)/2$

Domain (g) = range (f) = R.   (e)

Range (g) = domain (f) = R.

(d) $g(x) = (x-1)/2 \Rightarrow g'(x) = 1/2$

$\Rightarrow g'(3) = 1/2$ as in (b).

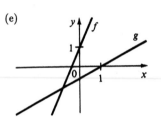

21. (a) $x_1 \neq x_2 \Rightarrow x_1^3 \neq x_2^3 \Rightarrow f(x_1) \neq f(x_2)$, so f is one-to-one.

(b) $f'(x) = 3x^2$ and $f(2) = 8 \Rightarrow g(8) = 2$, so $g'(8) = 1/f'(g(8))$

$= 1/f'(2) = 1/12$ (c) $y = x^3 \Rightarrow x = y^{1/3}$. Interchanging x and y

gives : $y = x^{1/3}$, so $f^{-1}(x) = x^{1/3}$.

Domain (g) = range (f) = R. Range (g) = domain (f) = R.

(d) $g(x) = x^{1/3} \Rightarrow$ (e)

$g'(x) = (1/3)x^{-2/3} \Rightarrow g'(8) =$

$(1/3)(1/4) = 1/12$ as in (b).

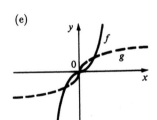

23. $f(x) = 9 - x^2$, $0 \leq x \leq 3$. (a) Since $x \geq 0$, $x_1 \neq x_2 \Rightarrow x_1^2 \neq x_2^2 \Rightarrow$

$9 - x_1^2 \neq 9 - x_2^2 \Rightarrow f(x_1) \neq f(x_2)$, so f is 1-1. (b) $f'(x) = -2x$ and

$f(1) = 8 \Rightarrow g(8) = 1$, so $g'(8) = 1/f'(g(8)) = 1/f'(1) = 1/(-2) = -1/2$.

(c) $y = 9 - x^2 \Rightarrow x^2 = 9 - y$　　　　　　　(e)

$\Rightarrow x = \sqrt{9-x}$. Interchange x and y :

$y = \sqrt{9-x}$, so $f^{-1}(x) = \sqrt{9-x}$.

Domain (g) = range (f) = $[0,9]$

Range (g) = domain (f) = $[0,3]$

(d) $g'(x) = -1/(2\sqrt{9-x}) \Rightarrow$

$g'(8) = -1/2$ as in (b).

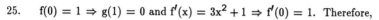

25.　$f(0) = 1 \Rightarrow g(1) = 0$ and $f'(x) = 3x^2 + 1 \Rightarrow f'(0) = 1$. Therefore,

$g'(1) = \dfrac{1}{f'(g(1))} = \dfrac{1}{f'(0)} = \dfrac{1}{1} = 1$.

27.　$f(0) = 3 \Rightarrow g(3) = 0$ and $f'(x) = 2x + \pi \cos \pi x \Rightarrow f'(0) = \pi$. Therefore,

$g'(3) = \dfrac{1}{f'(g(3))} = \dfrac{1}{f'(0)} = \dfrac{1}{\pi}$.

29.　$f(0) = 1 \Rightarrow g(1) = 0$ and $f'(x) = e^x \Rightarrow f'(0) = 1$. Therefore,

$g'(1) = \dfrac{1}{f'(g(1))} = \dfrac{1}{f'(0)} = \dfrac{1}{1} = 1$.

31.　$f(4) = 5 \Rightarrow g(5) = 4$. Therefore, $g'(5) = \dfrac{1}{f'(g(5))} = \dfrac{1}{f'(4)} = \dfrac{1}{2/3} = \dfrac{3}{2}$.

33.　$y = \sqrt[n]{x} \Rightarrow y^n = x \Rightarrow ny^{n-1}y' = 1 \Rightarrow y' = \dfrac{1}{ny^{n-1}} = \dfrac{1}{n\sqrt[n]{x^{n-1}}} = \frac{1}{n}x^{1/n-1}$

35.　Suppose that f is increasing. If $x_1 \neq x_2$, then either $x_1 < x_2$ or $x_2 < x_1$. If $x_1 < x_2$,

　　then $f(x_1) < f(x_2)$. If $x_2 < x_1$, then $f(x_2) < f(x_1)$. In either case, $x_1 \neq x_2 \Rightarrow$

　　$f(x_1) \neq f(x_2)$, so f is one-to-one.

EXERCISES 3.4

1.　$\log_2 64 = 6$ since $2^6 = 64$　　　　　　3.　$\log_8 2 = 1/3$ since $8^{1/3} = 2$

5.　$\log_3 \frac{1}{27} = -3$ since $3^{-3} = \frac{1}{27}$　　　　7.　$\ln e^{\sqrt{2}} = \sqrt{2}$

9.　$\log_{10} 1.25 + \log_{10} 80 = \log_{10} (1.25 \cdot 80) = \log_{10} 100 = 2$

11.　$\log_8 6 - \log_8 3 + \log_8 4 = \log_8 \frac{6 \cdot 4}{3} = \log_8 8 = 1$

13. $2^{(\log_2 3 + \log_2 5)} = 2^{\log_2 15} = 15$

15. $\log_5 a + \log_5 b - \log_5 c = \log_5 \frac{ab}{c}$

17. $2 \ln 4 - \ln 2 = \ln 4^2 - \ln 2 = \ln 16 - \ln 2 = \ln \frac{16}{2} = \ln 8$ [Or: $2\ln 4 - \ln 2 = 2\ln 2^2 - \ln 2$

$= 4\ln 2 - \ln 2 = 3\ln 2$.]

19. $(1/3) \ln x - 4 \ln (2x + 3) = \ln \left(x^{1/3} \right) - \ln (2x + 3)^4 = \ln \left[x^{1/3} / (2x + 3)^4 \right]$

21.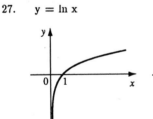

23. $y = \log_{1.1} x$

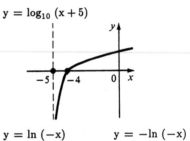

25. $y = \log_{10} x$

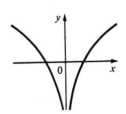

$y = \log_{10} (x + 5)$

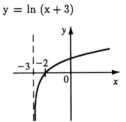

27. $y = \ln x$

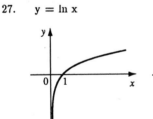

$y = -\ln x$

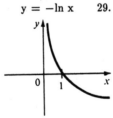

29. $y = \ln (-x)$

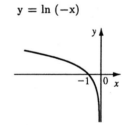

$y = -\ln (-x)$

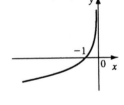

31. $y = \ln (x^2) = 2 \ln |x|$

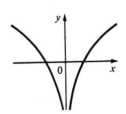

33. $y = \ln x$

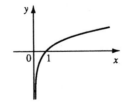

$y = \ln (x + 3)$

35. $\log_2 x = 3 \Rightarrow x = 2^3 = 8$

37. $e^x = 16 \Rightarrow \ln e^x = \ln 16 \Rightarrow x = \ln 16 \Rightarrow x = 4 \ln 2$

39. $\ln(2x-1) = 3 \Rightarrow e^{\ln(2x-1)} = e^3 \Rightarrow 2x-1 = e^3 \Rightarrow x = \frac{1}{2}(e^3+1)$

41. $3^{x+2} = m \Rightarrow \log_3 m = x+2 \Rightarrow x = \log_3 m - 2$

43. $\ln x = \ln 5 + \ln 8 = \ln 40 \Rightarrow x = 40$

45. $5 = \ln(e^{2x-1}) = 2x-1 \Rightarrow x = 3$

47. $\ln(\ln x) = 1 \Rightarrow e^{\ln(\ln x)} = e^1 \Rightarrow \ln x = e^1 = e \Rightarrow e^{\ln x} = e^e \Rightarrow x = e^e$

49. $2^{3^x} = 5 \Rightarrow 3^x = \log_2 5 \Rightarrow \log_3(\log_2 5) = x$. OR: $2^{3^x} = 5 \Rightarrow \ln 2^{3^x} = \ln 5 \Rightarrow 3^x \ln 2 = \ln 5$

    $\Rightarrow 3^x = (\ln 5)/\ln 2$. Hence $\ln 3^x = x \ln 3 = \ln[(\ln 5)/\ln 2] \Rightarrow x = \ln[(\ln 5)/\ln 2]/\ln 3$.

51. $\ln(x+6) + \ln(x-3) = \ln 5 + \ln 2 \Rightarrow \ln[(x+6)(x-3)] = \ln 10 \Rightarrow (x+6)(x-3) = 10$

    $\Rightarrow x^2 + 3x - 18 = 10 \Rightarrow x^2 + 3x - 28 = 0 \Rightarrow (x+7)(x-4) = 0 \Rightarrow x = -7, 4$. However

    $x = -7$ is not a solution since $\ln(-7+6)$ is not defined. So $x = 4$ is the only solution.

53. $\ln(x-5) = 3 \Rightarrow x-5 = e^3 \Rightarrow x = e^3 + 5 \approx 25.0855$

55. $e^{2-3x} = 20 \Rightarrow 2 - 3x = \ln 20 \Rightarrow x = \frac{1}{3}(2 - \ln 20) \approx -0.3319$

57. Let $\log_a x = r$ and $\log_a y = s$. Then $a^r = x$ and $a^s = y$

    (a) $xy = a^r a^s = a^{r+s} \Rightarrow \log_a(xy) = r + s = \log_a x + \log_a y$

    (b) $\frac{x}{y} = \frac{a^r}{a^s} = a^{r-s} \Rightarrow \log_a\left[\frac{x}{y}\right] = r - s = \log_a x - \log_a y$

    (c) $x^y = (a^r)^y = a^{ry} \Rightarrow \log_a(x^y) = ry = y \log_a x$

59. Take $a = e$ in Exercise 58(a) : $\log_e b \; \log_b c = \log_e c \Rightarrow$

    $\ln b \; \log_b c = \ln c \Rightarrow \log_b c = (\ln c)/(\ln b)$

61. If I is the intensity of the 1989 San Francisco earthquake, then $\log_{10}(I/S) = 6.9 \Rightarrow$
    $\log_{10}(25I/S) = \log_{10} 25 + \log_{10}(I/S) = \log_{10} 25 + 6.9 \approx 8.3$

63. $\lim_{x \to 5^+} \ln(x-5) = -\infty$ since $x - 5 \to 0^+$ as $x \to 5^+$

65. $\lim_{x \to \infty} \log_2(x^2 - x) = \infty$ since $x^2 - x \to \infty$ as $x \to \infty$

67. $\lim_{x \to \pi/2^-} \log_{10}(\cos x) = -\infty$ since $\cos x \to 0^+$ as $x \to \pi/2^-$

69. $\lim_{x \to \infty} \ln(1 + e^{-x^2}) = \ln(1 + \lim_{x \to \infty} e^{-x^2}) = \ln(1 + 0) = 0$

71. $f(x) = \log_{10}(1-x)$. Domain $= \{x \mid 1 - x > 0\} = \{x \mid x < 1\} = (-\infty, 1)$ Range $= \mathbb{R}$.

73. $F(t) = \sqrt{t} \ln(t^2 - 1)$ Domain $= \left\{ t \mid t \geq 0 \text{ and } t^2 - 1 > 0 \right\} = \left\{ t \mid t > 1 \right\} = (1, \infty)$.

Range $= \mathbb{R}$.

75. $y = \ln(x + 3) \Rightarrow e^y = e^{\ln(x+3)} = x + 3 \Rightarrow x = e^y - 3$.

Interchange x and y : the inverse function is $y = e^x - 3$.

77. $y = e^{\sqrt{x}} \Rightarrow \ln y = \ln e^{\sqrt{x}} = \sqrt{x} \Rightarrow x = (\ln y)^2$.

Interchange x and y : the inverse function is $y = (\ln x)^2$.

79. $y = \dfrac{10^x}{10^x + 1} \Rightarrow 10^x y + y = 10^x \Rightarrow 10^x(1 - y) = y \Rightarrow 10^x = \dfrac{y}{1-y} \Rightarrow$

$x = \log_{10}\left[\dfrac{y}{1-y}\right]$ Interchange x and y: $y = \log_{10}\left[\dfrac{x}{1-x}\right]$

81.   (a)   We have to show that $-f(x) = f(-x)$.

$-f(x) = -\ln(x + \sqrt{x^2 + 1}) = \ln[(x + \sqrt{x^2 + 1})^{-1}] = \ln[1/(x + \sqrt{x^2 + 1})]$

$= \ln\left[\dfrac{1}{x + \sqrt{x^2 + 1}} \cdot \dfrac{x - \sqrt{x^2 + 1}}{x - \sqrt{x^2 + 1}}\right] = \ln[(x - \sqrt{x^2 + 1})/(x^2 - x^2 - 1)]$

$= \ln(\sqrt{x^2 + 1} - x) = f(-x)$. Thus, f is an odd function.

(b)   Let $y = \ln(x + \sqrt{x^2 + 1})$, then $e^y = x + \sqrt{x^2 + 1} \Leftrightarrow (e^y - x)^2 = x^2 + 1$

$\Leftrightarrow e^{2y} - 2xe^y + x^2 = x^2 + 1 \Leftrightarrow 2xe^y = e^{2y} - 1 \Leftrightarrow x = (e^{2y} - 1)/(2e^y)$

$= \frac{1}{2}(e^y - e^{-y})$. Thus, the inverse function is $f^{-1}(x) = \frac{1}{2}(e^x - e^{-x})$.

83.   Let $x = \log_{10} 99$, $y = \log_9 82$. Then $10^x = 99 < 10^2 \Rightarrow x < 2$, and

$9^y = 82 > 9^2 \Rightarrow y > 2$. Therefore $y = \log_9 82$ is larger.

85.   (a)   Let $\epsilon > 0$ be given. We need N such that $|a^x - 0| < \epsilon$ when $x < N$.

But $a^x < \epsilon \Leftrightarrow x < \log_a \epsilon$. Let $N = \log_a \epsilon$. Then $x < N \Rightarrow$

$x < \log_a \epsilon \Rightarrow |a^x - 0| = a^x < \epsilon$, so $\lim\limits_{x \to -\infty} a^x = 0$.

(b)   Let $M > 0$ be given. We need N such that $a^x > M$ when $x > N$.

But $a^x > M \Leftrightarrow x > \log_a M$. Let $N = \log_a M$. Then $x > N \Rightarrow$

$x > \log_a M \Rightarrow a^x > M$, so $\lim\limits_{x \to \infty} a^x = \infty$.

87.   Notice that $4^x = (2^2)^x = 2^{2x} = (2^x)^2$, and $2^{x+3} = 2^3 2^x = 8 \cdot 2^x$, so that

$4^x - 2^{x+3} + 12 = 0 \Leftrightarrow (2^x)^2 - 8 \cdot 2^x + 12 = (2^x - 6)(2^x - 2) = 0 \Leftrightarrow 2^x = 6$ or $2^x = 2$.

Hence $x = \log_2 6$ or $x = 1$.

EXERCISES 3.5

1. $f(x) = \ln (x + 1) \Rightarrow f'(x) = 1/(x + 1)$. Dom (f) = dom (f') = $\left\{ x \mid x + 1 > 0 \right\}$

$= \left\{ x \mid x > -1 \right\} = (-1 , \infty)$. $\left[\text{Note that, in general, dom (f')} \subset \text{dom (f).}\right]$

3. $f(x) = \ln (\cos x) \Rightarrow f'(x) = \frac{1}{\cos x} (-\sin x) = -\tan x$

Dom (f) = dom (f') = $\left\{ x \mid \cos x > 0 \right\} = (-\pi/2 , \pi/2) \cup (3\pi/2 , 5\pi/2) \cup \cdots$

$= \left\{ x \mid (4n - 1)\pi/2 < x < (4n + 1)\pi/2 , n = 0 , \pm 1 , \pm 2 , ...\right\}$

5. $f(x) = \ln (2 - x - x^2) \Rightarrow f'(x) = (-1 - 2x) / (2 - x - x^2)$ Dom (f) = dom (f') =

$\left\{ x \mid 2 - x - x^2 > 0 \right\} = \left\{ x \mid (2 + x)(1 - x) > 0 \right\} = \left\{ x \mid -2 < x < 1 \right\} = (-2 , 1)$

7. $f(x) = x^2 \ln (1 - x^2) \Rightarrow f'(x) = 2x \ln (1 - x^2) + x^2 \left[1/(1 - x^2) \right] (-2x)$

$= 2x \ln (1 - x^2) - 2x^3 / (1 - x^2)$. Dom (f) = dom (f') = $\left\{ x \mid 1 - x^2 > 0 \right\}$

$= \left\{ x \mid |x| < 1 \right\} = (-1 , 1)$

9. $f(x) = \log_3 (x^2 - 4) \Rightarrow f'(x) = \frac{1}{x^2 - 4}(\log_3 e)(2x) = \frac{2x}{(\ln 3)(x^2 - 4)}$

Dom (f) = dom (f') = $\left\{ x \mid x^2 - 4 > 0 \right\} = \left\{ x \mid |x| > 2 \right\} = (-\infty , -2) \cup (2 , \infty)$

11. $y = x \ln x \Rightarrow y' = \ln x + x(1/x) = \ln x + 1 \Rightarrow y'' = 1/x$

13. $y = \log_{10} x \Rightarrow y' = \frac{1}{x} \log_{10} e \Rightarrow y'' = -\frac{1}{x^2} \log_{10} e$

$\left[\text{OR}: y' = 1/(x \ln 10), y'' = -1/(x^2 \ln 10)\right]$

15. $f(x) = \sqrt{x} \ln x \Rightarrow f'(x) = \frac{1}{2\sqrt{x}} \ln x + \sqrt{x} (1/x) = (\ln x + 2)/(2\sqrt{x})$

17. $g(x) = \ln \frac{a - x}{a + x} = \ln (a - x) - \ln (a + x) \Rightarrow g'(x) = \frac{-1}{a - x} - \frac{1}{a + x} = \frac{-2a}{a^2 - x^2}$

19. $F(x) = \ln \sqrt{x} = (1/2) \ln x \Rightarrow F'(x) = (1/2) (1/x) = 1/(2x)$

21. $f(t) = \log_2 (t^4 - t^2 + 1) \Rightarrow f'(t) = \left[1/(t^4 - t^2 + 1) \right] (\log_2 e) (4t^3 - 2t)$

$= (4t^3 - 2t)/[(\ln 2)(t^4 - t^2 + 1)]$

23. $h(y) = \ln (y^3 \sin y) = 3 \ln y + \ln \sin y \Rightarrow h'(y) = 3/y + (1/\sin y) (\cos y) = 3/y + \cot y$

25. $g(u) = \frac{1 - \ln u}{1 + \ln u} \Rightarrow g'(u) = \frac{(1 + \ln u)\,(-1/u) - (1 - \ln u)\,(1/u)}{(1 + \ln u)^2} = -2/[u\,(1 + \ln u)^2]$

27. $y = (\ln \sin x)^3 \Rightarrow y' = 3\,(\ln \sin x)^2\,\frac{\cos x}{\sin x} = 3\,(\ln \sin x)^2\,\cot x$

29. $y = \frac{\ln x}{1 + x^2} \Rightarrow y' = \frac{(1 + x^2)\,(1/x) - 2x \ln x}{(1 + x^2)^2} = \frac{1 + x^2 - 2x^2 \ln x}{x\,(1 + x^2)^2}$

31. $y = \ln\left[\frac{x + 1}{x - 1}\right]^{3/5} = \frac{3}{5}\left[\ln\,(x + 1) - \ln\,(x - 1)\right] \Rightarrow y' = \frac{3}{5}\left[\frac{1}{x + 1} - \frac{1}{x - 1}\right] = \frac{-6}{5\,(x^2 - 1)}$

33. $y = \ln\,|x^3 - x^2| \Rightarrow y' = \frac{1}{x^3 - x^2}\,(3x^2 - 2x) = \frac{x\,(3x - 2)}{x^2\,(x - 1)} = \frac{3x - 2}{x\,(x - 1)}$

35. $F(x) = e^x \ln x \Rightarrow F'(x) = e^x \ln x + e^x\,(1/x) = e^x\,(\ln x + 1/x)$

37. $f(t) = \pi^{-t} \Rightarrow f'(t) = \pi^{-t}\,(\ln \pi)\,(-1) = -\pi^{-t} \ln \pi$

39. $h(t) = t^3 - 3^t \Rightarrow h'(t) = 3t^2 - 3^t \ln 3$

41. $y = 2^{3^x} \Rightarrow y' = 2^{3^x}\,(\ln 2)\,3^x \ln 3 = (\ln 2)\,(\ln 3)\,3^x\,2^{3^x}$

43. $y = \ln\left[e^{-x}(1 + x)\right] = \ln\,(e^{-x}) + \ln\,(1 + x) = -x + \ln\,(1 + x) \Rightarrow$

$y' = -1 + 1/(1 + x) = -x/(1 + x)$

45. $y = x^{\sin x} \Rightarrow \ln y = \sin x \ln x \Rightarrow y'/y = \cos x \ln x + (\sin x)/x$

$\Rightarrow y' = x^{\sin x}\left[\cos x \ln x + (\sin x)/x\right]$

47. $y = x^{e^x} \Rightarrow \ln y = e^x \ln x \Rightarrow y'/y = e^x \ln x + e^x/x \Rightarrow y' = x^{e^x} e^x\,(\ln x + 1/x)$

49. $y = (\ln x)^x \Rightarrow \ln y = x \ln \ln x \Rightarrow y'/y = \ln \ln x + x \cdot \frac{1}{\ln x} \cdot \frac{1}{x} \Rightarrow$

$y' = (\ln x)^x\,(\ln \ln x + 1/\ln x)$

51. $y = x^{1/\ln x} \Rightarrow \ln y = (1/\ln x) \ln x = 1 \Rightarrow y = e \Rightarrow y' = 0$

53. $y = x^{1/x} \Rightarrow \ln y = (1/x) \ln x \Rightarrow y'/y = (-1/x^2) \ln x + (1/x)\,(1/x)$

$\Rightarrow y' = x^{1/x}\,(1 - \ln x)/x^2$

55. $y = (x^x)^x = x^{x^2} \Rightarrow \ln y = x^2 \ln x \Rightarrow y'/y = 2x \ln x + x^2\,(1/x) \Rightarrow$

$y' = x^{x^2} x\,(2 \ln x + 1) = x^{x^2 + 1}\,(2 \ln x + 1)$

57. $f(x) = \frac{x}{\ln x} \Rightarrow f'(x) = \frac{\ln x - x(1/x)}{(\ln x)^2} = \frac{\ln x - 1}{(\ln x)^2} \Rightarrow f'(e) = \frac{1-1}{1^2} = 0$

59. $y = f(x) = \ln \ln x \Rightarrow f'(x) = (1/\ln x)(1/x) \Rightarrow f'(e) = 1/e$, so the

    equation of the tangent at $(e, 0)$ is $y - 0 = \frac{1}{e}(x - e)$ or $x - ey = e$.

61. $y = f(x) = \ln(1 + e^x) \Rightarrow f'(x) = e^x/(1 + e^x) \Rightarrow f'(0) = 1/(1 + 1) = 1/2$ and so the

    equation of the tangent at $(0, \ln 2)$ is $y - \ln 2 = (1/2) x$ or $x - 2y + 2 \ln 2 = 0$

63. $y = \ln(x^2 + y^2) \Rightarrow y' = \frac{2x + 2yy'}{x^2 + y^2} \Rightarrow x^2 y' + y^2 y' = 2x + 2yy' \Rightarrow y' = \frac{2x}{x^2 + y^2 - 2y}$

65. $f(x) = \ln(x - 1) \Rightarrow f'(x) = 1/(x - 1) = (x - 1)^{-1} \Rightarrow f''(x) = -(x - 1)^{-2}$

    $\Rightarrow f'''(x) = 2(x - 1)^{-3} \Rightarrow f^{(4)}(x) = -2 \cdot 3(x - 1)^{-4} \Rightarrow \cdots \Rightarrow$

    $f^{(n)}(x) = (-1)^{n-1} 2 \cdot 3 \cdot 4 \cdots (n - 1)(x - 1)^{-n} = (-1)^{n-1}(n - 1)!/(x - 1)^n$

67. $y = (3x - 7)^4(8x^2 - 1)^3 \Rightarrow \ln|y| = 4 \ln|3x - 7| + 3 \ln|8x^2 - 1| \Rightarrow$

    $\frac{y'}{y} = \frac{12}{3x - 7} + \frac{48x}{8x^2 - 1} \Rightarrow y' = (3x - 7)^4(8x^2 - 1)^3 \left[\frac{12}{3x - 7} + \frac{48x}{8x^2 - 1} \right]$

69. $y = (x + 1)^4(x - 5)^3/(x - 3)^8 \Rightarrow \ln|y| = 4 \ln|x + 1| + 3 \ln|x - 5| - 8 \ln|x - 3|$

    $\Rightarrow \frac{y'}{y} = \frac{4}{x + 1} + \frac{3}{x - 5} - \frac{8}{x - 3} \Rightarrow y' = \frac{(x + 1)^4(x - 5)^3}{(x - 3)^8} \left[\frac{4}{x + 1} + \frac{3}{x - 5} - \frac{8}{x - 3} \right]$

71. $y = \frac{e^x \sqrt{x^5 + 2}}{(x + 1)^4(x^2 + 3)^2} \Rightarrow \ln y = x + \frac{1}{2}\ln(x^5 + 2) - 4 \ln|x + 1| - 2 \ln(x^2 + 3)$

    $\Rightarrow \frac{y'}{y} = 1 + \frac{5x^4}{2(x^5 + 2)} - \frac{4}{x + 1} - \frac{4x}{x^2 + 3}$

    $y' = \frac{e^x \sqrt{x^5 + 2}}{(x + 1)^4(x^2 + 3)^2} \left[1 + \frac{5x^4}{2(x^5 + 2)} - \frac{4}{x + 1} - \frac{4x}{x^2 + 3} \right]$

73. $f(x) = 2x + \ln x \Rightarrow f'(x) = 2 + 1/x$. If $g = f^{-1}$, then $f(1) = 2 \Rightarrow$

    $g(2) = 1$, so $g'(2) = 1/f'(g(2)) = 1/f'(1) = 1/3$.

75. Let $m = n/x$. Therefore, $n = xm$ and as $n \to \infty$, $m \to \infty$.

    Therefore, $\lim_{n \to \infty} \left[1 + \frac{x}{n} \right]^n = \lim_{m \to \infty} \left[1 + \frac{1}{m} \right]^{mx} = \left[\lim_{m \to \infty} \left[1 + \frac{1}{m} \right]^m \right]^x = e^x$ by (3.34).

EXERCISES 3.6

1. (a) By Theorem 3.36, $y(t) = y(0) e^{kt} = 100 e^{kt} \Rightarrow y(1/3) = 100 e^{k/3} = 200 \Rightarrow$

$k/3 = \ln(200/100) = \ln 2 \Rightarrow k = 3 \ln 2$. So $y(t) = 100 e^{(3 \ln 2)t} = 100 \cdot 2^{3t}$

(since $e^{\ln 2} = 2$).

(b) $y(10) = 100 \cdot 2^{30} \approx 1.07 \times 10^{11}$ cells

(c) $y(t) = 100 \cdot 2^{3t} = 10000 \Rightarrow 2^{3t} = 100 \Rightarrow 3t \ln 2 = \ln 100 \Rightarrow$

$t = (\ln 100)/(3 \ln 2) \approx 2.2$ h

3. (a) $y(t) = y(0) e^{kt} = 500 e^{kt} \Rightarrow y(3) = 500 e^{3k} = 8000 \Rightarrow e^{3k} = 16$

$\Rightarrow 3k = \ln 16 \Rightarrow y(t) = 500 e^{(\ln 16) t/3} = 500 \cdot 16^{t/3}$

(b) $y(4) = 500 \cdot 16^{4/3} \approx 20159$ (c) $y(t) = 500 \cdot 16^{t/3} = 30000 \Rightarrow$

$16^{t/3} = 60 \Rightarrow (t/3) \ln 16 = \ln 60 \Rightarrow t = 3 \ln 60 / \ln 16 \approx 4.4$ h

5. (a) We are given that $k = 0.05$ in (3.35) and so, by Theorem 3.36,

$y(t) = y(0)e^{0.05t} = 307000 e^{0.05t}$, where t is the number of years

after 1985. So $y(10) = 307000 e^{10(0.05)} \approx 506157$.

(b) $y(16) = 307000 e^{16(0.05)} \approx 683241$.

7. (a) If $y = [N_2O_5]$ then $\frac{dy}{dt} = -0.0005 y \Rightarrow y(t) = y(0) e^{-0.0005t}$

$= C e^{-0.0005t}$ (b) $y(t) = C e^{-0.0005t} = 0.9 C \Rightarrow e^{-0.0005t} = 0.9 \Rightarrow$

$-0.0005t = \ln(0.9) \Rightarrow t = -2000 \ln(0.9) \approx 211$ s.

9. (a) If $y(t)$ is the mass remaining after t days, then $y(t) = y(0) e^{kt}$

$= 50 e^{kt} \Rightarrow y(.00014) = 50 e^{0.00014k} = 25 \Rightarrow e^{0.00014k} = 1/2 \Rightarrow$

$k = -\ln 2 / 0.00014 \Rightarrow y(t) = 50 e^{-(\ln 2) t/0.00014} = 50 \cdot 2^{-t/0.00014}$

(b) $y(0.01) = 50 \cdot 2^{-0.01/0.00014} \approx 1.57 \times 10^{-20}$ mg

(c) $50 e^{-(\ln 2) t/0.00014} = 40 \Rightarrow -(\ln 2) t/0.00014 = \ln(0.8) \Rightarrow$

$t = -\ln(0.8) (0.00014)/(\ln 2) \approx 4.5 \times 10^{-5}$ s.

11. Let $y(t)$ be the level of radioactivity. Thus, $y(t) = y(0)e^{-kt}$ and k is determined by

using the half–life : $\frac{1}{2} = e^{-5730k} \Rightarrow k = -(\ln \frac{1}{2})/5730 = \ln 2/5730$. If $.74$ of the ^{14}C

remains, then we know that $.74 = e^{-t(\ln 2)/5730} \Rightarrow \ln .74 = -t(\ln 2)/5730$

$\Rightarrow t = -5730 \, (\ln 0.74)/\ln 2 \approx 2489 \approx 2500$ years.

13. Let $y(t) = $ temperature after t minutes . Then $\frac{dy}{dt} = -\frac{1}{10}(y(t) - 21)$.

If $u(t) = y(t) - 21$, then $\frac{du}{dt} = -\frac{1}{10}u \Rightarrow u(t) = u(0) \, e^{-t/10} = 12 \, e^{-t/10}$

$\Rightarrow y(t) = 21 + u(t) = 21 + 12 \, e^{-t/10}$.

15. (a) Let $y(t) = $ temperature after t minutes . Newton's Law of Cooling $\Rightarrow$

$\frac{dy}{dt} = k(y - 75)$. Let $u(t) = y(t) - 75$. Then $\frac{du}{dt} = ku$, so

$u(t) = u(0) \, e^{kt} = 110 \, e^{kt} \Rightarrow y(t) = 75 + 110 \, e^{kt} \Rightarrow y(30) = 75 + 110$

$e^{30k} = 150 \Rightarrow e^{30k} = \frac{75}{110} = \frac{15}{22} \Rightarrow k = \frac{1}{30}\ln\frac{15}{22}$,

so $y(t) = 75 + 110 \, e^{\frac{t}{30} \ln (\frac{15}{22})}$ and $y(45) = 75 + 110 \, e^{\frac{45}{30} \ln (\frac{15}{22})} \approx 137 \, °F$

(b) $y(t) = 75 + 110 \, e^{\frac{t}{30} \ln (\frac{15}{22})} = 100 \Rightarrow e^{\frac{t}{30} \ln (\frac{15}{22})} = \frac{25}{110}$

$\Rightarrow \frac{t}{30}\ln\frac{15}{22} = \ln\frac{25}{110} \Rightarrow t = \frac{30 \ln(25/110)}{\ln(15/22)} \approx 116$ min

17. (a) Let $P(h)$ be the pressure at altitude h . Then $dP/dh = kP \Rightarrow$

$P(h) = P(0) \, e^{kh} = 101.3 \, e^{kh} \Rightarrow P(1000) = 101.3 \, e^{1000k} = 87.14$

$\Rightarrow 1000k = \ln(87.14 / 101.3) \Rightarrow P(h) = 101.3 \, e^{\ln(87.14/101.3) \, h/1000}$

$P(3000) = 101.3 \, e^{\ln(87.14/101.3)(3)} \approx 64.5$ kPa

(b) $P(6187) = 101.3 \, e^{\ln(87.14/101.3)(6187/1000)} \approx 39.9$ kPa

19. With the notation of Example 4 , $A_0 = 3000$, $i = 0.09$, and $t = 5$.

(a) $n = 1$: $\quad A = 3000(1.09)^5 = \$4615.87$

(b) $n = 2$: $\quad A = 3000(1 + .09/2)^{10} = \$4658.91$

(c) $n = 12$: $\quad A = 3000(1 + .09/12)^{60} = \$4697.04$

(d) $n = 52$: $\quad A = 3000(1 + .09/52)^{5(52)} = \$4703.11$

(e) $n = 365$: $\quad A = 3000(1 + .09/365)^{5(365)} = \$4704.67$

(f) continuously : $A = 3000 \, e^{(0.09)5} = \$4704.94$

21. (a) If $y(t)$ is the amount of salt at time t, then $y(0) = 1500(0.3) = 450$ kg. The

rate of change of y is $\frac{dy}{dt} = -\left[\frac{y(t)}{1500} \frac{\text{kg}}{\text{L}}\right]\left[20 \frac{\text{L}}{\text{min}}\right] = -\frac{1}{75} y(t) \frac{\text{kg}}{\text{min}}$

so $y(t) = y(0) e^{-t/75} = 450 e^{-t/75} \Rightarrow y(30) = 450 e^{-0.4} \approx 301.6$ kg

(b) When the concentration is 0.2 kg/L, the amount of salt is

$(0.2)(1500) = 300$ kg. So $y(t) = 450 e^{-t/75} = 300 \Rightarrow e^{-t/75} = 2/3$

$\Rightarrow -t/75 = \ln(2/3) \Rightarrow t = -75 \ln(2/3) \approx 30.41$ min.

EXERCISES 3.7

1. $\cos^{-1}(-1) = \pi$ since $\cos \pi = -1$

3. $\tan^{-1} \sqrt{3} = \pi/3$ since $\tan \pi/3 = \sqrt{3}$

5. $\csc^{-1} \sqrt{2} = \pi/4$ since $\csc \pi/4 = \sqrt{2}$

7. $\cot^{-1}(-\sqrt{3}) = 5\pi/6$ since $\cot 5\pi/6 = -\sqrt{3}$

9. $\sin^{-1}(-1/\sqrt{2}) = -\pi/4$ since $\sin(-\pi/4) = -1/\sqrt{2}$

11. $\arctan(-\sqrt{3}/3) = -\pi/6$ since $\tan(-\pi/6) = -\sqrt{3}/3$

13. $\sin(\sin^{-1} 0.7) = 0.7$

15. $\tan(\tan^{-1} 10) = 10$

17. $\cos(\sin^{-1} \sqrt{3}/2) = \cos \pi/3 = 1/2$

19. Let $\theta = \cos^{-1} 4/5$, so $\cos \theta = 4/5$. Then $\sin(\cos^{-1} 4/5) = \sin \theta$
$= \sqrt{1 - (4/5)^2} = \sqrt{9/25} = 3/5$.

21. $\arcsin(\sin 5\pi/4) = \arcsin(-1/\sqrt{2}) = -\pi/4$

23. Let $\theta = \sin^{-1}(5/13)$. Then $\sin \theta = 5/13$, so $\cos(2 \sin^{-1}(5/13)) =$
$\cos 2\theta = 1 - 2 \sin^2 \theta = 1 - 2(5/13)^2 = 119/169$

25. Let $x = \sin^{-1}(1/3)$ and $y = \sin^{-1}(2/3)$. Then $\sin x = 1/3$, $\cos x =$

$\sqrt{1-(1/3)^2} = 2\sqrt{2}/3$, $\sin y = 2/3$, $\cos y = \sqrt{1-(2/3)^2} = \sqrt{5}/3$, so

$\sin [\sin^{-1} (1/3) + \sin^{-1} (2/3)] = \sin (x+y) = \sin x \cos y + \cos x \sin y =$

$(1/3) (\sqrt{5}/3) + (2\sqrt{2}/3) (2/3) = (\sqrt{5} + 4\sqrt{2}) /9$.

27. Let $y = \sin^{-1} x$. Then $-\pi/2 \le y \le \pi/2 \Rightarrow \cos y \ge 0$, so

$\cos (\sin^{-1} x) = \cos y = \sqrt{1 - \sin^2 y} = \sqrt{1 - x^2}$

29. Let $y = \tan^{-1} x$. Then $\tan y = x$,

so from the triangle we see that

$\sin (\tan^{-1} x) = \sin y = \dfrac{x}{\sqrt{1+x^2}}$

31. Let $y = \cos^{-1} x$. Then $\cos y = x$ and $0 \le y \le \pi \Rightarrow -\sin y \dfrac{dy}{dx} = 1 \Rightarrow$

$\dfrac{dy}{dx} = -\dfrac{1}{\sin y} = -\dfrac{1}{\sqrt{1 - \cos^2 y}} = -\dfrac{1}{\sqrt{1-x^2}}$ $\left[\textit{Note that } \sin y \ge 0 \textit{ for } 0 \le y \le \pi \right]$

33. Let $y = \cot^{-1} x$. Then $\cot y = x \Rightarrow -\csc^2 y \; dy/dx = 1 \Rightarrow$

$\dfrac{dy}{dx} = -\dfrac{1}{\csc^2 y} = -\dfrac{1}{1 + \cot^2 y} = -\dfrac{1}{1+x^2}$

35. Let $y = \csc^{-1} x$. Then $\csc y = x \Rightarrow -\csc y \cot y \; dy/dx = 1 \Rightarrow$

$\dfrac{dy}{dx} = -\dfrac{1}{\csc y \cot y} = -\dfrac{1}{\csc y \sqrt{\csc^2 y - 1}} = -\dfrac{1}{x\sqrt{x^2 - 1}}$

$\left[\text{Note that } \cot y \ge 0 \text{ on the domain of } \csc^{-1} x. \right]$

37. $g(x) = \tan^{-1} (x^3) \Rightarrow g'(x) = \dfrac{1}{1 + (x^3)^2} (3x^2) = \dfrac{3x^2}{1 + x^6}$

39. $y = \sin^{-1} (x^2) \Rightarrow y' = \left(1/\sqrt{1 - (x^2)^2}\right) (2x) = 2x/\sqrt{1 - x^4}$

41. $F(x) = \tan^{-1} (x/a) \Rightarrow F'(x) = \dfrac{1}{1 + (x/a)^2} \left[\dfrac{1}{a}\right] = \dfrac{a}{a^2 + x^2}$

43. $H(x) = (1+x^2) \arctan x \Rightarrow H'(x) = (2x) \arctan x + (1+x^2) \left[1/(1+x^2) \right]$

$= 1 + 2x \arctan x$

45. $g(t) = \sin^{-1} (4/t) \Rightarrow g'(t) = \left(1/\sqrt{1 - (4/t)^2}\right) (-4/t^2) = -4/\sqrt{t^4 - 16t^2}$

47. $G(t) = \cos^{-1} \sqrt{2t - 1} \Rightarrow G'(t) = -\dfrac{1}{\sqrt{1 - (2t - 1)}} \dfrac{2}{2\sqrt{2t - 1}} = -\dfrac{1}{\sqrt{2(-2t^2 + 3t - 1)}}$

49. $y = \sec^{-1}\sqrt{1+x^2} \Rightarrow y' = \dfrac{1}{\sqrt{1+x^2}\sqrt{(1+x^2)-1}} \ \dfrac{2x}{2\sqrt{1+x^2}} = \dfrac{x}{(1+x^2)\sqrt{x^2}}$

$= \dfrac{x}{(1+x^2)\,|x|}$

51. $y = \tan^{-1}(\sin x) \Rightarrow y' = \dfrac{\cos x}{1+\sin^2 x}$

53. $y = \dfrac{\sin^{-1}x}{\cos^{-1}x} \Rightarrow y' = \dfrac{(\cos^{-1}x)/\sqrt{1-x^2} + \sin^{-1}x/\sqrt{1-x^2}}{(\cos^{-1}x)^2} = \dfrac{\cos^{-1}x + \sin^{-1}x}{\sqrt{1-x^2}\,(\cos^{-1}x)^2}$

$= \pi/\left[2\sqrt{1-x^2}\,(\cos^{-1}x)^2\right]$

55. $y = (\tan^{-1}x)^{-1} \Rightarrow y' = -(\tan^{-1}x)^{-2}\left[1/(1+x^2)\right] = -1/[(1+x^2)(\tan^{-1}x)^2]$

57. $y = x^2\cot^{-1}(3x) \Rightarrow y' = 2x\cot^{-1}(3x) + x^2\left[-1/(1+(3x)^2)\right](3)$

$= 2x\cot^{-1}(3x) - 3x^2/(1+9x^2)$

59. $y = \arccos\left[\dfrac{b + a\cos x}{a + b\cos x}\right] \Rightarrow y' = $

$-\dfrac{1}{\sqrt{1-\left[(b+a\cos x)/(a+b\cos x)\right]^2}} \ \dfrac{(a+b\cos x)(-a\sin x) - (b+a\cos x)(-b\sin x)}{(a+b\cos x)^2}$

$= \dfrac{1}{\sqrt{a^2 + b^2\cos^2 x - b^2 - a^2\cos^2 x}} \ \dfrac{(a^2-b^2)\sin x}{|a+b\cos x|}$

$= \dfrac{1}{\sqrt{a^2-b^2}\sqrt{1-\cos^2 x}} \ \dfrac{(a^2-b^2)\sin x}{|a+b\cos x|} = \dfrac{\sqrt{a^2-b^2}}{|a+b\cos x|} \ \dfrac{\sin x}{|\sin x|}$ but $0 \le x \le \pi$ and so

$|\sin x| = \sin x$. Also $a > b > 0 \Rightarrow b\cos x \ge -b > -a$, so $a + b\cos x > 0$.

Thus, $y' = \dfrac{\sqrt{a^2-b^2}}{a+b\cos x}$

61. $g(x) = \sin^{-1}(3x+1) \Rightarrow g'(x) = 3/\sqrt{1-(3x+1)^2} = 3/\sqrt{-9x^2-6x}$

Dom $(g) = \left\{x \mid -1 \le 3x+1 \le 1\right\} = \left\{x \mid -2/3 \le x \le 0\right\} = [\,-2/3\,,\,0\,]$

Dom $(g') = \left\{x \mid -1 < 3x+1 < 1\right\} = (-2/3\,,\,0)$

63. $S(x) = \sin^{-1}(\tan^{-1}x) \Rightarrow S'(x) = 1/[\sqrt{1-(\tan^{-1}x)^2}\,(1+x^2)]$

Dom $(S) = \left\{x \mid -1 \le \tan^{-1}x \le 1\right\} = \left\{x \mid \tan(-1) \le x \le \tan 1\right\}$

$= [\,-\tan 1\,,\,\tan 1\,], \quad \text{dom}\,(S') = \left\{x \mid -1 < \tan^{-1}x < 1\right\} = (-\tan 1\,,\,\tan 1)$

65. $G(x) = \sqrt{\csc^{-1}x} \Rightarrow G'(x) = [1/(2\sqrt{\csc^{-1}x})][-1/(x\sqrt{x^2-1})]$

$= -1/(2x\sqrt{(x^2-1)}\csc^{-1}x)$. Dom $(G) = \left\{ x \mid |x| \geq 1 \right\} = (-\infty, -1] \cup [1, \infty)$

[Note : $\csc^{-1}x > 0$]   dom $(G') = \left\{ x \mid |x| > 1 \right\} = (-\infty, -1) \cup (1, \infty)$

67. $U(t) = 2^{\arctan t} \Rightarrow U'(t) = 2^{\arctan t}(\ln 2)/(1+t^2)$. Dom $(U) = $ dom $(U') = R$

69. $g(x) = x \sin^{-1}(x/4) + \sqrt{16-x^2} \Rightarrow g'(x) = \sin^{-1}(x/4) + x/(4\sqrt{1-(x/4)^2})$

$-x/\sqrt{16-x^2} = \sin^{-1}(x/4) \Rightarrow g'(2) = \sin^{-1}(1/2) = \pi/6$

71. $\lim\limits_{x \to -1^+} \sin^{-1}x = \sin^{-1}(-1) = -\pi/2$

73. $\lim\limits_{x \to -\infty} \cot^{-1}x = \pi$ since $\cot x \to -\infty$ as $x \to \pi^-$

75. $\lim\limits_{x \to -\infty} \csc^{-1}x = \pi$ since $\csc x \to -\infty$ as $x \to \pi^+$

77. $\lim\limits_{x \to \infty} (\tan^{-1}x)^2 = (\lim\limits_{x \to \infty} \tan^{-1}x)^2 = (\pi/2)^2 = \pi^2/4$

79. $\lim\limits_{x \to 1^-} \dfrac{\arcsin x}{\tan(\pi x/2)} = 0$ since $\lim\limits_{x \to 1^-} \arcsin x = \frac{\pi}{2}$ and $\lim\limits_{x \to 1^-} \tan(\pi x/2) = \infty$

81.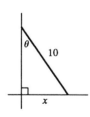

$\dfrac{dx}{dt} = 2$ ft/s , $\sin\theta = \frac{x}{10} \Rightarrow \theta = \sin^{-1}(x/10)$

$\dfrac{d\theta}{dx} = \dfrac{1/10}{\sqrt{1-(x/10)^2}}$, $\dfrac{d\theta}{dt} = \dfrac{d\theta}{dx}\dfrac{dx}{dt} = \dfrac{1/10}{\sqrt{1-(x/10)^2}}(2)$ rad/s

$\dfrac{d\theta}{dt}\Big|_{x=6} = \dfrac{2/10}{\sqrt{1-(6/10)^2}}$ rad/s $= \frac{1}{4}$ rad/s

83. $y = \sec^{-1}x$

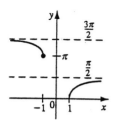

85.   Let $a = \arctan x$ and let $b = \arctan y$. Then by Equation 14(a) Appendix B,

$$\tan(a+b) = \frac{\tan a + \tan b}{1-(\tan a)(\tan b)} = \frac{\tan(\arctan x) + \tan(\arctan y)}{1-\tan(\arctan x)\tan(\arctan y)} \Rightarrow \tan(a+b) = \frac{x+y}{1-xy} \Rightarrow$$

$$\arctan x + \arctan y \ = a+b = \arctan\left[\frac{x+y}{1-xy}\right], \text{ since } -\pi/2 < \arctan x + \arctan y < \pi/2.$$

87. (a) Since $\left|\arctan(1/x)\right| < \pi/2$, we have $0 \le \left|x\arctan(1/x)\right| \le \frac{\pi}{2}|x| \to 0$ as $x \to 0$. So, by the

Squeeze Theorem, $\lim\limits_{x \to 0} f(x) = 0 = f(0)$, so f is continuous at 0.

(b) Here $\frac{f(x)-f(0)}{x-0} = \frac{x\arctan\frac{1}{x}-0}{x} = \arctan\frac{1}{x}$. So $f'_+(0) = \lim\limits_{x \to 0^+} \frac{f(x)-f(0)}{x-0}$

$= \lim\limits_{x \to 0^+} \arctan\frac{1}{x} = \lim\limits_{y \to \infty} \arctan y = \pi/2$, while $f'_-(0) = \lim\limits_{x \to 0^-} \frac{f(x)-f(0)}{x-0}$

$= \lim\limits_{x \to 0^-} \arctan\frac{1}{x} = \lim\limits_{y \to -\infty} \arctan y = -\pi/2$. So $f'(0)$ does not exist.

89. $y = \sec^{-1} x \Rightarrow \sec y = x \Rightarrow \sec y \tan y \frac{dy}{dx} = 1 \Rightarrow \frac{dy}{dx} = \frac{1}{\sec y \tan y}$

Now $\tan^2 y = \sec^2 y - 1 = x^2 - 1$, so $\tan y = \pm\sqrt{x^2-1}$. For $y \in [0, \pi/2)$,

$x \ge 1$, so $\sec y = x = |x|$ and $\tan y \ge 0 \Rightarrow \frac{dy}{dx} = \frac{1}{x\sqrt{x^2-1}} = \frac{1}{|x|\sqrt{x^2-1}}$

For $y \in (\pi/2, \pi]$, $x \le -1$, so $|x| = -x$ and $\tan y = -\sqrt{x^2-1} \Rightarrow$

$\frac{dy}{dx} = \frac{1}{\sec y \tan y} = \frac{1}{x(-\sqrt{x^2-1})} = \frac{1}{(-x)\sqrt{x^2-1}} = \frac{1}{|x|\sqrt{x^2-1}}$

EXERCISES 3.8

1. $\sinh(-x) = \frac{1}{2}\left(e^{-x} - e^{-(-x)}\right) = \frac{1}{2}\left(e^{-x} - e^{x}\right) = -\frac{1}{2}\left(e^{x} - e^{-x}\right) = -\sinh x$

3. $\cosh x + \sinh x = \frac{1}{2}\left(e^{x} + e^{-x}\right) + \frac{1}{2}\left(e^{x} - e^{-x}\right) = \frac{1}{2}\left(2e^{x}\right) = e^{x}$

5. $\sinh x \cosh y + \cosh x \sinh y$

$= (1/2)\left(e^{x} - e^{-x}\right)(1/2)\left(e^{y} + e^{-y}\right) + (1/2)\left(e^{x} + e^{-x}\right)(1/2)\left(e^{y} - e^{-y}\right)$

$$= (1/4) \left[(e^{x+y} + e^{x-y} - e^{-x+y} - e^{-x-y}) + (e^{x+y} - e^{x-y} + e^{-x+y} - e^{-x-y}) \right]$$

$$= (1/4)\, (2e^{x+y} - 2e^{-x-y}) = (1/2)\, (e^{x+y} - e^{-(x+y)}) = \sinh (x + y)$$

7. Divide both sides of the identity $\cosh^2 x - \sinh^2 x = 1$ by $\sinh^2 x$:

$$\frac{\cosh^2 x}{\sinh^2 x} - 1 = \frac{1}{\sinh^2 x} \qquad \text{or} \qquad \coth^2 x - 1 = \operatorname{csch}^2 x$$

9. By Exercise 5 , $\sinh 2x = \sinh (x + x) = \sinh x \cosh x + \cosh x \sinh x = 2 \sinh x \cosh x$.

11. By Exercise 10 , $\cosh 2y = \cosh^2 y + \sinh^2 y = 1 + 2 \sinh^2 y \Rightarrow \sinh^2 y = (\cosh 2y - 1)/2$.

Put $x = 2y$. Then $\sinh (x/2) = \pm \sqrt{(\cosh x - 1)/2}$.

13. $\tanh (\ln x) = \dfrac{\sinh (\ln x)}{\cosh (\ln x)} = \dfrac{(e^{\ln x} - e^{-\ln x})/2}{(e^{\ln x} + e^{-\ln x})/2} = \dfrac{x - 1/x}{x + 1/x} = \dfrac{x^2 - 1}{x^2 + 1}$

15. By Exercise 3 , $(\cosh x + \sinh x)^n = (e^x)^n = e^{nx} = \cosh nx + \sinh nx$

17. $\tanh x = 4/5 > 0$, so $x > 0$. $\coth x = 1/\tanh x = 5/4$, $\operatorname{sech}^2 x = 1 - \tanh^2 x$

$= 1 - (4/5)^2 = 9/25 \Rightarrow \operatorname{sech} x = 3/5$ (since $\operatorname{sech} x > 0$) , $\cosh x = 1/\operatorname{sech} x = 5/3$,

$\sinh x = \tanh x \cosh x = (4/5)\,(5/3) = 4/3$, and $\operatorname{csch} x = 1/\sinh x = 3/4$.

19. (a) $\displaystyle \lim_{x \to \infty} \tanh x = \lim_{x \to \infty} \frac{e^x - e^{-x}}{e^x + e^{-x}} = \lim_{x \to \infty} \frac{1 - e^{-2x}}{1 + e^{-2x}} = \frac{1 - 0}{1 + 0} = 1$

(b) $\displaystyle \lim_{x \to -\infty} \tanh x = \lim_{x \to -\infty} \frac{e^x - e^{-x}}{e^x + e^{-x}} = \lim_{x \to -\infty} \frac{e^{2x} - 1}{e^{2x} + 1} = \frac{0 - 1}{0 + 1} = -1$

(c) $\displaystyle \lim_{x \to \infty} \sinh x = \lim_{x \to \infty} \frac{e^x - e^{-x}}{2} = \infty$

(d) $\displaystyle \lim_{x \to -\infty} \sinh x = \lim_{x \to -\infty} \frac{e^x - e^{-x}}{2} = -\infty$

(e) $\displaystyle \lim_{x \to \infty} \operatorname{sech} x = \lim_{x \to \infty} \frac{2}{e^x + e^{-x}} = 0$

(f) $\displaystyle \lim_{x \to \infty} \coth x = \lim_{x \to \infty} \frac{e^x + e^{-x}}{e^x - e^{-x}} = \lim_{x \to \infty} \frac{1 + e^{-2x}}{1 - e^{-2x}} = \frac{1 + 0}{1 - 0} = 1$ [or use (a)]

(g) $\displaystyle \lim_{x \to 0^+} \coth x = \lim_{x \to 0^+} \frac{\cosh x}{\sinh x} = \infty$, since $\sinh x \to 0$ and $\coth x > 0$

(h) $\displaystyle \lim_{x \to 0^-} \coth x = \lim_{x \to 0^-} \frac{\cosh x}{\sinh x} = -\infty$, since $\sinh x \to 0$ and $\coth x < 0$

(i) $\displaystyle\lim_{x \to -\infty} \operatorname{csch} x = \lim_{x \to -\infty} \frac{2}{e^x - e^{-x}} = 0$

21. Let $y = \sinh^{-1} x$. Then $\sinh y = x$ and, by Example 1(a), $\cosh y = \sqrt{1 + \sinh^2 y}$

$= \sqrt{1 + x^2}$. So by Exercise 3, $e^y = \sinh y + \cosh y = x + \sqrt{1 + x^2} \;\Rightarrow\; y = \ln(x + \sqrt{1 + x^2})$.

23. (a) Let $y = \tanh^{-1} x$. Then $x = \tanh y = \dfrac{e^y - e^{-y}}{e^y + e^{-y}} = \dfrac{e^{2y} - 1}{e^{2y} + 1} \;\Rightarrow$

$xe^{2y} + x = e^{2y} - 1 \;\Rightarrow\; e^{2y} = \dfrac{1+x}{1-x} \;\Rightarrow\; 2y = \ln\left[\dfrac{1+x}{1-x}\right] \;\Rightarrow\; y = \tfrac{1}{2}\ln\left[\dfrac{1+x}{1-x}\right]$

(b) Let $y = \tanh^{-1} x$. Then $x = \tanh y$, so from Exercise 14 we have

$e^{2y} = \dfrac{1 + \tanh y}{1 - \tanh y} = \dfrac{1+x}{1-x} \;\Rightarrow\; 2y = \ln\left[\dfrac{1+x}{1-x}\right] \;\Rightarrow\; y = \tfrac{1}{2}\ln\left[\dfrac{1+x}{1-x}\right]$

25. (a) Let $y = \cosh^{-1} x$. Then $\cosh y = x$ and $y \geq 0 \;\Rightarrow\; \sinh y \dfrac{dy}{dx} = 1$

$\Rightarrow\; \dfrac{dy}{dx} = \dfrac{1}{\sinh y} = \dfrac{1}{\sqrt{\cosh^2 y - 1}} = \dfrac{1}{\sqrt{x^2 - 1}}$ (since $\sinh y \geq 0$ for $y \geq 0$)

[OR: Use Formula 3.55.]

(b) Let $y = \tanh^{-1} x$. Then $\tanh y = x \;\Rightarrow\; \operatorname{sech}^2 y \dfrac{dy}{dx} = 1 \;\Rightarrow$

$\dfrac{dy}{dx} = \dfrac{1}{\operatorname{sech}^2 y} = \dfrac{1}{1 - \tanh^2 y} = \dfrac{1}{1 - x^2}$ [OR: Use Formula 3.56.]

(c) Let $y = \operatorname{csch}^{-1} x$. Then $\operatorname{csch} y = x \;\Rightarrow\; -\operatorname{csch} y \coth y \dfrac{dy}{dx} = 1$

$\Rightarrow\; \dfrac{dy}{dx} = -\dfrac{1}{\operatorname{csch} y \coth y}$ By Exercise 7, $\coth y = \pm\sqrt{\operatorname{csch}^2 y + 1} =$

$\pm\sqrt{x^2 + 1}$. If $x > 0$, then $\coth y > 0$, so $\coth y = \sqrt{x^2 + 1}$. If $x < 0$,

then $\coth y < 0$, so $\coth y = -\sqrt{x^2 + 1}$. In either case we have

$\dfrac{dy}{dx} = -\dfrac{1}{\operatorname{csch} y \coth y} = -1 / |x| \sqrt{x^2 + 1}$.

(d) Let $y = \operatorname{sech}^{-1} x$. Then $\operatorname{sech} y = x \;\Rightarrow\; -\operatorname{sech} y \tanh y \dfrac{dy}{dx} = 1$

$\Rightarrow\; \dfrac{dy}{dx} = -\dfrac{1}{\operatorname{sech} y \tanh y} = -\dfrac{1}{\operatorname{sech} y \sqrt{1 - \operatorname{sech}^2 y}} = -\dfrac{1}{x \sqrt{1 - x^2}}$

[Note that $y > 0$ and so $\tanh y > 0$.]

(e) Let $y = \coth^{-1} x$. Then $\coth y = x \Rightarrow -\operatorname{csch}^2 y \dfrac{dy}{dx} = 1 \Rightarrow$

$$\dfrac{dy}{dx} = -\dfrac{1}{\operatorname{csch}^2 y} = \dfrac{1}{1 - \coth^2 y} = \dfrac{1}{1 - x^2} \quad \text{by Exercise 7.}$$

27. $f(x) = \tanh 3x \Rightarrow f'(x) = 3 \operatorname{sech}^2 3x$

29. $h(x) = \cosh (x^4) \Rightarrow h'(x) = \sinh (x^4) \cdot 4x^3$

31. $G(x) = x^2 \operatorname{sech} x \Rightarrow G'(x) = 2x \operatorname{sech} x - x^2 \operatorname{sech} x \tanh x$

33. $H(t) = \tanh (e^t) \Rightarrow H'(t) = \operatorname{sech}^2(e^t) (e^t)$

35. $y = x^{\cosh x} \Rightarrow \ln y = \cosh x \ln x \Rightarrow \dfrac{y'}{y} = \sinh x \ln x + \dfrac{\cosh x}{x} \Rightarrow$

$y' = x^{\cosh x} (\sinh x \ln x + (\cosh x)/x)$

37. $y = \cosh^{-1} (x^2) \Rightarrow y' = \left(1/\sqrt{(x^2)^2 - 1} \right) (2x) = 2x/\sqrt{x^4 - 1}$

39. $y = x \ln (\operatorname{sech} 4x) \Rightarrow y' = \ln (\operatorname{sech} 4x) + x \dfrac{-\operatorname{sech} 4x \tanh 4x}{\operatorname{sech} 4x} (4)$

$= \ln (\operatorname{sech} 4x) - 4x \tanh 4x$

41. $y = \tanh^{-1} (x/a) \Rightarrow y' = \dfrac{1}{1 - (x/a)^2} \left[\dfrac{1}{a}\right] = \dfrac{a}{a^2 - x^2}$

43. $y = \operatorname{csch}^{-1} (x^4) \Rightarrow y' = -4x^3 / \left[x^4 \sqrt{(x^4)^2 + 1} \right] = -4/\left(x \sqrt{x^8 + 1} \right)$

45. $y = \coth^{-1} \sqrt{x^2 + 1} \Rightarrow y' = \dfrac{1}{1 - (x^2 + 1)} \dfrac{2x}{2\sqrt{x^2 + 1}} = -\dfrac{1}{x \sqrt{x^2 + 1}}$

47. The tangent to $y = \cosh x$ has slope 1 when $y' = \sinh x = 1 \Rightarrow x = \sinh^{-1} 1 = \ln (1 + \sqrt{2})$,

by (6.80). Since $\sinh x = 1$ and $\cosh x = \sqrt{1 + \sinh^2 x}$, we have $\cosh x = \sqrt{2}$. The

point is $(\ln (1 + \sqrt{2}), \sqrt{2})$.

49. $\cosh x = \cosh[\ln(\sec\theta + \tan\theta)] = \dfrac{e^{\ln(\sec\theta + \tan\theta)} + e^{-\ln(\sec\theta + \tan\theta)}}{2}$

$= \dfrac{\sec\theta + \tan\theta + 1/(\sec\theta + \tan\theta)}{2} = \dfrac{\sec\theta + \tan\theta + \dfrac{\sec\theta - \tan\theta}{(\sec\theta + \tan\theta)(\sec\theta - \tan\theta)}}{2}$

$= \dfrac{\sec\theta + \tan\theta + \dfrac{\sec\theta - \tan\theta}{\sec^2\theta - \tan^2\theta}}{2} = \dfrac{\sec\theta + \tan\theta + \sec\theta - \tan\theta}{2} = \sec\theta$

EXERCISES 3.9

NOTE : The use of l'Hospital's Rule is indicated by H (above the equal sign).

1. $\displaystyle\lim_{x\to 2}\frac{x-2}{x^2-4} = \lim_{x\to 2}\frac{x-2}{(x-2)(x+2)} = \lim_{x\to 2}\frac{1}{x+2} = \frac{1}{4}$

3. $\displaystyle\lim_{x\to -1}\frac{x^6-1}{x^4-1} \overset{H}{=} \lim_{x\to -1}\frac{6x^5}{4x^3} = \frac{-6}{-4} = \frac{3}{2}$

5. $\displaystyle\lim_{x\to 0}\frac{e^x-1}{\sin x} \overset{H}{=} \lim_{x\to 0}\frac{e^x}{\cos x} = \frac{1}{1} = 1$

7. $\displaystyle\lim_{x\to 0}\frac{\sin x}{x^3} \overset{H}{=} \lim_{x\to 0}\frac{\cos x}{3x^2} = \infty$

9. $\displaystyle\lim_{x\to 0}\frac{\tan x}{x+\sin x} \overset{H}{=} \lim_{x\to 0}\frac{\sec^2 x}{1+\cos x} = \frac{1}{1+1} = \frac{1}{2}$

11. $\displaystyle\lim_{x\to\infty}\frac{\ln x}{x} \overset{H}{=} \lim_{x\to\infty}\frac{1/x}{1} = 0$

13. $\displaystyle\lim_{x\to\infty}\frac{e^x}{x^3} \overset{H}{=} \lim_{x\to\infty}\frac{e^x}{3x^2} \overset{H}{=} \lim_{x\to\infty}\frac{e^x}{6x} \overset{H}{=} \lim_{x\to\infty}\frac{e^x}{6} = \infty$

15. $\displaystyle\lim_{x\to a}\frac{x^{1/3}-a^{1/3}}{x-a} \overset{H}{=} \lim_{x\to a}\frac{(1/3)x^{-2/3}}{1} = \frac{1}{3\,a^{2/3}}$

17. $\displaystyle\lim_{x\to 0}\frac{e^x-1-x}{x^2}\overset{H}{=}\lim_{x\to 0}\frac{e^x-1}{2x}\overset{H}{=}\lim_{x\to 0}\frac{e^x}{2}=\tfrac{1}{2}$

19. $\displaystyle\lim_{x\to 0}\frac{\sin x}{e^x}=\tfrac{0}{1}=0$

21. $\displaystyle\lim_{x\to 0}\frac{1-\cos x}{x^2}\overset{H}{=}\lim_{x\to 0}\frac{\sin x}{2x}\overset{H}{=}\lim_{x\to 0}\frac{\cos x}{2}=\tfrac{1}{2}$

23. $\displaystyle\lim_{x\to 2^-}\frac{\ln x}{\sqrt{2-x}}=\infty$ since $\sqrt{2-x}\to 0$ but $\ln x\to\ln 2$.

25. $\displaystyle\lim_{x\to\infty}\frac{\ln\ln x}{\sqrt{x}}\overset{H}{=}\lim_{x\to\infty}\frac{1/(x\ln x)}{1/2\sqrt{x}}=\lim_{x\to\infty}\frac{2}{\sqrt{x}\ln x}=0$

27. $\displaystyle\lim_{x\to 0}\frac{\tan^{-1}(2x)}{3x}\overset{H}{=}\lim_{x\to 0}\frac{2/(1+4x^2)}{3}=\tfrac{2}{3}$

29. $\displaystyle\lim_{x\to 0}\frac{\tan\alpha x}{x}\overset{H}{=}\lim_{x\to 0}\frac{\alpha\sec^2\alpha x}{1}=\alpha$

31. $\displaystyle\lim_{x\to 0}\frac{\tan 2x}{\tanh 3x}\overset{H}{=}\lim_{x\to 0}\frac{2\sec^2 2x}{3\operatorname{sech}^2 3x}=\tfrac{2}{3}$

33. $\displaystyle\lim_{x\to 0}\frac{x+\sin 3x}{x-\sin 3x}\overset{H}{=}\lim_{x\to 0}\frac{1+3\cos 3x}{1-3\cos 3x}=\frac{1+3}{1-3}=-2$

35. $\displaystyle\lim_{x\to 0}\frac{e^{4x}-1}{\cos x}=\tfrac{0}{1}=0$

37. $\displaystyle\lim_{x\to 0}\frac{\tan x-\sin x}{x^3}\overset{H}{=}\lim_{x\to 0}\frac{\sec^2 x-\cos x}{3x^2}\overset{H}{=}\lim_{x\to 0}\frac{2\sec^2 x\tan x+\sin x}{6x}$

$\displaystyle\overset{H}{=}\lim_{x\to 0}\frac{4\sec^2 x\tan^2 x+2\sec^4 x+\cos x}{6}=\frac{0+2+1}{6}=\tfrac{1}{2}$

39. $\displaystyle\lim_{x\to 0^+}\sqrt{x}\ln x=\lim_{x\to 0^+}\frac{\ln x}{x^{-1/2}}\overset{H}{=}\lim_{x\to 0^+}\frac{1/x}{-(1/2)x^{-3/2}}=\lim_{x\to 0^+}(-2\sqrt{x})=0$

41. $\displaystyle\lim_{x\to\infty}e^{-x}\ln x=\lim_{x\to\infty}\frac{\ln x}{e^x}\overset{H}{=}\lim_{x\to\infty}\frac{1/x}{e^x}=\lim_{x\to\infty}\frac{1}{xe^x}=0$

43. $\displaystyle\lim_{x\to\infty}x^3 e^{-x^2}=\lim_{x\to\infty}\frac{x^3}{e^{x^2}}\overset{H}{=}\lim_{x\to\infty}\frac{3x^2}{2xe^{x^2}}=\lim_{x\to\infty}\frac{3x}{2e^{x^2}}\overset{H}{=}\lim_{x\to\infty}\frac{3}{4xe^{x^2}}=0$

45. $\displaystyle\lim_{x\to\pi}(x-\pi)\cot x = \lim_{x\to\pi}\frac{x-\pi}{\tan x} \overset{H}{=} \lim_{x\to\pi}\frac{1}{\sec^2 x} = \frac{1}{(-1)^2} = 1$

47. $\displaystyle\lim_{x\to 0}\left[\frac{1}{x^4}-\frac{1}{x^2}\right] = \lim_{x\to 0}\frac{1-x^2}{x^4} = \infty$

49. $\displaystyle\lim_{x\to 0}\left[\frac{1}{x}-\csc x\right] = \lim_{x\to 0}\left[\frac{1}{x}-\frac{1}{\sin x}\right] = \lim_{x\to 0}\frac{\sin x - x}{x \sin x}$

$\overset{H}{=} \displaystyle\lim_{x\to 0}\frac{\cos x - 1}{\sin x + x \cos x} \overset{H}{=} \lim_{x\to 0}\frac{-\sin x}{2\cos x - x \sin x} = \frac{0}{2} = 0$

51. $\displaystyle\lim_{x\to\infty}(x-\sqrt{x^2-1}) = \lim_{x\to\infty}(x-\sqrt{x^2-1})\,\frac{x+\sqrt{x^2-1}}{x+\sqrt{x^2-1}} = \lim_{x\to\infty}\frac{x^2-(x^2-1)}{x+\sqrt{x^2-1}}$

$= \displaystyle\lim_{x\to\infty}\frac{1}{x+\sqrt{x^2-1}} = 0$

53. $\displaystyle\lim_{x\to\infty}\left[\frac{x^3}{x^2-1}-\frac{x^3}{x^2+1}\right] = \lim_{x\to\infty}\frac{x^3(x^2+1)-x^3(x^2-1)}{(x^2-1)(x^2+1)} = \lim_{x\to\infty}\frac{2x^3}{x^4-1}$

$= \displaystyle\lim_{x\to\infty}\frac{2/x}{1-1/x^4} = 0$

55. $y = x^{\sin x} \Rightarrow \ln y = \sin x \ln x$, so $\displaystyle\lim_{x\to 0^+}\ln y = \lim_{x\to 0^+}\sin x \ln x$

$= \displaystyle\lim_{x\to 0^+}\frac{\ln x}{\csc x} \overset{H}{=} \lim_{x\to 0^+}\frac{1/x}{-\csc x \cot x} = -\lim_{x\to 0^+}\frac{\sin x}{x}\lim_{x\to 0^+}\tan x = -1\cdot 0$

$= 0 \Rightarrow \displaystyle\lim_{x\to 0^+}x^{\sin x} = \lim_{x\to 0^+}e^{\ln y} = e^0 = 1$

57. $x^{1/\ln x} = e^{(\ln x)(1/\ln x)} = e^1 = e$, so $\displaystyle\lim_{x\to 0^+}x^{1/\ln x} = \lim_{x\to 0^+}x^{1/\ln x} = \lim_{x\to 0^+}e = e$

59. $y = (1-2x)^{1/x} \Rightarrow \ln y = (1/x)\ln(1-2x) \Rightarrow \displaystyle\lim_{x\to 0}\ln y = \lim_{x\to 0}\frac{\ln(1-2x)}{x}$

$\overset{H}{=} \displaystyle\lim_{x\to 0}\frac{-2/(1-2x)}{1} = -2 \Rightarrow \lim_{x\to 0}(1-2x)^{1/x} = \lim_{x\to 0}e^{\ln y} = e^{-2}$

61. $y = (1+3/x+5/x^2)^x \Rightarrow \ln y = x\ln(1+3/x+5/x^2) \Rightarrow$

$\displaystyle\lim_{x\to\infty}\ln y = \lim_{x\to\infty}\frac{\ln(1+3/x+5/x^2)}{1/x} \overset{H}{=} \lim_{x\to\infty}\frac{(-3/x^2-10/x^3)/(1+3/x+5/x^2)}{-1/x^2}$

$$= \lim_{x \to \infty} \frac{3 + 10/x}{1 + 3/x + 5/x^2} = 3 , \text{ so } \lim_{x \to \infty} (1 + 3/x + 5/x^2)^x = \lim_{x \to \infty} e^{\ln y} = e^3$$

63. $y = x^{1/x} \Rightarrow \ln y = (1/x) \ln x \Rightarrow \lim_{x \to \infty} \ln y = \lim_{x \to \infty} \frac{\ln x}{x} \overset{H}{=} \lim_{x \to \infty} \frac{1/x}{1} = 0$

$\Rightarrow \lim_{x \to \infty} x^{1/x} = \lim_{x \to \infty} e^{\ln y} = e^0 = 1$

65. $y = (\cot x)^{\sin x} \Rightarrow \ln y = \sin x \ln (\cot x) \Rightarrow \lim_{x \to 0^+} \ln y =$

$\lim_{x \to 0^+} \frac{\ln (\cot x)}{\csc x} \overset{H}{=} \lim_{x \to 0^+} \frac{-\csc^2 x / \cot x}{-\csc x \cot x} = \lim_{x \to 0^+} \frac{\csc x}{\cot^2 x} = \lim_{x \to 0^+} \frac{\sin x}{\cos^2 x} = 0$

so $\lim_{x \to 0^+} (\cot x)^{\sin x} = \lim_{x \to 0^+} e^{\ln y} = e^0 = 1$

67. $y = \left[\frac{x}{x+1}\right]^x \Rightarrow \ln y = x \ln \left[\frac{x}{x+1}\right] \Rightarrow \lim_{x \to \infty} \ln y = \lim_{x \to \infty} x \ln \left[\frac{x}{x+1}\right]$

$= \lim_{x \to \infty} \frac{\ln x - \ln (x+1)}{1/x} \overset{H}{=} \lim_{x \to \infty} \frac{1/x - 1/(x+1)}{-1/x^2} = \lim_{x \to \infty} \left[-x + \frac{x^2}{x+1}\right]$

$= \lim_{x \to \infty} \frac{-x}{x+1} = -1 , \text{ so } \lim_{x \to \infty} \left[\frac{x}{x+1}\right]^x = \lim_{x \to \infty} e^{\ln y} = e^{-1} = \frac{1}{e}$

$\left[\text{OR} : \lim_{x \to \infty} \left[\frac{x}{x+1}\right]^x = \lim_{x \to \infty} \left[\left[\frac{x+1}{x}\right]^{-1}\right]^x = \left[\lim_{x \to \infty} \left[1 + \frac{1}{x}\right]^x\right]^{-1} = e^{-1}\right]$

69. $y = x^{x^x} \Rightarrow \ln y = x^x \ln x \Rightarrow \lim_{x \to 0^+} \ln y = \lim_{x \to 0^+} x^x \ln x = -\infty$ since

$\ln x \to -\infty$ and $x^x \to 1$ (from Example 9). So $\lim_{x \to 0^+} x^{x^x} = \lim_{x \to 0^+} e^{\ln y} = 0$

71. $\lim_{x \to \infty} \frac{x}{2x + 3 \sin x} = \lim_{x \to \infty} \frac{1}{2 + 3(\sin x / x)} = \frac{1}{2 + 3(0)} = \frac{1}{2}$

73. $\lim_{x \to 0^+} \frac{x + 1 - e^x}{x^3} \overset{H}{=} \lim_{x \to 0^+} \frac{1 - e^x}{3x^2} \overset{H}{=} \lim_{x \to 0^+} \frac{-e^x}{6x} = -\infty$ since $6x \to 0^+$ and $-e^x \to -1$

75. $\lim_{x \to 0} \frac{2x \sin x}{\sec x - 1} \overset{H}{=} \lim_{x \to 0} \frac{2 \sin x + 2x \cos x}{\sec x \tan x} \overset{H}{=} \lim_{x \to 0} \frac{4 \cos x - 2x \sin x}{\sec x \tan^2 x + \sec^3 x} = 4$

77. $\lim_{x \to 0} \frac{\cos x - 1 + x^2/2}{x^4} \overset{H}{=} \lim_{x \to 0} \frac{-\sin x + x}{4x^3} \overset{H}{=} \lim_{x \to 0} \frac{-\cos x + 1}{12x^2} \overset{H}{=} \lim_{x \to 0} \frac{\sin x}{24x} = \frac{1}{24}$

79. Since $\lim_{h \to 0} (f(x + h) - f(x - h)) = f(x) - f(x) = 0$ (*f is differentiable and hence*

continuous) and $\lim_{h \to 0} 2h = 0$, we use l'Hospital's Rule:

$$\lim_{h \to 0} \frac{f(x+h) - f(x-h)}{2h} \overset{H}{=} \lim_{h \to 0} \frac{f'(x+h) - f'(x-h)(-1)}{2} = \frac{f'(x) + f'(x)}{2} = \frac{2f'(x)}{2} = f'(x)$$

81. $\displaystyle\lim_{x \to \infty} \frac{e^x}{x^n} \overset{H}{=} \lim_{x \to \infty} \frac{e^x}{n\, x^{n-1}} \overset{H}{=} \lim_{x \to \infty} \frac{e^x}{n(n-1)\, x^{n-2}} \overset{H}{=} \cdots \overset{H}{=} \lim_{x \to \infty} \frac{e^x}{n!} = \infty$

83. $\displaystyle\lim_{x \to 0^+} x^\alpha \ln x = \lim_{x \to 0^+} \frac{\ln x}{x^{-\alpha}} \overset{H}{=} \lim_{x \to 0^+} \frac{1/x}{-\alpha\, x^{-\alpha-1}} = \lim_{x \to 0^+} \frac{x^\alpha}{-\alpha} = 0$ since $\alpha > 0$

REVIEW EXERCISES FOR CHAPTER 3

1. $y = 7^x$

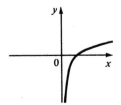

3. $y = e^{-x}$ $y = -e^{-x}$

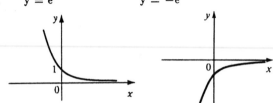

5. $y = \log_5 x$

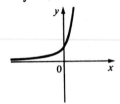

7. $= -\ln x$ $y = 2 - \ln x$

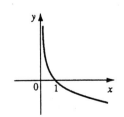

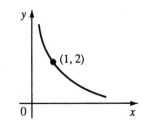

(1, 2)

9. $e^x = 5 \ \Rightarrow\ x = \ln (e^x) = \ln 5$

11. $\log_{10} (e^x) = 1 \ \Rightarrow\ e^x = 10 \ \Rightarrow\ x = \ln (e^x) = \ln 10$

       [OR : $1 = \log_{10} (e^x) = x \log_{10} e \ \Rightarrow\ x = 1/\log_{10} e$]

13. $2 = \ln (x^\pi) = \pi \ln x \ \Rightarrow\ \ln x = 2/\pi \ \Rightarrow\ x = e^{2/\pi}$

15. $\tan x = 4 \ \Rightarrow\ x = \tan^{-1} 4 + n\pi = \arctan 4 + n\pi$, n an integer

17. $y = \log_{10}(x^2 - x) \Rightarrow y' = \frac{1}{x^2 - x}(\log_{10} e)(2x - 1) = \frac{2x - 1}{(\ln 10)(x^2 - x)}$

19. $y = \frac{\sqrt{x+1}\,(2-x)^5}{(x+3)^7} \Rightarrow \ln|y| = \frac{1}{2}\ln(x+1) + 5\ln|2-x| - 7\ln(x+3) \Rightarrow$

$\frac{y'}{y} = \frac{1}{2(x+1)} + \frac{-5}{2-x} - \frac{7}{x+3} \Rightarrow y' = \frac{\sqrt{x+1}\,(2-x)^5}{(x+3)^7}\left[\frac{1}{2(x+1)} - \frac{5}{2-x} - \frac{7}{x+3}\right]$

21. $y = e^{cx}(c\sin x - \cos x) \Rightarrow$

$y' = ce^{cx}(c\sin x - \cos x) + e^{cx}(c\cos x + \sin x) = (c^2 + 1)e^{cx}\sin x$

23. $y = \ln(\sec^2 x) = 2\ln|\sec x| \Rightarrow y' = (2/\sec x)(\sec x \tan x) = 2\tan x$

25. $y = xe^{-1/x} \Rightarrow y' = e^{-1/x} + xe^{-1/x}(1/x^2) = e^{-1/x}(1 + 1/x)$

27. $y = (\cos^{-1} x)^{\sin^{-1} x} \Rightarrow \ln y = \sin^{-1} x \ln(\cos^{-1} x) \Rightarrow$

$\frac{y'}{y} = \frac{1}{\sqrt{1-x^2}}\ln(\cos^{-1} x) + \sin^{-1} x \frac{1}{\cos^{-1} x}\left[-\frac{1}{\sqrt{1-x^2}}\right]$

$\Rightarrow y' = (\cos^{-1} x)^{\sin^{-1} x}\left[\frac{\ln(\cos^{-1} x) - \sin^{-1} x/\cos^{-1} x}{\sqrt{1-x^2}}\right]$

29. $y = e^{e^x} \Rightarrow y' = e^{e^x} e^x = e^{x+e^x}$

31. $y = \ln(1/x) + 1/\ln x = -\ln x + (\ln x)^{-1} \Rightarrow y' = -1/x - 1/[x(\ln x)^2]$

33. $y = 7^{\sqrt{2x}} \Rightarrow y' = 7^{\sqrt{2x}}(\ln 7)[1/(2\sqrt{2x})](2) = 7^{\sqrt{2x}}(\ln 7)/\sqrt{2x}$

35. $y = xe^{\sec^{-1} x} \Rightarrow y' = e^{\sec^{-1} x} + xe^{\sec^{-1} x}[1/(x\sqrt{x^2-1})] = e^{\sec^{-1} x}(1 + 1/\sqrt{x^2-1})$

37. $y = \ln(\cosh 3x) \Rightarrow y' = (1/\cosh 3x)(\sinh 3x)(3) = 3\tanh 3x$

39. $y = \cosh^{-1}(\sinh x) \Rightarrow y' = \cosh x/\sqrt{\sinh^2 x - 1}$

41. $y = \ln\sin x - \frac{1}{2}\sin^2 x \Rightarrow y' = \frac{\cos x}{\sin x} - \sin x \cos x = \cot x - \sin x \cos x$

43. $y = \sin^{-1}\left[\frac{x-1}{x+1}\right] \Rightarrow y' = \frac{1}{\sqrt{1-[(x-1)/(x+1)]^2}} \cdot \frac{(x+1)-(x-1)}{(x+1)^2}$

$= \frac{1}{\sqrt{(x+1)^2-(x-1)^2}} \cdot \frac{2}{x+1} = \frac{2}{\sqrt{4x}\,(x+1)} = \frac{1}{\sqrt{x}\,(x+1)}$

[*Note that the domain of y is $x \geq 0$.*]

45. $y = \frac{1}{4}\left[\ln(x^2+x+1) - \ln(x^2-x+1)\right] + \frac{1}{2\sqrt{3}}\left[\tan^{-1}\left[\frac{2x+1}{\sqrt{3}}\right] + \tan^{-1}\left[\frac{2x-1}{\sqrt{3}}\right]\right] \Rightarrow$

$y' = \frac{1}{4}\left[\frac{2x+1}{x^2+x+1} - \frac{2x-1}{x^2-x+1}\right] + \frac{1}{2\sqrt{3}}\left[\frac{2/\sqrt{3}}{1+[(2x+1)/\sqrt{3}]^2} + \frac{2/\sqrt{3}}{1+[(2x-1)/\sqrt{3}]^2}\right]$

$= \frac{1}{4}\left[\frac{2x+1}{x^2+x+1} - \frac{2x-1}{x^2-x+1}\right] + \frac{1}{4(x^2+x+1)} + \frac{1}{4(x^2-x+1)}$

$= \frac{1}{2}\left[\frac{x+1}{x^2+x+1} - \frac{x-1}{x^2-x+1}\right] = \frac{1}{x^4+x^2+1}$

47. $f(x) = 2^x \Rightarrow f'(x) = 2^x \ln 2 \Rightarrow f''(x) = 2^x (\ln 2)^2 \Rightarrow \cdots \Rightarrow f^{(n)}(x) = 2^x (\ln 2)^n$

49. We first show it is true for $n = 1 : f'(x) = e^x + xe^x = (x+1)e^x$.

We now assume it is true for $n = k : f^{(k)}(x) = (x+k)e^x$. We must now show it is true

for $n = k+1 : f^{(k+1)}(x) = \frac{d}{dx}[f^{(k)}(x)] = \frac{d}{dx}[(x+k)e^x] = e^x + (x+k)e^x = [x + (k+1)]e^x$.

Thus, $f^{(n)}(x) = (x+n)e^x$ by mathematical induction.

51. $y = f(x) = \ln(e^x + e^{2x}) \Rightarrow f'(x) = (e^x + 2e^{2x})/(e^x + e^{2x}) \Rightarrow f'(0) = 3/2$,

so the tangent line at $(0, \ln 2)$ is $y - \ln 2 = \frac{3}{2}x$ or $3x - 2y + \ln 4 = 0$.

53. $y = [\ln(x+4)]^2 \Rightarrow y' = 2\ln(x+4)/(x+4) = 0 \Leftrightarrow \ln(x+4) = 0 \Leftrightarrow$

$x + 4 = 1 \Leftrightarrow x = -3$, so the tangent is horizontal at $(-3, 0)$.

55. $\lim\limits_{x \to -\infty} 10^{-x} = \infty$ since $-x \to \infty$ as $x \to -\infty$

57. $\lim\limits_{x \to 0^+} \ln(\tan x) = -\infty$ since $\tan x \to 0^+$ as $x \to 0^+$

59. $\lim\limits_{x \to -4^+} e^{1/(x+4)} = \infty$ since $\dfrac{1}{x+4} \to \infty$ as $x \to -4^+$

61. $\lim\limits_{x \to \infty} \dfrac{e^x}{e^{2x} + e^{-x}} = \lim\limits_{x \to \infty} \dfrac{e^{-x}}{1 + e^{-3x}} = \dfrac{0}{1+0} = 0$

63. $\lim\limits_{x \to 1} \cos^{-1}\left[\dfrac{x}{x+1}\right] = \cos^{-1}\left[\dfrac{1}{2}\right] = \dfrac{\pi}{3}$

65. $\lim\limits_{x \to \pi} \dfrac{\sin x}{x^2 - \pi^2} \overset{H}{=} \lim\limits_{x \to \pi} \dfrac{\cos x}{2x} = \dfrac{-1}{2\pi}$

67. $\lim\limits_{x \to \infty} \dfrac{\ln(\ln x)}{\ln x} \overset{H}{=} \lim\limits_{x \to \infty} \dfrac{1/(x \ln x)}{1/x} = \lim\limits_{x \to \infty} \dfrac{1}{\ln x} = 0$

69. $\lim\limits_{x \to 0} \dfrac{\ln(1-x) + x + x^2/2}{x^3} \overset{H}{=} \lim\limits_{x \to 0} \dfrac{-1/(1-x) + 1 + x}{3x^2} \overset{H}{=} \lim\limits_{x \to 0} \dfrac{-1/(1-x)^2 + 1}{6x}$

$\overset{H}{=} \lim\limits_{x \to 0} \dfrac{-2/(1-x)^3}{6} = -\dfrac{2}{6} = -\dfrac{1}{3}$

71. $\lim\limits_{x \to 0^+} \sin x \, (\ln x)^2 = \lim\limits_{x \to 0^+} \dfrac{(\ln x)^2}{\csc x} \overset{H}{=} \lim\limits_{x \to 0^+} \dfrac{2 \ln x \, / \, x}{-\csc x \cot x}$

$= -2 \lim\limits_{x \to 0} \dfrac{\sin x}{x} \lim\limits_{x \to 0} \dfrac{\ln x}{\cot x} = -2 \lim\limits_{x \to 0} \dfrac{\ln x}{\cot x} \overset{H}{=} -2 \lim\limits_{x \to 0} \dfrac{1/x}{-\csc^2 x}$

$= 2 \lim\limits_{x \to 0} \dfrac{\sin^2 x}{x} = 2 \lim\limits_{x \to 0} \dfrac{\sin x}{x} \lim\limits_{x \to 0} \sin x = 2 \cdot 1 \cdot 0 = 0$

73. $\lim\limits_{x \to 1} (\ln x)^{\sin x} = (\ln 1)^{\sin 1} = 0^{\sin 1} = 0$

75. $\lim\limits_{x \to 0^+} \dfrac{x^{1/3} - 1}{x^{1/4} - 1} = \dfrac{0-1}{0-1} = 1$

77. (a) $y(t) = y(0)e^{kt} = 1000e^{kt} \Rightarrow y(2) = 1000e^{2k} = 9000 \Rightarrow e^{2k} = 9$

$\Rightarrow 2k = \ln 9 \Rightarrow k = (1/2)\ln 9 = \ln 3 \Rightarrow y(t) = 1000e^{(\ln 3)\, t}$

$= 1000 \cdot 3^t$

(b) $y(3) = 1000 \cdot 3^3 = 27000$ (c) $1000 \cdot 3^t = 2000$

$\Rightarrow 3^t = 2 \Rightarrow t \ln 3 = \ln 2 \Rightarrow t = (\ln 2)/(\ln 3) \approx 0.63 \text{ h.}$

79. Using the formula in Example 4, Section 3.6, $A(t) = A_0(1 + \frac{i}{n})^{nt}$ where $A_0 = 10000.00$ and $i = .10$ then for :

(a) $n = 1, \ A(4) = 10,000(1 + .10)^{1 \cdot 4} = \$14,641.00$

(b) $n = 2, \ A(4) = 10,000(1 + \frac{10}{2})^{2 \cdot 4} = \$14,774.55$

(c) $n = 4, \ A(4) = 10,000(1 + \frac{10}{4})^{4 \cdot 4} = \$14,845.06$

(d) $n = 12, \ A(4) = 10,000(1 + \frac{10}{12})^{12 \cdot 4} = \$14,893.54$

(e) $n = 365, \ A(4) = 10,000(1 + \frac{10}{365})^{365 \cdot 4} = \$14,917.44$

(f) Using the formula for continuous interest, $A(t) = A_0 e^{it}$, we have

$A(4) = 10000.00 e^{.10 \times 4} = \$14,918.25$

81. (a) $C'(t) = -kC(t) \Rightarrow C(t) = C(0)e^{-kt}$ by Theorem (3.36). But $C(0) = C_0$. Thus,

$C(t) = C_0 e^{-kt}$.

(b) $C(30) = .5C_0$ since the concentration is reduced by half. Thus, $.5C_0 = C_0 e^{-k30} \Rightarrow$

$\ln .5 = -30k \Rightarrow k = -\frac{1}{30} \ln .5 = \frac{\ln 2}{30}$. Since 10% of the original concentration remains

if 90% is eliminated, we want the value of t such that $C(t) = .1C_0$. Therefore,

$.1C_0 = C_0 e^{-t(\ln 2/30)} \Rightarrow t = -\frac{30}{\ln 2} \ln .1 \approx 100 \text{ h.}$

83. $s(t) = Ae^{-ct} \cos(\omega t + \delta) \Rightarrow v(t) = s'(t)$

$= -cAe^{-ct} \cos(\omega t + \delta) + Ae^{-ct}[-\omega \sin(\omega t + \delta)] = -Ae^{-ct}[c \cos(\omega t + \delta) + \omega \sin(\omega t + \delta)]$

$a(t) = v'(t) = cAe^{-ct}[c \cos(\omega t + \delta) + \omega \sin(\omega t + \delta)]$

$$- Ae^{-ct}[-\omega c \sin (\omega t + \delta) + \omega^2 \cos (\omega t + \delta)]$$

$$= Ae^{-ct} [(c^2 - \omega^2) \cos (\omega t + \delta) + 2c\omega \sin (\omega t + \delta)]$$

85. $f(x) = e^{g(x)} \Rightarrow f'(x) = e^{g(x)} g'(x)$

87. $f(x) = \ln |g(x)| \Rightarrow f'(x) = g'(x)/g(x)$

89. $f(x) = \ln g(e^x) \Rightarrow f'(x) = \dfrac{1}{g(e^x)} g'(e^x)e^x$

91. $f(x) = \ln x + \tan^{-1} x \Rightarrow f(1) = \ln 1 + \tan^{-1} 1 = \pi/4 \Rightarrow g(\pi/4) = 1$

       $f'(x) = 1/x + 1/(1+x^2)$, so $g'(\pi/4) = 1/f'(1) = 1/(3/2) = 2/3$

Applications Plus (page 252)

1. The equation of the parabola is of the form $y = kx^2$ and we know that the point $(-100, 100)$,

    where the car starts, is on the parabola, so $100 = k(-100)^2$. Thus $k = 1/100$ and the equation of

    the parabola is $y = x^2/100$. Let $P(a, a^2/100)$ be the point where the headlights illuminate the

    statue. Then the tangent line at P will pass through the point $Q(100, 50)$ where the statue is

    located. Equating slopes, we get

    $$\tfrac{a}{50} = \text{slope of tangent at } P = m_{PQ} = \frac{(a^2/100) - 50}{a - 100}$$

    $$\Rightarrow a^2 - 100a = \tfrac{1}{2}a^2 - 2500 \;\Rightarrow\; a^2 - 200a + 5000 = 0 \;\Rightarrow\; a = \frac{200 \pm \sqrt{20{,}000}}{2} = 100 \pm 50\sqrt{2}.$$

    But $a < 100$, so $a = 100 + 50\sqrt{2}$ and the point is $(100 + 50\sqrt{2}, \, 150 - 100\sqrt{2})$, which is about 29

    m east and 9 m north of the origin.

3. (a) $\frac{dC}{dt} = r - kC = k\left(\frac{r}{k} - C\right)$. Let $u(t) = \frac{r}{k} - C(t)$. Then $\frac{du}{dt} = -\frac{dC}{dt}$. Therefore $\frac{du}{dt} = -ku$.

    $\frac{du}{u} = -k\,dt \Rightarrow \ln|u| = -kt + C$. When $t = 0$, $C = C_0$, so $u = \frac{r}{k} - C_0$ and $\ln\left|\frac{r}{k} - C_0\right| = C$.

    Therefore $\ln|u| = -kt + \ln\left|\frac{r}{k} - C_0\right|$. Now $\ln\left|\frac{r}{k} - C\right| = -kt + \ln\left|\frac{r}{k} - C_0\right|$

    $$\Rightarrow \ln\left|\frac{\frac{r}{k} - C}{\frac{r}{k} - C_0}\right| = -kt \;\Rightarrow\; \left|\frac{\frac{r}{k} - C}{\frac{r}{k} - C_0}\right| = e^{-kt} \;\Rightarrow\; \frac{\frac{r}{k} - C}{\frac{r}{k} - C_0} = \pm e^{-kt}.$$

    Taking $t = 0$ gives $1 = \pm 1$, so the sign is $+$. Thus $\frac{r}{k} - C(t) = \left(\frac{r}{k} - C_0\right)e^{-kt}$

    $$\Rightarrow C(t) = \frac{r}{k} - \left(\frac{r}{k} - C_0\right)e^{-kt} = C_0 e^{-kt} + \frac{r}{k}(1 - e^{-kt}).$$

    (b) If $C_0 < r/k$, then the first formula for $C(t)$ shows that $C(t)$ increases monotonically

    and $\lim_{t \to \infty} C(t) = r/k$. The second expression for $C(t)$ shows how the role of C_0 steadily

    diminishes as that of r/k increases.

5. $V = \frac{4}{3}\pi r^3 \Rightarrow \frac{dV}{dt} = 4\pi r^2 \frac{dr}{dt}$. But $\frac{dV}{dt}$ is proportional to the surface area so that $\frac{dV}{dt} = k \cdot 4\pi r^2$ for

some constant k. Therefore

$4\pi r^2 \frac{dr}{dt} = k \cdot 4\pi r^2 \Rightarrow \frac{dr}{dt} = k = \text{constant} \Rightarrow r = kt + r_0$. To find k we use the fact that when

$t = 3$, $r = 3k + r_0$ and $V = \frac{1}{2}V_0 \Rightarrow \frac{4}{3}\pi(3k + r_0)^3 = \frac{1}{2}\frac{4}{3}\pi r_0^3 \Rightarrow (3k + r_0)^3 = \frac{1}{2}r_0^3$

$\Rightarrow (3k + r_0) = \frac{1}{\sqrt[3]{2}}r_0 \Rightarrow k = \frac{r_0}{3}\left(\frac{1}{\sqrt[3]{2}} - 1\right)$. Therefore $r = \frac{r_0}{3}\left(\frac{1}{\sqrt[3]{2}} - 1\right)t + r_0$. When the snowball

has melted completely we have that $r = 0 \Rightarrow \frac{r_0}{3}\left(\frac{1}{\sqrt[3]{2}} - 1\right)t + r_0 = 0$ which gives $t = \frac{3\sqrt[3]{2}}{\sqrt[3]{2} - 1}$.

Therefore it takes $\frac{3\sqrt[3]{2}}{\sqrt[3]{2} - 1} - 3 = \frac{3}{\sqrt[3]{2} - 1} \approx 11$ hours and 33 minutes longer.

7.

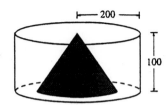

(a) We are given that $V = \frac{1}{3}\pi r^2 h$, $\frac{dV}{dt} = 60{,}000\pi$ ft^3/h, and $r = 1.5h = \frac{3}{2}h$. So

$V = \frac{1}{3}\pi\left(\frac{3}{2}h\right)^2 h = \frac{3}{4}\pi h^3 \Rightarrow \frac{dV}{dt} = \frac{3}{4}\pi \cdot 3h^2 \frac{dh}{dt} = \frac{9}{4}\pi h^2 \frac{dh}{dt}$.

Therefore, $\frac{dh}{dt} = \frac{4\frac{dV}{dt}}{9\pi h^2} = \frac{240{,}000\pi}{9\pi h^2} = \frac{80{,}000}{3h^2}$ (*)

$\Rightarrow 3h^2 dh = 80{,}000\, dt \Rightarrow h^3 = 80{,}000\, t + C$.

When $t = 0$, $h = 60$. Therefore, $C = (60)^3 = 216{,}000$, so $h^3 = 80{,}000\, t + 216{,}000$.

Let $h = 100$. Then $(100)^3 = 1{,}000{,}000 = 80{,}000\, t + 216{,}000 \Rightarrow$

$80{,}000\, t = 784{,}000$, so the time required is $t = 9.8$ hours.

(b) Floor area of silo: $\pi(200)^2 = 40{,}000\pi$ ft^2

Area of base of pile: $A = \pi r^2 = \pi\left(\frac{3}{2}h\right)^2 = \frac{9\pi h^2}{4} = \frac{9\pi (60)^2}{4} = 8100\pi \text{ ft}^2$

Floor area remaining: $40{,}000\pi - 8100\pi = 31{,}900\pi \text{ ft}^2 \approx 100{,}000 \text{ ft}^2$

$A = \frac{9\pi h^2}{4} \Rightarrow \frac{dA}{dt} = \frac{18\pi h}{4}\frac{dh}{dt}$ and from $(*)$ in (a) we know that

when $h = 60$, $\frac{dh}{dt} = \frac{80{,}000}{3(60)^2} = \frac{200}{27}$ ft/h.

Therefore, $\frac{dA}{dt} = \frac{18\pi(60)}{4}\left(\frac{200}{27}\right) = 2000\pi \approx 6283 \text{ ft}^2/\text{h}.$

(c) At $h = 90$ ft, $\frac{dV}{dt} = 60{,}000\pi - 20{,}000\pi = 40{,}000\pi \text{ ft}^3/\text{h}$. From $(*)$ in **(a)**,

$\frac{dh}{dt} = \frac{4\frac{dV}{dt}}{9\pi h^2} = \frac{4(40{,}000\pi)}{9\pi h^2} = \frac{160{,}000}{9h^2} \Rightarrow 9h^2\,dh = 160{,}000\,dt \Rightarrow 3h^3 = 160{,}000\,t + C.$

When $t = 0$, $h = 90$; therefore, $C = 3 \cdot 729{,}000 = 2{,}187{,}000$. Therefore,

$3h^3 = 160{,}000\,t + 2{,}187{,}000$. At the top, $h = 100$, so $3(100)^3 = 160{,}000\,t + 2{,}187{,}000 \Rightarrow$

$t = \frac{813{,}000}{160{,}000} \approx 5.1$. The pile reaches the top after about 5.1 h.

9. $2yy' = 4p \Rightarrow y' = \frac{2p}{y} \Rightarrow$ slope of tangent at P is $\frac{2p}{y_1}$. Slope of FP is $\frac{y_1}{x_1 - p}$, so using the formula

from Exercise 64(b) in Section 2 of Review and Preview, we have

$\tan\alpha = \frac{-2p/y_1 + y_1/(x_1 - p)}{1 + (2p/y_1)\,y_1/(x_1 - p)} = \frac{-2p(x_1 - p) + y_1^2}{y_1(x_1 - p) + 2py_1}$

$= \frac{-2px_1 + 2p^2 + 4px_1}{y_1(p + x_1)} = \frac{2p(p + x_1)}{y_1(p + x_1)} = \frac{2p}{y_1} =$ slope of tangent at P $= \tan\beta.$

Since $0 \le \alpha, \beta \le \frac{\pi}{2}$, this proves that $\alpha = \beta.$

CHAPTER FOUR

EXERCISES 4.1

1. Absolute maximum at e; absolute minimum at d; local maximum at b, e; local minimum at d, s.

3. $f(x) = 1 + 2x, x \geq -1$. Absolute minimum $f(-1) = -1$; no local minimum. No local or absolute maximum.

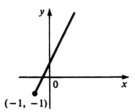

5. $f(x) = |x|, -2 \leq x \leq 1$. Absolute maximum $f(-2) = 2$; no local maximum. Absolute and local minimum $f(0) = 0$.

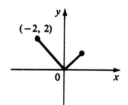

7. $f(x) = 1 - x^2, 0 < x < 1$. No extrema.

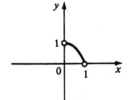

9. $f(x) = 1 - x^2, 0 \leq x < 1$. Absolute maximum $f(0) = 1$; no local maximum. No absolute or local minimum.

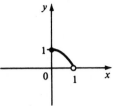

11. $f(x) = 1 - x^2,\ -2 \le x \le 1.$ Absolute and local maximum $f(0) = 1.$

Absolute minimum $f(-2) = -3;$ no local minimum.

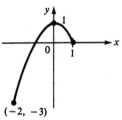

$(-2, -3)$

13. $f(t) = 1/t,\ 0 < t < 1.$ No extrema.

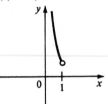

15. $f(\theta) = \sin\theta,\ -2\pi \le \theta \le 2\pi.$ Absolute and local maxima $f(-3\pi/2) = f(\pi/2) = 1.$

Absolute and local minima $f(-\pi/2) = f(3\pi/2) = -1.$

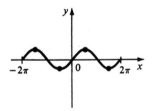

17. $f(\theta) = \cos(\theta/2),\ -\pi < \theta < \pi.$ Local and absolute maximum $f(0) = 1.$

No local or absolute minimum.

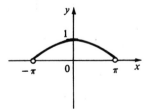

19. $f(x) = x^5$. No extrema.

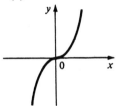

21. $f(x) = 2x$ if $0 \le x < 1$, $f(x) = 2 - x$ if $1 \le x \le 2$. Absolute minima $f(0) = f(2) = 0$; no local minima. No absolute or local maximum.

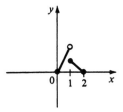

23. $f(x) = 2x - 3x^2 \Rightarrow f'(x) = 2 - 6x = 0 \Leftrightarrow x = 1/3$. So the critical number is $1/3$.

25. $f(x) = x^3 - 3x + 1 \Rightarrow f'(x) = 3x^2 - 3 = 3(x^2 - 1) = 3(x + 1)(x - 1)$.
    So the critical numbers are $x = \pm 1$.

27. $f(t) = 2t^3 + 3t^2 + 6t + 4 \Rightarrow f'(t) = 6t^2 + 6t + 6$. But $t^2 + t + 1 = 0$ has no real solutions since $b^2 - 4ac = 1 - 4(1)(1) = -3 < 0$. No critical numbers.

29. $s(t) = 2t^3 + 3t^2 - 6t + 4 \Rightarrow s'(t) = 6t^2 + 6t - 6 = 6(t^2 + t - 1)$.
    By the quadratic formula, the critical numbers are $t = (-1 \pm \sqrt{5})/2$.

31. $g(x) = \sqrt[9]{x} = x^{1/9} \Rightarrow g'(x) = \frac{1}{9}x^{-8/9} = 1/\sqrt[9]{x^8} \ne 0$, but $g'(0)$ does not exist, so $x = 0$ is a critical number.

33. $g(t) = 5t^{2/3} + t^{5/3} \Rightarrow g'(t) = (10/3)t^{-1/3} + (5/3)t^{2/3}$. $g'(0)$ does not exist, so $t = 0$ is a critical number. $g'(t) = (5/3)t^{-1/3}(2 + t) = 0 \Leftrightarrow t = -2$, so $t = -2$ is also a critical number.

35. $f(r) = \dfrac{r}{r^2 + 1} \Rightarrow f'(r) = \dfrac{1(r^2 + 1) - r(2r)}{(r^2 + 1)^2} = \dfrac{-r^2 + 1}{(r^2 + 1)^2} = 0 \Leftrightarrow r^2 = 1 \Leftrightarrow r = \pm 1$, so these are the critical numbers.

37. $F(x) = x^{4/5}(x - 4)^2 \Rightarrow F'(x) = (4/5)x^{-1/5}(x - 4)^2 + 2x^{4/5}(x - 4)$
    $= (x - 4)(7x - 8)/(5x^{1/5}) = 0$ when $x = 4, 8/7$ and $F'(0)$ does not exist. Critical numbers are $0, 8/7, 4$.

39. $V(x) = x\sqrt{x - 2}$. $V'(x) = \sqrt{x - 2} + \dfrac{x}{2\sqrt{x - 2}} \Rightarrow V'(2)$ does not exist. For $x > 2$

131

(the domain of $V'(x)$), $V'(x) > 0$, so 2 is the only critical number.

41. $f(\theta) = \sin^2(2\theta) \Rightarrow f'(\theta) = 2\sin(2\theta)\cos(2\theta)(2) = 2\sin 4\theta = 0 \Leftrightarrow \sin(4\theta) = 0$
$\Leftrightarrow 4\theta = n\pi$, n an integer. So $\theta = n\pi/4$ are the critical numbers.

43. $f(x) = x\ln x \Rightarrow f'(x) = \ln x + 1 = 0 \Leftrightarrow \ln x = -1 \Leftrightarrow x = e^{-1} = \frac{1}{e}$. So $x = \frac{1}{e}$ is the only critical number.

45. $f(x) = x^2 - 2x + 2$, $[0,3]$. $f'(x) = 2x - 2 = 0 \Leftrightarrow x = 1$. $f(0) = 2$, $f(1) = 1$, $f(3) = 5$.
So $f(3) = 5$ is the absolute maximum and $f(1) = 1$ is the absolute minimum.

47. $f(x) = x^3 - 12x + 1$, $[-3,5]$. $f'(x) = 3x^2 - 12 = 3(x^2 - 4) = 3(x+2)(x-2)$
$= 0 \Leftrightarrow x = \pm 2$. $f(-3) = 10$, $f(-2) = 17$, $f(2) = -15$, $f(5) = 66$. So $f(2) = -15$ is
the absolute minimum and $f(5) = 66$ is the absolute maximum.

49. $f(x) = 2x^3 + 3x^2 + 4$, $[-2,1]$. $f'(x) = 6x^2 + 6x = 6x(x+1) = 0 \Leftrightarrow x = -1, 0$.
$f(-2) = 0$, $f(-1) = 5$, $f(0) = 4$, $f(1) = 9$. So $f(1) = 9$ is the absolute maximum and
$f(-2) = 0$ is the absolute minimum.

51. $f(x) = x^4 - 4x^2 + 2$, $[-3,2]$. $f'(x) = 4x^3 - 8x = 4x(x^2 - 2) = 0 \Leftrightarrow x = 0, \pm\sqrt{2}$.
$f(-3) = 47$, $f(-\sqrt{2}) = -2$, $f(0) = 2$, $f(\sqrt{2}) = -2$, $f(2) = 2$, so $f(\pm\sqrt{2}) = -2$ is the
absolute minimum and $f(-3) = 47$ is the absolute maximum.

53. $f(x) = x^2 + \frac{2}{x}$, $[1/2, 2]$. $f'(x) = 2x - \frac{2}{x^2} = 2(x^3 - 1)/x^2 = 0 \Leftrightarrow x = 1$. $f(1/2) = 17/4$,
$f(1) = 3$, $f(2) = 5$. So $f(1) = 3$ is the absolute minimum and $f(2) = 5$ is the absolute
maximum.

55. $f(x) = x^{4/5}$, $[-32,1]$. $f'(x) = (4/5)x^{-1/5} \Rightarrow f'(x) \neq 0$ but $f'(0)$ does not exist, so 0 is
the only critical number. $f(-32) = 16$, $f(0) = 0$, $f(1) = 1$. So $f(0) = 0$ is the absolute
minimum and $f(-32) = 16$ is the absolute maximum.

57. $f(x) = |x - 1| - 1$, $[-1,2]$. $f'(x) = -1$ if $x < 1$, $f'(x) = 1$ if $x > 1$ and $f'(1)$ does not
exist, so 1 is the only critical number. $f(-1) = 1$, $f(1) = -1$, $f(2) = 0$. So $f(1) = -1$
is the absolute minimum and $f(-1) = 1$ is the absolute maximum.

59. $f(x) = \sin x + \cos x$, $[0, \pi/3]$. $f'(x) = \cos x - \sin x = 0 \Leftrightarrow x = \pi/4$. $f(0) = 1$,
$f(\pi/4) = \sqrt{2}$, $f(\pi/3) = (\sqrt{3} + 1)/2$. So $f(0) = 1$ is the absolute minimum and
$f(\pi/4) = \sqrt{2}$ the absolute maximum.

61. $f(x) = xe^{-x}$, $[0,1]$. $f'(x) = e^{-x} - xe^{-x} = e^{-x}(1-x) = 0 \Leftrightarrow x = 1$. $f(0) = 0$,
$f(1) = e^{-1}$. So $f(1) = e^{-1}$ is the absolute maximum and $f(0) = 0$ is the absolute
minimum.

63. $f(x) = [\![x]\!]$ is discontinuous at every integer n (*See Example* 11 *and Exercise* 71 *in Section* 1.3),
       so that $f'(n)$ does not exist. For all other real numbers a, $[\![x]\!]$ is constant on an open interval
       containing a, and so $f'(a) = 0$. Therefore every real number is a critical number of $f(x) = [\![x]\!]$.

65. $f(x) = x^5$. $f'(x) = 5x^4 \Rightarrow f'(0) = 0$ so 0 is a critical number. But $f(0) = 0$ and f
       takes both positive and negative values in any open interval containing 0, so f does not
       have a local extremum at 0.

67. $f(x) = x^{101} + x^{51} + x + 1 \Rightarrow f'(x) = 101x^{100} + 51x^{50} + 1 \geq 1$ for all x, so $f'(x) = 0$ has no
       solutions.   Thus f(x) has no critical numbers, so f(x) can have no local extrema.

69.

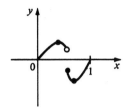

71.   If f has a local minimum at c, then $g(x) = -f(x)$ has a local maximum at c, so $g'(c) = 0$
       by the case of Fermat's Theorem proved in the text.   Thus $f'(c) = -g'(c) = 0$.

EXERCISES 4.2

1. $f(x) = x^3 - x$, $[-1, 1]$. f, being a polynomial, is continuous on $[-1, 1]$ and differentiable
       on $(-1, 1)$.   Also $f(-1) = 0 = f(1)$. $f'(c) = 3c^2 - 1 = 0 \Rightarrow c = \pm 1/\sqrt{3}$.

3. $f(x) = \cos 2x$, $[0, \pi]$. f is continuous on $[0, \pi]$ and differentiable on $(0, \pi)$.   Also
       $f(0) = 1 = f(\pi)$. $f'(c) = -2\sin 2c = 0 \Rightarrow \sin 2c = 0 \Rightarrow 2c = \pi \Rightarrow c = \pi/2$
       (since $c \in (0, \pi)$).

5. $f(x) = 1 - x^{2/3}$. $f(-1) = 1 - (-1)^{2/3} = 1 - 1 = 0 = f(1)$.
       $f'(x) = -(2/3)x^{-1/3} \Rightarrow f'(c) = 0$ has no solutions.   This does not contradict Rolle's
       Theorem since $f'(0)$ does not exist.

7. $f(x) = 1 - x^2$, $[0, 3]$. f, being a polynomial, is continuous on $[0, 3]$ and differentiable on $(0, 3)$. $\frac{f(3) - f(0)}{3 - 0} = \frac{-8 - 1}{3} = -3$ and $-3 = f'(c) = -2c \Rightarrow c = 3/2$.

9. $f(x) = x^3 - 2x + 1$, $[-2, 3]$. f, being a polynomial, is continuous on $[-2, 3]$ and differentiable on $(-2, 3)$. $\frac{f(3) - f(-2)}{3 - (-2)} = \frac{22 - (-3)}{5} = 5$ and $5 = f'(c) = 3c^2 - 2 \Rightarrow 3c^2 = 7 \Rightarrow c = \pm\sqrt{7/3}$.

11. $f(x) = 1/x$, $[1, 2]$. f, being a rational function, is continuous on $[1, 2]$ and differentiable on $(1, 2)$. $\frac{f(2) - f(1)}{2 - 1} = \frac{(1/2) - 1}{1} = -1/2$ and $-1/2 = f'(c) = -1/c^2$ $\Rightarrow c^2 = 2 \Rightarrow c = \sqrt{2}$ (since c must lie in $[1, 2]$).

13. 1 and $x - 1$ are continuous on R by 1.20, $\sqrt[3]{x}$ is continuous on R by 1.22; therefore $f(x) = 1 + \sqrt[3]{x - 1}$ is continuous on R by 1.24 and 1.19(1), and hence continuous on $[2, 9]$. $f'(x) = \frac{1}{3}(x - 1)^{-2/3}$, so that f is differentiable for all $x \neq 1$ and so f is differentiable on $(2, 9)$. By the Mean Value Theorem, there exists a number c such that $f'(c) = \frac{1}{3}(c - 1)^{-2/3} = \frac{f(9) - f(2)}{9 - 2} = \frac{3 - 2}{7} = \frac{1}{7} \Rightarrow \frac{1}{3}(c - 1)^{-2/3} = \frac{1}{7}$ $\Rightarrow (c - 1)^2 = (\frac{7}{3})^3 \Rightarrow c = \pm(\frac{7}{3})^{3/2} + 1 \Rightarrow c = (\frac{7}{3})^{3/2} + 1 \approx 4.564$ since c must be in $[2, 9]$.

15. $f(x) = |x - 1|$. $f(3) - f(0) = |3 - 1| - |0 - 1| = 1$. Since $f'(c) = -1$ if $c < 1$ and $f'(c) = 1$ if $c > 1$, $f'(c)(3 - 0) = \pm 3$ and so is never $= 1$. This does not contradict the Mean Value Theorem since $f'(1)$ does not exist.

17. $f(x) = x^5 + 10x + 3 = 0$. Since f is continuous and $f(-1) = -8$ and $f(0) = 3$, the equation has at least one root in $(-1, 0)$ by the Intermediate Value Theorem. Suppose that the equation has more than one root, say a and b are both roots with $a < b$. Then $f(a) = 0 = f(b)$ so by Rolle's Theorem $f'(x) = 5x^4 + 10 = 0$ has a root in (a, b). But this is impossible since clearly $f'(x) \geq 10 > 0$ for all real x.

19. $f(x) = x^5 - 6x + c = 0$. Suppose that $f(x)$ has two roots a and b with $-1 \leq a < b \leq 1$. Then $f(a) = 0 = f(b)$, so by Rolle's Theorem there is a number d in (a, b) with $f'(d) = 0$. Now $0 = f'(d) = 5d^4 - 6 \Rightarrow d = \pm\sqrt[4]{6/5}$ neither of which is in the interval $[-1, 1]$ and hence not in (a, b). Thus $f(x)$ can have at most one root in $[-1, 1]$.

21. (a) Suppose that a cubic polynomial $P(x)$ has roots $a_1 < a_2 < a_3 < a_4$, so $P(a_1) = P(a_2) = P(a_3) = P(a_4)$. By Rolle's Theorem there are numbers c_1, c_2, c_3 with $a_1 < c_1 < a_2$, $a_2 < c_2 < a_3$ and $a_3 < c_3 < a_4$ and $P'(c_1) = P'(c_2) = P'(c_3) = 0$. Thus the second degree polynomial $P'(x)$ has 3

distinct real roots, a contradiction.

(b)   We prove by induction that a polynomial of degree n has at most n real roots. This is certainly true for $n = 1$. Suppose that the result is true for all polynomials of degree n and let $P(x)$ be a polynomial of degree $n + 1$. Suppose that $P(x)$ has more than $n + 1$ real roots, say $a_1 < a_2 < a_3 < \cdots < a_{n+1} < a_{n+2}$. Then $P(a_1) = P(a_2) = \cdots = P(a_{n+2}) = 0$. By Rolle's Theorem there are real numbers $c_1, \cdots, c_{n+1}$ with $a_1 < c_1 < a_2, \cdots, a_{n+1} < c_{n+1} < a_{n+2}$ and $P'(c_1) = \cdots = P'(c_{n+1}) = 0$. Thus the nth degree polynomial $P'(x)$ has at least $n + 1$ roots. This contradiction shows that $P(x)$ has at most $n + 1$ real roots.

23.   Suppose that such a function f exists. By the Mean Value Theorem there is a number $0 < c < 2$ with $f'(c) = \dfrac{f(2) - f(0)}{2 - 0} = 5/2$. But this is impossible since $f'(x) \le 2 < 5/2$ for all x, so no such function can exist.

25.   Let $f(x) = \sin x$ and let $b < a$. Then $f(x)$ is continuous on $[b, a]$ and differentiable on (b, a). By the Mean Value Theorem, there is a number $c \in (b, a)$ with $\sin a - \sin b = f(a) - f(b) = f'(c)(a - b) = (\cos c)(a - b)$. Thus $|\sin a - \sin b| \le |\cos c||b - a| \le |a - b|$. If $a < b$, then $|\sin a - \sin b| = |\sin b - \sin a| \le |b - a| = |a - b|$. If $a = b$, both sides of the inequality collapse to 0.

27.   For $x > 0$, $f(x) = g(x)$, so $f'(x) = g'(x)$. For $x < 0$, $f'(x) = (1/x)' = -1/x^2$ and $g'(x) = (1 + 1/x)' = -1/x^2$, so again $f'(x) = g'(x)$. However, the domain of $g(x)$ is not an interval (it is $(-\infty, 0) \cup (0, \infty)$) so we cannot conclude that $f - g$ is constant (in fact it is not).

29.   Let $f(x) = 2\sin^{-1}x - \cos^{-1}(1 - 2x^2)$. Then $f'(x) = \dfrac{2}{\sqrt{1 - x^2}} - \dfrac{4x}{\sqrt{1 - (1 - 2x^2)^2}}$

$= \dfrac{2}{\sqrt{1 - x^2}} - \dfrac{4x}{2x\sqrt{1 - x^2}} = 0$ (since $x \ge 0$). Thus $f'(x) = 0$ for all $x \in [0, 1)$. Thus (by 4.15 or 4.17) $f(x) = C$. To find C let $x = 0$. Thus $2\sin^{-1}(0) - \cos^{-1}(1) = 0 = C$. Therefore we see that $f(x) = 2\sin^{-1}x - \cos^{-1}(1 - 2x^2) = 0 \Rightarrow 2\sin^{-1}x = \cos^{-1}(1 - 2x^2)$.

31.   Let $g(t)$ and $h(t)$ be the position functions of the two runners and let $f(t) = g(t) - h(t)$. By hypothesis $f(0) = g(0) - h(0) = 0$ and $f(b) = g(b) - h(b) = 0$ where b is the finishing time. Then by Rolle's Theorem, there is a time $0 < c < b$ with $0 = f'(c) = g'(c) - h'(c)$. Hence $g'(c) = h'(c)$, so at time c, both runners have the same velocity $g'(c) = h'(c)$.

EXERCISES 4.3

1. $f(x) = 20 - x - x^2$, $f'(x) = -1 - 2x = 0 \Rightarrow x = -1/2$ (the only critical number)

    (a) $f'(x) > 0 \Leftrightarrow -1 - 2x > 0 \Leftrightarrow x < -1/2$, $f'(x) < 0 \Leftrightarrow x > -1/2$, so f is increasing on $(-\infty, -1/2]$ and decreasing on $[-1/2, \infty)$.

    (b) By the First Derivative Test, (c)
        $f(-1/2) = 20.25$ is a local
        maximum.

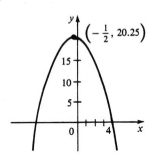

3. $f(x) = x^3 + x + 1$. $f'(x) = 3x^2 + 1 > 0$ for all $x \in R$.

    (a) f is increasing on R. (c)

    (b) f has no local extrema.

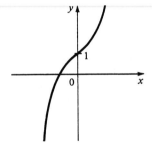

5. $f(x) = x^3 - 2x^2 + x$. $f'(x) = 3x^2 - 4x + 1 = (3x - 1)(x - 1)$. So the critical numbers are $x = 1/3, 1$.

    (a) $f'(x) > 0 \Leftrightarrow (3x - 1)(x - 1) > 0 \Leftrightarrow x < 1/3$ or $x > 1$ and $f'(x) < 0 \Leftrightarrow 1/3 < x < 1$. So f is increasing on $(-\infty, 1/3]$ and $[1, \infty)$ and f is decreasing on $[1/3, 1]$.

    (b) The local maximum is $f(1/3) = 4/27$ (c)
        and the local minimum is $f(1) = 0$

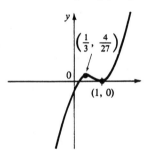

Section 4.3

7. $f(x) = 2x^2 - x^4$. $f'(x) = 4x - 4x^3 = 4x(1 - x^2) = 4x(1 + x)(1 - x)$, so the critical numbers are $x = -1, 0, 1$.

(a)

Interval	4x	1 + x	1 − x	f'(x)	f
x < −1	−	−	+	+	increasing on $(-\infty, -1]$
−1 < x < 0	−	+	+	−	decreasing on $[-1, 0]$
0 < x < 1	+	+	+	+	increasing on $[0, 1]$
x > 1	+	+	−	−	decreasing on $[1, \infty)$

(b) Local maximum $f(-1) = 1$,

local minimum $f(0) = 0$,

local maximum $f(1) = 1$.

(c)

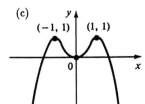

9. $f(x) = x^4 + 4x + 1$. $f'(x) = 4x^3 + 4 = 4(x^3 + 1) = 4(x + 1)(x^2 - x + 1) \Rightarrow x = -1$ is the only critical number.

(a) $f'(x) > 0 \Leftrightarrow 4(x^3 + 1) > 0 \Leftrightarrow x > -1$.

$f'(x) < 0 \Leftrightarrow x < -1$. So f is increasing on $[-1, \infty)$ and decreasing on $(-\infty, -1]$.

(b) Local minimum $f(-1) = -2$.

(c)

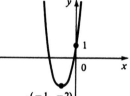

11. $f(x) = x^3(x - 4)^4$. $f'(x) = 3x^2(x - 4)^4 + x^3[4(x - 4)^3] = x^2(x - 4)^3(7x - 12)$. The critical numbers are $x = 0, 4, 12/7$.

(a) $x^2(x - 4)^2 \geq 0$ so $f'(x) \geq 0 \Leftrightarrow (x - 4)(7x - 12) \geq 0 \Leftrightarrow x \leq 12/7$ or $x \geq 4$.

$f'(x) \leq 0 \Leftrightarrow 12/7 \leq x \leq 4$. So f is increasing on $(-\infty, 12/7]$ and $[4, \infty)$ and decreasing on $[12/7, 4]$.

(b) Local maximum $f(12/7)$

$= 12^3 \cdot 16^4/7^7 \approx 137.5$,

local minimum $f(4) = 0$.

(c)

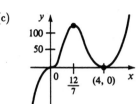

13. $f(x) = x\sqrt{6 - x}$. $f'(x) = \sqrt{6 - x} + x\left[-1/(2\sqrt{6 - x})\right] = 3(4 - x)/(2\sqrt{6 - x})$. Critical numbers are $x = 4, 6$.

137

(a) $f'(x) > 0 \Leftrightarrow 4 - x > 0$ (and $x < 6$)

$\Leftrightarrow x < 4$ and $f'(x) < 0 \Leftrightarrow 4 - x < 0$

(and $x < 6$) $\Leftrightarrow 4 < x < 6$. So f is

increasing on $(-\infty, 4]$ and

decreasing on $[4, 6]$.

(b) Local maximum $f(4) = 4\sqrt{2}$.

(c)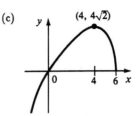

15. $f(x) = x^{1/5}(x + 1)$. $f'(x) = (1/5)x^{-4/5}(x + 1) + x^{1/5} = (1/5)x^{-4/5}(6x + 1)$.

The critical numbers are $x = 0, -1/6$.

(a) $f'(x) > 0 \Leftrightarrow 6x + 1 > 0$ $(x \neq 0)$ $\Leftrightarrow x > -1/6$ $(x \neq 0)$ and $f'(x) < 0 \Leftrightarrow x < -1/6$.

So f is increasing on $[-1/6, \infty)$ and decreasing on $(-\infty, -1/6]$.

(b) Local minimum $f(-1/6)$

$= -5/6^{6/5} \approx -.58$.

(c)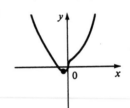

17. $f(x) = x\sqrt{x - x^2}$. The domain of f is $\{x | x(1 - x) \geq 0\} = [0, 1]$.

$f'(x) = \sqrt{x - x^2} + x(1 - 2x)/\left(2\sqrt{x - x^2}\right) = x(3 - 4x)/\left(2\sqrt{x - x^2}\right)$.

So the critical numbers are $x = 0, 3/4, 1$.

(a) $f'(x) > 0 \Leftrightarrow 3 - 4x > 0 \Leftrightarrow 0 < x < 3/4$. $f'(x) < 0 \Leftrightarrow 3/4 < x < 1$. So f is

increasing on $[0, 3/4]$ and decreasing on $[3/4, 1]$.

(b) Local maximum $f(3/4) = 3\sqrt{3}/16$.

(c)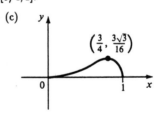

19. $f(x) = x - 2\sin x$, $0 \leq x \leq 2\pi$. $f'(x) = 1 - 2\cos x$. So

$f'(x) = 0 \Leftrightarrow \cos x = 1/2 \Leftrightarrow x = \pi/3$ or $5\pi/3$.

(a) $f'(x) > 0 \Leftrightarrow 1 - 2\cos x > 0 \Leftrightarrow \frac{1}{2} > \cos x \Leftrightarrow \frac{\pi}{3} < x < \frac{5}{3}\pi$. $f'(x) < 0 \Leftrightarrow 0 \leq x < \pi/3$

or $5\pi/3 < x \leq 2\pi$. So f is increasing on $[\pi/3, 5\pi/3]$ and decreasing on $[0, \pi/3]$ and

$[5\pi/3, 2\pi]$.

(b) Local minimum $f(\pi/3) = \pi/3 - \sqrt{3}$ (c)
     $\approx -.68$, local maximum $f(5\pi/3)$
     $= \sqrt{3} + 5\pi/3 \approx 6.97$.

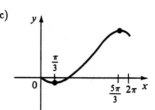

21. $f(x) = \sin^4 x + \cos^4 x$, $0 \le x \le 2\pi$. $f'(x) = 4\sin^3 x \cos x - 4\cos^3 x \sin x$
     $= -4\sin x \cos x(\cos^2 x - \sin^2 x) = -2\sin 2x \cos 2x = -\sin 4x$.
     $f'(x) = 0 \Leftrightarrow \sin 4x = 0 \Leftrightarrow 4x = n\pi \Leftrightarrow x = \pi n/4$.   So the critical numbers are 0, $\pi/4$,
     $\pi/2$, $3\pi/4$, π, $5\pi/4$, $3\pi/2$, $7\pi/4$, 2π.

     (a) $f'(x) > 0 \Leftrightarrow \sin 4x < 0 \Leftrightarrow \pi/4 < x < \pi/2$ or $3\pi/4 < x < \pi$ or $5\pi/4 < x < 3\pi/2$
         or $7\pi/4 < x < 2\pi$.   f is increasing on these intervals.   f is decreasing on $[0, \pi/4]$,
         $[\pi/2, 3\pi/4]$, $[\pi, 5\pi/4]$, $[3\pi/2, 7\pi/4]$.

     (b)   Local maxima $f(\pi/2) = f(\pi)$ (c)
         $= f(3\pi/2) = 1$, local minima
         $f(\pi/4) = f(3\pi/4) = f(5\pi/4)$
         $= f(7\pi/4) = 1/2$.

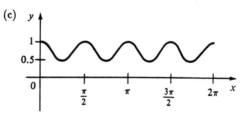

23. $f(x) = x^3 + 2x^2 - x + 1$. $f'(x) = 3x^2 + 4x - 1 = 0 \Rightarrow x = (-4 \pm \sqrt{28})/6 = (-2 \pm \sqrt{7})/3$.
     Now $f'(x) > 0$ for $x < (-2 - \sqrt{7})/3$ or $x > (-2 + \sqrt{7})/3$ and $f'(x) < 0$ for
     $(-2 - \sqrt{7})/3 < x < (-2 + \sqrt{7})/3$.   f is increasing on $(-\infty, (-2 - \sqrt{7})/3]$ and
     $[(-2 + \sqrt{7})/3, \infty)$ and decreasing on $[(-2 - \sqrt{7})/3, (-2 + \sqrt{7})/3]$.

25. $f(x) = x^6 + 192x + 17$. $f'(x) = 6x^5 + 192 = 6(x^5 + 32)$.   So
     $f'(x) > 0 \Leftrightarrow x^5 > -32 \Leftrightarrow x > -2$ and $f'(x) < 0 \Leftrightarrow x < -2$.   So f is increasing on
     $[-2, \infty)$ and decreasing on $(-\infty, -2]$.

27. $f(x) = x e^x \Rightarrow f'(x) = e^x + x e^x = e^x(1 + x) > 0 \Leftrightarrow 1 + x > 0 \Leftrightarrow x > -1$, so
     f is increasing on $[-1, \infty)$ and decreasing on $(-\infty, -1]$.

29. $f(x) = (\ln x)/\sqrt{x} \Rightarrow f'(x) = \dfrac{\sqrt{x}\,(1/x) - (\ln x)[1/(2\sqrt{x})]}{x} = \dfrac{2 - \ln x}{2x^{3/2}} \Rightarrow$
     $f'(x) > 0$ when $\ln x < 2 \Leftrightarrow x < e^2$.
     So f is increasing on $(0, e^2]$ and decreasing on $[e^2, \infty)$.

31. $f(x) = x^3 - 3x^2 + 6x - 2$, $-1 \le x \le 1$. $f'(x) = 3x^2 - 6x + 6 = 3(x^2 - 2x + 2)$. Since
$f'(x)$ has no real roots, $f'(x) > 0$ for all $x \in R$, so f is increasing and has no local
extremum. The absolute minimum is $f(-1) = -12$ and the absolute maximum is
$f(1) = 2$.

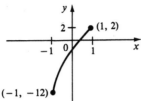

33. $f(x) = x + \sqrt{1-x}$, $0 \le x \le 1$. $f'(x) = 1 - 1/(2\sqrt{1-x}) = (2\sqrt{1-x} - 1)/(2\sqrt{1-x}) = 0$ when
$2\sqrt{1-x} - 1 = 0 \Rightarrow \sqrt{1-x} = 1/2 \Rightarrow 1 - x = 1/4 \Rightarrow x = 3/4$. For $0 < x < 3/4$,
$f'(x) > 0$ and for $1 > x > 3/4$, $f'(x) < 0$. So the local maximum is
$f(3/4) = 5/4$. Also $f(0) = 1$ and $f(1) = 1$ are the absolute minima and $f(3/4) = 5/4$
is the absolute maximum.

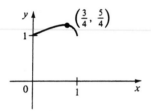

35. $g(x) = \dfrac{x}{x^2+1}$, $-5 \le x \le 5$. $g'(x) = \dfrac{(x^2+1) - x(2x)}{(x^2+1)^2} = \dfrac{1-x^2}{(x^2+1)^2}$. The critical
numbers are $x = \pm 1$. $g'(x) > 0 \Leftrightarrow x^2 < 1 \Leftrightarrow -1 < x < 1$ and $g'(x) < 0 \Leftrightarrow x < -1$
or $x > 1$. So $g(-1) = -1/2$ is a local minimum and $g(1) = 1/2$ is a
local maximum. Also $g(-5) = -5/26$ and $g(5) = 5/26$. So $g(-1) = -1/2$ is the
absolute minimum and $g(1) = 1/2$ is the absolute maximum.

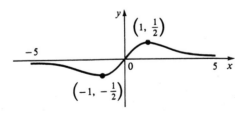

37. Let $f(x) = x + 1/x$, so $f'(x) = 1 - 1/x^2 = (x^2 - 1)/x^2$. Thus $f'(x) > 0$ for $x > 1 \Rightarrow f$ is increasing on $[1, \infty)$. Hence for $1 < a < b$, $a + \frac{1}{a} = f(a) < f(b) = b + \frac{1}{b}$.

39. Let $f(x) = 2\sqrt{x} - 3 + \frac{1}{x}$. Then $f'(x) = \frac{1}{\sqrt{x}} - \frac{1}{x^2} > 0$ for $x > 1$ since for $x > 1$, $x^2 > x > \sqrt{x}$. Hence f is increasing, so for $x > 1$, $f(x) > f(1) = 0$ or $2\sqrt{x} - 3 + 1/x > 0$ for $x > 1$. Hence $2\sqrt{x} > 3 - 1/x$ for $x > 1$.

41. Let $f(x) = \sin x - x + \frac{x^3}{6}$. Then $f'(x) = \cos x - 1 + \frac{x^2}{2}$. By Exercise 40, $f'(x) > 0$ for $x > 0$, so f is increasing for $x > 0$. Thus $f(x) > f(0) = 0$ for $x > 0$. Hence $\sin x - x + x^3/6 > 0$ or $\sin x > x - x^3/6$ for $x > 0$.

43. (a) Let $f(x) = x + \frac{1}{x}$, so $f'(x) = 1 - \frac{1}{x^2} > 0 \Leftrightarrow x^2 < 1 \Leftrightarrow 0 < x < 1$ (*since $x > 0$*), and $f'(x) > 0$ for $x > 1$. By the First Derivative Test, there is an absolute minimum for $f(x)$ on $(0, \infty)$ where $x = 1$. Thus $f(x) = x + \frac{1}{x} \geq f(1) = 2$ for $x > 0$.

(b) Let $y = \frac{1}{x}$, then $(\sqrt{x} - \sqrt{y})^2 \geq 0 \Rightarrow (\sqrt{x} - \frac{1}{\sqrt{x}})^2 \geq 0 \Rightarrow x - 2 + \frac{1}{x} \geq 0 \Rightarrow x + \frac{1}{x} \geq 2$.

45. $f(x) = ax^3 + bx^2 + cx + d \Rightarrow f(1) = a + b + c + d = 0$ and $f(-2) = -8a + 4b - 2c + d = 3$. Also $f'(1) = 3a + 2b + c = 0$ and $f'(-2) = 12a - 4b + c = 0$ by Fermat's Theorem. Solving these four equations, we get $a = 2/9$, $b = 1/3$, $c = -4/3$, $d = 7/9$, so the function is $f(x) = (1/9)(2x^3 + 3x^2 - 12x + 7)$.

47.

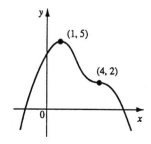

49. Let $x_1, x_2 \in I$ with $x_1 < x_2$. Then $f(x_1) < f(x_2)$ and $g(x_1) < g(x_2)$ (*since f and g are increasing on I*), so $(f + g)(x_1) = f(x_1) + g(x_1) < f(x_2) + g(x_2) = (f + g)(x_2)$. Therefore $f + g$ is increasing on I.

51. Let x_1 and x_2 be any two numbers in $[a, b]$ with $x_1 < x_2$. Then f is continuous on $[x_1, x_2]$ and differentiable on (x_1, x_2), so by the Mean Value Theorem there is a number c between x_1 and x_2 such that $f(x_2) - f(x_1) = f'(c)(x_2 - x_1)$. Now $f'(c) < 0$ by

assumption and $x_2 - x_1 > 0$ because $x_1 < x_2$. Thus $f(x_2) - f(x_1) = f'(c)(x_2 - x_1)$ is negative, so $f(x_2) - f(x_1) < 0$ or $f(x_2) < f(x_1)$. This shows that f is decreasing on $[a, b]$.

53. (a) Let $f(x) = e^x - 1 - x$ where $x \geq 0 \Rightarrow f'(x) = e^x - 1 \geq 0 \Leftrightarrow x \geq 0$. Thus, $f(x)$ is

increasing on $[0, \infty)$ and $f(0) = 0$. Therefore, $f(x) \geq 0$ for $x \geq 0$

$\Rightarrow e^x - 1 - x \geq 0 \Rightarrow e^x \geq 1 + x$.

(b) Let $f(x) = e^x - 1 - x - \frac{1}{2}x^2$. Thus, $f'(x) = e^x - 1 - x \geq 0$ for $x \geq 0$ by part (a).

Thus, $f(x)$ is increasing on $[0, \infty)$ so that $0 = f(0) \leq f(x) = e^x - 1 - x - \frac{x^2}{2}$

$\Rightarrow e^x \geq 1 + x + \frac{1}{2}x^2$.

(c) By part (a), the result holds for $n = 1$. Suppose that

$$e^x \geq 1 + x + \frac{x^2}{2!} + \cdots + \frac{x^k}{k!} \text{ for } x \geq 0. \text{ Let } f(x) = e^x - 1 - x - \frac{x^2}{2!} - \cdots - \frac{x^k}{k!} - \frac{x^{k+1}}{(k+1)!}.$$

Then $f'(x) = e^x - 1 - x - \cdots - \frac{x^k}{k!} \geq 0$ by assumption. Hence $f(x)$ is increasing on

$[0, \infty)$. So $0 \leq x$ implies that $0 = f(0) \leq f(x) = e^x - 1 - x - \cdots - \frac{x^k}{k!} - \frac{x^{k+1}}{(k+1)!}$, and

hence $e^x \geq 1 + x + \cdots + \frac{x^k}{k!} + \frac{x^{k+1}}{(k+1)!}$ for $x \geq 0$. Therefore, for $x \geq 0$,

$$e^x \geq 1 + x + \frac{x^2}{2!} + \cdots + \frac{x^n}{n!} \text{ for every positive integer n by mathemetical induction.}$$

EXERCISES 4.4

1. (a) $f(x) = x^3 - x \Rightarrow f'(x) = 3x^2 - 1 = 0 \Leftrightarrow x^2 = 1/3 \Leftrightarrow x = \pm 1/\sqrt{3}$.

$f'(x) > 0 \Leftrightarrow x^2 > 1/3 \Leftrightarrow |x| > 1/\sqrt{3} \Leftrightarrow x > 1/\sqrt{3}$ or $x < -1/\sqrt{3}$.

$f'(x) < 0 \Leftrightarrow |x| < 1/\sqrt{3} \Leftrightarrow -1/\sqrt{3} < x < 1/\sqrt{3}$. So f is increasing on

$(-\infty, -1/\sqrt{3}]$ and $[1/\sqrt{3}, \infty)$, decreasing on $[-1/\sqrt{3}, 1/\sqrt{3}]$.

(b) Local maximum $f(-1/\sqrt{3}) = 2/(3\sqrt{3}) \approx .38$, local minimum

$f(1/\sqrt{3}) = -2/(3\sqrt{3}) \approx -.38$.

(c) $f''(x) = 6x \Rightarrow f''(x) > 0 \Leftrightarrow x > 0$,

so f is CU on $(0, \infty)$ and CD on $(-\infty, 0)$.

(d) Point of inflection at x = 0. $\left(\frac{-1}{\sqrt{3}}, \frac{2}{3\sqrt{3}}\right)$

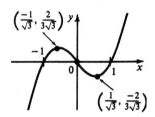

3. (a) $f(x) = x^3 - x^2 - x + 1 \Rightarrow f'(x) = 3x^2 - 2x - 1 = (3x+1)(x-1) = 0 \Leftrightarrow x = -1/3$

   or 1. $f'(x) > 0 \Leftrightarrow x < -1/3$ or $x > 1$; $f'(x) < 0 \Leftrightarrow -1/3 < x < 1$, so f is

   increasing on $(-\infty, -1/3]$ and $[1, \infty)$ and decreasing on $[-1/3, 1]$.

   (b) Local maximum $f(-1/3) = 32/27$, local minimum $f(1) = 0$.

   (c) $f''(x) = 6x - 2 > 0 \Leftrightarrow x > 1/3$, so f is CU on $(1/3, \infty)$ and CD on $(-\infty, 1/3)$.

   (d) Inflection point at x = 1/3.

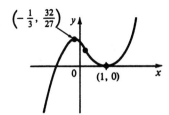

5. (a) $g(x) = x^4 - 3x^3 + 3x^2 - x \Rightarrow g'(x) = 4x^3 - 9x^2 + 6x - 1 = (x-1)^2(4x-1) = 0$

   when x = 1 or 1/4. $g'(x) \geq 0 \Leftrightarrow 4x - 1 \geq 0 \Leftrightarrow x \geq 1/4$ and

   $g'(x) \leq 0 \Leftrightarrow x \leq 1/4$, so g is increasing on $[1/4, \infty)$ and decreasing on $(-\infty, 1/4]$.

   (b) Local minimum $g(1/4) = -27/256$.

   (c) $g''(x) = 12x^2 - 18x + 6 = 6(2x-1)(x-1) > 0 \Leftrightarrow x < 1/2$ or $x > 1$,

   $g'(x) < 0 \Leftrightarrow 1/2 < x < 1$, so g is CU on $(-\infty, 1/2)$ and $(1, \infty)$ and CD on $(1/2, 1)$.

   (d) Inflection points at x = 1/2 and 1.

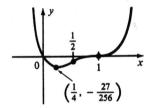

7. (a) $h(x) = 3x^5 - 5x^3 + 3 \Rightarrow h'(x) = 15x^4 - 15x^2 = 15x^2(x^2 - 1) = 0$ when $x = 0, \pm 1$.

    $h'(x) > 0 \Leftrightarrow x^2 > 1 \Leftrightarrow x > 1$ or $x < -1$, so h is increasing on $(-\infty, -1]$ and $[1, \infty)$

    and decreasing on $[-1, 1]$.

    (b) Local maximum $h(-1) = 5$, local minimum $h(1) = 1$.

    (c) $h''(x) = 60x^3 - 30x = 30x(2x^2 - 1) = 60x(x + 1/\sqrt{2})(x - 1/\sqrt{2}) \Rightarrow h''(x) > 0$ when

    $x > 1/\sqrt{2}$ or $-1/\sqrt{2} < x < 0$, so h is CU on $(1/\sqrt{2}, \infty)$ and $(-1/\sqrt{2}, 0)$ and CD on

    $(-\infty, -1/\sqrt{2})$ and $(0, 1/\sqrt{2})$.

    (d) Inflection points at $x = \pm 1/\sqrt{2}$ and 0.

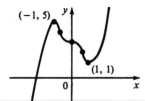

9. (a) $G(x) = 8 - \sqrt[3]{x} \Rightarrow G'(x) = -(1/3)x^{-2/3} < 0$ $(x \neq 0)$ so G is decreasing on

    $(-\infty, \infty)$.

    (b) No extrema.

    (c) $G''(x) = (2/9)x^{-5/3} > 0$ if $x > 0$, so G is CU on $(0, \infty)$ and CD on $(-\infty, 0)$.

    (d) IP at $x = 0$.

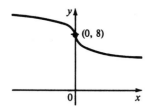

11. (a) $P(x) = x\sqrt{x^2 + 1} \Rightarrow P'(x) = \sqrt{x^2 + 1} + x^2/\sqrt{x^2 + 1} = (2x^2 + 1)/\sqrt{x^2 + 1} > 0$,

    so P is increasing on R.

    (b) No extrema.

    (c) $P''(x) = \dfrac{4x\sqrt{x^2 + 1} - (2x^2 + 1)(x/\sqrt{x^2 + 1})}{x^2 + 1} = \dfrac{x(2x^2 + 3)}{(x^2 + 1)^{3/2}} > 0 \Leftrightarrow x > 0$

    so P is CU on $(0, \infty)$ and CD on $(-\infty, 0)$.

(d) IP at x = 0.

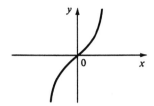

13. (a) $Q(x) = x^{1/3}(x+3)^{2/3} \Rightarrow Q'(x) = \frac{1}{3}x^{-2/3}(x+3)^{2/3} + x^{1/3}(\frac{2}{3})(x+3)^{-1/3}$

   $= (x+1)/\left[x^{2/3}(x+3)^{1/3}\right]$. The critical numbers are -3, -1, and 0. Note that

   $x^{2/3} \geq 0$ for all x. So $Q'(x) > 0$ when $x < -3$ or $x > -1$ and $Q'(x) < 0$ when

   $-3 < x < -1 \Rightarrow Q$ is increasing on $(-\infty, -3]$ and $[-1, \infty)$ and decreasing on

   $[-3, -1]$.

   (b) $Q(-3) = 0$ is a local maximum and $Q(-1) = -4^{1/3} \approx -1.6$ is a local minimum.

   (c) $Q''(x) = -2/\left[x^{5/3}(x+3)^{4/3}\right] \Rightarrow Q''(x) > 0 \Leftrightarrow x < 0$, so Q is CU on $(-\infty, -3)$ and

   $(-3, 0)$ and CD on $(0, \infty)$.

   (d) IP at x = 0.

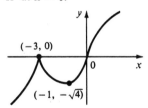

15. (a) $f(\theta) = \sin^2\theta \Rightarrow f'(\theta) = 2\sin\theta\cos\theta = \sin 2\theta > 0 \Leftrightarrow 2\theta \in (2n\pi, (2n+1)\pi) \Leftrightarrow \theta \in$

   $(n\pi, n\pi + \pi/2)$, n an integer. So f is increasing on $[n\pi, n\pi + \pi/2]$ and decreasing on

   $[n\pi + \pi/2, (n+1)\pi]$.

   (b) Local minima $f(n\pi) = 0$, local maxima $f(n\pi + \pi/2) = 1$.

   (c) $f''(\theta) = 2\cos 2\theta > 0 \Leftrightarrow 2\theta \in (2n\pi - \pi/2, 2n\pi + \pi/2) \Leftrightarrow \theta \in (n\pi - \pi/4, n\pi + \pi/4)$,

   so f is CU on these intervals and CD on $(n\pi + \pi/4, n\pi + 3\pi/4)$.

(d)   IP at $\theta = n\pi \pm \pi/4$, n an integer.

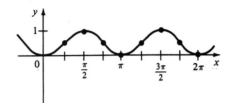

17.   (a) $f(t) = t + \sin t \Rightarrow f'(t) = 1 + \cos t \geq 0$ for all $t \Rightarrow$ f is increasing on R.

(b)   Thus there is no local extremum.

(c) $f''(t) = -\sin t > 0$ when $\sin t < 0$.   So f is CU on $((2n-1)\pi, 2n\pi)$ and CD on

$(2n\pi, (2n+1)\pi)$, n an integer.

(d)   IP at $x = n\pi$.

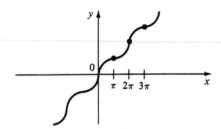

19. $f(x) = 6x^2 - 2x^3 - x^4 \Rightarrow f'(x) = 12x - 6x^2 - 4x^3 \Rightarrow f''(x) = 12 - 12x - 12x^2 = 0$

$\Leftrightarrow x^2 + x - 1 = 0 \Rightarrow x = \dfrac{-1 \pm \sqrt{5}}{2}$.   For $x < \dfrac{-1-\sqrt{5}}{2}$, $f''(x) < 0$.

For $\dfrac{-1-\sqrt{5}}{2} < x < \dfrac{-1+\sqrt{5}}{2}$, $f''(x) > 0$, and if $x > \dfrac{-1+\sqrt{5}}{2}$ then $f''(x) < 0$.

Therefore f is CU on $(\dfrac{-1-\sqrt{5}}{2}, \dfrac{-1+\sqrt{5}}{2})$.

21. $f(x) = x(1+x)^{-2} \Rightarrow f'(x) = (1+x)^{-2} - 2x(1+x)^{-3} = (1+x)^{-3}(1-x)$

$\Rightarrow f''(x) = -3(1+x)^{-4}(1-x) - (1+x)^{-3} = (1+x)^{-4}(2x-4) > 0$

$\Leftrightarrow (2x-4) > 0 \Leftrightarrow x > 2$.   Therefore f is CU on $(2, \infty)$.

23. $f(x) = xe^x \Rightarrow f'(x) = e^x + xe^x = e^x(1+x) > 0 \Rightarrow$

$f''(x) = e^x(1+x) + e^x = e^x(2+x) > 0 \Leftrightarrow 2+x > 0 \Leftrightarrow x > -2$, so f is CU

on $(-2, \infty)$ and CD on $(-\infty, -2)$.

25. $f(x) = (\ln x)/\sqrt{x} \Rightarrow f'(x) = \dfrac{\sqrt{x}\,(1/x) - (\ln x)[1/(2\sqrt{x})]}{x} = \dfrac{2 - \ln x}{2x^{3/2}} \Rightarrow$

$f''(x) = \dfrac{2x^{3/2}\,(-1/x) - (2 - \ln x)\,(3x^{1/2})}{4x^3} = \dfrac{3\ln x - 8}{4x^{5/2}} > 0 \Leftrightarrow$

$\ln x > 8/3 \Leftrightarrow x > e^{8/3}$, so f is CU on $(e^{8/3}, \infty)$.

27.

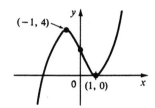

29.

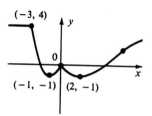

31. $f(x) = ax^3 + bx^2 + cx + d$, $a \neq 0$, $\Rightarrow f'(x) = 3ax^2 + 2bx + c \Rightarrow f''(x) = 6ax + 2b$. f is

CU when $6ax + 2b > 0$ and CD when $6ax + 2b < 0$. So there is exactly one inflection

point, namely when $x = -b/(3a)$.

33. By hypothesis $g = f'$ is differentiable on an open interval containing c. Since $(c, f(c))$ is

a point of inflection, the concavity changes at $x = c$, so $f'(x)$ changes signs at $x = c$.

Hence, by the First Derivative Test, f' has a local extremum at $x = c$. Thus, by

Fermat's Theorem, $f''(c) = 0$.

35. Using the fact that $|x| = \sqrt{x^2}$ (*see Exercises 79–81 in section 2.5*), we have that

$g(x) = x\sqrt{x^2} \Rightarrow g'(x) = \sqrt{x^2} + \sqrt{x^2} = 2\sqrt{x^2} = 2|x| \Rightarrow g''(x) = 2x(x^2)^{-1/2} = \dfrac{2x}{|x|} < 0$ for

$x < 0$ and $g''(x) > 0$ for $x > 0$, so $(0, 0)$ is an inflection point. But $g''(0)$ does not exist.

37. If f and g are CU on I, then $f'' > 0$ and $g'' > 0$ on I, so $(f + g)'' = f'' + g'' > 0$ on

$I \Rightarrow f + g$ is CU on I.

39. Since f and g are positive, increasing, and CU on I, we have $f > 0$, $f' > 0$, $f'' > 0$, $g > 0$,

$g' > 0$, $g'' > 0$ on I. Then $(fg)' = f'g + fg' \Rightarrow (fg)'' = f''g + 2f'g' + fg'' > 0$

$\Rightarrow fg$ is CU on I.

EXERCISES 4.5

Abbreviations: D: the domain of f,   VA: vertical asymptote(s),

HA: horizontal asymptote,   IP: inflection point(s)

1. $y = f(x) = 1 - 3x + 5x^2 - x^3$.   A. $D = R$.   B. y–intercept $= f(0) = 1$.   C. No

symmetry.   D. No asymptotes.   E. $f'(x) = -3 + 10x - 3x^2 = -(3x - 1)(x - 3) > 0$

$\Leftrightarrow (3x - 1)(x - 3) < 0 \Leftrightarrow \frac{1}{3} < x < 3$. $f'(x) < 0 \Leftrightarrow x < \frac{1}{3}$ or $x > 3$. So f is increasing

on $[1/3, 3]$ and decreasing on $(-\infty, 1/3]$ and $[3, \infty)$.   F. The critical numbers occur

when $f'(x) = -(3x - 1)(x - 3) = 0$ H.

$\Leftrightarrow x = \frac{1}{3}, 3$.

The local minimum is $f(\frac{1}{3}) = \frac{14}{27}$ and the local

maximum is $f(3) = 10$.

G. $f''(x) = 10 - 6x > 0$

$\Leftrightarrow x < \frac{5}{3}$, so f is CU on $(-\infty, 5/3)$ and

CD on $(5/3, \infty)$.   IP $(5/3, 142/27)$.

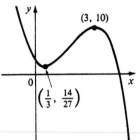

3. $y = f(x) = x^4 - 6x^2$.   A. $D = R$.   B. y–intercept $= f(0) = 0$, x–intercepts occur

when $f(x) = 0 \Rightarrow x^4 - 6x^2 = 0 \Leftrightarrow x^2(x^2 - 6) = 0 \Leftrightarrow x = 0, \pm\sqrt{6}$.

C. Since $f(-x) = (-x)^4 - 6(-x^2) = x^4 - 6x^2 = f(x)$, f is an even function and the

curve is symmetric about the y–axis.   H.

D. No asymptotes.   E. $f'(x) = 4x^3 - 12x$

$= 4x(x^2 - 3) = 0$ when $x = 0, \pm\sqrt{3}$.

$f'(x) < 0$ for $x < -\sqrt{3}$ and $0 < x < \sqrt{3}$.

$f'(x) > 0$ for $-\sqrt{3} < x < 0$ and $x > \sqrt{3}$,

so that f is increasing on $[-\sqrt{3}, 0]$ and $[\sqrt{3}, \infty)$

and decreasing on $(-\infty, -\sqrt{3}]$ and $[0, \sqrt{3}]$.

F. Local minima $f(\pm\sqrt{3}) = -9$, local maximum $f(0) = 0$.   G. $f''(x) = 12x^2 - 12$

$= 12(x^2 - 1) > 0 \Leftrightarrow x^2 > 1 \Leftrightarrow |x| > 1 \Leftrightarrow x > 1$ or $x < -1$, so f is CU on

$(-\infty, -1)$, $(1, \infty)$ and CD on $(-1, 1)$.   IP $(1, -5)$ and $(-1, -5)$.

5. $y = f(x) = 1/(x - 1)$.   A. $D = \{x \mid x \neq 1\} = (-\infty, 1) \cup (1, \infty)$.

B. y–intercept $= f(0) = -1$, no x–intercept.   C. No symmetry.   D. $\lim\limits_{x \to \pm\infty} \frac{1}{x - 1}$

$= 0$, so $y = 0$ is a HA. $\lim\limits_{x \to 1^+} \frac{1}{x - 1} = \infty$ and $\lim\limits_{x \to 1^-} \frac{1}{x - 1} = -\infty$, so $x = 1$ is

a VA.   E. $f'(x) = -1/(x - 1)^2 < 0$ $(x \neq 1)$, so f is decreasing on $(-\infty, 1)$ and $(1, \infty)$.

F. No extrema

G. $f''(x) = 2/(x-1)^3 \Rightarrow f''(x) > 0$

$\Leftrightarrow x > 1$, so f is CU on $(1, \infty)$ and

CD on $(-\infty, 1)$. No IP.

H.

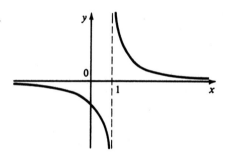

7.　$y = f(x) = 1/(x^2 - 9)$. A. $D = \{x \mid x \neq \pm 3\} = (-\infty, -3) \cup (-3, 3) \cup (3, \infty)$.

B. y–intercept $= f(0) = -1/9$, no x–intercept. C. $f(-x) = f(x) \Rightarrow$ f is even, the

curve is symmetric about the y–axis. D. $\lim\limits_{x \to \pm\infty} \dfrac{1}{x^2 - 9} = 0$, so $y = 0$ is a HA.

$\lim\limits_{x \to 3^-} \dfrac{1}{x^2 - 9} = -\infty$, $\lim\limits_{x \to 3^+} \dfrac{1}{x^2 - 9} = \infty$, $\lim\limits_{x \to -3^-} \dfrac{1}{x^2 - 9} = \infty$, $\lim\limits_{x \to -3^+} \dfrac{1}{x^2 - 9} = -\infty$, so

$x = 3$ and $x = -3$ are VA. E. $f'(x) = -2x/(x^2 - 9)^2 > 0 \Leftrightarrow x < 0 \ (x \neq -3)$ so f is

increasing on $(-\infty, -3)$ and $(-3, 0]$ and decreasing on $[0, 3)$ and $(3, \infty)$.

F. Local maximum $f(0) = -1/9$.

G. $y'' = \dfrac{-2(x^2 - 9)^2 + (2x)\, 2\, (x^2 - 9)\, (2x)}{(x^2 - 9)^4}$

$= \dfrac{6\, (x^2 + 3)}{(x^2 - 9)^3} > 0 \Leftrightarrow x^2 > 9$

$\Leftrightarrow x > 3$ or $x < -3$, so f is

CU on $(-\infty, -3)$ and $(3, \infty)$ and

CD on $(-3, 3)$. No IP.

H.

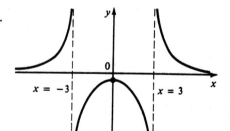

9.　$y = f(x) = x/(2x - 3)^2$. A. $D = \{x \mid x \neq 3/2\} = (-\infty, 3/2) \cup (3/2, \infty)$. B. Both

intercepts are 0. C. No symmetry. D. $\lim\limits_{x \to \pm\infty} \dfrac{x}{(2x - 3)^2} = 0$, so $y = 0$ is a HA.

$\lim\limits_{x \to 3/2} \dfrac{x}{(2x - 3)^2} = \infty$, so $x = 3/2$ is a VA. E. $f'(x) = \dfrac{(2x - 3)^2 - x[2(2x - 3)(2)]}{(2x - 3)^4}$

$= -\dfrac{2x + 3}{(2x - 3)^3} \Rightarrow f'(x) > 0 \Leftrightarrow -3/2 < x < 3/2$, so f is increasing on $[-3/2, 3/2)$ and decreasing on $(-\infty, -3/2]$ and $(3/2, \infty)$.

F. $f(-3/2) = -1/24$ is a local

minimum.

G. $y'' = \dfrac{-2(2x-3)^3 + (2x+3)\,6\,(2x-3)^2}{(2x-3)^6}$

$= \dfrac{8(x+3)}{(2x-3)^4} > 0 \Leftrightarrow x > -3 \ (x \neq 3/2)$

So f is CU on $(-3, 3/2)$ and $(3/2, \infty)$

and CD on $(-\infty, -3)$.   IP $= (-3, -1/27)$.

H.

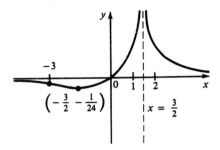

11. $y = f(x) = (x-3)/(x+3)$.   A. $D = \{x \mid x \neq -3\} = (-\infty, -3) \cup (-3, \infty)$.

B.   x–intercept is 3, y–intercept $= f(0) = -1$.   C.   No symmetry.

D. $\displaystyle\lim_{x \to \pm\infty} \frac{x-3}{x+3} = \lim_{x \to \pm\infty} \frac{1 - 3/x}{1 + 3/x} = 1$,   so $y = 1$ is a HA. $\displaystyle\lim_{x \to -3^-} \frac{x-3}{x+3} = \infty$ and

$\displaystyle\lim_{x \to -3^+} \frac{x-3}{x+3} = -\infty$, so $x = -3$ is a VA.

E. $f'(x) = \dfrac{(x+3) - (x-3)}{(x+3)^2} = \dfrac{6}{(x+3)^2}$

$\Rightarrow f'(x) > 0 \ (x \neq -3)$ so f is

increasing on $(-\infty, -3)$ and $(-3, \infty)$

F.   No extrema

G. $f''(x) = \dfrac{-12}{(x+3)^3} > 0 \Leftrightarrow$

$x < -3$, so f is CU on $(-\infty, -3)$

and CD on $(-3, \infty)$.   No IP.

H.

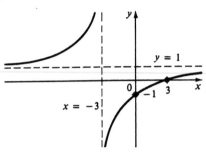

13. $y = f(x) = 1/[(x-1)(x+2)]$.   A. $D = \{x \mid x \neq 1, -2\} = (-\infty, -2) \cup (-2, 1) \cup (1, \infty)$.

B.   No x–intercept, y–intercept $= f(0) = -1/2$.   C.   No symmetry.

D. $\displaystyle\lim_{x \to \pm\infty} \frac{1}{(x-1)(x+2)} = 0$, so $y = 0$ is a HA. $\displaystyle\lim_{x \to 1^-} \frac{1}{(x-1)(x+2)} = -\infty$,

$\displaystyle\lim_{x \to 1^+} \frac{1}{(x-1)(x+2)} = \infty$, $\displaystyle\lim_{x \to -2^-} \frac{1}{(x-1)(x+2)} = \infty$, $\displaystyle\lim_{x \to -2^+} \frac{1}{(x-1)(x+2)} = -\infty$.   So

$x = 1$ and $x = -2$ are VA.   E. $f'(x) = -(2x+1)/[(x-1)(x+2)]^2 \Rightarrow f'(x) > 0$

$\Leftrightarrow x < -1/2 \ (x \neq -2)$, so f is increasing on $(-\infty, -2)$ and $(-2, -1/2]$ and decreasing

on $[-1/2, 1)$ and $(1, \infty)$.   F. $f(-1/2) = -4/9$ is a local maximum.

G. $f''(x) = \dfrac{6(x^2 + x + 1)}{[(x-1)(x+2)]^3}.$

Now $x^2 + x + 1 > 0$ for all x, so

$f''(x) > 0 \Leftrightarrow (x-1)(x+2) > 0 \Leftrightarrow$

$x < -2$ or $x > 1$.   Thus f is CU on

$(-\infty, -2)$ and $(1, \infty)$ and CD on $(-2, 1)$.   No IP.

H.

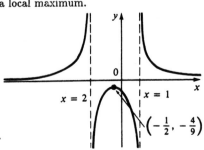

15. $y = f(x) = \dfrac{1+x^2}{1-x^2} = -1 + \dfrac{2}{1-x^2}$. A. $D = \{x \mid x \neq \pm 1\}$. B. No x–intercept,

y–intercept $= f(0) = 1$. C. $f(-x) = f(x)$, so f is even and the curve is symmetric

about the y-axis. D. $\displaystyle\lim_{x \to \pm\infty} \dfrac{1+x^2}{1-x^2} = \lim_{x \to \pm\infty} \dfrac{(1/x^2)+1}{(1/x^2)-1} = -1$, so $y = -1$ is a HA.

$\displaystyle\lim_{x \to 1^-} \dfrac{1+x^2}{1-x^2} = \infty,\ \lim_{x \to 1^+} \dfrac{1+x^2}{1-x^2} = -\infty,\ \lim_{x \to -1^-} \dfrac{1+x^2}{1-x^2} = -\infty,$

$\displaystyle\lim_{x \to -1^+} \dfrac{1+x^2}{1-x^2} = \infty$. So $x = 1$ and $x = -1$ are VA. E. $f'(x) = 4x/(1-x^2)^2 > 0$

$\Leftrightarrow x > 0$ $(x \neq 1)$, so f increases on $[0,1)$, $(1,\infty)$, decreases on $(-\infty,-1)$, $(-1,0]$.

F. $f(0) = 1$ is a local minimum. H.

G. $y'' = \dfrac{4(1-x^2)^2 - 4x \cdot 2(1-x^2)(-2x)}{(1-x^2)^4}$

$= \dfrac{4(1+3x^2)}{(1-x^2)^3} > 0 \Leftrightarrow x^2 < 1$

$\Leftrightarrow -1 < x < 1$, so f is CU on $(-1,1)$

and CD on $(-\infty,-1)$ and $(1,\infty)$.

No IP.

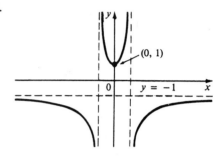

(0, 1)

$y = -1$

17. $y = f(x) = 1/(x^3 - x) = 1/[x(x-1)(x+1)]$. A. $D = \{x \mid x \neq 0, \pm 1\}$. B. No

intercepts. C. $f(-x) = -f(x)$, symmetric about $(0,0)$.

D. $\displaystyle\lim_{x \to \pm\infty} \dfrac{1}{x^3 - x} = 0$, so $y = 0$ is a HA. $\displaystyle\lim_{x \to 0^-} \dfrac{1}{x^3 - x} = \infty,\ \lim_{x \to 0^+} \dfrac{1}{x^3 - x} = -\infty,$

$\displaystyle\lim_{x \to 1^-} \dfrac{1}{x^3 - x} = -\infty,\ \lim_{x \to 1^+} \dfrac{1}{x^3 - x} = \infty,\ \lim_{x \to -1^-} \dfrac{1}{x^3 - x} = -\infty,$

$\displaystyle\lim_{x \to -1^+} \dfrac{1}{x^3 - x} = \infty$. So $x = 0$, $x = 1$, and $x = -1$ are VA.

E. $f'(x) = (1 - 3x^2)/(x^3 - x)^2 \Rightarrow f'(x) > 0 \Leftrightarrow x^2 < 1/3 \Leftrightarrow$

$-1/\sqrt{3} < x < 1/\sqrt{3}$ $(x \neq 0)$, so f is increasing on $[-1/\sqrt{3},0)$, $(0,1/\sqrt{3}]$

and decreasing on $(-\infty,-1)$, $(-1,-1/\sqrt{3}]$, $[1/\sqrt{3},1)$, and $(1,\infty)$.

F. Local minimum $f(-1/\sqrt{3}) = 3\sqrt{3}/2$. H.

local maximum $f(1/\sqrt{3}) = -3\sqrt{3}/2$

G. $f''(x) = \dfrac{2(6x^4 - 3x^2 + 1)}{(x^3 - x)^3}$. Since

$6x^4 - 3x^2 + 1$ has negative discriminant

as a quadratic in x^2, it is $> 0 \Rightarrow$

$f''(x) > 0 \Leftrightarrow x^3 - x > 0 \Leftrightarrow x > 1$ or $-1 < x < 0$.

f is CU on $(-1,0)$ and $(1,\infty)$, and CD on

$(-\infty,-1)$ and $(0,1)$. No IP.

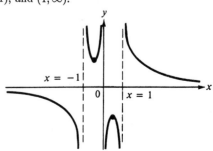

$x = -1$

$x = 1$

19. $y = f(x) = \sqrt{2-x}$.   A. $D = \{x \mid x \le 2\} = (-\infty, 2]$.   B.   x-intercept $= 2$,

y-intercept $= f(0) = \sqrt{2}$.   C.   No symmetry.   D. $\lim\limits_{x \to -\infty} \sqrt{2-x} = \infty$, no asymptotes.

E. $f'(x) = -1/(2\sqrt{2-x}) < 0$, so f is decreasing on $(-\infty, 2]$.

F.   No extrema.

H.

G. $f''(x) = -1/[4(2-x)^{3/2}] < 0$

so f is CD on $(-\infty, 2)$

No IP.

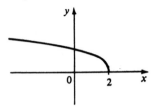

21. $y = f(x) = x + \sqrt{x}$.   A. $D = \{x \mid x \ge 0\} = [0, \infty)$.   B.   Both intercepts are 0.

C.   No symmetry.   D. $\lim\limits_{x \to \infty} (x + \sqrt{x}) = \infty$,   H.

no asymptotes.   E. $f'(x) = 1 + 1/(2\sqrt{x})$

> 0, so f is increasing on $[0, \infty)$.

F.   No extrema.

G. $f''(x) = -(1/4) x^{-3/2} < 0$, so

f is CD on $(0, \infty)$.   No IP.

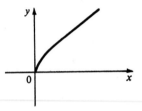

23. $y = f(x) = \sqrt{x^2 + 1} - x$.   A. $D = R$.   B.   No x-intercept, y-intercept $= 1$.

C.   No symmetry.   D. $\lim\limits_{x \to -\infty} (\sqrt{x^2 + 1} - x) = \infty$ and $\lim\limits_{x \to \infty} (\sqrt{x^2 + 1} - x)$

$= \lim\limits_{x \to \infty} (\sqrt{x^2 + 1} - x) \dfrac{\sqrt{x^2 + 1} + x}{\sqrt{x^2 + 1} + x} = \lim\limits_{x \to \infty} \dfrac{1}{\sqrt{x^2 + 1} + x} = 0$,   so $y = 0$ is a HA.

E. $f'(x) = \dfrac{x}{\sqrt{x^2 + 1}} - 1 = \dfrac{x - \sqrt{x^2 + 1}}{\sqrt{x^2 + 1}}$ H.

$\Rightarrow f'(x) < 0$, so f is decreasing on R.

F.   No extrema.

G. $f''(x) = 1/(x^2 + 1)^{3/2} > 0$, so f is

CU on R.   No IP.

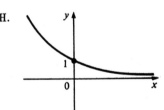

25. $y = f(x) = \sqrt[4]{x^2 - 25}$.   A. $D = \{x \mid x^2 \ge 25\} = (-\infty, -5] \cup [5, \infty)$.   B.   x-intercepts

are ± 5, no y-intercept.   C. $f(-x) = f(x)$, so the curve is symmetric about the y-axis.

D. $\lim\limits_{x \to \pm\infty} \sqrt[4]{x^2 - 25} = \infty$, no asymptotes.   E. $f'(x) = (1/4)(x^2 - 25)^{-3/4}(2x)$

$= x/[2(x^2 - 25)^{3/4}] > 0$ if $x > 5$, so f is increasing on $[5, \infty)$ and decreasing on

$(-\infty, -5]$. F. No local extrema.

G. $y'' = \dfrac{2(x^2 - 25)^{3/4} - 3x^2(x^2 - 25)^{-1/4}}{4(x^2 - 25)^{3/2}}$

$= -\dfrac{x^2 + 50}{4(x^2 - 25)^{7/4}} < 0 \ \text{so } f \text{ is CD}$

on $(-\infty, -5)$ and $(5, \infty)$. No IP.

H.

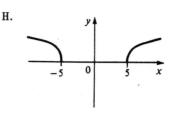

27. $y = f(x) = x\sqrt{x^2 - 9}$. A. $D = \{x \mid x^2 \geq 9\} = (-\infty, -3] \cup [3, \infty)$. B. x–intercepts

are ± 3, no y–intercept. C. $f(-x) = -f(x)$, so the curve is symmetric about the origin.

D. $\lim\limits_{x \to \infty} x\sqrt{x^2 - 9} = \infty$, $\lim\limits_{x \to -\infty} x\sqrt{x^2 - 9} = -\infty$, no asymptotes.

E. $f'(x) = \sqrt{x^2 - 9} + x^2/\sqrt{x^2 - 9} > 0$ for $x \in D$, so f is increasing on $(-\infty, -3]$ and

$[3, \infty)$. F. No extrema.

G. $f''(x) = \dfrac{x}{\sqrt{x^2 - 9}} + \dfrac{2x\sqrt{x^2 - 9} - x^2(x/\sqrt{x^2 - 9})}{x^2 - 9} = \dfrac{x(2x^2 - 27)}{(x^2 - 9)^{3/2}} > 0$

$\Leftrightarrow x > 3\sqrt{3/2}$ or $-3\sqrt{3/2} < x < 0$, so f is CU on $(3\sqrt{3/2}, \infty)$

and $(-3\sqrt{3/2}, -3)$ and CD on $(-\infty, -3\sqrt{3/2})$ and $(3, 3\sqrt{3/2})$. IP $(\pm 3\sqrt{3/2}, \pm 9\sqrt{3}/2)$.

H.

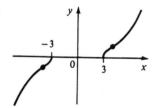

29. $y = f(x) = \sqrt{1 - x^2}/x$. A. $D = \{x \mid |x| \leq 1, x \neq 0\} = [-1, 0) \cup (0, 1]$.

B. x–intercepts ± 1, no y–intercept. C. $f(-x) = -f(x)$, so the curve is symmetric

about $(0, 0)$. D. $\lim\limits_{x \to 0^+} \dfrac{\sqrt{1 - x^2}}{x} = \infty$, $\lim\limits_{x \to 0^-} \dfrac{\sqrt{1 - x^2}}{x} = -\infty$, so $x = 0$ is a VA.

E. $f'(x) = \dfrac{(-x^2/\sqrt{1 - x^2}) - \sqrt{1 - x^2}}{x^2} = -1/(x^2\sqrt{1 - x^2}) < 0$, so f is decreasing on $[-1, 0)$

153

and $(0, 1]$.   F.   No extrema

H.

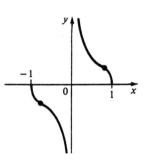

G. $f''(x) = \dfrac{2 - 3x^2}{x^3(1 - x^2)^{3/2}} > 0 \Leftrightarrow$

$-1 < x < -\sqrt{2/3}$ or $0 < x < \sqrt{2/3}$, so f is

CU on $(-1, -\sqrt{2/3})$ and $(0, \sqrt{2/3})$ and

CD on $(-\sqrt{2/3}, 0)$ and $(\sqrt{2/3}, 1)$.

IP $(\pm\sqrt{2/3}, \pm 1/\sqrt{2})$.

31. $y = f(x) = x + 3x^{2/3}$.   A. $D = R$.   B. $y = x + 3x^{2/3} = x^{2/3}(x^{1/3} + 3) = 0$ if $x = 0$

or -27 (x–intercepts), y–intercept $= f(0) = 0$.   C.   No symmetry.

D. $\lim\limits_{x \to \infty}(x + 3x^{2/3}) = \infty$, $\lim\limits_{x \to -\infty}(x + 3x^{2/3}) = \lim\limits_{x \to -\infty}x^{2/3}(x^{1/3} + 3) = -\infty$, no

asymptotes.   E. $f'(x) = 1 + 2x^{-1/3} = (x^{1/3} + 2)/x^{1/3} > 0 \Leftrightarrow x > 0$ or $x < -8$, so f

increases on $(-\infty, -8]$, $[0, \infty)$ and decreases on $[-8, 0]$.

F.   Local maximum $f(-8) = 4$

local minimum $f(0) = 0$

G. $f''(x) = -(2/3)x^{-4/3} < 0$ $(x \neq 0)$ so f is CD on $(-\infty, 0)$ and $(0, \infty)$.   No IP.

H.

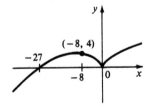

33. $y = f(x) = \cos x - \sin x$.   A. $D = R$.   B. $y = 0 \Leftrightarrow \cos x = \sin x \Leftrightarrow x = n\pi + \pi/4$, n an

integer, (x–intercepts), y–intercept $= f(0) = 1$.   C.   Periodic with period 2π.

D.   No asymptotes.   E. $f'(x) = -\sin x - \cos x = 0 \Leftrightarrow \cos x = -\sin x$

$\Leftrightarrow x = 2n\pi + 3\pi/4$ or $2n\pi + 7\pi/4$. $f'(x) > 0 \Leftrightarrow \cos x < -\sin x$

$\Leftrightarrow 2n\pi + 3\pi/4 < x < 2n\pi + 7\pi/4$, so f is increasing on $[2n\pi + 3\pi/4, 2n\pi + 7\pi/4]$ and

decreasing on $[2n\pi - \pi/4, 2n\pi + 3\pi/4]$.   F.   Local maximum $f(2n\pi - \pi/4) = \sqrt{2}$, local minimum $f(2n\pi + 3\pi/4) = -\sqrt{2}$.

G. $f''(x) = -\cos x + \sin x > 0$

$\Leftrightarrow \sin x > \cos x \Leftrightarrow$

$x \in (2n\pi + \pi/4, 2n\pi + 5\pi/4)$, so f is

CU on these intervals and CD on

$(2n\pi - 3\pi/4, 2n\pi + \pi/4)$.

IP $(n\pi + \pi/4, 0)$.

H.

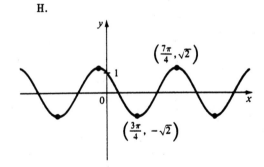

35. $y = f(x) = \sin x - \tan x$. A. $D = \{x \mid x \neq (2n+1)\pi/2\}$. B. $y = 0 \Leftrightarrow \sin x = \tan x$

$= \sin x/\cos x \Leftrightarrow \sin x = 0$ or $\cos x = 1 \Leftrightarrow x = n\pi$ (*x–intercepts*),

y–intercept $= f(0) = 0$. C. $f(-x) = -f(x)$, so the curve is symmetric about $(0,0)$.

Also periodic with period 2π. D. $\lim\limits_{x \to \pi/2^-} (\sin x - \tan x) = -\infty$ and

$\lim\limits_{x \to \pi/2^+} (\sin x - \tan x) = \infty$, so $x = n\pi + \pi/2$ are VA. E. $f'(x) = \cos x - \sec^2 x \leq 0$, so f

decreases on each interval in its domain, i.e., on $((2n-1)\pi/2, (2n+1)\pi/2)$.

F. No extrema.

G. $f''(x) = -\sin x - 2\sec^2 x \tan x = -\sin x(1 + 2\sec^3 x)$. Note that $1 + 2\sec^3 x \neq 0$

since $\sec^3 x \neq -1/2$. $f''(x) > 0$ for $-\pi/2 < x < 0$ and $3\pi/2 < x < 2\pi$ so f is CU on

$(n\pi - \pi/2, n\pi)$ and CD on $(n\pi, n\pi + \pi/2)$. IP $(n\pi, 0)$. Note also that $f'(0) = 0$ but

$f'(\pi) = -2$.

H.

37. $y = f(x) = x \tan x$, $-\pi/2 < x < \pi/2$. A. $D = (-\pi/2, \pi/2)$. B. Intercepts 0.

C. $f(-x) = f(x)$, so the curve is symmetric about the y–axis. D. $\lim\limits_{x \to \pi/2^-} x \tan x = \infty$

and $\lim\limits_{x \to -\pi/2^+} x \tan x = \infty$, so $x = \pi/2$ and $x = -\pi/2$ are VA.

E. $f'(x) = \tan x + x \sec^2 x > 0 \Leftrightarrow 0 < x < \pi/2$, so f increases on $[0, \pi/2)$ and decreases on

$(-\pi/2, 0]$. F. Absolute minimum $f(0) = 0$. G. $y'' = 2\sec^2 x + 2x \tan x \sec^2 x > 0$ for

155

$-\pi/2 < x < \pi/2$, so f is CU on $(-\pi/2, \pi/2)$.

H.

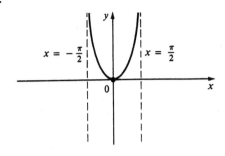

39. $y = f(x) = x/2 - \sin x$, $0 < x < 3\pi$. A. $D = (0, 3\pi)$. B. No y-intercept (*x-intercept*

could be found approximately by Newton's Method, see Exercise 2.10.11). C. No

symmetry. D. No asymptotes. E. $f'(x) = 1/2 - \cos x > 0 \Leftrightarrow \cos x < 1/2$

$\Leftrightarrow \pi/3 < x < 5\pi/3$ or $7\pi/3 < x < 3\pi$, so f is increasing on $[\pi/3, 5\pi/3]$ and $[7\pi/3, 3\pi)$

and decreasing on $(0, \pi/3]$ and $[5\pi/3, 7\pi/3]$.   H.

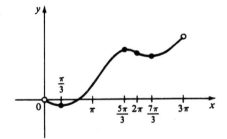

F. $f(\pi/3) = \pi/6 - \sqrt{3}/2$ is a local

minimum, $f(5\pi/3) = 5\pi/6 + \sqrt{3}/2$ is

a local maximum, $f(7\pi/3) = 7\pi/6 - \sqrt{3}/2$

is a local minimum.

G. $f''(x) = \sin x > 0 \Leftrightarrow 0 < x < \pi$ or

$2\pi < x < 3\pi$, so f is CU on $(0, \pi)$ and

$(2\pi, 3\pi)$ and CD on $(\pi, 2\pi)$.

IP $(\pi, \pi/2)$ and $(2\pi, \pi)$.

41. $y = f(x) = 2\cos x + \sin^2 x$. A. $D = R$. B. y-intercept $= f(0) = 2$.

C. $f(-x) = f(x)$, so the curve is symmetric about the y-axis. Periodic with period 2π.

D. No asymptotes. E. $f'(x) = -2\sin x + 2\sin x \cos x = 2\sin x (\cos x - 1) > 0$

$\Leftrightarrow \sin x < 0 \Leftrightarrow (2n-1)\pi < x < 2n\pi$, so f is increasing on $[(2n-1)\pi, 2n\pi]$ and decreasing on $[2n\pi, (2n+1)\pi]$.

H.

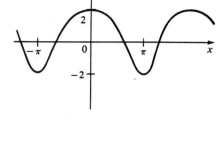

F. $f(2n\pi) = 2$ is a local maximum.

$f((2n+1)\pi) = -2$ is a local minimum.

G. $f''(x) = -2\cos x + 2\cos 2x$

$= 2(2\cos^2 x - \cos x - 1)$

$= 2(2\cos x + 1)(\cos x - 1) > 0$

$\Leftrightarrow \cos x < -1/2 \Leftrightarrow x \in (2n\pi + 2\pi/3, 2n\pi + 4\pi/3)$,

so f is CU on these intervals and CD on

$(2n\pi - 2\pi/3, 2n\pi + 2\pi/3)$. IP when $x = 2n\pi \pm 2\pi/3$.

43.　$y = f(x) = \sin 2x - 2\sin x$.　A.　$D = \mathbb{R}$.　B.　y−intercept $= f(0) = 0$.

$y = 0 \Leftrightarrow 2\sin x = \sin 2x = 2\sin x \cos x \Leftrightarrow \sin x = 0$　or　$\cos x = 1 \Leftrightarrow x = n\pi$ (the

x−intercepts).　C.　$f(-x) = -f(x)$,　so the curve is symmetric about $(0,0)$.　f is periodic

with period 2π, so we do parts D−G for $-\pi \le x \le \pi$.　D.　No asymptotes.

E.　$f'(x) = 2\cos 2x - 2\cos x$.　As in Exercise 41G, we see that

$f'(x) > 0 \Leftrightarrow -\pi < x < -2\pi/3$　or　$2\pi/3 < x < \pi$,　so f is increasing on $[-\pi, -2\pi/3]$

and $[2\pi/3, \pi]$ and decreasing on $[-2\pi/3, 2\pi/3]$.　F.　$f(-2\pi/3) = 3\sqrt{3}/2$ is a local

maximum, $f(2\pi/3) = -3\sqrt{3}/2$ is a local minimum.

G.　$f''(x) = -4\sin 2x + 2\sin x$

$= 2\sin x(1 - 4\cos x) = 0$　when

$x = 0, \pm\pi$　or　$\cos x = 1/4$.

If $\alpha = \cos^{-1}(1/4)$, then f is CU

on $(-\alpha, 0)$ and (α, π) and CD on

$(-\pi, -\alpha)$ and $(0, \alpha)$.　IP $(0,0)$,

$(\pi, 0)$, $(\alpha, -3\sqrt{15}/8)$, $(-\alpha, 3\sqrt{15}/8)$.

H.

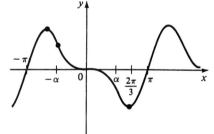

45.　$y = f(x) = e^{-x^2}$　A.　$D = \mathbb{R}$　B.　No x−intercept , y−intercept $= 1$

C.　$f(-x) = f(x)$,　so the curve is symmetric about the y−axis.

D. $\lim\limits_{x\to\pm\infty} e^{-x^2} = 0$, so $y = 0$ is HA. No VA. E. $f'(x) = -2xe^{-x^2} \Rightarrow$

$f'(x) > 0 \Leftrightarrow x < 0$, so f is increasing on $(-\infty, 0]$, decreasing on $[0, \infty)$.

F. $f(0) = 1$ is a local and

absolute maximum

G. $f''(x) = 2e^{-x^2}(2x^2 - 1) > 0 \Leftrightarrow$

$x^2 > 1/2 \Leftrightarrow |x| > 1/\sqrt{2}$, so f is CU

on $(-\infty, -1/\sqrt{2})$ and $(1/\sqrt{2}, \infty)$ and

CD on $(-1/\sqrt{2}, 1/\sqrt{2})$. IP $(\pm 1/\sqrt{2}, 1/\sqrt{e})$

H.

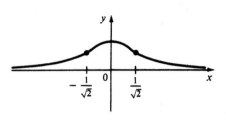

47. $y = f(x) = e^{-1/(x+1)}$ A. $D = \left\{x \mid x \neq -1\right\} = (-\infty, -1) \cup (-1, \infty)$

B. No x–intercept, y–intercept $= f(0) = e^{-1}$ C. No symmetry

D. $\lim\limits_{x\to\pm\infty} e^{-1/(x+1)} = 1$ since $-1/(x+1) \to 0$, so $y = 1$ is a HA.

$\lim\limits_{x\to -1^+} e^{-1/(x+1)} = 0$ since $-1/(x+1) \to -\infty$, $\lim\limits_{x\to -1^-} e^{-1/(x+1)} = \infty$

since $-1/(x+1) \to \infty$, so $x = -1$ is a VA E. $f'(x) = e^{-1/(x+1)}/(x+1)^2$

$\Rightarrow f'(x) > 0$ for all x $(\neq -1)$, so f is increasing on $(-\infty, -1)$ and $(-1, \infty)$

F. No extrema

G $f''(x) = e^{-1/(x+1)}/(x+1)^4 + e^{-1/(x+1)}(-2)/(x+1)^3$

$= -e^{-1/(x+1)}(2x+1)/(x+1)^4 \Rightarrow f''(x) > 0 \Leftrightarrow 2x + 1 < 0 \Leftrightarrow x < -1/2$

So f is CU on $(-\infty, -1)$, $(-1, -1/2)$, and CD on $(-1/2, \infty)$. IP $(-1/2, e^{-2})$

H.

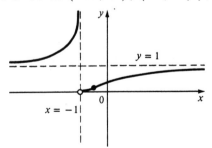

49. $y = f(x) = e^{1/x^2}$ A. $D = \left\{x \mid x \neq 0\right\} = (-\infty, 0) \cup (0, \infty)$ B. No intercepts

C. $f(-x) = f(x)$, so the curve is symmetric about the y–axis.

D. $\lim\limits_{x\to\pm\infty} e^{1/x^2} = e^0 = 1$, so $y = 1$ is a HA. $\lim\limits_{x\to 0} e^{1/x^2} = \infty$, so $x = 0$

is a VA. E. $f'(x) = e^{1/x^2}(-2/x^3) > 0 \Leftrightarrow x < 0$, so f is increasing

on $(-\infty, 0)$ and decreasing on $(0, \infty)$.   F. No extrema

G. $f''(x) = e^{1/x^2}(4/x^6) + e^{1/x^2}(6/x^4) > 0$, so f is CU

on $(-\infty, 0)$ and $(0, \infty)$.

H.

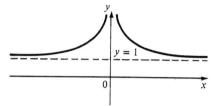

51. $y = f(x) = \ln(\cos x)$ A. $D = \Big\{x \mid \cos x > 0\Big\} = (-\pi/2, \pi/2) \cup$

$(3\pi/2, 5\pi/2) \cup \cdots = \Big\{x \mid 2n\pi - \pi/2 < x < 2n\pi + \pi/2, \ n = 0, \pm 1, \pm 2, \cdots\Big\}$

B. x–intercepts occur when $\ln(\cos x) = 0 \Leftrightarrow \cos x = 1 \Leftrightarrow x = 2n\pi$,

y–intercept $= f(0) = 0$ C. $f(-x) = f(x)$, so the curve is symmetric

about the y–axis . $f(x + 2\pi) = f(x)$, f has period 2π, so in D–G we

consider only $-\pi/2 < x < \pi/2$.   D. $\lim\limits_{x\to\pi/2^-} \ln(\cos x) = -\infty$

$\lim\limits_{x\to -\pi/2^+} \ln(\cos x) = -\infty$, so $x = \pi/2$ and $x = -\pi/2$ are VA. No HA.

E. $f'(x) = (1/\cos x)(-\sin x) = -\tan x > 0 \Leftrightarrow -\pi/2 < x < 0$

so f is increasing on $(-\pi/2, 0]$

and decreasing on $[0, \pi/2)$.

F. $f(0) = 0$ is a local maximum

G. $f''(x) = -\sec^2 x < 0 \Rightarrow$ f is

CD on $(-\pi/2, \pi/2)$. No IP .

H.

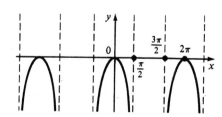

53. $y = f(x) = \ln(x + \sqrt{1 + x^2})$ A. $x + \sqrt{1 + x^2} > 0$ for all x since $1 + x^2 > x^2 \Rightarrow$

$\sqrt{1 + x^2} > |x|$, so $D = R$. B. y–intercept $= f(0) = 0$, x–intercept occurs when

$x + \sqrt{1 + x^2} = 1 \Rightarrow \sqrt{1 + x^2} = 1 - x \Rightarrow 1 + x^2 = 1 - 2x + x^2 \Rightarrow x = 0$.

C. $\ln(-x + \sqrt{1 + x^2}) = -\ln(x + \sqrt{1 + x^2})$ since $(\sqrt{1 + x^2} - x)(\sqrt{1 + x^2} + x)$

$= 1$, so the curve is symmetric about the origin.   D. $\lim\limits_{x\to\infty} \ln(x + \sqrt{1 + x^2}) = \infty$,

$$\lim_{x \to -\infty} \ln(x + \sqrt{1+x^2}) = \lim_{x \to -\infty} \ln \frac{1}{\sqrt{1+x^2} - x} = -\infty, \text{ no HA.}$$

E. $f'(x) = \dfrac{1}{x + \sqrt{1+x^2}} \left[1 + \dfrac{x}{\sqrt{1+x^2}}\right] = \dfrac{1}{\sqrt{1+x^2}} > 0$, so f is increasing on

$(-\infty, \infty)$. F. No extrema

H.

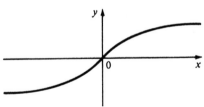

G. $f''(x) = -\dfrac{x}{(1+x^2)^{3/2}}$

$\Rightarrow f''(x) > 0 \Leftrightarrow x < 0$, so

f is CU on $(-\infty, 0)$, CD on $(0, \infty)$

and the IP is $(0, 0)$.

55. $y = f(x) = \ln(1+x^2)$ A. D = R B. Both intercepts are 0.

C. $f(-x) = f(x)$, so the curve is symmetric about the y-axis.

D. $\lim\limits_{x \to \pm\infty} \ln(1+x^2) = \infty$, no asymptotes E. $f'(x) = \dfrac{2x}{1+x^2} > 0 \Leftrightarrow x > 0$

so f is increasing on $[0, \infty)$ and decreasing on $(-\infty, 0]$. F. $f(0) = 0$

is a local and absolute minimum.

H.

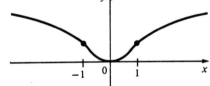

G. $f''(x) = \dfrac{2(1+x^2) - 2x(2x)}{(1+x^2)^2}$

$= \dfrac{2(1-x^2)}{(1+x^2)^2} > 0 \Leftrightarrow |x| < 1$, so f is

CU on $(-1, 1)$, CD on $(-\infty, -1)$ and

$(1, \infty)$. IP $(1, \ln 2)$ and $(-1, \ln 2)$

57. $y = f(x) = \sin^{-1}(x/(x+1))$ A. $D = \left\{x \mid -1 \le x/(x+1) \le 1\right\}$ For $x > -1$

we have $-x-1 \le x \le x+1 \Leftrightarrow 2x \ge -1 \Leftrightarrow x \ge -1/2$, so $D = [-1/2, \infty)$.

B. Intercepts are 0. C. No symmetry D. $\lim\limits_{x \to \infty} \sin^{-1}\left[\dfrac{x}{x+1}\right] =$

$\lim\limits_{x \to \infty} \sin^{-1}\left[\dfrac{1}{1+1/x}\right] = \sin^{-1} 1 = \pi/2$, so $y = \pi/2$ is a HA.

E. $f'(x) = \dfrac{1}{\sqrt{1 - (x/(x+1))^2}} \dfrac{(x+1) - x}{(x+1)^2} = \dfrac{1}{(x+1)\sqrt{2x+1}} > 0$, so f is

increasing on $[-1/2, \infty)$.

F. No local maximum or minimum.

$f(-1/2) = \sin^{-1}(-1) = -\pi/2$

is an absolute minimum.

H.

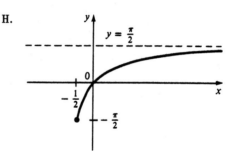

G. $f''(x) = \dfrac{\sqrt{2x+1} + (x+1)/\sqrt{2x+1}}{(x+1)^2(2x+1)}$

$= -\dfrac{3x+2}{(x+1)^2(2x+1)^{3/2}} < 0$ on D,

so f is CD on $(-1/2, \infty)$.

59. $y = f(x) = x\,e^{-x}$ A. $D = R$ B. Intercepts are 0   C. No symmetry

D. $\displaystyle\lim_{x\to\infty} xe^{-x} = \lim_{x\to\infty}\frac{x}{e^x} \overset{H}{=} \lim_{x\to\infty}\frac{1}{e^x} = 0$, so $y = 0$ is a HA . $\displaystyle\lim_{x\to-\infty} xe^{-x} = -\infty$

E. $f'(x) = e^{-x} - xe^{-x} = e^{-x}(1-x) > 0 \Leftrightarrow x < 1$, so f is increasing on

$(-\infty, 1]$ and decreasing on $[1, \infty)$.

F. Absolute maximum $f(1) = 1/e$

H.

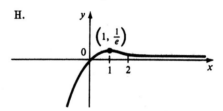

G. $f''(x) = e^{-x}(x-2) > 0 \Leftrightarrow x > 2$

so f is CU on $(2, \infty)$ and CD on $(-\infty, 2)$.

IP is $(2, 2/e^2)$

61. $y = f(x) = x \ln x$ A. $D = (0, \infty)$ B. x–intercept when $\ln x = 0 \Leftrightarrow x = 1$, no

y–intercept   C. No symmetry   D. $\displaystyle\lim_{x\to\infty} x \ln x = \infty$, $\displaystyle\lim_{x\to 0^+} x \ln x = \lim_{x\to 0^+}\frac{\ln x}{1/x}$

$\overset{H}{=} \displaystyle\lim_{x\to 0^+}\frac{1/x}{-1/x^2} = \lim_{x\to 0^+}(-x) = 0$, no asymptotes

E. $f'(x) = \ln x + 1 = 0$ when $\ln x = -1 \Leftrightarrow x = e^{-1}$. $f'(x) > 0 \Leftrightarrow$

$\ln x > -1 \Leftrightarrow x > e^{-1}$, so f is

increasing on $[1/e, \infty)$ and

decreasing on $(0, 1/e]$.

F. $f(1/e) = -1/e$ is an absolute

minimum.   G. $f''(x) = 1/x > 0$, so

f is CU on $(0, \infty)$. No IP.

H.

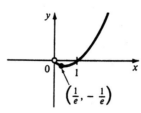

63. $y = f(x) = x^2 \ln x$ A. $D = (0, \infty)$ B. x–intercept when $\ln x = 0 \Leftrightarrow$

$x = 1$, no y–intercept   C. No symmetry   D. $\displaystyle\lim_{x\to\infty} x^2 \ln x = \infty$,

$\lim\limits_{x \to 0^+} x^2 \ln x = \lim\limits_{x \to 0^+} \dfrac{\ln x}{1/x^2} \overset{H}{=} \lim\limits_{x \to 0^+} \dfrac{1/x}{-2/x^3} = \lim\limits_{x \to 0^+} \left(-\dfrac{x^2}{2}\right) = 0$, no asymptote

E. $f'(x) = 2x \ln x + x = x(2 \ln x + 1) > 0 \Leftrightarrow \ln x > -1/2 \Leftrightarrow$

$x > e^{-1/2}$, so f is increasing on $[1/\sqrt{e}, \infty)$, decreasing on $(0, 1/\sqrt{e}]$.

F. $f(1/\sqrt{e}) = -1/(2e)$ is an absolute

minimum   G. $f''(x) = 2 \ln x + 3 > 0$

$\Leftrightarrow \ln x > -3/2 \Leftrightarrow x > e^{-3/2}$, so f

is CU on $(e^{-3/2}, \infty)$ and CD on

$(0, e^{-3/2})$. IP is $(e^{-3/2}, -3/(2e^3))$

H.

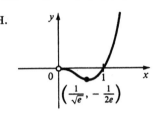

$\left(\dfrac{1}{\sqrt{e}}, -\dfrac{1}{2e}\right)$

65. $y = f(x) = x e^{-x^2}$ A. $D = R$ B. Intercepts are 0   C. $f(-x) = -f(x)$,

so the curve is symmetric about the origin.

D. $\lim\limits_{x \to \pm\infty} x e^{-x^2} = \lim\limits_{x \to \pm\infty} \dfrac{x}{e^{x^2}} \overset{H}{=} \lim\limits_{x \to \pm\infty} \dfrac{1}{2xe^{x^2}} = 0$, so $y = 0$ is a HA.

E. $f'(x) = e^{-x^2} - 2x^2 e^{-x^2} = e^{-x^2}(1 - 2x^2) > 0 \Leftrightarrow x^2 < 1/2 \Leftrightarrow |x| < 1/\sqrt{2}$

so f is increasing on $[-1/\sqrt{2}, 1/\sqrt{2}]$ and decreasing on $(-\infty, -1/\sqrt{2}]$ and

$[1/\sqrt{2}, \infty)$.   F. $f(1/\sqrt{2}) = 1/\sqrt{2e}$ is a local maximum, $f(-1/\sqrt{2}) = -1/\sqrt{2e}$

is a local minimum.

H.

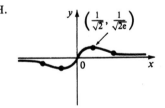

$\left(\dfrac{1}{\sqrt{2}}, \dfrac{1}{\sqrt{2e}}\right)$

G. $f''(x) = -2x e^{-x^2}(1 - 2x^2) - 4x e^{-x^2}$

$= 2x e^{-x^2}(2x^2 - 3) > 0 \Leftrightarrow x > \sqrt{3/2}$ or

$-\sqrt{3/2} < x < 0$, so f is CU on $(\sqrt{3/2}, \infty)$

and $(-\sqrt{3/2}, 0)$ and CD on $(-\infty, -\sqrt{3/2})$

and $(0, \sqrt{3/2})$. IP are $(0, 0)$ and

$(\pm\sqrt{3/2}, \pm\sqrt{3/2} \, e^{-3/2})$

67. $y = f(x) = e^x/x$ A. $D = \left\{x \mid x \neq 0\right\}$ B. No intercepts   C. No symmetry

D. $\lim\limits_{x \to \infty} \dfrac{e^x}{x} \overset{H}{=} \lim\limits_{x \to \infty} \dfrac{e^x}{1} = \infty$, $\lim\limits_{x \to -\infty} \dfrac{e^x}{x} = 0$, so $y = 0$ is a HA. $\lim\limits_{x \to 0^+} \dfrac{e^x}{x} = \infty$,

$\lim\limits_{x \to 0^-} \dfrac{e^x}{x} = -\infty$, so $x = 0$ is a VA. E. $f'(x) = \dfrac{xe^x - e^x}{x^2} > 0 \Leftrightarrow (x - 1) e^x > 0$

$\Leftrightarrow x > 1$, so f is increasing on $[1, \infty)$, decreasing on $(-\infty, 0)$ and $(0, 1]$

F. $f(1) = e$ is a local minimum

H.

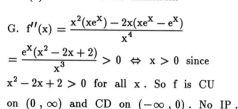

G. $f''(x) = \dfrac{x^2(xe^x) - 2x(xe^x - e^x)}{x^4}$

$= \dfrac{e^x(x^2 - 2x + 2)}{x^3} > 0 \Leftrightarrow x > 0$ since

$x^2 - 2x + 2 > 0$ for all x. So f is CU

on $(0, \infty)$ and CD on $(-\infty, 0)$. No IP.

69. $y = f(x) = xe^{1/x}$ A. $D = \left\{ x \mid x \neq 0 \right\}$ B. No intercepts   C. No symmetry

D. $\lim\limits_{x \to \infty} xe^{1/x} = \infty$, $\lim\limits_{x \to -\infty} xe^{1/x} = -\infty$, no HA. $\lim\limits_{x \to 0^+} xe^{1/x} = \lim\limits_{x \to 0^+} \dfrac{e^{1/x}}{1/x}$

$\overset{H}{=} \lim\limits_{x \to 0^+} \dfrac{e^{1/x}(-1/x^2)}{-1/x^2} = \lim\limits_{x \to 0^+} e^{1/x} = \infty$, so $x = 0$ is a VA.

Also $\lim\limits_{x \to 0^-} xe^{1/x} = 0$ since $\frac{1}{x} \to -\infty \Rightarrow e^{1/x} \to 0$

E. $f'(x) = e^{1/x} + xe^{1/x}(-1/x^2) = e^{1/x}(1 - 1/x) > 0 \Leftrightarrow 1/x < 1 \Leftrightarrow x < 0$

or $x > 1$, so f is increasing on $(-\infty, 0)$ and $[1, \infty)$, decreasing on $(0, 1]$

F. $f(1) = e$ is a local minimum

H.

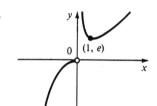

G. $f''(x) = e^{1/x}(-1/x^2)(1 - 1/x)$

$+ e^{1/x}(1/x^2) = e^{1/x}/x^3 > 0 \Leftrightarrow$

$x > 0$, so f is CU on $(0, \infty)$ and

CD on $(-\infty, 0)$. No IP.

71. $y = f(x) = x - \ln(1 + x)$ A. $D = \left\{ x \mid x > -1 \right\} = (-1, \infty)$ B. Intercepts

are 0.   C. No symmetry   D. $\lim\limits_{x \to -1^+} (x - \ln(1 + x)) = \infty$, so $x = -1$ is a

VA. $\lim\limits_{x \to \infty} (x - \ln(1 + x)) = \lim\limits_{x \to \infty} x\left[1 - \dfrac{\ln(1 + x)}{x}\right] = \infty$, since $\lim\limits_{x \to \infty} \dfrac{\ln(1 + x)}{x}$

$\overset{H}{=} \lim\limits_{x \to \infty} \dfrac{1/(1 + x)}{1} = 0$ E. $f'(x) = 1 - \dfrac{1}{1 + x} = \dfrac{x}{1 + x} > 0 \Leftrightarrow x > 0$ since

$x + 1 > 0$. So f is increasing on

$[0, \infty)$ and decreasing on $(-1, 0]$.

F. $f(0) = 0$ is an absolute minimum.

G. $f''(x) = 1/(1 + x)^2 > 0$, so f is

CU on $(-1, \infty)$.

H.

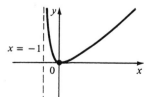

73. $y = f(x) = x^3/(x^2 - 1)$.   A. $D = \{x \mid x \neq \pm 1\} = (-\infty, -1) \cup (-1, 1) \cup (1, \infty)$.

B. x-intercept $= 0$, y-intercept $= 0$.   C. $f(-x) = -f(x) \Rightarrow f$ is odd, so the curve is symmetric about the origin.   D. $\lim\limits_{x\to\infty} \dfrac{x^3}{x^2-1} = \infty$ but long division gives

$\dfrac{x^3}{x^2-1} = x + \dfrac{x}{x^2-1}$ so $f(x) - x = \dfrac{x}{x^2-1} \to 0$ as $x \to \pm\infty \Rightarrow y = x$ is a slant asymptote.

$\lim\limits_{x\to 1^-} \dfrac{x^3}{x^2-1} = -\infty, \ \lim\limits_{x\to 1^+} \dfrac{x^3}{x^2-1} = \infty, \ \lim\limits_{x\to -1^-} \dfrac{x^3}{x^2-1} = -\infty, \ \lim\limits_{x\to -1^+} \dfrac{x^3}{x^2-1} = \infty,$ so

$x = 1$ and $x = -1$ are VA.   E. $f'(x) = \dfrac{3x^2(x^2-1) - x^3(2x)}{(x^2-1)^2} = \dfrac{x^2(x^2-3)}{(x^2-1)^2}$

$\Rightarrow f'(x) > 0 \Leftrightarrow x^2 > 3 \Leftrightarrow x > \sqrt{3}$ or $x < -\sqrt{3}$, so f is increasing on $(-\infty, -\sqrt{3}]$ and

$[\sqrt{3}, \infty)$ and decreasing on $[-\sqrt{3}, -1), (-1, 1),$ and $(1, \sqrt{3}]$.

F. $f(-\sqrt{3}) = -3\sqrt{3}/2$ is a local

maximum and $f(\sqrt{3}) = 3\sqrt{3}/2$ is a local minimum.

G. $y'' = \dfrac{2x(x^2+3)}{(x^2-1)^3} > 0 \Leftrightarrow x > 1$ or

$-1 < x < 0,$ so f is CU on $(1, \infty)$ and

$(-1, 0)$ and CD on $(-\infty, -1)$ and

$(0, 1)$.   IP is $(0, 0)$.

H.

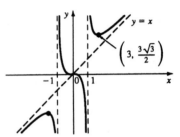

$\left(3, \dfrac{3\sqrt{3}}{2}\right)$

75. $y = f(x) = (x^2 + 4)/x = x + 4/x.$ A. $D = \{x \mid x \neq 0\} = (-\infty, 0) \cup (0, \infty).$

B. No intercepts.   C. $f(-x) = -f(x) \Rightarrow$ symmetry about the origin.

D. $\lim\limits_{x\to\infty}(x + \tfrac{4}{x}) = \infty$ but $f(x) - x = \tfrac{4}{x} \to 0$ as $x \to \pm\infty,$ so $y = x$ is a slant asymptote.

$\lim\limits_{x\to 0^+}(x + \tfrac{4}{x}) = \infty$ and $\lim\limits_{x\to 0^-}(x + \tfrac{4}{x}) = -\infty,$ so $x = 0$ is a VA.   E. $f'(x) = 1 - \tfrac{4}{x^2} > 0$

$\Leftrightarrow x^2 > 4 \Leftrightarrow x > 2$ or $x < -2,$ so f is increasing on $(-\infty, -2]$ and $[2, \infty)$ and

decreasing on $[-2, 0)$ and $(0, 2]$.   F. $f(-2) = -4$ is a local maximum and $f(2) = 4$ is

a local minimum.   G. $f''(x) = \tfrac{8}{x^3} > 0 \Leftrightarrow x > 0$ so f is CU on $(0, \infty)$ and CD on

$(-\infty, 0)$.   No IP.   H.

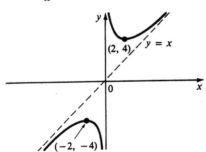

77. $y = \dfrac{1}{x-1} - x.$ A. $D = \{x \mid x \neq 1\}.$ B. $y = 0 \Leftrightarrow x = 1/(x-1) \Leftrightarrow x^2 - x - 1 = 0$

$\Rightarrow x = (1 \pm \sqrt{5})/2$ (x–intercepts), y–intercept $= f(0) = -1.$ C. No symmetry.

D. $y - (-x) = 1/(x-1) \to 0$ as $x \to \pm\infty,$ so $y = -x$ is a slant asymptote.

$\lim\limits_{x \to 1^+}\left[\frac{1}{x-1} - x\right] = \infty$ and $\lim\limits_{x \to 1^-}\left[\frac{1}{x-1} - x\right] = -\infty$, so $x = 1$ is a VA.

E. $f'(x) = -1 - 1/(x-1)^2 < 0$ for all $x \neq 1$, so f is decreasing on $(-\infty, 1)$ and $(1, \infty)$.

F.   No local extrema.

G. $f''(x) = \dfrac{2}{(x-1)^3} > 0 \Leftrightarrow x > 1$

so f is CU on $(1, \infty)$ and CD on $(-\infty, 1)$.

No IP.

H.

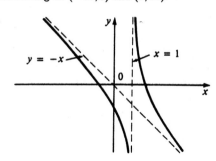

79. $y = f(x) = e^x + x$ A. $D = R$ B. x–intercept between -1 and 0,

y–intercept $= f(0) = 1$ C. No symmetry   D. $\lim\limits_{x \to \infty} (e^x + x) = \infty$,

$\lim\limits_{x \to -\infty} (e^x + x) = -\infty$, so no HA. But $f(x) - x = e^x \to 0$ as $x \to -\infty$, so

$y = x$ is a slant asymptote.

E. $f'(x) = e^x + 1 > 0$ for all x,

so f is increasing on R.

F.   No extrema

G. $f''(x) = e^x > 0$ for all x,

so f is CU on R.

H.

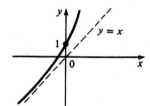

81. $\dfrac{x^2}{a^2} - \dfrac{y^2}{b^2} = 1 \Rightarrow y = \pm\frac{b}{a}\sqrt{x^2 - a^2}$.   Now $\lim\limits_{x \to \infty}\left[\frac{b}{a}\sqrt{x^2 - a^2} - \frac{b}{a}x\right]$

$= \frac{b}{a} \cdot \lim\limits_{x \to \infty}\left[\left(\sqrt{x^2 - a^2} - x\right)\dfrac{\sqrt{x^2 - a^2} + x}{\sqrt{x^2 - a^2} + x}\right] = \frac{b}{a} \cdot \lim\limits_{x \to \infty}\dfrac{-a^2}{\sqrt{x^2 - a^2} + x} = 0$, which shows that

$y = \frac{b}{a}x$ is a slant asymptote.   Similarly $\lim\limits_{x \to \infty}\left[-\frac{b}{a}\sqrt{x^2 - a^2} - \left(-\frac{b}{a}x\right)\right]$

$= -\frac{b}{a} \cdot \lim\limits_{x \to \infty}\dfrac{-a^2}{\sqrt{x^2 - a^2} + x} = 0$, which shows that $y = -\frac{b}{a}x$ is a slant asymptote.

EXERCISES 4.6

1.   If x is one number, the other is $100 - x$.   Maximize $f(x) = x(100 - x) = 100x - x^2$.

$f'(x) = 100 - 2x = 0 \Rightarrow x = 50$.   Now $f''(x) = -2 < 0$, so there is an absolute maximum at x = 50. The numbers are 50 and 50.

3. The two numbers are x and $100/x$ where $x > 0$. Minimize $f(x) = x + 100/x$.
$f'(x) = 1 - 100/x^2 = (x^2 - 100)/x^2$. The critical number is $x = 10$. Since $f'(x) < 0$
for $0 < x < 10$ and $f'(x) > 0$ for $x > 10$, there is an absolute minimum at $x = 10$.
The numbers are 10 and 10.

5. Let p be the perimeter and x and y the sides, so $p = 2x + 2y \Rightarrow y = \frac{1}{2}p - x$. The area
is $A(x) = x(\frac{1}{2}p - x) = \frac{1}{2}px - x^2$. Now $0 = A'(x) = \frac{1}{2}p - 2x \Rightarrow x = p/4$. Since
$A''(x) = -2 < 0$, there is an absolute maximum where $x = p/4$. The sides of the
rectangle are $p/4$ and $p/2 - p/4 = p/4$, so the rectangle is a square.

7. Here $5x + 2y = 750$ so $y = (750 - 5x)/2$. Maximize $A = xy = x(750 - 5x)/2$
$= 375x - (5/2)x^2$. Now $A'(x) = 375 - 5x = 0 \Rightarrow x = 75$. Since $A''(x) = -5 < 0$
there is an absolute maximum when $x = 75$. Then $y = 375/2$. The largest area is
$75(375/2) = 14,062.5$ ft^2.

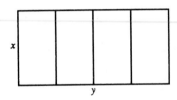

9. Let b be the base of the box and h the height. The surface area is
$1200 = b^2 + 4hb \Rightarrow h = (1200 - b^2)/4b$. The volume is
$V = b^2h = b^2(1200 - b^2)/4b = 300b - b^3/4 \Rightarrow V'(b) = 300 - (3/4)b^2$.
$V'(b) = 0 \Rightarrow b = \sqrt{400} = 20$. Since $V'(b) > 0$ for $0 < b < 20$ and $V'(b) < 0$ for
$b > 20$, there is an absolute maximum when $b = 20$. Then $h = 10$, so the largest
possible volume is $(20)^2(10) = 4000$ cm^3.

11. $10 = (2w)(w)h = 2w^2h$, so $h = 5/w^2$.
The cost is $C(w) = 10(2w^2)$
$+ 6[2(2wh) + 2hw] + 6(2w^2) = 32w^2 + 36wh$
$= 32w^2 + 180/w$. $C'(w) = 64w - 180/w^2$
$= 4(16w^3 - 45)/w^2 \Rightarrow w = \sqrt[3]{45/16}$ is the
critical number. $C'(w) < 0$ for
$0 < w < \sqrt[3]{45/16}$ and $C'(w) > 0$ for $w > \sqrt[3]{45/16}$. The minimum cost is
$C(\sqrt[3]{45/16}) = 32(2.8125)^{2/3} + 180/\sqrt[3]{2.8125} \approx \$191.28$.

13. For (x, y) on the line $y = 2x - 3$, the distance to the origin is $\sqrt{(x - 0)^2 + (2x - 3)^2}$. We

minimize the square of the distance $= x^2 + (2x - 3)^2 = 5x^2 - 12x + 9 = D(x)$.
$D'(x) = 10x - 12 = 0 \Rightarrow x = 6/5$. Since there is a point closest to the origin, $x = 6/5$
and hence $y = -3/5$. So the point is $(6/5, -3/5)$.

15. By symmetry, the points are (x, y) and $(x, -y)$, where $y > 0$. The square of the
distance is $D(x) = (x - 2)^2 + y^2 = (x - 2)^2 + 4 + x^2 = 2x^2 - 4x + 8$. So
$D'(x) = 4x - 4 = 0 \Rightarrow x = 1$ and $y = \pm\sqrt{4 + 1} = \pm\sqrt{5}$. The points are $(1, \pm\sqrt{5})$.

17.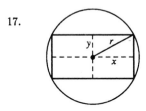

Area of rectangle is $4xy$. Also $r^2 = x^2 + y^2$

so $y = \sqrt{r^2 - x^2}$, so the area is

$A(x) = 4x\sqrt{r^2 - x^2}$. Now

$A'(x) = 4\left[\sqrt{r^2 - x^2} - x^2/\sqrt{r^2 - x^2}\right]$

$= 4(r^2 - 2x^2)/\sqrt{r^2 - x^2}$. The critical number is $x = r/\sqrt{2}$. Clearly this gives a
maximum. The dimensions are $2x = \sqrt{2}r$ and $2y = \sqrt{2}r$.

19.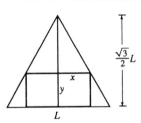

$\frac{(\sqrt{3}/2)L - y}{x} = \frac{(\sqrt{3}/2)L}{L/2} = \sqrt{3}$ (similar

triangles) $\Rightarrow \sqrt{3}x = (\sqrt{3}/2)L - y \Rightarrow$

$y = (\sqrt{3}/2)(L - 2x)$. The area of the

inscribed rectangle is

$A(x) = (2x)y = \sqrt{3}x(L - 2x)$ where

$0 \le x \le L/2$. Now

$0 = A'(x) = \sqrt{3}L - 4\sqrt{3}x \Rightarrow x = \sqrt{3}L/4\sqrt{3} = L/4$. Since $A(0) = A(L/2) = 0$, the
maximum occurs when $x = L/4$, and $y = (\sqrt{3}/2)L - \sqrt{3}L/4 = \sqrt{3}L/4$, so the dimensions
are $L/2$ and $\sqrt{3}L/4$.

21.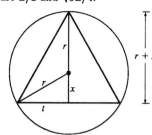

The area of the triangle is $A(x) =$
$\frac{1}{2}(2t)(r + x) = t(r + x) = \sqrt{r^2 - x^2}\,(r + x)$.

Then $0 = A'(x) = r\,\dfrac{-2x}{2\sqrt{r^2 - x^2}} + \sqrt{r^2 - x^2}$

$+ x\,\dfrac{-2x}{2\sqrt{r^2 - x^2}} = -\dfrac{x^2 + rx}{\sqrt{r^2 - x^2}} + \sqrt{r^2 - x^2}$

$\Rightarrow \dfrac{x^2 + rx}{\sqrt{r^2 - x^2}} = \sqrt{r^2 - x^2} \Rightarrow x^2 + rx = r^2 - x^2 \Rightarrow 0 = 2x^2 + rx - r^2$

$= (2x - r)(x + r) \Rightarrow x = r/2$ or $x = -r$. Now $A(r) = 0 = A(-r) \Rightarrow$ the maximum

occurs where $x = r/2$, so the triangle has height $r + r/2 = 3r/2$ and base

$2\sqrt{r^2 - (r/2)^2} = 2\sqrt{3r^2/4} = \sqrt{3}r$.

23.

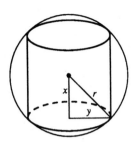

The cylinder has volume $V = \pi y^2(2x)$.
Also $x^2 + y^2 = r^2 \Rightarrow y^2 = r^2 - x^2$, so
$V(x) = \pi(r^2 - x^2)(2x) = 2\pi(r^2 x - x^3)$,
where $0 \le x \le r$. $V'(x) = 2\pi(r^2 - 3x^2) = 0$
$\Rightarrow x = r/\sqrt{3}$. Now $V(0) = V(r) = 0$,
so there is a maximum when $x = r/\sqrt{3}$ and
$V(r/\sqrt{3}) = 4\pi r^3/(3\sqrt{3})$.

25.

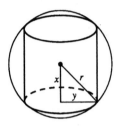

The cylinder has surface area
$2\pi y^2 + 2\pi y(2x)$. Now $x^2 + y^2 = r^2$
$\Rightarrow y = \sqrt{r^2 - x^2}$, so the surface area is
$S(x) = 2\pi(r^2 - x^2) + 4\pi x\sqrt{r^2 - x^2}$, $0 \le x \le r$.
$S'(x) = -4\pi x + 4\pi\sqrt{r^2 - x^2} - 4\pi x^2/\sqrt{r^2 - x^2}$
$= 4\pi(r^2 - 2x^2 - x\sqrt{r^2 - x^2})/\sqrt{r^2 - x^2} = 0 \Rightarrow$

$(\star)\ x\sqrt{r^2 - x^2} = r^2 - 2x^2 \Rightarrow x^2(r^2 - x^2) = r^4 - 4r^2 x^2 + 4x^4 \Rightarrow 5x^4 - 5r^2 x^2 + r^4 = 0$.
By the quadratic formula, $x^2 = [(5 \pm \sqrt{5})/10]\,r^2$, but we reject the root with the $+$ sign
since it doesn't satisfy $(\star)$. So $x = \sqrt{(5 - \sqrt{5})/10}\,r$. Since $S(0) = S(r) = 0$, the
maximum occurs at the critical number and $x^2 = [(5 - \sqrt{5})/10]\,r^2 \Rightarrow y^2 = [(5 + \sqrt{5})/10]\,r^2$
$\Rightarrow$ the surface area is $2\pi[(5 + \sqrt{5})/10]\,r^2 + 4\pi\sqrt{(5 - \sqrt{5})/10}\sqrt{(5 + \sqrt{5})/10}\,r^2 = \pi r^2(1 + \sqrt{5})$.

27.

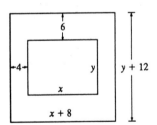

$xy = 384 \Rightarrow y = 384/x$. Total area
is $A(x) = (8 + x)(12 + 384/x)$
$= 12(40 + x + 256/x)$
$A'(x) = 12(1 - 256/x^2) = 0 \Rightarrow x = 16$.
There is an absolute minimum when
$x = 16$ since $A'(x) < 0$ for $0 < x < 16$ and
$A'(x) > 0$ for $x > 16$. When $x = 16$, $y = 384/16 = 24$, so the dimensions are 24 cm
and 36 cm.

29.

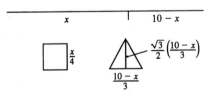

The total area is
$A(x) = x^2/16 + (\sqrt{3}/36)(10 - x)^2$, $0 \le x \le 10$.
$A'(x) = x/8 - (\sqrt{3}/18)(10 - x) = 0$
$\Rightarrow x = 40\sqrt{3}/(9 + 4\sqrt{3})$. Now
$A(0) = (\sqrt{3}/36)\,100 \approx 4.81$,
$A(10) = 100/16 = 6.25$ and

$A(40\sqrt{3}/(9 + 4\sqrt{3})) \approx 2.72$, so (a) the maximum occurs when $x = 10$ m and (b) the
minimum when $x = 40\sqrt{3}/(9 + 4\sqrt{3}) \approx 4.35$ m.

31.

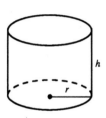

The volume is $V = \pi r^2 h$ and the

surface area is $S(r) = \pi r^2 + 2\pi rh$

$= \pi r^2 + 2\pi r(V/\pi r^2) = \pi r^2 + 2V/r$.

$S'(r) = 2\pi r - 2V/r^2 = 0 \Rightarrow 2\pi r^3 = 2V$

$\Rightarrow r = \sqrt[3]{V/\pi}$. This gives an absolute

minimum since $S'(r) < 0$ for $0 < r < \sqrt[3]{V/\pi}$

and $S'(r) > 0$ for $r > \sqrt[3]{V/\pi}$. When $r = \sqrt[3]{V/\pi}$, $h = V/(\pi r^2) = V/[\pi(V/\pi)^{2/3}] = \sqrt[3]{V/\pi}$.

33.

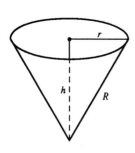

$h^2 + r^2 = R^2 \Rightarrow V = (\pi/3)r^2 h$

$= (\pi/3)(R^2 - h^2)h = (\pi/3)(R^2 h - h^3)$.

$V'(h) = (\pi/3)(R^2 - 3h^2) = 0$ when

$h = R^2/\sqrt{3}$. This gives an absolute

maximum since $V'(h) > 0$ for $0 < h < R/\sqrt{3}$

and $V'(h) < 0$ for $h > R/\sqrt{3}$. Maximum

volume is $V(R/\sqrt{3}) = 2\pi R^3/(9\sqrt{3})$.

35. $F(\theta) = \dfrac{\mu W}{\mu \sin\theta + \cos\theta} \Rightarrow F'(\theta) = \dfrac{-\mu W(\mu\cos\theta - \sin\theta)}{(\mu\sin\theta + \cos\theta)^2} = 0 \Leftrightarrow \mu\cos\theta - \sin\theta = 0$

$\Leftrightarrow \mu = \tan\theta$. For $\tan\theta < \mu$, $F'(\theta) < 0$, and for $\tan\theta > \mu$, $F'(\theta) > 0$, so

$\tan\theta = \mu$ minimizes F.

37. Here $T(x) = \dfrac{\sqrt{x^2 + 25}}{6} + \dfrac{5 - x}{8}$, $0 \le x \le 5$, $\Rightarrow T'(x) = \dfrac{x}{6\sqrt{x^2 + 25}} - \dfrac{1}{8} = 0$

$\Leftrightarrow 8x = 6\sqrt{x^2 + 25} \Leftrightarrow 16x^2 = 9(x^2 + 25) \Leftrightarrow x = \dfrac{15}{\sqrt{7}}$. But $\dfrac{15}{\sqrt{7}} > 5$, so T has no critical

number. Since $T(0) \approx 1.46$ and $T(5) \approx 1.18$, he should row directly to B.

39.

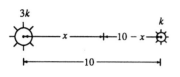

The total illumination is

$I(x) = \dfrac{3k}{x^2} + \dfrac{k}{(10 - x)^2}$, $0 < x < 10$. Then

$I'(x) = \dfrac{-6k}{x^3} + \dfrac{2k}{(10 - x)^3} = 0 \Rightarrow$

$6k(10 - x)^3 = 2kx^3 \Rightarrow \sqrt[3]{3}(10 - x) = x$

$\Rightarrow x = 10\sqrt[3]{3}/(1 + \sqrt[3]{3}) \approx 5.9$ ft.

This gives a minimum since there is clearly no maximum.

41.

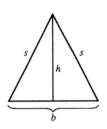

Here $s^2 = h^2 + b^2/4$ so $h^2 = s^2 - b^2/4$.

The area is $A = \frac{1}{2}b\sqrt{s^2 - b^2/4}$. Let the

perimeter be p so $2s + b = p$ or $s = (p-b)/2$

$\Rightarrow A(b) = \frac{1}{2}b\sqrt{(p-b)^2/4 - b^2/4}$

$= b\sqrt{p^2 - 2pb}/4$. Now $A'(b) = \frac{1}{4}\sqrt{p^2 - 2pb}$

$- \frac{1}{4}bp/\sqrt{p^2 - 2pb} = (-3pb + p^2)/(4\sqrt{p^2 - 2pb})$.

Now $A'(b) = 0 \Rightarrow -3pb + p^2 = 0 \Rightarrow b = p/3$. Since $A'(b) > 0$ for $b < p/3$ and

$A'(b) < 0$ for $b > p/3$, there is an absolute maximum when $b = p/3$. But then

$2s + p/3 = p$ so $s = p/3 \Rightarrow s = b \Rightarrow$ the triangle is equilateral.

43.

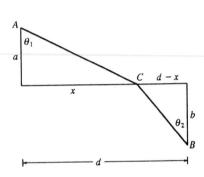

The total time $T(x) = $ (time from A

to C) + (time from C to B)

$= \sqrt{a^2 + x^2}/v_1 + \sqrt{b^2 + (d-x)^2}/v_2$

$0 < x < d$.

$T'(x) = \dfrac{x}{v_1\sqrt{a^2 + x^2}} - \dfrac{d-x}{v_2\sqrt{b^2 + (d-x)^2}}$

$= \dfrac{\sin\theta_1}{v_1} - \dfrac{\sin\theta_2}{v_2}$

The minimum occurs when $T'(x) = 0$

$\Rightarrow \dfrac{\sin\theta_1}{v_1} = \dfrac{\sin\theta_2}{v_2}$.

45.

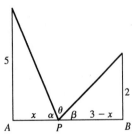

$x = 5\cot\alpha$, $3 - x = 2\cot\beta \Rightarrow$

$\theta = \pi - \cot^{-1}(x/5) - \cot^{-1}((3-x)/2)$

$\dfrac{d\theta}{dx} = \dfrac{1}{1 + (x/5)^2}\left[\frac{1}{5}\right] + \dfrac{1}{1 + ((3-x)/2)^2}\left[-\frac{1}{2}\right] = 0$

$\Rightarrow 5(1 + x^2/25) = 2[1 + (9 - 6x + x^2)/4] \Rightarrow$

$50 + 2x^2 = 65 - 30x + 5x^2 \Rightarrow x^2 - 10x + 5 = 0$

$\Rightarrow x = 5 \pm 2\sqrt{5}$. We reject the root with the $+$ sign, since it is > 3.

$d\theta/dx > 0$ for $x < 5 - 2\sqrt{5}$ and $d\theta/dx < 0$ for $x > 5 - 2\sqrt{5}$, so θ is

maximized when $|AP| = x = 5 - 2\sqrt{5}$.

47.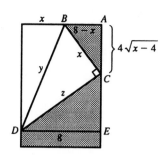

$y^2 = x^2 + z^2$, but triangles CDE and BCA
are similar, so $z/8 = x/(4\sqrt{x-4})$.
Thus we minimize $f(x) = y^2$
$$= x^2 + 4x^2/(x-4) = x^3/(x-4), \ 4 < x \le 8.$$
$$f'(x) = \frac{3x^2(x-4) - x^3}{(x-4)^2} = \frac{2x^2(x-6)}{(x-4)^2} = 0$$
when $x = 6$. $f'(x) < 0$ when $x < 6$,

$f'(x) > 0$ when $x > 6$, so the minimum occurs when $x = 6$ in.

49.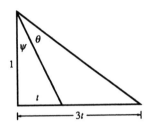

It suffices to maximize $\tan\theta$. Now

$$\tan(\psi + \theta) = \tfrac{3t}{1} = \frac{\tan\psi + \tan\theta}{1 - \tan\psi\tan\theta} \ (Equation$$

14(a), Appendix B) $= \dfrac{t + \tan\theta}{1 - t\tan\theta}$. So

$$3t(1 - t\tan\theta) = t + \tan\theta$$

$$\Rightarrow 2t = (1 + 3t^2)\tan\theta \Rightarrow \tan\theta = \frac{2t}{1+3t^2}.$$

Let $f(t) = \tan\theta = \dfrac{2t}{1+3t^2} \Rightarrow f'(t) = \dfrac{2(1+3t^2) - 2t(6t)}{(1+3t^2)^2}$

$= \dfrac{2(1-3t^2)}{(1+3t^2)^2} = 0 \Leftrightarrow 1 - 3t^2 = 0 \Leftrightarrow t = \tfrac{1}{\sqrt{3}}$ since $t \ge 0$. Since $f'(t) > 0$ for $0 \le t < \tfrac{1}{\sqrt{3}}$

and $f'(t) < 0$ for $t > \tfrac{1}{\sqrt{3}}$, there is an absolute maximum when $t = \tfrac{1}{\sqrt{3}}$ where

$\tan\theta = \dfrac{2(1/\sqrt{3})}{1 + 3(1/\sqrt{3})^2} = \tfrac{1}{\sqrt{3}} \Rightarrow \theta = \tfrac{\pi}{6}$

51. $a = W\sin\theta$, $c = W\cos\theta$, $b = L\cos\theta$, $d = L\sin\theta$, so the area of the circumscribed
rectangle is $A(\theta) = (a+b)(c+d) = (W\sin\theta + L\cos\theta)(W\cos\theta + L\sin\theta)$
$= LW\sin^2\theta + LW\cos^2\theta + (L^2 + W^2)\sin\theta\cos\theta = LW + (1/2)(L^2 + W^2)\sin 2\theta$,
$0 \le \theta \le \pi/2$. This expression shows without calculus, that the maximum value of $A(\theta)$
occurs when $\sin 2\theta = 1 \Leftrightarrow 2\theta = \pi/2 \Leftrightarrow x = \pi/4$. So the maximum area is
$A(\pi/4) = LW + (1/2)(L^2 + W^2) = (L + W)^2/2$.

EXERCISES 4.7

1. (a) $C(x) = 10000 + 25x + x^2$, $C(1000) = \$1035000$. $c(x) = \dfrac{C(x)}{x} = \dfrac{10000}{x} + 25 + x$,
   $c(1000) = \$1035$. $C'(x) = 25 + 2x$, $C'(1000) = \$2025/\text{unit}$.

   (b) We must have $c(x) = C'(x) \Rightarrow 10000/x + 25 + x = 25 + 2x \Rightarrow 10000/x = x$
   $\Rightarrow x^2 = 10000 \Rightarrow x = 100$. This is a minimum since $c''(x) = 20000/x^3 > 0$.

(c) The minimum average cost is $c(100) = \$225$.

3. (a) $C(x) = 45 + \frac{x}{2} + \frac{x^2}{560}$, $C(1000) = \$2330.71$. $c(x) = \frac{45}{x} + \frac{1}{2} + \frac{x}{560}$, $c(1000) = \$2.33$.
    $C'(x) = \frac{1}{2} + \frac{x}{280}$, $C'(1000) = \$4.07/\text{unit}$

    (b) We must have $C'(x) = c(x) \Rightarrow \frac{1}{2} + \frac{x}{280} = \frac{45}{x} + \frac{1}{2} + \frac{x}{560} \Rightarrow \frac{45}{x} = \frac{x}{560}$
    $\Rightarrow x^2 = (45)(560) \Rightarrow x = \sqrt{25200} \approx 159$. This is a minimum since
    $c''(x) = 90/x^2 > 0$.

    (c) The minimum average cost is $c(159) = \$1.07$.

5. (a) $C(x) = 2\sqrt{x} + \frac{x^2}{8000}$, $C(1000) = \$188.25$. $c(x) = \frac{2}{\sqrt{x}} + \frac{x}{8000}$, $c(1000) = \$.19$.
    $C'(x) = \frac{1}{\sqrt{x}} + \frac{x}{4000}$, $C'(1000) = \$.28/\text{unit}$.

    (b) We must have $C'(x) = c(x) \Rightarrow \frac{1}{\sqrt{x}} + \frac{x}{4000} = \frac{2}{\sqrt{x}} + \frac{x}{8000} \Rightarrow \frac{x}{8000} = 1/\sqrt{x}$
    $\Rightarrow x^{3/2} = 8000 \Rightarrow x = (8000)^{2/3} = 400$. This is a minimum since
    $c''(x) = (3/2)x^{-5/2} > 0$.

    (c) The minimum average cost is $c(400) = \$.15$.

7. $C(x) = 680 + 4x + 0.01x^2$, $p(x) = 12 \Rightarrow R(x) = xp(x) = 12x$. If the profit is
    maximum, then $R'(x) = C'(x) \Rightarrow 12 = 4 + 0.02x \Rightarrow 0.02x = 8 \Rightarrow x = 400$. Now
    $R''(x) = 0 < .02 = C''(x)$, so $x = 400$ gives a maximum.

9. $C(x) = 1200 + 25x - .0001x^2$, $p(x) = 55 - x/1000$. Then $R(x) = xp(x)$
    $= 55x - x^2/1000$. If the profit is maximum, then $R'(x) = C'(x)$ or
    $55 - x/500 = 25 - .0002x \Rightarrow 30 = 0.0018x \Rightarrow x = 30/0.0018 \approx 16667$. Now
    $R''(x) = -\frac{1}{500} < -.0002 = C''(x)$, so $x = 16667$ gives a maximum.

11. $C(x) = 1450 + 36x - x^2 + 0.001x^3$, $p(x) = 60 - 0.01x$. Then
    $R(x) = xp(x) = 60x - .01x^2$. If the profit is maximum, then $R'(x) = C'(x)$ or

    $60 - .02x = 36 - 2x + .003x^2 \Rightarrow .003x^2 - 1.98x - 24 = 0$. By the quadratic formula,
    $x = \frac{1.98 \pm \sqrt{(-1.98)^2 + 4(.003)(24)}}{2(.003)} = \frac{1.98 \pm \sqrt{4.2084}}{.006}$. Since $x > 0$,
    $x \approx (1.98 + 2.05)/.006 \approx 672$. Now $R''(x) = -.02$ and
    $C''(x) = -2 + .006x \Rightarrow C''(672) = 2.032 \Rightarrow R''(672) < C''(672) \Rightarrow$ there is a maximum
    at $x = 672$.

13. $C(x) = .001x^3 - 0.3x^2 + 6x + 900$. The marginal cost is $C'(x) = .003x^2 - .6x + 6$.
    $C'(x)$ is increasing when $C''(x) > 0$ or $.006x - .6 > 0$ or $x > .6/.006 = 100$. So $C'(x)$
    starts to increase when $x = 100$.

15. (a) We are given that the demand function p is linear and $p(27000) = 10$,

172

$p(33000) = 8$, so the slope is $\dfrac{10-8}{27000-33000} = -\dfrac{1}{3000}$ and the equation of the graph is $y - 10 = (-1/3000)(x - 27000) \Rightarrow p(x) = 19 - x/3000$.

(b) The revenue is $R(x) = xp(x) = 19x - x^2/3000 \Rightarrow R'(x) = 19 - x/1500 = 0$ when $x = 28500$. Since $R''(x) = -1/1500 < 0$, the maximum revenue occurs when $x = 28500 \Rightarrow$ the price is $p(28500) = \$9.50$.

17. (a) $p(x) = 450 - (1/10)(x - 1000) = 550 - x/10$.

(b) $R(x) = xp(x) = 500x - x^2/10$. $R'(x) = 550 - x/5 = 0$ when $x = 5(550) = 2750$. $p(2750) = 275$, so the rebate should be $450 - 275 = \$175$.

(c) $P(x) = R(x) - C(x) = 550x - x^2/10 - 6800 - 150x = 400x - x^2/10 - 6800$. $P'(x) = 400 - x/5 = 0$ when $x = 2000$. $p(2000) = 550 - 200 = 350$. Therefore the rebate to maximize profits should be $450 - 350 = \$100$.

EXERCISES 4.8

1. $f(x) = 12x^2 + 6x - 5 \Rightarrow F(x) = 12(x^3/3) + 6(x^2/2) - 5x + C = 4x^3 + 3x^2 - 5x + C$.

3. $f(x) = 6x^9 - 4x^7 + 3x^2 + 1 \Rightarrow F(x) = 6(x^{10}/10) - 4(x^8/8) + 3(x^3/3) + x + C$
$= (3/5)x^{10} - (1/2)x^8 + x^3 + x + C$.

5. $f(x) = \sqrt{x} + \sqrt[3]{x} = x^{1/2} + x^{1/3} \Rightarrow F(x) = x^{3/2}/(3/2) + x^{4/3}/(4/3) + C$
$= (2/3)x^{3/2} + (3/4)x^{4/3} + C$.

7. $f(x) = 6/x^5 = 6x^{-5} \Rightarrow F(x) = 6x^{-4}/(-4) + C_1 = -3/(2x^4) + C_1$, if $x > 0$;
$F(x) = -3/(2x^4) + C_2$, if $x < 0$.

9. $f(x) = \sqrt{x} + 1/\sqrt{x} = x^{1/2} + x^{-1/2} \Rightarrow F(x) = x^{3/2}/(3/2) + x^{1/2}/(1/2) + C$
$= (2/3)x^{3/2} + 2x^{1/2} + C$.

11. $g(t) = (t^3 + 2t^2)/\sqrt{t} = t^{5/2} + 2t^{3/2} \Rightarrow G(t) = t^{7/2}/(7/2) + 2t^{5/2}/(5/2) + C$
$= (2/7)t^{7/2} + (4/5)t^{5/2} + C$.

13. $h(x) = \sin x - 2\cos x \Rightarrow H(x) = -\cos x - 2\sin x + C$.

15. $f(t) = \sec^2 t + t^2 \Rightarrow F(t) = \tan t + t^3/3 + C_n$ on the interval $((2n - 1)\pi/2, (2n + 1)\pi/2)$.

17. $f(x) = 2x + 5/\sqrt{1 - x^2} \Rightarrow F(x) = x^2 + 5\sin^{-1} x + C$.

19. $f'(x) = x^4 - 2x^2 + x - 1 \Rightarrow f(x) = x^5/5 - 2x^3/3 + x^2/2 - x + C$.

21. $f''(x) = x^2 + x^3 \Rightarrow f'(x) = x^3/3 + x^4/4 + C \Rightarrow f(x) = x^4/12 + x^5/20 + Cx + D$.

23. $f''(x) = 1 \Rightarrow f'(x) = x + C \Rightarrow f(x) = x^2/2 + Cx + D.$

25. $f'''(x) = 24x \Rightarrow f''(x) = 12x^2 + C \Rightarrow f'(x) = 4x^3 + Cx + D$

    $\Rightarrow f(x) = x^4 + Cx^2/2 + Dx + E.$

27. $f'(x) = 4x + 3 \Rightarrow f(x) = 2x^2 + 3x + C \Rightarrow -9 = f(0) = C \Rightarrow f(x) = 2x^2 + 3x - 9.$

29. $f'(x) = 3\sqrt{x} - 1/\sqrt{x} = 3x^{1/2} - x^{-1/2} \Rightarrow f(x) = 3x^{3/2}/(3/2) - x^{1/2}/(1/2) + C$

    $\Rightarrow 2 = f(1) = 2 - 2 + C = C \Rightarrow f(x) = 2x^{3/2} - 2x^{1/2} + 2.$

31. $f'(x) = 3\cos x + 5\sin x \Rightarrow f(x) = 3\sin x - 5\cos x + C \Rightarrow 4 = f(0) = -5 + C \Rightarrow C = 9$

    $\Rightarrow f(x) = 3\sin x - 5\cos x + 9.$

33. $f'(x) = 2 + x^{3/5} \Rightarrow f(x) = 2x + (5/8)x^{8/5} + C \Rightarrow 3 = f(1) = 2 + (5/8) + C \Rightarrow C = 3/8$

    $\Rightarrow f(x) = 2x + (5/8)x^{8/5} + 3/8.$

35. $f'(x) = \frac{2}{x}, \; x < 0 \Rightarrow f(x) = 2\ln|x| + C \Rightarrow 7 = f(-1) = 2\ln 1 + C = C \Rightarrow C = 7$

    $\Rightarrow f(x) = 2\ln|x| + 7.$

37. $f''(x) = -8 \Rightarrow f'(x) = -8x + C \Rightarrow 5 = f'(0) = C \Rightarrow f'(x) = -8x + 5$

    $\Rightarrow f(x) = -4x^2 + 5x + D \Rightarrow 6 = f(0) = D \Rightarrow f(x) = -4x^2 + 5x + 6.$

39. $f''(x) = 20x^3 - 10 \Rightarrow f'(x) = 5x^4 - 10x + C \Rightarrow -5 = f'(1) = 5 - 10 + C \Rightarrow C = 0$

    $\Rightarrow f'(x) = 5x^4 - 10x \Rightarrow f(x) = x^5 - 5x^2 + D \Rightarrow 1 = f(1) = 1 - 5 + D$

    $\Rightarrow D = 5 \Rightarrow f(x) = x^5 - 5x^2 + 5.$

41. $f''(x) = x^2 + 3\cos x \Rightarrow f'(x) = x^3/3 + 3\sin x + C \Rightarrow 3 = f'(0) = C$

    $\Rightarrow f'(x) = x^3/3 + 3\sin x + 3 \Rightarrow f(x) = x^4/12 - 3\cos x + 3x + D \Rightarrow 2 = f(0) = -3 + D$

    $\Rightarrow D = 5 \Rightarrow f(x) = x^4/12 - 3\cos x + 3x + 5.$

43. $f''(x) = 6x + 6 \Rightarrow f'(x) = 3x^2 + 6x + C \Rightarrow f(x) = x^3 + 3x^2 + Cx + D \Rightarrow 4 = f(0) = D$

    and $3 = f(1) = 1 + 3 + C + D = 4 + C + 4 \Rightarrow C = -5 \Rightarrow f(x) = x^3 + 3x^2 - 5x + 4.$

45. $f''(x) = x^{-3} \Rightarrow f'(x) = -(1/2)x^{-2} + C \Rightarrow f(x) = (1/2)x^{-1} + Cx + D$

    $\Rightarrow 0 = f(1) = (1/2) + C + D$ and $0 = f(2) = (1/4) + 2C + D.$ Solving these equations,

    we get $C = 1/4, \; D = -3/4,$ so $f(x) = 1/(2x) + x/4 - 3/4.$

47. $f''(x) = x^{-2}, \; x > 0, \; \Rightarrow f'(x) = -1/x + C \Rightarrow f(x) = -\ln x + Cx + D.$

    $0 = f(1) = C + D$ and $0 = f(2) = -\ln 2 + 2C + D = -\ln 2 + 2C - C$

    $= -\ln 2 + C \Rightarrow C = \ln 2$ and $D = -\ln 2.$ So $f(x) = -\ln x + (\ln 2)x - \ln 2.$

49. We have that $f'(x) = 2x + 1 \Rightarrow f(x) = x^2 + x + C.$ But f passes through $(1,6)$ so that

    $6 = f(1) = 1^2 + 1 + C \Rightarrow C = 4.$

    Therefore $f(x) = x^2 + x + 4 \Rightarrow f(2) = 2^2 + 2 + 4 = 10.$

51. $v(t) = s'(t) = 3 - 2t \Rightarrow s(t) = 3t - t^2 + C \Rightarrow 4 = s(0) = C \Rightarrow s(t) = 3t - t^2 + 4.$

53. $a(t) = v'(t) = 3t + 8 \Rightarrow v(t) = (3/2) t^2 + 8t + C \Rightarrow -2 = v(0) = C$

    $\Rightarrow v(t) = (3/2) t^2 + 8t - 2 \Rightarrow s(t) = (1/2) t^3 + 4t^2 - 2t + D \Rightarrow 1 = s(0) = D$

    $\Rightarrow s(t) = (1/2) t^3 + 4t^2 - 2t + 1.$

55. $a(t) = v'(t) = t^2 - t \Rightarrow v(t) = (1/3) t^3 - (1/2) t^2 + C$

    $\Rightarrow s(t) = (1/12) t^4 - (1/6) t^3 + Ct + D \Rightarrow 0 = s(0) = D$ and

    $12 = s(6) = 108 - 36 + 6C + 0 \Rightarrow C = -10 \Rightarrow s(t) = (1/12) t^4 - (1/6) t^3 - 10t.$

57. (a) $v'(t) = a(t) = -9.8 \Rightarrow v(t) = -9.8t + C$, but $C = v(0) = 0$, so

    $v(t) = -9.8t \Rightarrow s(t) = -4.9t^2 + D \Rightarrow D = s(0) = 450 \Rightarrow s(t) = 450 - 4.9t^2$

    (b) It reaches the ground when

    $0 = s(t) = 450 - 4.9t^2 \Rightarrow t^2 = 450/4.9 \Rightarrow t = \sqrt{450/4.9} \approx 9.58$ s.

    (c) $v = -9.8\sqrt{450/4.9} \approx -93.9$ m/s.

59. (a) $v'(t) = -9.8 \Rightarrow v(t) = -9.8t + C \Rightarrow 5 = v(0) = C$, so $v(t) = 5 - 9.8t$

    $\Rightarrow s(t) = 5t - 4.9t^2 + D \Rightarrow D = s(0) = 450 \Rightarrow s(t) = 450 + 5t - 4.9t^2.$

    (b) It reaches the ground when $450 + 5t - 4.9t^2 = 0$. By the quadratic formula, the

    positive root of this equation is $t = (5 + \sqrt{8845})/9.8 \approx 10.1$ s.

    (c) $v = 5 - 9.8(5 + \sqrt{8845})/9.8 \approx -94.0$ m/s.

61. By Exercise 60, $s(t) = -4.9t^2 + v_0 t + s_0$ and $v(t) = s'(t) = -9.8t + v_0$.

    So $[v(t)]^2 = (9.8)^2 t^2 - 19.6 v_0 t + v_0^2$ and

    $v_0^2 - 19.6[s(t) - s_0] = v_0^2 - 19.6[-4.9t^2 + v_0 t] = v_0^2 + (9.8)^2 t^2 - 19.6 v_0 t = [v(t)]^2.$

63. Marginal cost $= 1.92 - 0.002x = C'(x) \Rightarrow C(x) = 1.92x - 0.001x^2 + K.$ But

    $C(1) = 1.92 - 0.001 + K = 562 \Rightarrow K = 560.081.$ Therefore

    $C(x) = 1.92x - 0.001x^2 + 560.081 \Rightarrow C(100) = 1.92(100) - 0.001(100)^2 + 560.081$

    $= 742.081$, so the cost of producing 100 items is \$742.08.

65. Taking the upward direction to be positive we have that for $0 \le t \le 10$

    (using the subscript 1 to refer to $0 \le t \le 10$)

    $a_1(t) = -9 + 0.9t = v_1'(t) \Rightarrow v_1(t) = -9t + 0.45t^2 + v_0,$

    but $v_1(0) = v_0 = -10 \Rightarrow v_1(t) = -9t + 0.45t^2 - 10 = s_1'(t)$

    $\Rightarrow s_1(t) = -\frac{9}{2}t^2 + .15t^3 - 10t + s_0.$ But $s_1(0) = 500 = s_0$

    $\Rightarrow s_1(t) = -\frac{9}{2}t^2 + .15t^3 - 10t + 500.$ Now for $t > 10,$

    $a(t) = 0 = v'(t) \Rightarrow v(t) = \text{constant} = v_1(10) = -9(10) + 0.45(10)^2 - 10 = -55$

    $\Rightarrow v(t) = -55 = s'(t) \Rightarrow s(t) = -55t + s_{10}.$ But $s(10) = s_1(10)$

    $\Rightarrow -55(10) + s_{10} = 100 \Rightarrow s_{10} = 650 \Rightarrow s(t) = -55t + 650$

When the raindrop hits the ground we have that

$s(t) = 0 \Rightarrow -55t + 650 = 0 \Rightarrow t = \frac{650}{55} = \frac{130}{11} \approx 11.8$ s.

67. $a(t) = a$ and the initial velocity is 30 mi/h $= 30 \cdot 5280/(60)^2 = 44$ ft/s and final velocity

50 mi/h $= 50 \cdot 5280/(60)^2 = 220/3$ ft/s.

So $v(t) = at + 44 \Rightarrow 220/3 = v(5) = 5a + 44 \Rightarrow a = 88/15 \approx 5.87$ ft/s^2.

69.   The height at time t is $s(t) = -16t^2 + h$, where $h = s(0)$ is the height of the cliff.

$v(t) = -32t = -120$ when $t = 3.75$, so

$0 = s(3.75) = -16(3.75)^2 + h \Rightarrow h = 16(3.75)^2 = 225$ ft.

REVIEW EXERCISES FOR CHAPTER 4

1.   False. For example take $f(x) = x^3$, then $f'(x) = 3x^2$ and $f'(0) = 3(0)^2 = 0$, but

$f(0) = 0$ is not a maximum or minimum; $(0,0)$ is an inflection point.

3.   False. For example, $f(x) = x$ is continuous on $(0,1)$ but attains neither a maximum nor

a minimum value on $(0,1)$

5.   True, by the Test for Monotonic Functions.

7.   False. $f(x) = g(x) + C$ by Corollary (4.17). For example $f(x) = x + 2$,

$g(x) = x + 1 \Rightarrow f'(x) = g'(x) = 1$, but $f(x) \neq g(x)$.

9.   True. Let $x_1 < x_2$ where $x_1, x_2 \in I$. Then $f(x_1) < f(x_2)$ and $g(x_1) < g(x_2)$ (since f and

g are increasing on I), so $(f+g)(x_1) = f(x_1) + g(x_1) < f(x_2) + g(x_2) = (f+g)(x_2)$.

11.   False. Take $f(x) = x$ and $g(x) = x - 1$. Then both f and g are increasing on $[0,1]$. But

$f(x)g(x) = x(x-1)$ is not increasing on $[0,1]$.

13.   True. Let $x_1, x_2 \in I$ and $x_1 < x_2$, then $f(x_1) < f(x_2)$ (since f is increasing)

$\Rightarrow \frac{1}{f(x_1)} > \frac{1}{f(x_2)}$ (since f is positive) $\Rightarrow g(x_1) > g(x_2) \Rightarrow g(x) = \frac{1}{f(x)}$ is decreasing on I.

15. $f(x) = x^3 - 12x + 5$, $-5 \leq x \leq 3$. $f'(x) = 3x^2 - 12 = 0 \Rightarrow x^2 = 4 \Rightarrow x = \pm 2$.

$f''(x) = 6x \Rightarrow f''(-2) = -12 < 0$, so $f(-2) = 21$ is a local maximum, and

$f''(2) = 12 > 0$, so $f(2) = -11$ is a local minimum. Also $f(-5) = -60$ and $f(3) = -4$,

so $f(-2) = 21$ is the absolute maximum and $f(-5) = -60$ is the absolute minimum.

17. $f(x) = \frac{x-2}{x+2}$, $0 \leq x \leq 4$. $f'(x) = \frac{(x+2)-(x-2)}{(x+2)^2} = \frac{4}{(x+2)^2} > 0 \Rightarrow f$ is increasing on

$[0,4]$, so f has no local extrema and $f(0) = -1$ is the absolute minimum and $f(4) = 1/3$

is the absolute maximum.

19. $f(x) = x - \sqrt{2}\sin x,\ 0 \le x \le \pi.\quad f'(x) = 1 - \sqrt{2}\cos x = 0 \Rightarrow \cos x = 1/\sqrt{2} \Rightarrow x = \pi/4.$
$f''(\pi/4) = \sqrt{2}\sin(\pi/4) = 1 > 0,$ so $f(\pi/4) = (\pi/4) - 1$ is a local minimum. Also
$f(0) = 0$ and $f(\pi) = \pi,$ so the absolute minimum is $f(\pi/4) = \frac{\pi}{4} - 1,$ the absolute
maximum is $f(\pi) = \pi.$

21.

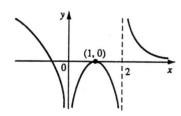

23. $y = f(x) = 1 + x + x^3.$ A. $D = R.$ B. y–intercept $= 1,$ x–intercept requires the
solution of a cubic, so we don't find it. C. No symmetry. D. $\lim\limits_{x \to \infty}(1 + x + x^3) = \infty,$
$\lim\limits_{x \to -\infty}(1 + x + x^3) = -\infty,$ no asymptotes. H.
E. $f'(x) = 1 + 3x^2 \Rightarrow f'(x) > 0,$ so f is
increasing on R. F. No local extrema.
G. $f''(x) = 6x \Rightarrow f''(x) > 0$ if $x > 0$
and $f''(x) < 0$ if $x < 0,$ so f is CU on
$(0, \infty)$ and CD on $(-\infty, 0).$ IP $(0, 1).$

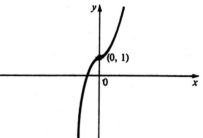

25. $y = f(x) = 1/[x(x-3)]^2.$ A. $D = \{x \mid x \ne 0, 3\} = (-\infty, 0) \cup (0, 3) \cup (3, \infty).$

B. No intercepts. C. No symmetry. D. $\lim\limits_{x \to \pm\infty} \dfrac{1}{x(x-3)^2} = 0,$ so $y = 0$ is a HA.

$\lim\limits_{x \to 0^+} \dfrac{1}{x(x-3)^2} = \infty,\ \lim\limits_{x \to 0^-} \dfrac{1}{x(x-3)^2} = -\infty,\ \lim\limits_{x \to 3} \dfrac{1}{x(x-3)^2} = \infty,$ so $x = 0$ and
$x = 3$ are VA. E. $f'(x) = -\dfrac{(x-3)^2 + 2x(x-3)}{x^2(x-3)^4} = \dfrac{3(1-x)}{x^2(x-3)^3} \Rightarrow f'(x) > 0$

$\Leftrightarrow 1 < x < 3,$ so f is increasing on $[1, 3)$ and decreasing on $(-\infty, 0),\ (0, 1],$ and $(3, \infty).$

F. $f(1) = 1/4$ is a local minimum. H.

G. $f''(x) = \dfrac{6(2x^2 - 4x + 3)}{x^3(x-3)^4}.$ Note that
$2x^2 - 4x + 3 > 0$ for all x since it has
negative discriminant. So $f''(x) > 0$
$\Leftrightarrow x > 0 \Rightarrow$ f is CU on $(0, 3)$ and
$(3, \infty)$ and CD on $(-\infty, 0).$ No IP.

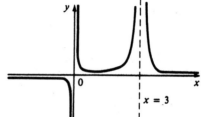

27. $y = f(x) = x\sqrt{5-x}$. A. $D = \{x \mid x \le 5\} = (-\infty, 5]$. B. x–intercepts $0, 5$;

y–intercept $= f(0) = 0$. C. No symmetry. D. $\lim\limits_{x \to -\infty} x\sqrt{5-x} = -\infty$, no asymptote.

E. $f'(x) = \sqrt{5-x} - \dfrac{x}{2\sqrt{5-x}} = \dfrac{10-3x}{2\sqrt{5-x}} > 0 \Leftrightarrow x < \dfrac{10}{3}$. So f is increasing on

$(-\infty, 10/3]$ and decreasing on $[10/3, 5]$.

F. $f(10/3) = 10\sqrt{5}/(3\sqrt{3})$ is a local and absolute maximum.

G. $f''(x) = \dfrac{-6\sqrt{5-x} - (10-3x)(-1/\sqrt{5-x})}{4(5-x)}$ H.

$= \dfrac{3x - 20}{4(5-x)^{3/2}} < 0$ for all x in D,

so f is CD on $(-\infty, 5)$.

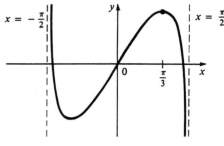

29. $y = f(x) = 4x - \tan x$, $-\pi/2 < x < \pi/2$. A. $D = (-\pi/2, \pi/2)$.

B. y–intercept $= f(0) = 0$. C. $f(-x) = -f(x)$, so the curve is symmetric about $(0, 0)$.

D. $\lim\limits_{x \to \pi/2^-} (4x - \tan x) = -\infty$, $\lim\limits_{x \to -\pi/2^+} (4x - \tan x) = \infty$, so $x = \pi/2$ and $x = -\pi/2$

are VA. E. $f'(x) = 4 - \sec^2 x > 0 \Leftrightarrow \sec x < 2 \Leftrightarrow \cos x > 1/2$

$\Leftrightarrow -\pi/3 < x < \pi/3$, so f is increasing on $[-\pi/3, \pi/3]$ and decreasing on

$(-\pi/2, -\pi/3]$ and $[\pi/3, \pi/2)$.   H.

F. $f(\pi/3) = 4\pi/3 - \sqrt{3}$ is a local

maximum, $f(-\pi/3) = \sqrt{3} - 4\pi/3$ is

a local minimum. G. $f''(x) = -2\sec^2 x \tan x > 0$

$\Leftrightarrow \tan x < 0 \Leftrightarrow -\pi/2 < x < 0$, so f is

CU on $(-\pi/2, 0)$ and CD on $(0, \pi/2)$. IP $(0, 0)$.

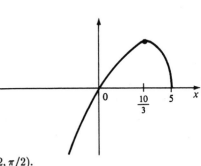

31. $y = f(x) = e^x + e^{-3x}$ A. $D = R$ B. No x–intercept , y–intercept $= f(0) = 2$

C. No symmetry D. $\lim\limits_{x \to \pm\infty} (e^x + e^{-3x}) = \infty$, no asymptote

E. $f'(x) = e^x - 3e^{-3x} = e^{-3x}(e^{4x} - 3) > 0 \Leftrightarrow e^{4x} > 3 \Leftrightarrow 4x > \ln 3 \Leftrightarrow$

$x > \frac{1}{4} \ln 3$, so f is increasing

on $[(\ln 3)/4 , \infty)$ and decreasing on

$(-\infty , (\ln 3)/4]$. F. Absolute

minimum $f((\ln 3)/4) = 3^{1/4} + 3^{-3/4}$

G. $f''(x) = e^x + 9e^{-3x} > 0$, so f is

CU on $(-\infty , \infty)$.

H.

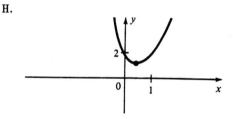

33. $y = f(x) = 2x^2 - \ln x$ A. $D = (0 , \infty)$ B. No intercepts C. No

symmetry D. $\lim\limits_{x\to 0^+} (2x^2 - \ln x) = \infty$, so $x = 0$ is a VA .

$\lim\limits_{x\to\infty} (2x^2 - \ln x) = \lim\limits_{x\to\infty} x^2 \left[2 - \frac{\ln x}{x^2}\right] = \infty$, since $\lim\limits_{x\to\infty} \frac{\ln x}{x^2} \overset{H}{=} \lim\limits_{x\to\infty} \frac{1/x}{2x}$

$= \lim\limits_{x\to\infty} \frac{1}{2x^2} = 0$ E. $f'(x) = 4x - \frac{1}{x} = \frac{4x^2 - 1}{x} > 0 \Leftrightarrow x^2 > 1/4$

$\Leftrightarrow x > 1/2$ (since $x > 0$) , so f is

increasing on $[1/2 , \infty)$ and

decreasing on $(0 , 1/2]$. F. Absolute

minimum $f(1/2) = \ln 2 + 1/2$

G. $f''(x) = 4 + 1/x^2 > 0$, so f is

CU on $(0 , \infty)$.

H.

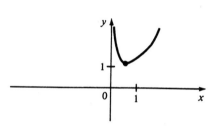

35. $f(x) = x^{101} + x^{51} + x - 1 = 0$. Since f is continuous and $f(0) = -1$ and $f(1) = 2$, the

equation has at least one root in $(0, 1)$ by the Intermediate Value Theorem. Suppose

the equation has two roots, a and b, with $a < b$. Then $f(a) = 0 = f(b)$ so by Rolle's

Theorem $f'(x) = 101x^{100} + 51x^{50} + 1 = 0$ has a root in (a, b). But this is impossible

since $f'(x) \geq 1$ for all x.

37. Since f is continuous on $[32, 33]$ and differentiable on $(32, 33)$, then by the Mean Value

Theorem there exists a number c in $(32, 33)$ such that $f'(c) = \frac{1}{5}c^{-4/5} = \frac{\sqrt[5]{33} - \sqrt[5]{32}}{33 - 32}$

$= \sqrt[5]{33} - 2$, but $\frac{1}{5}c^{-4/5} > 0 \Rightarrow \sqrt[5]{33} - 2 > 0 \Rightarrow \sqrt[5]{33} > 2$. Also f' is decreasing, so that

$f'(c) < f'(32) = \frac{1}{5}(32)^{-4/5} = 0.0125 \Rightarrow 0.0125 > f'(c) = \sqrt[5]{33} - 2 \Rightarrow \sqrt[5]{33} < 2.0125$.

Therefore $2 < \sqrt[5]{33} < 2.0125$.

39.

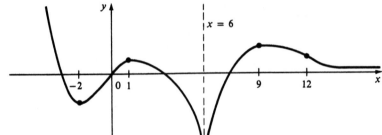

41. For $(1,6)$ to be on the curve, we have that $6 = y(1) = 1^3 + a(1)^2 + b(1) + 1$

$= a + b + 2 \Rightarrow b = 4 - a.$ Now $y' = 3x^2 + 2ax + b$ and $y'' = 6x + 2a.$

Also, for $(1,6)$ to be an inflection point it must be true that $y''(1) = 6(1) + 2a = 0$

$\Rightarrow a = -3 \Rightarrow b = 4 - (-3) = 7.$

43. If $B = 0,$ the line is vertical and the distance from $x = -C/A$ to (x_1, y_1) is

$|x_1 + C/A| = |Ax_1 + By_1 + C|/\sqrt{A^2 + B^2},$ so assume $B \neq 0.$ The square of the distance

from (x_1, y_1) to the line is $f(x) = (x - x_1)^2 + (y - y_1)^2$ where $Ax + By + C = 0,$ so we

minimize $f(x) = (x - x_1)^2 + (-\frac{A}{B}x - \frac{C}{B} - y_1)^2 \Rightarrow f'(x) = 2(x - x_1) + 2(-\frac{A}{B}x - \frac{C}{B} - y_1)(-\frac{A}{B}).$

$f'(x) = 0 \Rightarrow x = (B^2x_1 - ABy_1 - AC)/(A^2 + B^2)$ and this gives a minimum since

$f''(x) = 2(1 + A^2/B^2) > 0.$ Substituting this value of x and simplifying gives

$f(x) = (Ax_1 + By_1 + C)^2/(A^2 + B^2)$ so the minimum distance is

$|Ax_1 + By_1 + C|/\sqrt{A^2 + B^2}.$

45.

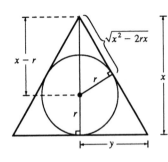

By similar triangles, $\frac{y}{x} = r/\sqrt{x^2 - 2rx},$ so

the area of the triangle is

$A(x) = (1/2)(2y)x = xy = rx^2/\sqrt{x^2 - 2rx} \Rightarrow$

$A'(x) = \dfrac{2rx\sqrt{x^2 - 2rx} - rx^2(x - r)/\sqrt{x^2 - 2rx}}{x^2 - 2rx}$

$= \dfrac{rx^2(x - 3r)}{(x^2 - 2rx)^{3/2}} = 0$ when $x = 3r.$

$A'(x) < 0$ when $2r < x < 3r,$ $A'(x) > 0$

when $x > 3r.$ So $x = 3r$ gives a minimum and $A(3r) = r(9r^2)/(\sqrt{3}r) = 3\sqrt{3}r^2.$

47.

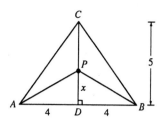

We minimize $L(x) = |PA| + |PB| + |PC|$
$= 2\sqrt{x^2 + 16} + (5 - x)$, $0 \le x \le 5$.
$L'(x) = 2x/\sqrt{x^2 + 16} - 1 = 0 \Leftrightarrow$
$2x = \sqrt{x^2 + 16} \Leftrightarrow 4x^2 = x^2 + 16$
$\Leftrightarrow x = 4/\sqrt{3}$. $L(0) = 13$,
$L(4/\sqrt{3}) \approx 11.9$, $L(5) \approx 12.8$, so the
minimum occurs when $x = 4/\sqrt{3}$.

49.

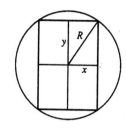

The strength is $S(x) = k(2x)(2y)^2 = 8kxy^2$
$= 8kx(R^2 - x^2) = 8k(R^2x - x^3)$, $0 \le x \le R$.
$S'(x) = 8k(R^2 - 3x^2) = 0 \Rightarrow x = R/\sqrt{3}$.
$S(0) = 0 = S(R)$, so the maximum occurs
when $x = R/\sqrt{3}$. Then the width is
$2x = 2R/\sqrt{3}$ and the depth is
$2y = 2\sqrt{R^2 - R^2/3} = 2\sqrt{2}R/\sqrt{3}$.

51.   Let x = selling price of ticket. Then $12 - x$ is the amount the ticket price has been
lowered, so the number of tickets sold is $11000 + 1000(12 - x) = 23000 - 1000x$. The
revenue is $R(x) = x(23000 - 1000x) = 23000x - 1000x^2$, so $R'(x) = 23000 - 2000x = 0$
when $x = 11.5$. Since $R''(x) = -2000 < 0$, the maximum revenue occurs when the
ticket prices are \$11.50.

53. $f'(x) = x - \sqrt[4]{x} = x - x^{1/4} \Rightarrow f(x) = (1/2)x^2 - (4/5)x^{5/4} + C$.

55. $f'(x) = (1 + x)/\sqrt{x} = x^{-1/2} + x^{1/2} \Rightarrow f(x) = 2x^{1/2} + (2/3)x^{3/2} + C$
$\Rightarrow 0 = f(1) = 2 + 2/3 + C \Rightarrow C = -8/3 \Rightarrow f(x) = 2x^{1/2} + (2/3)x^{3/2} - 8/3$.

57. $f'(x) = \dfrac{2}{1 + x^2} \Rightarrow f(x) = 2\tan^{-1}x + C \Rightarrow -1 = f(0) = 2\tan^{-1}0 + C = C \Rightarrow C = -1$
$\Rightarrow f(x) = 2\tan^{-1}x - 1$.

59. $f''(x) = x^3 + x \Rightarrow f'(x) = x^4/4 + x^2/2 + C \Rightarrow 1 = f'(0) = C \Rightarrow f'(x) = x^4/4 + x^2/2 + 1$
$\Rightarrow f(x) = x^5/20 + x^3/6 + x + D \Rightarrow -1 = f(0) = D \Rightarrow f(x) = x^5/20 + x^3/6 + x - 1$.

61.   Choosing the positive direction to be upward, we have

$a(t) = -9.8 \Rightarrow v(t) = -9.8t + v_0$, but $v(0) = 0 = v_0 \Rightarrow v(t) = -9.8t = s'(t)$

$\Rightarrow s(t) = -4.9t^2 + s_0$, but $s(0) = s_0 = 500 \Rightarrow s(t) = -4.9t^2 + 500$. When $s = 0$,

$-4.9t^2 + 500 = 0 \Rightarrow t = \sqrt{\dfrac{500}{4.9}} \Rightarrow v = -9.8\sqrt{\dfrac{500}{4.9}} \approx -98.995$ m/s. Therefore the

canister will not burst.

PROBLEMS PLUS (page 312)

1. Let $y = f(x) = e^{-x^2}$. The area of the rectangle under the curve from $-x$ to x is $A(x) = 2xe^{-x^2}$ where $x \geq 0$. We maximize $A(x)$.
$A'(x) = 2e^{-x^2} - 4x^2e^{-x^2} = 2e^{-x^2}(1 - 2x^2) = 0 \Rightarrow x = 1/\sqrt{2}$. This gives a maximum since $A'(x) > 0$ for $0 \leq x < 1/\sqrt{2}$ and $A'(x) < 0$ for $x > 1/\sqrt{2}$. We next determine the points of inflection of $f(x)$. Now $f'(x) = -2xe^{-x^2} = -A(x)$. So $f''(x) = -A'(x)$. So $f''(x) < 0$ for $-1/\sqrt{2} < x < 1/\sqrt{2}$ and $f''(x) > 0$ for $x < -1/\sqrt{2}$ and $x > 1/\sqrt{2}$. So $f(x)$ changes concavity at $x = \pm 1/\sqrt{2}$. So the two vertices of the rectangle of largest area are at the inflection points.

3. First notice that if we can prove the simpler inequality
$$\frac{x^2 + 1}{x} \geq 2 \qquad (x > 0)$$
then the given inequality follows because
$$\frac{(x^2 + 1)(y^2 + 1)(z^2 + 1)}{xyz} = \left(\frac{x^2 + 1}{x}\right)\left(\frac{y^2 + 1}{y}\right)\left(\frac{z^2 + 1}{z}\right) \geq 2 \cdot 2 \cdot 2 = 8$$
So we let $f(x) = (x^2 + 1)/x = x + \frac{1}{x}$, $x > 0$. Then $f'(x) = 1 - 1/x^2 = 0$ if $x = 1$, and $f'(x) < 0$ for $0 < x < 1$, $f'(x) > 0$ for $x > 1$. Thus the absolute minimum value of $f(x)$ for $x > 0$ is $f(1) = 2$. Therefore $\frac{x^2 + 1}{x} \geq 2$ for all positive x.

[Or, without calculus, $\frac{x^2 + 1}{x} \geq 2 \Leftrightarrow x^2 + 1 \geq 2x \Leftrightarrow x^2 - 2x + 1 \geq 0 \Leftrightarrow (x - 1)^2 \geq 0$, which is true.]

5. Differentiating $x^2 + xy + y^2 = 12$ implicitly with respect to x gives
$2x + y + x\frac{dy}{dx} + 2y\frac{dy}{dx} = 0$, so $\frac{dy}{dx} = -\frac{2x + y}{x + 2y}$. At a highest or lowest point, $\frac{dy}{dx} = 0$, so $y = -2x$. Substituting in the equation gives $x^2 + x(-2x) + (-2x)^2 = 12$, so $3x^2 = 12$ and $x = \pm 2$. If $x = 2$, then $y = -2x = -4$, and if $x = -2$ then $y = 4$. Thus the highest and lowest points are $(-2, 4)$ and $(2, -4)$.

7. Such a function cannot exist. $f'(x) > 3$ for all x means that f is differentiable (and hence continuous) for all x. So by Part 2 of the Fundamental Theorem,
$$\int_1^4 f'(x)\,dx = f(4) - f(1) = 7 - (-1) = 8. \text{ However, if } f'(x) > 3 \text{ for all } x,$$
then $\int_1^4 f'(x)\,dx \geq 3 \cdot (4 - 1) = 9$ by Property 4.14.8.

OR: By the Mean Value Theorem there exists a number $c \in (1, 4)$ such that
$$f'(c) = \frac{f(4) - f(1)}{4 - 1} = \frac{7 - (-1)}{3} = \frac{8}{3} \Rightarrow 8 = 3f'(c). \text{ But } f'(x) > 3 \Rightarrow 3f'(c) > 9,$$

so such a function cannot exist.

9. $f(x) = \dfrac{1}{1+|x|} + \dfrac{1}{1+|x-2|} = \begin{cases} \dfrac{1}{1-x} + \dfrac{1}{1-(x-2)} & \text{if } x < 0 \\[2mm] \dfrac{1}{1+x} + \dfrac{1}{1-(x-2)} & \text{if } 0 \le x < 2 \\[2mm] \dfrac{1}{1+x} + \dfrac{1}{1+(x-2)} & \text{if } x \ge 2 \end{cases}$

so $f'(x) = \begin{cases} \dfrac{1}{(1-x)^2} + \dfrac{1}{(3-x)^2} & \text{if } x < 0 \\[2mm] \dfrac{-1}{(1+x)^2} + \dfrac{1}{(3-x)^2} & \text{if } 0 < x < 2 \\[2mm] \dfrac{-1}{(1+x)^2} - \dfrac{1}{(x-1)^2} & \text{if } x > 2 \end{cases}$

Clearly $f'(x) > 0$ for $x < 0$ and $f'(x) < 0$ for $x > 2$. For $0 < x < 2$, we have,

$f'(x) = \dfrac{1}{(3-x)^2} - \dfrac{1}{(x+1)^2} = \dfrac{(x^2+2x+1) - (x^2-6x+9)}{(3-x)^2(x+1)^2} = \dfrac{8(x-1)}{(3-x)^2(x+1)^2}$, so

$f'(x) < 0$ for $x < 1$, $f'(1) = 0$ and $f'(x) > 0$ for $x > 1$. We have shown that $f'(x) > 0$ for

$x < 0$, $f'(x) < 0$ for $0 < x < 1$, $f'(x) > 0$ for $1 < x < 2$ and $f'(x) < 0$ for $x > 2$. Therefore by

the First Derivative Test, the local maxima of f are at $x = 0$ and $x = 2$, where f takes the value

of $\frac{4}{3}$. Therefore $\frac{4}{3}$ is the absolute maximum value of f.

11. We must find a value x_0 such that the normal lines to the parabola $y = x^2$ at $x = \pm x_0$

intersect at a point one unit from the points $(\pm x_0, x_0^2)$. The normal to $y = x^2$ at $x = \pm x_0$

has slope $\dfrac{-1}{\pm 2x_0}$ and passes through $(\pm x_0, x_0^2)$, so the normals have the equations

$y - x_0^2 = \dfrac{-1}{2x_0}(x - x_0)$ and $y - x_0^2 = \dfrac{1}{2x_0}(x + x_0)$. The common y-intercept is $x_0^2 + \frac{1}{2}$. We

want to find the value of x_0 for which the distance from $(0, x_0^2 + \frac{1}{2})$ to (x_0, x_0^2) equals 1. The

square of the distance is $(x_0 - 0)^2 + [x_0^2 - (x_0^2 + \frac{1}{2})]^2 = x_0^2 + \frac{1}{4} = 1 \Leftrightarrow x_0 = \pm\dfrac{\sqrt{3}}{2}$.

For these values of x_0, the y-intercept is $x_0^2 + \frac{1}{2} = \frac{5}{4}$, so the center of the circle is at $(0, \frac{5}{4})$.

ALTERNATE SOLUTION: Let the center of the circle be $(0, a)$. Then the equation of the

circle is $x^2 + (y-a)^2 = 1$. Solving with the equation of the parabola, $y = x^2$, we get

$x^2 + (x^2 - a)^2 = 1 \Leftrightarrow x^2 + x^4 - 2ax^2 + a^2 = 1 \Leftrightarrow x^4 + (1-2a)x^2 + a^2 - 1 = 0$. The

parabola and the circle will be tangent to each other when this quadratic equation in x^2 has

equal roots, that is, the discriminant is 0. Thus $(1-2a)^2 - 4(a^2 - a) = 0 \Leftrightarrow$

$1 - 4a + 4a^2 - 4a^2 + 4 = 0 \Leftrightarrow 4a = 5$, so $a = \frac{5}{4}$. The center of the circle is $(0, \frac{5}{4})$.

13. Let $\theta_1 = \text{arccot } x$, so $\cot \theta_1 = x = \frac{x}{1}$. So $\sin(\text{arccot } x) = \sin \theta_1 = \frac{1}{\sqrt{x^2+1}}$. Let

$\theta_2 = \arctan\left(\frac{1}{\sqrt{x^2+1}}\right)$, so $\tan \theta_2 = \frac{1}{\sqrt{x^2+1}}$. Hence

$\cos\{\arctan[\sin(\text{arccot } x)]\} = \cos \theta_2 = \frac{\sqrt{x^2+1}}{\sqrt{x^2+2}} = \sqrt{\frac{x^2+1}{x^2+2}}$.

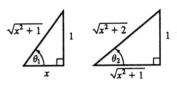

15. (a) $\{[f(x)]^2 + [g(x)]^2\}' = 2f(x)f'(x) + 2g(x)g'(x) = 2f(x)g(x) + 2g(x)[-f(x)] = 0$ for all x,

so $[f(x)]^2 + [g(x)]^2$ is a constant function. $[f(0)]^2 + [g(0)]^2 = 0^2 + 1^2 = 1$, so

$[f(x)]^2 + [g(x)]^2 = 1$ for all x.

(b) Let $h(x) = [f(x) - \sin x]^2 + [g(x) - \cos x]^2$. Then $h(x) = [f(x)]^2 + [g(x)]^2 + \sin^2 x$

$+ \cos^2 x - 2f(x)\sin x - 2g(x)\cos x = 2 - 2f(x)\sin x - 2g(x)\cos x$ (*using part (a)*). So

$h'(x) = -2f'(x)\sin x - 2f(x)\cos x - 2g'(x)\cos x + 2g(x)\sin x$

$= -2g(x)\sin x - 2f(x)\cos x + 2f(x)\cos x + 2g(x)\sin x = 0$. Therefore $h(x)$ is a constant

function, and since $h(0) = [0 - \sin 0]^2 + [1 - \cos 0]^2 = 0$, we have $h(x) = 0$ for all x.

(c) By part (b), $[f(x) - \sin x]^2 + [g(x) - \cos x]^2 = 0$ for all x. If, for some x, $f(x) \neq \sin x$ or

$g(x) \neq \cos x$, then one or both of the squared terms is positive, so the left hand side cannot

equal zero, contradicting (b). Therefore $f(x) = \sin x$ and $g(x) = \cos x$ for all x.

17. $f(x) = [\![x]\!] + \sqrt{x - [\![x]\!]}$. On each interval of the form $[n, n+1)$, where n is an integer, we have

$f(x) = n + \sqrt{x - n}$. It is easy to see that this function is continuous and increasing

on $[n, n+1)$. Also, the left hand limit $\lim\limits_{x \to (n+1)-} f(x) = \lim\limits_{x \to (n+1)-}\left[[\![x]\!] + \sqrt{x - [\![x]\!]}\right]$

$= \lim\limits_{x \to (n+1)-}[\![x]\!] + \sqrt{\lim\limits_{x \to (n+1)-} x - \lim\limits_{x \to (n+1)-}[\![x]\!]} = n + \sqrt{n+1-n} = n+1$

$= f(n+1) = \lim\limits_{x \to (n+1)+} f(x)$, so f is continuous and increasing everywhere.

19. Let $L = \lim\limits_{x \to \infty}\left(\frac{x+a}{x-a}\right)^x$, so $\ln L = \lim\limits_{x \to \infty}\ln\left(\frac{x+a}{x-a}\right)^x = \lim\limits_{x \to \infty} x\ln\left(\frac{x+a}{x-a}\right)$

$= \lim\limits_{x \to \infty}\frac{\ln(x+a) - \ln(x-a)}{1/x} = (\text{by } L'Hospital's Rule)} \lim\limits_{x \to \infty}\frac{\frac{1}{x+a} - \frac{1}{x-a}}{-1/x^2}$

$= -\lim\limits_{x \to \infty}\frac{(x-a)x^2 - (x+a)x^2}{(x+a)(x-a)} = -\lim\limits_{x \to \infty}\frac{-2ax^2}{x^2-a^2} = \lim\limits_{x \to \infty}\frac{2a}{1-a^2/x^2} = 2a$.

Hence $\ln L = 2a$, so $L = e^{2a}$. Hence $L = e^1 \Rightarrow 2a = 1$ so $a = \frac{1}{2}$.

21. $A = (x_1, x_1^2)$ and $B = (x_2, x_2^2)$, where x_1 and x_2 are the solutions of the quadratic equation

$x^2 = mx + b$. Let $P = (x, x^2)$ and set $A_1 = (x_1, 0)$, $B_1 = (x_2, 0)$, and $P_1 = (x, 0)$. Let $f(x)$

denote the area of triangle PAB. Then $f(x)$ can be expressed in terms of the areas of three

trapezoids as follows: $f(x) = \text{area}(A_1ABB_1) - \text{area}(A_1APP_1) - \text{area}(B_1BPP_1)$

$= \frac{1}{2}(x_2 - x_1)(x_1^2 + x_2^2) - \frac{1}{2}(x - x_1)(x_1^2 + x^2) - \frac{1}{2}(x_2 - x)(x^2 + x_2^2)$. After expansion, cancelling

of terms, and factoring, we find that $f(x) = \frac{1}{2}(x_2 - x_1)(x - x_1)(x_2 - x)$.

{[NOTE: Another way to get an expression for $f(x)$ is to use the formula for an area of a

triangle in terms of the coordinates of the vertices:

$f(x) = \frac{1}{2}[(x_2x_1^2 - x_1x_2^2) + (x_1x^2 - xx_1^2) + (xx_2^2 - x_2x^2)]\}$

From this it follows that $f'(x) = \frac{1}{2}(x_2 - x_1)(x_1 + x_2 - 2x)$ and $f''(x) = -(x_2 - x_1) < 0$.

Thus the area $f(x)$ is maximal when $x = \frac{1}{2}(x_1 + x_2)$. Its maximum value is

$f(\frac{1}{2}(x_1 + x_2)) = \frac{1}{2}(x_2 - x_1)\frac{1}{2}(x_2 - x_1)\frac{1}{2}(x_2 - x_1) = \frac{1}{8}(x_2 - x_1)^3$.

In terms of m and b, $x_1 = \frac{1}{2}(m - \sqrt{m^2 + 4b})$ and $x_2 = \frac{1}{2}(m + \sqrt{m^2 + 4b})$, so the maximal area

is $\frac{1}{8}(m^2 + 4b)^{3/2}$ and it is attained at the point $P(m/2, m^2/4)$.

23.   Since $[\![x]\!] \le x < [\![x]\!] + 1$, we have $1 \le \frac{x}{[\![x]\!]} \le 1 + \frac{1}{[\![x]\!]}$ for $x > 0$. As $x \to \infty$, $[\![x]\!] \to \infty$, so

$\frac{1}{[\![x]\!]} \to 0$ and $1 + \frac{1}{[\![x]\!]} \to 1$. Thus $\lim\limits_{x \to \infty} \frac{x}{[\![x]\!]} = 1$ by the Squeeze Theorem.

CHAPTER FIVE

EXERCISES 5.1

1. $\displaystyle\sum_{i=1}^{5} \sqrt{i} = \sqrt{1} + \sqrt{2} + \sqrt{3} + \sqrt{4} + \sqrt{5}$

3. $\displaystyle\sum_{i=4}^{6} 3^i = 3^4 + 3^5 + 3^6$

5. $\displaystyle\sum_{k=0}^{4} \frac{2k-1}{2k+1} = -1 + \frac{1}{3} + \frac{3}{5} + \frac{5}{7} + \frac{7}{9}$

7. $\displaystyle\sum_{i=1}^{n} i^{10} = 1^{10} + 2^{10} + 3^{10} + \cdots + n^{10}$

9. $\displaystyle\sum_{j=0}^{n-1} (-1)^j = 1 - 1 + 1 - 1 + \cdots + (-1)^{n-1}$

11. $1 + 2 + 3 + 4 + \cdots + 10 = \displaystyle\sum_{i=1}^{10} i$

13. $\frac{1}{2} + \frac{2}{3} + \frac{3}{4} + \frac{4}{5} + \cdots + \frac{19}{20} = \displaystyle\sum_{i=1}^{19} \frac{i}{i+1}$

15. $2 + 4 + 6 + 8 + \cdots + 2n = \displaystyle\sum_{i=1}^{n} 2i$

17. $1 + 2 + 4 + 8 + 16 + 32 = \displaystyle\sum_{i=0}^{5} 2^i$

19. $x + x^2 + x^3 + \cdots + x^n = \displaystyle\sum_{i=1}^{n} x^i$

21. $\displaystyle\sum_{i=4}^{8} (3i - 2) = 10 + 13 + 16 + 19 + 22 = 80$

23. $\displaystyle\sum_{j=1}^{6} 3^{j+1} = 3^2 + 3^3 + 3^4 + 3^5 + 3^6 + 3^7 = 9 + 27 + 81 + 243 + 729 + 2187 = 3276$

[For a more general method, see #51.]

25. $\displaystyle\sum_{n=1}^{20} (-1)^n = -1 + 1 - 1 + 1 - 1 + 1 - 1 + 1 - 1 + 1 - 1 + 1 - 1 + 1 - 1 + 1 - 1$
$+ 1 - 1 + 1 = 0$

26. $\displaystyle\sum_{i=1}^{100} 4 = 4 + 4 + 4 + \cdots + 4 = 100 \cdot 4 = 400$
    (100 *summands*)

27. $\displaystyle\sum_{i=0}^{4} (2^i + i^2) = (1 + 0) + (2 + 1) + (4 + 4) + (8 + 9) + (16 + 16) = 61$

29. $\displaystyle\sum_{i=1}^{n} \left(\frac{1}{i} - \frac{1}{i+1}\right) = \left(1 - \frac{1}{2}\right) + \left(\frac{1}{2} - \frac{1}{3}\right) + \left(\frac{1}{3} - \frac{1}{4}\right) + \cdots + \left(\frac{1}{n-1} - \frac{1}{n}\right) + \left(\frac{1}{n} - \frac{1}{n+1}\right)$
$= 1 - \frac{1}{n+1} = \frac{n}{n+1}$

31. $\displaystyle\sum_{i=1}^{n} 2i = 2\sum_{i=1}^{n} i = n(n+1)$

33. $\displaystyle\sum_{i=1}^{n} (i^2 + 3i + 4) = \sum_{i=1}^{n} i^2 + 3\sum_{i=1}^{n} i + \sum_{i=1}^{n} 4 = \frac{n(n+1)(2n+1)}{6} + \frac{3n(n+1)}{2} + 4n$

$= \frac{1}{6}\left[(2n^3 + 3n^2 + n) + (9n^2 + 9n) + 24n\right] = \frac{1}{6}(2n^3 + 12n^2 + 34n) = n(n^2 + 6n + 17)/3$

35. $\displaystyle\sum_{i=1}^{n} (i+1)(i+2) = \sum_{i=1}^{n} (i^2 + 3i + 2) = \sum_{i=1}^{n} i^2 + 3\sum_{i=1}^{n} i + \sum_{i=1}^{n} 2$

$= \frac{n(n+1)(2n+1)}{6} + \frac{3n(n+1)}{2} + 2n = \frac{n(n+1)}{6}\left[(2n+1) + 9)\right] + 2n$

$= \frac{n(n+1)}{3}(n+5) + 2n = \frac{n}{3}\left[(n+1)(n+5) + 6\right] = n(n^2 + 6n + 11)/3$

37. $\displaystyle\sum_{i=1}^{n} (i^3 - i - 2) = \sum_{i=1}^{n} i^3 - \sum_{i=1}^{n} i - \sum_{i=1}^{n} 2 = \left[\frac{n(n+1)}{2}\right]^2 - \frac{n(n+1)}{2} - 2n$

$= \frac{n(n+1)}{4}\left[n(n+1) - 2\right] - 2n = \frac{n(n+1)(n+2)(n-1)}{4} - 2n$

$= \frac{n}{4}\left[(n+1)(n-1)(n+2) - 8\right] = \frac{n}{4}\left[(n^2 - 1)(n+2) - 8\right] = n(n^3 + 2n^2 - n - 10)/4$

39. By Theorem 5.2(a) and Example 3, $\displaystyle\sum_{i=1}^{n} c = c\sum_{i=1}^{n} 1 = cn$

41. Let S_n be the statement that $\displaystyle\sum_{i=1}^{n} i^3 = \left[\frac{n(n+1)}{2}\right]^2$.

1. S_1 is true because $1^3 = \left[\frac{1 \cdot 2}{2}\right]^2$.

2. Assume S_k is true. Then $\displaystyle\sum_{i=1}^{k} i^3 = \left[\frac{k(k+1)}{2}\right]^2$, so $\displaystyle\sum_{i=1}^{k+1} i^3 = \left[\frac{k(k+1)}{2}\right]^2 + (k+1)^3$

$= \frac{(k+1)^2}{4}\left[k^2 + 4(k+1)\right] = \frac{(k+1)^2}{4}(k+2)^2 = \left[\frac{(k+1)\left[(k+1)+1\right]}{2}\right]^2$, showing

that S_{k+1} is true. Therefore S_n is true for all n by mathematical induction.

43. The area of G_i is $\displaystyle\left(\sum_{k=1}^{i} k\right)^2 - \left(\sum_{k=1}^{i-1} k\right)^2 = \left[\frac{i(i+1)}{2}\right]^2 - \left[\frac{(i-1)i}{2}\right]^2$

$= \frac{i^2}{4}\left[(i+1)^2 - (i-1)^2\right] = \frac{i^2}{4}\left[(i^2 + 2i + 1) - (i^2 - 2i + 1)\right] = \frac{i^2}{4}(4i) = i^3$.

Thus the area of ABCD equals $\displaystyle\sum_{i=1}^{n} i^3$ and also equals $\left[n(n+1)/2\right]^2$.

45. (a) $\displaystyle\sum_{i=1}^{n} \left(i^4 - (i-1)^4\right) = (1^4 - 0^4) + (2^4 - 1^4) + (3^4 - 2^4) + \cdots$

$+ \left[n^4 - (n-1)^4\right] = n^4 - 0 = n^4$.

(b) $\displaystyle\sum_{i=1}^{100}(5^i - 5^{i-1}) = (5^1 - 5^0) + (5^2 - 5^1) + (5^3 - 5^2) + \cdots + (5^{100} - 5^{99}) = 5^{100} - 5^0$

$= 5^{100} - 1.$

(c) $\displaystyle\sum_{i=3}^{99}\left(\frac{1}{i} - \frac{1}{i+1}\right) = \left(\frac13 - \frac14\right) + \left(\frac14 - \frac15\right) + \left(\frac15 - \frac16\right) + \cdots + \left(\frac{1}{99} - \frac{1}{100}\right) = \frac13 - \frac{1}{100} = \frac{97}{300}.$

(d) $\displaystyle\sum_{i=1}^{n}(a_i - a_{i-1}) = (a_1 - a_0) + (a_2 - a_1) + (a_3 - a_2) + \cdots + (a_n - a_{n-1}) = a_n - a_0.$

47. $\displaystyle\lim_{n\to\infty}\sum_{i=1}^{n}\frac{1}{n}\left(\frac{i}{n}\right)^2 = \lim_{n\to\infty}\frac{1}{n^3}\sum_{i=1}^{n}i^2 = \lim_{n\to\infty}\frac{1}{n^3}\frac{n(n+1)(2n+1)}{6} = \lim_{n\to\infty}\frac16\left(1 + \frac1n\right)\left(2 + \frac1n\right)$

$= \frac16 \cdot 1 \cdot 2 = \frac13$

49. $\displaystyle\lim_{n\to\infty}\sum_{i=1}^{n}\frac{2}{n}\left[\left(\frac{2i}{n}\right)^3 + 5\left(\frac{2i}{n}\right)\right] = \lim_{n\to\infty}\sum_{i=1}^{n}\left[\frac{16}{n^4}i^3 + \frac{20}{n^2}i\right]$

$= \lim_{n\to\infty}\left[\frac{16}{n^4}\sum_{i=1}^{n}i^3 + \frac{20}{n^2}\sum_{i=1}^{n}i\right] = \lim_{n\to\infty}\left[\frac{16}{n^4}\frac{n^2(n+1)^2}{4} + \frac{20}{n^2}\frac{n(n+1)}{2}\right]$

$= \lim_{n\to\infty}\left[\frac{4(n+1)^2}{n^2} + \frac{10n(n+1)}{n^2}\right] = \lim_{n\to\infty}\left[4\left(1 + \frac1n\right)^2 + 10\left(1 + \frac1n\right)\right] = 4 \cdot 1 + 10 \cdot 1 = 14$

51. Let $S = \displaystyle\sum_{i=1}^{n}ar^{i-1} = a + ar + ar^2 + \cdots + ar^{n-1}$. Then $rS = ar + ar^2 + \cdots + ar^{n-1} + ar^n$.

Subtracting the first equation from the second, we find $(r - 1)S = ar^n - a = a(r^n - 1)$, so

$S = \dfrac{a(r^n - 1)}{r - 1}.$

53. $\displaystyle\sum_{i=1}^{n}(2i + 2^i) = 2\sum_{i=1}^{n}i + \sum_{i=1}^{n}2 \cdot 2^{i-1} = 2\frac{n(n+1)}{2} + \frac{2(2^n - 1)}{2 - 1} = 2^{n+1} + n^2 + n - 2.$ For the

first sum we have used Theorem 5.3(c), and for the second, Exercise 51 with $a = r = 2$.

55. Using (4.3(c)) we have that, $\displaystyle\sum_{i=1}^{n}i = \frac{n(n+1)}{2} = 78 \Leftrightarrow n(n+1) = 156$

$\Leftrightarrow n^2 + n - 156 = 0 \Leftrightarrow (n + 13)(n - 12) = 0 \Leftrightarrow n = 12$ or -13. But $n = -13$ would

produce a negative answer for the sum, so $n = 12$.

57. $\sin x \sin y = \frac12\left[\cos(x - y) - \cos(x + y)\right]$ by (18c) of Appendix B, so

$2\sin u \sin v = \cos(u - v) - \cos(u + v)$ $(\star)$. Taking $u = \frac12 x$ and $v = ix$, we get

$2\sin\frac12 x \sin ix = \cos\left(\frac12 - i\right)x - \cos\left(\frac12 + i\right)x = \cos\left(i - \frac12\right)x - \cos\left(i + \frac12\right)x.$

Thus $2\sin\frac12 x\displaystyle\sum_{i=1}^{n}\sin ix = \sum_{i=1}^{n}2\sin\frac12 x \sin ix$

$$= \sum_{i=1}^{n} \left[\cos\left(i - \tfrac{1}{2}\right)x - \cos\left(i + \tfrac{1}{2}\right)x \right] = -\sum_{i=1}^{n} \left[\cos\left(i + \tfrac{1}{2}\right)x - \cos\left(i - \tfrac{1}{2}\right)x \right]$$

$$= -\left[\cos(n + \tfrac{1}{2})x - \cos\tfrac{1}{2}x \right] \quad \text{(telescoping sum)}$$

$$= \cos\left[\tfrac{1}{2}(n+1)x - \tfrac{1}{2}nx\right] - \cos\left[\tfrac{1}{2}(n+1)x + \tfrac{1}{2}nx\right] = 2\sin\tfrac{1}{2}(n+1)x \sin\tfrac{1}{2}nx$$

$\left[by\ (\star)\ with\ u = \tfrac{1}{2}(n+1)x\ and\ v = \tfrac{1}{2}nx \right]$. If x is not an integer multiple of 2π, then

$\sin\tfrac{1}{2}x \neq 0$, so we can divide by $2\sin\tfrac{1}{2}x$ and get $\displaystyle\sum_{i=1}^{n} \sin ix = \frac{\sin\tfrac{1}{2}nx \sin\tfrac{1}{2}(n+1)x}{\sin\tfrac{1}{2}x}$.

EXERCISES 5.2

1. (a) $\|P\| = \max\{1, 1, 1, 1\} = 1$ (c)

    (b) $\displaystyle\sum_{i=1}^{n} f(x_i^*)\Delta x_i = \sum_{i=1}^{4} f(i - 1) \cdot 1$

    $= 16 + 15 + 12 + 7 = 50$

3. (a) $\|P\| = \max\{1, 1, 1, 1\} = 1$ (c)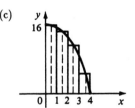

    (b) $\displaystyle\sum_{i=1}^{n} f(x_i^*)\Delta x_i = \sum_{i=1}^{4} f(i - \tfrac{1}{2}) \cdot 1$

    $= 15.75 + 13.75 + 9.75 + 3.75 = 43$

5. (a) $\|P\| = \max\{1, 1, 1, 1, 1\} = 1$ (c)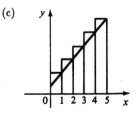

    (b) $\displaystyle\sum_{i=1}^{n} f(x_i^*)\Delta x_i = \sum_{i=1}^{5} f(i) \cdot 1$

    $= 3 + 5 + 7 + 9 + 11 = 35$

7. (a) $\|P\| = \max\{.5, .5, .5, .5, .5, .5\} = .5$

(b) $\displaystyle\sum_{i=1}^{n} f(x_i^*)\Delta x_i = [f(-.5) + f(0)$

$+ f(.5) + f(1) + f(1.5) + f(2)](0.5)$

$= \frac{1}{2}[1.875 + 2 + 2.125 + 3 + 5.375 + 10]$

$= \frac{1}{2}(24.375) = 12.1875$

(c)

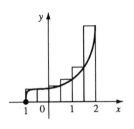

9. (a) $\|P\| = \max\left\{\frac{\pi}{4}, \frac{\pi}{4}, \frac{\pi}{4}, \frac{\pi}{4}\right\} = \frac{\pi}{4}$

(b) $\displaystyle\sum_{i=1}^{n} f(x_i^*)\Delta x_i = \sum_{i=1}^{4} f(x_i^*)\frac{\pi}{4}$

$= \frac{\pi}{4}\left[f(\frac{\pi}{6}) + f(\frac{\pi}{3}) + f(\frac{2\pi}{3}) + f(\frac{5\pi}{6})\right]$

$= \frac{\pi}{4}\left[1 + \sqrt{3} + \sqrt{3} + 1\right] = \frac{\pi}{2}(1 + \sqrt{3})$

(c)

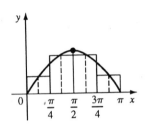

11. $f(x) = x^2 + 1$ on $[0, 2]$ with partition points $x_i = \frac{2i}{n}$ ($i = 0, 1, 2, \ldots, n$), so

$\Delta x_1 = \Delta x_2 = \cdots = \Delta x_n = \frac{2}{n}$. $\|P\| = \max\{\Delta x_i\} = \frac{2}{n}$, so $\|P\| \to 0$ is equivalent to

$n \to \infty$. Taking x_i^* to be the midpoint of $[x_{i-1}, x_i] = [2(i-1)/n, 2i/n]$, we get

$x_i^* = (2i-1)/n$. Thus $A = \displaystyle\lim_{\|P\| \to 0} \sum_{i=1}^{n} f(x_i^*)\Delta x_i = \lim_{n \to \infty} \sum_{i=1}^{n}\left[\left(\frac{2i-1}{n}\right)^2 + 1\right]\frac{2}{n}$

$= \displaystyle\lim_{n \to \infty} \sum_{i=1}^{n}\left[\frac{8i^2}{n^3} - \frac{8i}{n^3} + \frac{2}{n^3} + \frac{2}{n}\right]$

$= \displaystyle\lim_{n \to \infty}\left[\frac{8}{n^3}\sum_{i=1}^{n} i^2 - \frac{8}{n^3}\sum_{i=1}^{n} i + \left(\frac{2}{n^3} + \frac{2}{n}\right)\sum_{i=1}^{n} 1\right]$

$= \displaystyle\lim_{n \to \infty}\left[\frac{8}{n^3}\frac{n(n+1)(2n+1)}{6} - \frac{8}{n^3}\frac{n(n+1)}{2} + \left(\frac{2}{n^3} + \frac{2}{n}\right)n\right]$

$= \displaystyle\lim_{n \to \infty}\left[\frac{4}{3}\cdot 1\left(1 + \frac{1}{n}\right)\left(2 + \frac{1}{n}\right) - \frac{4}{n}\cdot 1\left(1 + \frac{1}{n}\right) + \frac{2}{n^2} + 2\right]$

$= \frac{4}{3}\cdot 1 \cdot 1 \cdot 2 - 0 \cdot 1 \cdot 1 + 0 + 2 = \frac{8}{3} + 2 = \frac{14}{3}$

13. $f(x) = 16 - x^2$ on $[-4, 4]$ with partition points $x_i = -4 + \frac{8i}{n}$ ($i = 0, 1, \ldots, n$), so $\Delta x_i = \frac{8}{n}$ for

all i. $\|P\| = \frac{8}{n}$, so $\|P\| \to 0$ is equivalent to $n \to \infty$.

(a) $x_i^* = x_{i-1} = -4 + \frac{8(i-1)}{n}$. $A = \lim\limits_{n \to \infty} \sum\limits_{i=1}^{n} \left[16 - \left(-4 + \frac{8(i-1)}{n} \right)^2 \right] \frac{8}{n}$

$= \lim\limits_{n \to \infty} \sum\limits_{i=1}^{n} \left[\frac{64(i-1)}{n} - \frac{64(i-1)^2}{n^2} \right] \frac{8}{n} = \lim\limits_{n \to \infty} \sum\limits_{i=0}^{n-1} \left[\frac{64i}{n} - \frac{64i^2}{n^2} \right] \frac{8}{n}$

$= \lim\limits_{n \to \infty} \sum\limits_{i=1}^{n-1} \left[\frac{64i}{n} - \frac{64}{n^2}i^2 \right] \frac{8}{n} = \lim\limits_{n \to \infty} \left[\frac{64 \cdot 8}{n^2} \sum\limits_{i=1}^{n-1} i - \frac{64 \cdot 8}{n^3} \sum\limits_{i=1}^{n-1} i^2 \right]$

$= \lim\limits_{n \to \infty} \left[\frac{64 \cdot 8}{n^2} \cdot \frac{(n-1)n}{2} - \frac{64 \cdot 8}{n^3} \cdot \frac{(n-1)n(2n-1)}{6} \right]$

$= \lim\limits_{n \to \infty} \left[256 \left(1 - \frac{1}{n} \right) - \frac{256}{3} \left(1 - \frac{1}{n} \right) \left(2 - \frac{1}{n} \right) \right] = 256 \cdot 1 - \frac{256}{3} \cdot 1 \cdot 2 = \frac{256}{3}$

(b) $x_i^* = x_i = -4 + \frac{8i}{n}$. The only difference is that $i-1$ is replaced by i, so from (a) we see

that $A = \lim\limits_{n \to \infty} \sum\limits_{i=1}^{n} \left[\frac{64i}{n} - \frac{64i^2}{n^2} \right] \frac{8}{n} = \lim\limits_{n \to \infty} \left[\frac{64 \cdot 8}{n^2} \sum\limits_{i=1}^{n} i - \frac{64 \cdot 8}{n^3} \sum\limits_{i=1}^{n} i^2 \right]$

$= \lim\limits_{n \to \infty} \left[\frac{64 \cdot 8}{n^2} \cdot \frac{n(n+1)}{2} - \frac{64 \cdot 8}{n^3} \cdot \frac{n(n+1)(2n+1)}{6} \right]$

$= \lim\limits_{n \to \infty} \left[256 \left(1 + \frac{1}{n} \right) - \frac{256}{3} \left(1 + \frac{1}{n} \right) \left(2 + \frac{1}{n} \right) \right] = 256 - \frac{256}{3} \cdot 2 = \frac{256}{3}$

(c) $x_i^* = -4 + \frac{8i-4}{n} = 4 \left[\frac{2i-1}{n} - 1 \right] = 4 \left[\frac{2i}{n} - \frac{1}{n} - 1 \right]$.

$A = \lim\limits_{n \to \infty} \sum\limits_{i=1}^{n} \left[16 - 16 \left(\frac{2i}{n} - \frac{1}{n} - 1 \right)^2 \right] \frac{8}{n}$

$= \lim\limits_{n \to \infty} \sum\limits_{i=1}^{n} \left[16 - 16 \left(\frac{4i^2}{n^2} - \frac{4i}{n^2} - \frac{4i}{n} + \frac{1}{n^2} + \frac{2}{n} + 1 \right) \right] \frac{8}{n}$

$= \lim\limits_{n \to \infty} \left[-64 \frac{8}{n^3} \sum\limits_{i=1}^{n} i^2 + 64 \frac{8}{n^3} \sum\limits_{i=1}^{n} i + 64 \frac{8}{n^2} \sum\limits_{i=1}^{n} i - 16 \frac{8}{n^3} - \frac{16^2}{n^2} - 16 \frac{8}{n} \right]$

$= \lim\limits_{n \to \infty} \left[-64 \frac{8}{n^3} \frac{n(n+1)(2n+1)}{6} + 64 \frac{8}{n^3} \frac{n(n+1)}{2} + 64 \frac{8}{n^2} \frac{n(n+1)}{2} \right]$
(*The last three terms all* $\to 0$)

$= \lim\limits_{n \to \infty} \left[-64 \cdot \frac{4}{3} \left(1 + \frac{1}{n} \right) \left(2 + \frac{1}{n} \right) + 64 \cdot \frac{4}{3} \left(1 + \frac{1}{n} \right) + 64 \cdot 4 \left(1 + \frac{1}{n} \right) \right]$

$= -64 \cdot \frac{4}{3} \cdot 1 \cdot 2 + 64 \cdot 0 \cdot 1 + 64 \cdot 4 \cdot 1 = -\frac{512}{3} + 256 = \frac{256}{3}$

(a) 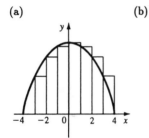 (b) 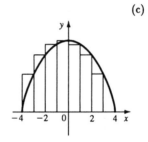 (c)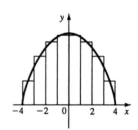

15. $f(x) = 5$ on $[-2, 2]$ with partition points $x_i = -2 + \frac{4i}{n}$ $(i = 0, 1, \ldots, n)$.

$$\Delta x_i = \frac{4}{n} \text{ for } i = 1, \ldots, n \Rightarrow \|P\| = \frac{4}{n}. \quad A = \lim_{n \to \infty} \sum_{i=1}^{n} f(x_i^*)\Delta x_i$$

$$= \lim_{n \to \infty} \sum_{i=1}^{n} 5 \cdot \frac{4}{n} = \lim_{n \to \infty} \frac{20}{n} \sum_{i=1}^{n} 1 = \lim_{n \to \infty} \frac{20}{n} \cdot n = 20.$$

The approximation is exact in this instance.

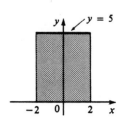

17. $f(x) = x^2 + 3x - 2$ on $[1, 4]$. $x_i^* = x_i = 1 + \frac{3i}{n}$ for $i = 1, 2, \ldots, n$. $\Delta x_i = \frac{3}{n}$ for all i.

$$A = \lim_{n \to \infty} \sum_{i=1}^{n} \left[\left(1 + \frac{3i}{n}\right)^2 + 3\left(1 + \frac{3i}{n}\right) - 2 \right] \frac{3}{n} = \lim_{n \to \infty} \sum_{i=1}^{n} \left[\frac{9i^2}{n^2} + \frac{15i}{n} + 2 \right] \frac{3}{n}$$

$$= \lim_{n \to \infty} \left[\frac{27}{n^3} \sum_{i=1}^{n} i^2 + \frac{45}{n^2} \sum_{i=1}^{n} i + \frac{6}{n} \sum_{i=1}^{n} 1 \right]$$

$$= \lim_{n \to \infty} \left[\frac{27}{n^3} \frac{n(n+1)(2n+1)}{6} + \frac{45}{n^2} \frac{n(n+1)}{2} + \frac{6}{n} n \right]$$

$$= \lim_{n \to \infty} \left[\frac{9}{2} \cdot 1 \left(1 + \frac{1}{n}\right) \cdot \left(2 + \frac{1}{n}\right) + \frac{45}{2} \cdot 1 \left(1 + \frac{1}{n}\right) + 6 \right]$$

$$= \frac{9}{2} \cdot 1 \cdot 1 \cdot 2 + \frac{45}{2} \cdot 1 \cdot 1 + 6 = 9 + \frac{45}{2} + 6 = \frac{75}{2} = 37.5$$

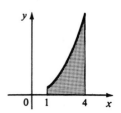

19. $f(x) = x^3 + 2x$ on $[0, 2]$. $x_i^* = x_i = \frac{2i}{n}$ for $i = 1, 2, \ldots, n$. $A = \lim_{n \to \infty} \sum_{i=1}^{n} \left[\left(\frac{2i}{n}\right)^3 + \frac{4i}{n} \right] \frac{2}{n}$

$$= \lim_{n \to \infty} \left[\frac{16}{n^4} \sum_{i=1}^{n} i^3 + \frac{8}{n^2} \sum_{i=1}^{n} i \right] = \lim_{n \to \infty} \left[\frac{16}{n^4} \frac{n^2(n+1)^2}{4} + \frac{8}{n^2} \frac{n(n+1)}{2} \right]$$

$$= \lim_{n \to \infty} \left[4\left(1 + \frac{1}{n}\right)^2 + 4\left(1 + \frac{1}{n}\right) \right]$$

$$= 4 \cdot 1 + 4 \cdot 1 = 8$$

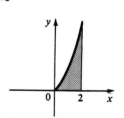

21. $f(x) = 1 - 2x^2 + x^4$ on $[-1, 1]$. $x_i^* = x_i = -1 + \frac{2i}{n}$, $\Delta x_i = \frac{2}{n}$.

$$A = \lim_{n \to \infty} \sum_{i=1}^{n} \left[1 - 2\left(-1 + \frac{2i}{n}\right)^2 + \left(-1 + \frac{2i}{n}\right)^4 \right] \frac{2}{n}$$

$$= \lim_{n \to \infty} \sum_{i=1}^{n} \left[1 - 2 + \frac{8i}{n} - \frac{8i^2}{n^2} + 1 - \frac{8i}{n} + \frac{24i^2}{n^2} - \frac{32i^3}{n^3} + \frac{16i^4}{n^4} \right] \frac{2}{n}$$

$$= \lim_{n \to \infty} \left[\frac{32}{n^5} \sum_{i=1}^{n} i^4 - \frac{64}{n^4} \sum_{i=1}^{n} i^3 + \frac{32}{n^3} \sum_{i=1}^{n} i^2 \right]$$

$$= \lim_{n \to \infty} \left[\frac{32}{n^5} \frac{n(n+1)(2n+1)(3n^2+3n-1)}{30} - \frac{64}{n^2} \frac{n^2(n+1)^2}{4} + \frac{32}{n^3} \frac{n(n+1)(2n+1)}{6} \right]$$

$$= \lim_{n \to \infty} \left[\frac{16}{15} \cdot 1 \left(1 + \frac{1}{n}\right)\left(2 + \frac{1}{n}\right)\left(3 + \frac{3}{n} - \frac{1}{n^2}\right) - 16 \cdot 1^2 \left(1 + \frac{1}{n}\right)^2 + \frac{16}{3} \cdot 1 \left(1 + \frac{1}{n}\right)\left(2 + \frac{1}{n}\right) \right]$$

$$= \frac{16}{15} \cdot 1^2 \cdot 2 \cdot 3 - 16 \cdot 1^4 + \frac{16}{3} \cdot 1^2 \cdot 2 = \frac{32}{5} - 16 + \frac{32}{3} = \frac{16}{15}.$$

23. $f(x) = \sin x$. $\Delta x_i = \frac{\pi}{n}$ and $x_i^* = x_i = \frac{i\pi}{n}$. So $A = \lim_{n \to \infty} \sum_{i=1}^{n} \left(\sin \frac{i\pi}{n}\right)\left(\frac{\pi}{n}\right)$

$$= \lim_{n \to \infty} \frac{\pi}{n} \sum_{i=1}^{n} \sin \frac{i\pi}{n} = \lim_{n \to \infty} \frac{\pi}{n} \frac{\sin \left(\frac{1}{2} n \frac{\pi}{n}\right) \sin \left(\frac{1}{2}(n+1)\frac{\pi}{n}\right)}{\sin \left(\frac{1}{2}\frac{\pi}{n}\right)} \quad (using\ 5.1\ \#57)$$

$$= \lim_{n \to \infty} \frac{\pi}{n} \frac{\sin \left(\frac{\pi}{2} + \frac{\pi}{2n}\right)}{\sin \frac{\pi}{2n}} = \lim_{n \to \infty} 2 \frac{\pi/(2n)}{\sin [\pi/(2n)]} \cdot \lim_{n \to \infty} \cos \left(\frac{\pi}{2n}\right) = 2 \cdot 1 \cdot 1 = 2. \quad \text{Here we}$$

have used the identity $\sin \left(\frac{\pi}{2} + x\right) = \cos x$.

EXERCISES 5.3

1. $f(x) = 7 - 2x$. (a) $\|P\| = \max\{.6, .6, .8, 1.2, .8\} = 1.2$

 (b) $\sum_{i=1}^{5} f(x_i^*)\Delta x_i = f(1.3)(.6) + f(1.9)(.6) + f(2.6)(.8) + f(3.6)(1.2) + f(4.6)(.8)$
 $= (4.4)(.6) + (3.2)(.6) + (1.8)(.8) + (-.2)(1.2) + (-2.2)(.8) = 4$

3. $f(x) = 2 - x^2$. (a) $\|P\| = \max\{.6, .4, 1, .8, .6, .6\} = 1$

 (b) $\sum_{i=1}^{6} f(x_i^*)\Delta x_i = f(-1.4)(.6) + f(-1)(.4) + f(0)(1) + f(.8)(.8) + f(1.4)(.6) + f(2)(.6)$
 $= (.04)(.6) + (1)(.4) + (2)(1) + (1.36)(.8) + (.04)(.6) + (-2)(.6) = 2.336$

5. $f(x) = x^3$. (a) $\|P\| = \max\{.5, .5, .5, .5\} = .5$

(b) $\sum_{i=1}^{n} f(x_i^*)\Delta x_i = \frac{1}{2}\sum_{i=1}^{4} f(x_i^*) = \frac{1}{2}\left[(-1)^3 + (-.4)^3 + (.2)^3 + 1^3\right] = -.028$

7. The width of the intervals is $\Delta x = (5-0)/5 = 1$ so the partition points are 0, 1, 2, 3, 4, 5 and the midpoints are .5, 1.5, 2.5, 3.5, 4.5.

$\int_0^5 x^3 dx \approx \sum_{i=1}^{5} f(\overline{x}_i)\Delta x = (0.5)^3 + (1.5)^3 + (2.5)^3 + (3.5)^3 + (4.5)^3 = 153.125.$

9. $\Delta x = (2-1)/10 = 0.1$ so the partition points are 1.0, 1.1, 1.2, 1.3, 1.4, 1.5, 1.6, 1.7, 1.8, 1.9, 2.0 and the midpoints are 1.05, 1.15, 1.25, 1.35, 1.45, 1.55, 1.65, 1.75, 1.85, 1.95.

$\int_1^2 \sqrt{1+x^2}\, dx \approx \sum_{i=1}^{10} f(\overline{x}_i)\Delta x = 0.1\left(\sqrt{1+(1.05)^2} + \sqrt{1+(1.15)^2} + \cdots + \sqrt{1+(1.95)^2}\right)$

≈ 1.8100

11. $\int_a^b c\, dx = \lim_{n\to\infty} \frac{b-a}{n}\sum_{i=1}^{n} c = \lim_{n\to\infty} \frac{b-a}{n}nc \;\left[By\ Theorem\ 5.3\right] = \lim_{n\to\infty} (b-a)c = (b-a)c$

13. $\int_1^4 (x^2 - 2)\, dx = \lim_{n\to\infty} \frac{3}{n}\sum_{i=1}^{n}\left[\left(1+\frac{3i}{n}\right)^2 - 2\right] = \lim_{n\to\infty} \frac{3}{n}\sum_{i=1}^{n}\left[\frac{9i^2}{n^2} + \frac{6i}{n} - 1\right]$

$= \lim_{n\to\infty}\left[\frac{27}{n^3}\sum_{i=1}^{n} i^2 + \frac{18}{n^2}\sum_{i=1}^{n} i - \frac{3}{n}\sum_{i=1}^{n} 1\right] = \lim_{n\to\infty}\left[\frac{27}{n^3}\frac{n(n+1)(2n+1)}{6} + \frac{18}{n^2}\frac{n(n+1)}{2} - \frac{3}{n}n\right]$

$= \lim_{n\to\infty}\left[\frac{9}{2}\cdot 1\left(1+\frac{1}{n}\right)\left(2+\frac{1}{n}\right) + 9\cdot 1\left(1+\frac{1}{n}\right) - 3\right] = \frac{9}{2}\cdot 2 + 9 - 3 = 15$

15. $\int_{-3}^0 (2x^2 - 3x - 4)\, dx = \lim_{n\to\infty} \frac{3}{n}\sum_{i=1}^{n}\left[2\left(-3+\frac{3i}{n}\right)^2 - 3\left(-3+\frac{3i}{n}\right) - 4\right]$

$= \lim_{n\to\infty} \frac{3}{n}\sum_{i=1}^{n}\left[\frac{18i^2}{n^2} - \frac{45i}{n} + 23\right] = \lim_{n\to\infty}\left[\frac{54}{n^3}\sum_{i=1}^{n} i^2 - \frac{135}{n^2}\sum_{i=1}^{n} i + \frac{69}{n}\sum_{i=1}^{n} 1\right]$

$= \lim_{n\to\infty}\left[\frac{54}{n^3}\frac{n(n+1)(2n+1)}{6} - \frac{135}{n^2}\frac{n(n+1)}{2} + \frac{69}{n}n\right]$

$= \lim_{n\to\infty}\left[9\cdot 1\left(1+\frac{1}{n}\right)\left(2+\frac{1}{n}\right) - \frac{135}{2}\cdot 1\left(1+\frac{1}{n}\right) + 69\right] = 9\cdot 2 - \frac{135}{2} + 69 = 19.5$

17. $\int_{-1}^1 (t^3 - t^2 + 1)\, dt = \lim_{n\to\infty} \frac{2}{n}\sum_{i=1}^{n}\left[\left(-1+\frac{2i}{n}\right)^3 - \left(-1+\frac{2i}{n}\right)^2 + 1\right]$

$= \lim_{n\to\infty} \frac{2}{n}\sum_{i=1}^{n}\left[\left(\frac{8i^3}{n^3} - \frac{12i^2}{n^2} + \frac{6i}{n} - 1\right) - \left(\frac{4i^2}{n^2} - \frac{4i}{n} + 1\right) + 1\right]$

$= \lim_{n\to\infty} \frac{2}{n}\sum_{i=1}^{n}\left[\frac{8i^3}{n^3} - \frac{16i^2}{n^2} + \frac{10i}{n} - 1\right]$

$= \lim_{n\to\infty}\left[\frac{16}{n^4}\sum_{i=1}^{n} i^3 - \frac{32}{n^3}\sum_{i=1}^{n} i^2 + \frac{20}{n^2}\sum_{i=1}^{n} i - \frac{2}{n}\sum_{i=1}^{n} 1\right]$

$$= \lim_{n \to \infty} \left[\frac{16}{n^4} \frac{n^2(n+1)^2}{4} - \frac{32}{n^3} \frac{n(n+1)(2n+1)}{6} + \frac{20}{n^2} \frac{n(n+1)}{2} - \frac{2}{n} n \right]$$

$$= \lim_{n \to \infty} \left[4 \cdot 1^2 \left(1 + \tfrac{1}{n}\right)^2 - \tfrac{16}{3} \cdot 1 \left(1 + \tfrac{1}{n}\right)\left(2 + \tfrac{1}{n}\right) + 10 \cdot 1 \left(1 + \tfrac{1}{n}\right) - 2 \right]$$

$$= 4 - \tfrac{32}{3} + 10 - 2 = \tfrac{4}{3}$$

19. $\int_0^b (x^3 + 4x)\,dx = \lim\limits_{n \to \infty} \dfrac{b}{n} \sum\limits_{i=1}^n \left[\left(\dfrac{bi}{n}\right)^3 + 4\left(\dfrac{bi}{n}\right) \right] = \lim\limits_{n \to \infty} \left[\dfrac{b^4}{n^4} \sum\limits_{i=1}^n i^3 + 4\dfrac{b^2}{n^2} \sum\limits_{i=1}^n i \right]$

$$= \lim_{n \to \infty} \left[\frac{b^4}{n^4} \frac{n^2(n+1)^2}{4} + \frac{4b^2}{n^2} \frac{n(n+1)}{2} \right] = \lim_{n \to \infty} \left[\frac{b^4}{4} \cdot 1^2 \left(1 + \tfrac{1}{n}\right)^2 + 2b^2 \cdot 1 \left(1 + \tfrac{1}{n}\right) \right]$$

$$= \frac{b^4}{4} + 2b^2.$$

21. $\int_0^2 (x^4 - x + 1)\,dx = \lim\limits_{n \to \infty} \dfrac{2}{n} \sum\limits_{i=1}^n \left[\left(\dfrac{2i}{n}\right)^4 - \left(\dfrac{2i}{n}\right) + 1 \right]$

$$= \lim_{n \to \infty} \left[\frac{32}{n^5} \sum_{i=1}^n i^4 - \frac{4}{n^2} \sum_{i=1}^n i + \frac{2}{n} \sum_{i=1}^n 1 \right]$$

$$= \lim_{n \to \infty} \left[\frac{32}{n^5} \frac{n(n+1)(2n+1)(3n^2 + 3n - 1)}{30} - \frac{4}{n^2} \frac{n(n+1)}{2} + \frac{2}{n} n \right]$$

$$= \lim_{n \to \infty} \left[\tfrac{16}{15} 1 \left(1 + \tfrac{1}{n}\right)\left(2 + \tfrac{1}{n}\right)\left(3 + \tfrac{3}{n} - \tfrac{1}{n^2}\right) - 2 \cdot 1 \left(1 + \tfrac{1}{n}\right) + 2 \right] = \tfrac{16}{15} \cdot 2 \cdot 3 - 2 + 2 = \tfrac{32}{5}$$

23. $\int_a^b x\,dx = \lim\limits_{n \to \infty} \dfrac{b-a}{n} \sum\limits_{i=1}^n \left[a + \dfrac{b-a}{n} i \right] = \lim\limits_{n \to \infty} \left[\dfrac{a(b-a)}{n} \sum\limits_{i=1}^n 1 + \dfrac{(b-a)^2}{n^2} \sum\limits_{i=1}^n i \right]$

$$= \lim_{n \to \infty} \left[\frac{a(b-a)}{n} n + \frac{(b-a)^2}{n^2} \cdot \frac{n(n+1)}{2} \right] = a(b-a) + \lim_{n \to \infty} \frac{(b-a)^2}{2} \left(1 + \tfrac{1}{n}\right)$$

$$= a(b-a) + \tfrac{1}{2}(b-a)^2 = (b-a)(a + \tfrac{1}{2}b - \tfrac{1}{2}a) = (b-a)\tfrac{1}{2}(b+a) = \tfrac{1}{2}(b^2 - a^2)$$

25. $\lim\limits_{\|P\| \to 0} \sum\limits_{i=1}^n \left[2(x_i^*)^2 - 5x_i^* \right] \Delta x_i = \int_0^1 (2x^2 - 5x)\,dx$

27. $\lim\limits_{\|P\| \to 0} \sum\limits_{i=1}^n \cos x_i \Delta x_i = \int_0^\pi \cos x\,dx$

29. $\lim\limits_{n \to \infty} \sum\limits_{i=1}^n \dfrac{i^4}{n^5} = \lim\limits_{n \to \infty} \dfrac{1}{n} \sum\limits_{i=1}^n \left(\dfrac{i}{n}\right)^4 = \int_0^1 x^4\,dx$

31. $\lim\limits_{n \to \infty} \sum\limits_{i=1}^n \left[3\left(1 + \tfrac{2i}{n}\right)^5 - 6 \right]\tfrac{2}{n} = \int_1^3 (3x^5 - 6)\,dx$

(To get started, notice that $\tfrac{2}{n} = \Delta x = \tfrac{b-a}{n}$ and $1 + \tfrac{2i}{n} = a + \tfrac{b-a}{n} i$.)

33. By the definition in Note 6, $\int_9^4 \sqrt{t}\,dt = -\int_4^9 \sqrt{t}\,dt = -\tfrac{38}{3}$

35.   (a) $f(x) = x^2 \sin x$ is continuous on $[0, 2]$ and hence integrable by (5.11).

      (b) $f(x) = \sec x$ is unbounded on $[0, 2]$, so f is not integrable (see the remarks following Theorem 5.11).

      (c) $f(x)$ is piecewise continuous on $[0, 2]$ with a single jump discontinuity at $x = 1$, so f is integrable.

      (d) $f(x)$ has an infinite discontinuity at $x = 1$, so f is not integrable on $[0, 2]$.

37.   f is bounded since $|f(x)| \leq 1$ for all x in $[a, b]$. To see that f is not integrable on $[a, b]$, notice that $\sum_{i=1}^{n} f(x_i^*)\Delta x_i = 0$ if $x_1^*, \ldots, x_n^*$ are all chosen to be rational numbers, but

$$\sum_{i=1}^{n} f(x_i^*)\Delta x_i = \sum_{i=1}^{n} \Delta x_i = b-a \text{ if } x_1^*, \ldots, x_n^* \text{ are all chosen to be irrational numbers.}$$

This is true no matter how small $\|P\|$ is, since every interval $[x_{i-1}, x_i]$ with $x_{i-1} < x_i$ contains both rational and irrational numbers. $\sum_{i=1}^{n} f(x_i^*)\Delta x_i$ cannot simultaneously approach both 0 and $b - a$ as $\|P\| \to 0$, so it has no limit as $\|P\| \to 0$.

39.   Choose $x_i = 1 + \frac{i}{n}$ and $x_i^* = \sqrt{x_{i-1}x_i} = \sqrt{\left(1 + \frac{i-1}{n}\right)\left(1 + \frac{i}{n}\right)}$.

Then $\int_1^2 x^{-2}\, dx = \lim_{n\to\infty} \frac{1}{n} \sum_{i=1}^{n} \frac{1}{\left(1 + \frac{i-1}{n}\right)\left(1 + \frac{i}{n}\right)} = \lim_{n\to\infty} n \sum_{i=1}^{n} \frac{1}{(n+i-1)(n+i)}$

$= \lim_{n\to\infty} n \sum_{i=1}^{n} \left[\frac{1}{n+i-1} - \frac{1}{n+1}\right]$ [by the hint] $= \lim_{n\to\infty} n \left[\sum_{i=0}^{n-1} \frac{1}{n+i} - \sum_{i=1}^{n} \frac{1}{n+i}\right]$

$= \lim_{n\to\infty} n \left[\frac{1}{n} - \frac{1}{2n}\right] = \lim_{n\to\infty} \left[1 - \frac{1}{2}\right] = \frac{1}{2}$.

EXERCISES 5.4

1. $\int_2^6 3\, dx = 3(6 - 2) = 12$ [Property 1]

3. $\int_{-4}^{-1} \sqrt{3}\, dx = \sqrt{3}(-1 + 4) = 3\sqrt{3}$

5. $\int_0^2 (5x + 3)\, dx = \int_0^2 5x\, dx + \int_0^2 3\, dx$ [Property 2] $= 5\int_0^2 x\, dx + \int_0^2 3\, dx$ [Property 3]
    $= 5 \cdot \frac{1}{2}(2^2 - 0^2) + 3(2 - 0) = 16$ [Property 1]

7. $\int_1^4 (2x^2 - 3x + 1)\, dx = 2\int_1^4 x^2\, dx - 3\int_1^4 x\, dx + \int_1^4 1\, dx$
    $= 2 \cdot \frac{1}{3}(4^3 - 1^3) - 3 \cdot \frac{1}{2}(4^2 - 1^2) + 1(4 - 1) = \frac{45}{2} = 22.5$

9. $\int_0^{\pi/3} (1 - 2\cos x)\,dx = \int_0^{\pi/3} 1\,dx - 2\int_0^{\pi/3} \cos x\,dx = 1(\frac{\pi}{3} - 0) - 2\sin\frac{\pi}{3} = \frac{\pi}{3} - \sqrt{3}$

11. $\int_{-2}^{2} |x + 1|\,dx = \int_{-2}^{-1} |x + 1|\,dx + \int_{-1}^{2} |x + 1|\,dx$ [Property 5]

$= \int_{-2}^{-1} (-x - 1)\,dx + \int_{-1}^{2} (x + 1)\,dx = -\int_{-2}^{-1} x\,dx - \int_{-2}^{-1} 1\,dx + \int_{-1}^{2} x\,dx + \int_{-1}^{2} 1\,dx$

$= -\frac{1}{2}\left[(-1)^2 - (-2)^2\right] - 1\left[-1 - (-2)\right] + \frac{1}{2}\left[2^2 - (-1)^2\right] + 1\left[2 - (-1)\right] = 5$

13. Using Property 5 we have $\int_{-2}^{5} [\![x]\!]\,dx$

$= \int_{-2}^{-1} [\![x]\!]\,dx + \int_{-1}^{0} [\![x]\!]\,dx + \int_{0}^{1} [\![x]\!]\,dx + \int_{1}^{2} [\![x]\!]\,dx + \int_{2}^{3} [\![x]\!]\,dx + \int_{3}^{4} [\![x]\!]\,dx + \int_{4}^{5} [\![x]\!]\,dx$

$= \int_{-2}^{-1} (-2)\,dx + \int_{-1}^{0} (-1)\,dx + \int_{0}^{1} 0\,dx + \int_{1}^{2} 1\,dx + \int_{2}^{3} 2\,dx + \int_{3}^{4} 3\,dx + \int_{4}^{5} 4\,dx$

$= -2(-1 + 2) - 1(0 + 1) + 0(1 - 0) + 1(2 - 1) + 2(3 - 2) + 3(4 - 3) + 4(5 - 4) = 7$

15. $\int_{-1}^{1} f(x)\,dx = \int_{-1}^{0} f(x)\,dx + \int_{0}^{1} f(x)\,dx = \int_{-1}^{0} (-2x)\,dx + \int_{0}^{1} 3x^2\,dx$

$= -2\int_{-1}^{0} x\,dx + 3\int_{0}^{1} x^2\,dx = -2 \cdot \frac{1}{2}\left[0^2 - (-1)^2\right] + 3 \cdot \frac{1}{3}\left[1^3 - 0^3\right] = 2$

17. $\int_{0}^{4} x^2\,dx + \int_{4}^{10} x^2\,dx = \int_{0}^{10} x^2\,dx = \frac{1}{3}\left[10^3 - 0^3\right] = \frac{1000}{3}$

19. $\int_{1}^{3} f(x)\,dx + \int_{3}^{6} f(x)\,dx + \int_{6}^{12} f(x)\,dx = \int_{1}^{6} f(x)\,dx + \int_{6}^{12} f(x)\,dx = \int_{1}^{12} f(x)\,dx$

21. $\int_{2}^{10} f(x)\,dx - \int_{2}^{7} f(x)\,dx = \int_{2}^{7} f(x)\,dx + \int_{7}^{10} f(x)\,dx - \int_{2}^{7} f(x)\,dx = \int_{7}^{10} f(x)\,dx$

23. $x \geq x^2$ on $[0, 1]$, so $\int_{0}^{1} x\,dx \geq \int_{0}^{1} x^2\,dx$ [Property 7]

25. $x^2 - 1 \geq 0$ on $[2, 6]$, so $\int_{2}^{6} (x^2 - 1)\,dx \geq 0$ [Property 6]

27. $0 \leq \sin x < 1$ on $[0, \pi/4]$, so $\sin^3 x \leq \sin^2 x$ on $[0, \pi/4]$. Hence $\int_0^{\pi/4} \sin^3 x\,dx \leq \int_0^{\pi/4} \sin^2 x\,dx$

  [Property 7]

29. $5 - x \geq 3 \geq x + 1$ on $[1, 2]$, so $\sqrt{5 - x} \geq \sqrt{x + 1}$ and $\int_{1}^{2} \sqrt{5 - x}\,dx \geq \int_{1}^{2} \sqrt{x + 1}\,dx$

31. $4 \leq x^2 \leq 16$ for $2 \leq x \leq 4$, so $4(4 - 2) \leq \int_{2}^{4} x^2\,dx \leq 16(4 - 2)$ [Property 8]; that is,

  $8 \leq \int_{2}^{4} x^2\,dx \leq 32$

33. If $-1 \leq x \leq 1$, then $0 \leq x^2 \leq 1$ and $1 \leq 1 + x^2 \leq 2$, so $1 \leq \sqrt{1 + x^2} \leq \sqrt{2}$ and

  $1[1 - (-1)] \leq \int_{-1}^{1} \sqrt{1 + x^2}\,dx \leq \sqrt{2}[1 - (-1)]$ [Property 8]; that is,

  $2 \leq \int_{-1}^{1} \sqrt{1 + x^2}\,dx \leq 2\sqrt{2}$

35. $1 \leq x^3 \leq 27$ for $1 \leq x \leq 3$, so $1(3 - 1) \leq \int_{1}^{3} x^3\,dx \leq 27(3 - 1)$. Thus $2 \leq \int_{1}^{3} x^3\,dx \leq 54$

37. If $1 \leq x \leq 2$, then $\frac{1}{2} \leq \frac{1}{x} \leq 1$, so $\frac{1}{2}(2 - 1) \leq \int_{1}^{2} \frac{1}{x}\,dx \leq 1(2 - 1)$ or $\frac{1}{2} \leq \int_{1}^{2} \frac{1}{x}\,dx \leq 1$

39. If $f(x) = x^2 + 2x$, $-3 \leq x \leq 0$, then $f'(x) = 2x + 2 = 0$ when $x = -1$, and $f(-1) = -1$. At

the endpoints, $f(-3) = 3$, $f(0) = 0$. Thus the absolute minimum is $m = -1$ and the absolute maximum is M = 3. Thus $-1\big[0-(-3)\big]\le\int_{-3}^{0}(x^2+2x)\,dx\le 3\big[0-(-3)\big]$ or $-3\le\int_{-3}^{0}(x^2+2x)\,dx\le 9$.

41. For $-1\le x\le 1$, $0\le x^4\le 1$ and $1\le\sqrt{1+x^4}\le\sqrt{2}$, so
$$1\big[1-(-1)\big]\le\int_{-1}^{1}\sqrt{1+x^4}\,dx\le\sqrt{2}\big[1-(-1)\big] \text{ or } 2\le\int_{-1}^{1}\sqrt{1+x^4}\,dx\le 2\sqrt{2}$$

43. $\sqrt{x^4+1}\ge\sqrt{x^4}=x^2$, so $\int_{1}^{3}\sqrt{x^4+1}\,dx\ge\int_{1}^{3}x^2\,dx=\frac{1}{3}(3^3-1^3)=\frac{26}{3}$

45. $0\le\sin x\le 1$ for $0\le x\le\frac{\pi}{2}$, so $x\sin x\le x$
$$\Rightarrow\int_{0}^{\pi/2}x\sin x\,dx\le\int_{0}^{\pi/2}x\,dx=\frac{1}{2}\Big[\big(\frac{\pi}{2}\big)^2-0^2\Big]=\frac{\pi^2}{8}$$

47. $\int_{a}^{b}\big[c\,f(x)+d\,g(x)\big]\,dx=\int_{a}^{b}c\,f(x)\,dx+\int_{a}^{b}d\,g(x)\,dx$ [Property 2]
$=c\int_{a}^{b}f(x)\,dx+d\int_{a}^{b}g(x)\,dx$ [Property 3]

49. By Property 7, $f(x)\le 0$ on $[a,b]$ implies that $\int_{a}^{b}f(x)\,dx\le\int_{a}^{b}0\,dx=0$ [Property 1]

51. By Property 7, the inequalities $-|f(x)|\le f(x)\le|f(x)|$ imply that

$\int_{a}^{b}(-|f(x)|)\,dx\le\int_{a}^{b}f(x)\,dx\le\int_{a}^{b}|f(x)|\,dx$. By Property 3, the left-hand integral equals

$-\int_{a}^{b}|f(x)|\,dx$. Thus $-M\le\int_{a}^{b}f(x)\,dx\le M$, where $M=\int_{a}^{b}|f(x)|\,dx$.

$\big[$Notice that $M\ge 0$ by Property 6.$\big]$ It follows that $\Big|\int_{a}^{b}f(x)\,dx\Big|\le M=\int_{a}^{b}|f(x)|\,dx$.

EXERCISES 5.5

1.
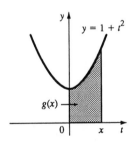

$y=1+t^2$

$g(x)$

$0 \qquad x \qquad t$

(a) By part 1 of the Fundamental Theorem,
$$g(x)=\int_{0}^{x}(1+t^2)\,dt$$
$$\Rightarrow g'(x)=f(x)=1+x^2.$$

(b) By part 2 of the Fundamental Theorem,
$$g(x)=\int_{0}^{x}(1+t^2)\,dt=\Big[t+\tfrac{1}{3}t^3\Big]_{0}^{x}$$
$$=(x+\tfrac{1}{3}x^3)-(0+\tfrac{1}{3}0^3)=x+\tfrac{1}{3}x^3$$
$$\Rightarrow g'(x)=1+x^2.$$

3. $g(x)=\int_{1}^{x}(t^2-1)^{20}\,dt\Rightarrow g'(x)=(x^2-1)^{20}$

5. $g(u)=\int_{\pi}^{u}\frac{1}{1+t^4}\,dt\Rightarrow g'(u)=\frac{1}{1+u^4}$

7. $F(x)=\int_{x}^{2}\cos(t^2)\,dt=-\int_{2}^{x}\cos(t^2)\,dt\Rightarrow F'(x)=-\cos(x^2)$

9. Let $u = \frac{1}{x}$. Then $\frac{du}{dx} = -\frac{1}{x^2}$, so $\frac{d}{dx}\int_2^{1/x} \sin^4 t\, dt = \frac{d}{du}\int_2^u \sin^4 t\, dt \cdot \frac{du}{dx} = \sin^4 u \frac{du}{dx}$

$= -\sin^4\left(\frac{1}{x}\right)/x^2$

11. Let $u = \tan x$. Then $\frac{du}{dx} = \sec^2 x$, so $\frac{d}{dx}\int_{\tan x}^{17} \sin(t^4)\, dt = -\frac{d}{dx}\int_{17}^{\tan x} \sin(t^4)\, dt$

$= -\frac{d}{du}\int_{17}^u \sin(t^4)\, dt \cdot \frac{du}{dx} = -\sin(u^4)\frac{du}{dx} = -\sin(\tan^4 x)\sec^2 x$

13. Let $t = 5x + 1$. Then $\frac{dt}{dx} = 5$, so $\frac{d}{dx}\int_0^{5x+1} \frac{1}{u^2-5}\, du = \frac{d}{dt}\int_0^t \frac{1}{u^2-5}\, du \cdot \frac{dt}{dx}$

$= \frac{1}{t^2-5}\frac{dt}{dx} = \frac{5}{(25x^2 + 10x - 4)}$

15. $\int_{-1}^8 6\, dx = 6x\big]_{-1}^8 = 48 - (-6) = 54$

17. $\int_0^1 x^{99}\, dx = \frac{1}{100}x^{100}\big|_0^1 = \frac{1^{100}}{100} - \frac{0^{100}}{100} = \frac{1}{100}$

19. $\int_0^1 (1 - 2x - 3x^2)\, dx = \left[x - 2\frac{x^2}{2} - 3\frac{x^3}{3}\right]_0^1 = \left[x - x^2 - x^3\right]_0^1 = (1 - 1 - 1) - 0 = -1$

21. $\int_{-3}^0 (5y^4 - 6y^2 + 14)\, dy = 5\frac{y^5}{5} - 6\frac{y^3}{3} + 14y\,]_{-3}^0 = y^5 - 2y^3 + 14y\,]_{-3}^0$

$= 0 - (-243 + 54 - 42) = 231$

23. $\int_0^4 \sqrt{x}\, dx = \int_0^4 x^{1/2}\, dx = \left[\frac{x^{3/2}}{3/2}\right]_0^4 = \left[\frac{2}{3}x^{3/2}\right]_0^4 = \frac{2}{3}(4)^{3/2} - 0 = \frac{16}{3}$

25. $\int_1^3 \left[\frac{1}{t^2} - \frac{1}{t^4}\right] dt = \int_1^3 (t^{-2} - t^{-4})\, dt = \left[\frac{t^{-1}}{-1} - \frac{t^{-3}}{-3}\right]_1^3 = \left[\frac{1}{3t^3} - \frac{1}{t}\right]_1^3$

$= \left[\frac{1}{81} - \frac{1}{3}\right] - \left[\frac{1}{3} - 1\right] = \frac{28}{81}$

27. $\int_1^2 \frac{x^2+1}{\sqrt{x}}\, dx = \int_1^2 (x^{3/2} + x^{-1/2})\, dx = \left[\frac{x^{5/2}}{5/2} + \frac{x^{1/2}}{1/2}\right]_1^2 = \frac{2}{5}x^{5/2} + 2x^{1/2}\,]_1^2$

$= \left(\frac{2}{5}4\sqrt{2} + 2\sqrt{2}\right) - \left(\frac{2}{5} + 2\right) = \frac{18\sqrt{2} - 12}{5} = \frac{6(3\sqrt{2} - 2)}{5}$

29. $\int_0^1 u(\sqrt{u} + \sqrt[3]{u})\, du = \int_0^1 (u^{3/2} + u^{4/3})\, du = \left[\frac{u^{5/2}}{5/2} + \frac{u^{7/3}}{7/3}\right]_0^1 = \frac{2}{5}u^{5/2} + \frac{3}{7}u^{7/3}\,]_0^1$

$= \frac{2}{5} + \frac{3}{7} = \frac{29}{35}$

31. $\int_{-2}^3 |x^2 - 1|\, dx = \int_{-2}^{-1} (x^2 - 1)\, dx + \int_{-1}^1 (1 - x^2)\, dx + \int_1^3 (x^2 - 1)\, dx$

$= \left[\frac{x^3}{3} - x\right]_{-2}^{-1} + \left[x - \frac{x^3}{3}\right]_{-1}^1 + \left[\frac{x^3}{3} - x\right]_1^3$

$= (-\frac{1}{3} + 1) - (-\frac{8}{3} + 2) + (1 - \frac{1}{3}) - (-1 + \frac{1}{3}) + (9 - 3) - (\frac{1}{3} - 1) = \frac{28}{3}$

33. $\int_3^3 \sqrt{x^5 + 2}\, dx = 0$

35. $\int_{-4}^{2} \frac{2}{x^6}\,dx$ does not exist since $f(x) = \frac{2}{x^6}$ has an infinite discontinuity at 0.

37. $\int_{1}^{4} (\sqrt{t} - \frac{2}{\sqrt{t}})\,dt = \int_{1}^{4} (t^{1/2} - 2t^{-1/2})\,dt = \left[\frac{t^{3/2}}{3/2} - 2\frac{t^{1/2}}{1/2}\right]_{1}^{4} = \frac{2}{3}t^{3/2} - 4t^{1/2}]_{1}^{4}$

$= \left[\frac{2}{3}8 - 4\cdot 2\right] - \left[\frac{2}{3} - 4\right] = \frac{2}{3}$

39. $\int_{-1}^{0} (x+1)^3\,dx = \int_{-1}^{0} (x^3 + 3x^2 + 3x + 1)\,dx = \frac{x^4}{4} + 3\frac{x^3}{3} + 3\frac{x^2}{2} + x]_{-1}^{0}$

$= \left[\frac{x^4}{4} + x^3 + \frac{3}{2}x^2 + x\right]_{-1}^{0} = 0 - \left[\frac{1}{4} - 1 + \frac{3}{2} - 1\right] = 2 - \frac{7}{4} = \frac{1}{4}$

41. $\int_{\pi/4}^{\pi/3} \sin t\,dt = -\cos t]_{\pi/4}^{\pi/3} = -\cos\frac{\pi}{3} + \cos\frac{\pi}{4} = -\frac{1}{2} + \frac{1}{\sqrt{2}} = \frac{\sqrt{2}-1}{2}$

43. $\int_{\pi/2}^{\pi} \sec x \tan x\,dx$ does not exist since $\sec x \tan x$ has an infinite discontinuity at $\pi/2$.

45. $\int_{\pi/6}^{\pi/3} \csc^2\theta\,d\theta = -\cot\theta]_{\pi/6}^{\pi/3} = -\cot\frac{\pi}{3} + \cot\frac{\pi}{6} = -\frac{1}{3}\sqrt{3} + \sqrt{3} = \frac{2}{3}\sqrt{3}$

47. $\int_{4}^{8} \frac{1}{x}\,dx = \ln x]_{4}^{8} = \ln 8 - \ln 4 = \ln\frac{8}{4} = \ln 2$

49. $\int_{8}^{9} 2^t\,dt = \left[\frac{1}{\ln 2}2^t\right]_{8}^{9} = \frac{1}{\ln 2}(2^9 - 2^8) = \frac{2^8}{\ln 2}$

51. $\int_{1}^{\sqrt{3}} \frac{6}{1+x^2}\,dx = 6\left[\tan^{-1}x\right]_{1}^{\sqrt{3}} = 6\tan^{-1}\sqrt{3} - 6\tan^{-1}1 = 6\frac{\pi}{3} - 6\frac{\pi}{4} = \frac{\pi}{2}$

53. $\int_{1}^{e} \frac{x^2 + x + 1}{x}\,dx = \int_{1}^{e}\left[x + 1 + \frac{1}{x}\right]\,dx = \frac{x^2}{2} + x + \ln x]_{1}^{e} = \left[\frac{e^2}{2} + e + \ln e\right] - \left[\frac{1}{2} + 1 + \ln 1\right]$

$= \frac{e^2}{2} + e - \frac{1}{2}$

55. $\int_{0}^{1}\left[\sqrt[4]{x^5} + \sqrt[5]{x^4}\right]\,dx = \int_{0}^{1} (x^{5/4} + x^{4/5})\,dx = \left[\frac{x^{9/4}}{9/4} + \frac{x^{9/5}}{9/5}\right]_{0}^{1} = \frac{4}{9}x^{9/4} + \frac{5}{9}x^{9/5}]_{0}^{1}$

$= \frac{4}{9} + \frac{5}{9} - 0 = 1$

57. $\int_{-1}^{2} (x - 2|x|)\,dx = \int_{-1}^{0} 3x\,dx + \int_{0}^{2} (-x)\,dx = 3\left[\frac{x^2}{2}\right]_{-1}^{0} - \left[\frac{x^2}{2}\right]_{0}^{2} = \left[3\cdot 0 - 3\frac{1}{2}\right] - (2 - 0)$

$= -\frac{7}{2} = -3.5$

59. $\int_{0}^{2} f(x)\,dx = \int_{0}^{1} x^4\,dx + \int_{1}^{2} x^5\,dx = \left[\frac{x^5}{5}\right]_{0}^{1} + \left[\frac{x^6}{6}\right]_{1}^{2} = \left(\frac{1}{5} - 0\right) + \left(\frac{64}{6} - \frac{1}{6}\right) = 10.7$

61. area $= \int_{0}^{2} (4x^2 - 4x + 3)\,dx = 4\frac{x^3}{3} - 4\frac{x^2}{2} + 3x]_{0}^{2} = 4\frac{8}{3} - 4\cdot 2 + 3\cdot 2 - 0 = \frac{26}{3}$

63. area $= \int_{0}^{27} x^{1/3}\,dx = \frac{3}{4}x^{4/3}]_{0}^{27} = \frac{3}{4}81 - 0 = \frac{243}{4}$

65. area $= \int_{0}^{\pi} \sin x\,dx = -\cos x]_{0}^{\pi} = -\cos\pi + \cos 0 = -(-1) + 1 = 2$

67. $\text{area} = \int_{-1}^{2} x^{0.8}\,dx = \left[\frac{x^{1.8}}{1.8}\right]_{-1}^{2} = \frac{2^{1.8}}{1.8} - \frac{-1}{1.8} = \frac{1+2^{1.8}}{1.8}$ or $\frac{5}{9}(1+2^{9/5})$

69. $\frac{d}{dx}\left[\frac{x}{a^2\sqrt{a^2-x^2}} + C\right] = \frac{1}{a^2}\frac{\sqrt{a^2-x^2} - x(-x)/\sqrt{a^2-x^2}}{a^2-x^2}$

   $= \frac{1}{a^2}\frac{(a^2-x^2)+x^2}{(a^2-x^2)^{3/2}} = \frac{1}{\sqrt{(a^2-x^2)^3}}$

71. $\frac{d}{dx}\left(\frac{x}{2} - \frac{\sin 2x}{4} + C\right) = \frac{1}{2} - \frac{1}{4}(\cos 2x)(2) + 0 = \frac{1}{2} - \frac{1}{2}\cos 2x = \frac{1}{2} - \frac{1}{2}(1 - 2\sin^2 x) = \sin^2 x$

73. $\int x\sqrt{x}\,dx = \int x^{3/2}\,dx = \frac{2}{5}x^{5/2} + C$

75. $\int (2-\sqrt{x})^2\,dx = \int (4 - 4\sqrt{x} + x)\,dx = 4x - 4\frac{x^{3/2}}{3/2} + \frac{x^2}{2} + C = 4x - \frac{8}{3}x^{3/2} + \frac{1}{2}x^2 + C$

77. $\int (2x + \sec x \tan x)\,dx = x^2 + \sec x + C$

79. (a) $\text{displacement} = \int_0^3 (3t-5)\,dt = \frac{3}{2}t^2 - 5t\big]_0^3 = \frac{27}{2} - 15 = -\frac{3}{2}\,m$

   (b) $\text{distance traveled} = \int_0^3 |3t-5|\,dt = \int_0^{5/3} (5-3t)\,dt + \int_{5/3}^3 (3t-5)\,dt$

   $= \left[5t - \frac{3}{2}t^2\right]_0^{5/3} + \left[\frac{3}{2}t^2 - 5t\right]_{5/3}^3 = \frac{25}{3} - \frac{3}{2}\cdot\frac{25}{9} + \frac{27}{2} - 15 - \left(\frac{3}{2}\cdot\frac{25}{9} - \frac{25}{3}\right) = \frac{41}{6}\,m$

81. (a) $v'(t) = a(t) = t+4 \Rightarrow v(t) = (1/2)t^2 + 4t + C \Rightarrow 5 = v(0) = C \Rightarrow$

   $v(t) = (1/2)t^2 + 4t + 5\,m/s$ [OR: $v(t) - v(0) = \int_0^t a(u)\,du$

   $= \int_0^t (u+4)\,du = \frac{u^2}{2} + 4u\big]_0^t = \frac{t^2}{2} + 4t \Rightarrow v(t) = \frac{t^2}{2} + 4t + 5\,m/s.$]

   (b) $\text{distance traveled} = \int_0^{10} |v(t)|\,dt = \int_0^{10}\left|\frac{t^2}{2} + 4t + 5\right|dt = \int_0^{10}\left[\frac{t^2}{2} + 4t + 5\right]dt$

   $= \left[\frac{t^3}{6} + 2t^2 + 5t\right]_0^{10} = \frac{500}{3} + 200 + 50 = \frac{1250}{3}\,m$

83. Since $m'(x) = \rho(x)$, $m = \int_0^4 \rho(x)dx = \int_0^4 (9 + 2\sqrt{x})dx = \left[9x + \frac{4}{3}x^{3/2}\right]_0^4$

   $= 36 + \frac{32}{3} - 0 = \frac{140}{3} = 46\frac{2}{3}\,kg.$

85. $g(x) = \int_{2x}^{3x} \frac{u-1}{u+1}\,du = \int_{2x}^0 \frac{u-1}{u+1}\,du + \int_0^{3x} \frac{u-1}{u+1}\,du = -\int_0^{2x} \frac{u-1}{u+1}\,du + \int_0^{3x} \frac{u-1}{u+1}\,du$

   $\Rightarrow g'(x) = -\frac{2x-1}{2x+1}\cdot\frac{d}{dx}(2x) + \frac{3x-1}{3x+1}\cdot\frac{d}{dx}(3x) = -\frac{2x-1}{2x+1}\cdot 2 + \frac{3x-1}{3x+1}\cdot 3$

87. $y = \int_{\sqrt{x}}^{x^3} \sqrt{t}\sin t\,dt = \int_{\sqrt{x}}^1 \sqrt{t}\sin t\,dt + \int_1^{x^3} \sqrt{t}\sin t\,dt = -\int_1^{\sqrt{x}} \sqrt{t}\sin t\,dt + \int_1^{x^3} \sqrt{t}\sin t\,dt$

   $\Rightarrow y' = -\sqrt[4]{x}(\sin\sqrt{x})\cdot\frac{d}{dx}(\sqrt{x}) + x^{3/2}\sin(x^3)\cdot\frac{d}{dx}(x^3)$

   $= -\sqrt[4]{x}(\sin\sqrt{x})/(2\sqrt{x}) + x^{3/2}\sin(x^3)(3x^2) = 3x^{7/2}\sin(x^3) - (\sin\sqrt{x})/(2\sqrt[4]{x})$

89. $F(x) = \int_1^x f(t)\,dt \Rightarrow F'(x) = f(x) = \int_1^{x^2} \frac{\sqrt{1+u^4}}{u}\,du$

$\Rightarrow F''(x) = f'(x) = \frac{\sqrt{1+(x^2)^4}}{x^2} \cdot \frac{d}{dx}(x^2) = \frac{2\sqrt{1+x^8}}{x}$. So $F''(2) = \sqrt{1+2^8} = \sqrt{257}$.

91. $\lim\limits_{n \to \infty} \sum\limits_{i=1}^{n} \frac{i^3}{n^4} = \lim\limits_{n \to \infty} \frac{1-0}{n} \sum\limits_{i=1}^{n} \left(\frac{i}{n}\right)^3 = \int_0^1 x^3\,dx = \frac{1}{4}x^4\Big|_0^1 = \frac{1}{4}$.

93. By (5.18), $\dfrac{g(x+h) - g(x)}{h} = \dfrac{1}{h}\int_x^{x+h} f$ for $h \neq 0$. Suppose $h < 0$. Since f is continuous on $[x+h, x]$, the Extreme Value Theorem says that there are numbers u and v in $[x+h, x]$ such that $f(u) = m$ and $f(v) = M$, where m and M are the absolute minimum and maximum values of f on $[x+h, x]$. By Property 8 of integrals,

$m(-h) \leq \int_{x+h}^x f \leq M(-h)$; that is, $f(u)(-h) \leq -\int_x^{x+h} f \leq f(v)(-h)$. Since $-h > 0$, we can

divide this inequality by $-h$: $f(u) \leq \dfrac{1}{h}\int_x^{x+h} f \leq f(v)$. Using (5.18) to replace the middle part

of the inequality, we obtain $f(u) \leq \dfrac{g(x+h) - g(x)}{h} \leq f(v)$ which is (5.19) in the case

where $h < 0$.

95. (a) Let $f(x) = \sqrt{x} \Rightarrow f'(x) = \frac{1}{2\sqrt{x}} > 0$ for $x > 0 \Rightarrow f$ is increasing on $[0, \infty)$. If $x \geq 0$,

then $x^3 \geq 0$, so $1 + x^3 \geq 1$ and since f is increasing, this means that

$f(1 + x^3) \geq f(1) \Rightarrow \sqrt{1+x^3} \geq 1$ for $x \geq 0$.   Next let

$g(t) = t^2 - t \Rightarrow g'(t) = 2t - 1 \Rightarrow g'(t) > 0$ when $t \geq 1$. Thus g is increasing on $[1, \infty)$.

And since $g(1) = 0$, $g(t) \geq 0$ when $t \geq 1$. Now let $t = \sqrt{1+x^3}$, where $x \geq 0$.

$\sqrt{1+x^3} \geq 1$ *(from above)* $\Rightarrow t \geq 1 \Rightarrow g(t) \geq 0 \Rightarrow (1+x^3) - \sqrt{1+x^3} \geq 0$ for $x \geq 0$.

Therefore $1 \leq \sqrt{1+x^3} \leq 1 + x^3$ for $x \geq 0$.

(b) From part (a) and Property 7: $\int_0^1 1\,dx \leq \int_0^1 \sqrt{1+x^3}\,dx \leq \int_0^1 (1+x^3)\,dx$

$\Leftrightarrow [x]_0^1 \leq \int_0^1 \sqrt{1+x^3}\,dx \leq \left[x + \frac{1}{4}x^4\right]_0^1 \Leftrightarrow 1 \leq \int_0^1 \sqrt{1+x^3}\,dx \leq 1 + \frac{1}{4} = 1.25$.

EXERCISES 5.6

1. Let $u = x^2 - 1$. Then $du = 2x\,dx$, so $\int x(x^2-1)^{99}\,dx = \int u^{99}\,\frac{1}{2}\,du = \frac{1}{2}\frac{u^{100}}{100} + C$

$= (x^2 - 1)^{100}/200 + C$

3. Let $u = 4x$. Then $du = 4\,dx$, so $\int \sin 4x\,dx = \int \sin u\,(1/4)\,du = (1/4)(-\cos u) + C$

$$= -(1/4)\cos 4x + C$$

5. Let $u = x^2 + 6x$. Then $du = 2(x+3)\,dx$, so $\displaystyle\int \frac{x+3}{(x^2+6x)^2}\,dx = \frac{1}{2}\int \frac{du}{u^2} = \frac{1}{2}\int u^{-2}\,du$

$$= -\tfrac{1}{2}u^{-1} + C = -1/\left[2(x^2+6x)\right] + C$$

7. Let $u = x^2 + x + 1$. Then $du = (2x+1)\,dx$, so $\displaystyle\int (2x+1)(x^2+x+1)^3\,dx = \int u^3\,du$

$$= u^4/4 + C = (x^2+x+1)^4/4 + C$$

9. Let $u = x - 1$. Then $du = dx$, so $\int \sqrt{x-1}\,dx = \int u^{1/2}\,du = \frac{2}{3}u^{3/2} + C$

$$= \tfrac{2}{3}(x-1)^{3/2} + C$$

11. Let $u = 2 + x^4$. Then $du = 4x^3\,dx$, so $\int x^3\sqrt{2+x^4}\,dx = \int u^{1/2}\frac{1}{4}\,du = \frac{1}{4}\frac{u^{3/2}}{3/2} + C$

$$= \tfrac{1}{6}(2+x^4)^{3/2} + C$$

13. Let $u = 3x^2 - 2x + 1$. Then $du = 2(3x-1)\,dx$, so $\displaystyle\int \frac{3x-1}{(3x^2-2x+1)^4}\,dx = \int u^{-4}\frac{1}{2}\,du$

$$= \tfrac{1}{2}\cdot\frac{u^{-3}}{-3} + C = -1/\left[6(3x^2-2x+1)^3\right] + C$$

15. Let $u = t + 1$. Then $du = dt$, so $\displaystyle\int \frac{2}{(t+1)^6}\,dt = 2\int u^{-6}\,du = -\frac{2}{5}u^{-5} + C$

$$= -2/\left[5(t+1)^5\right] + C$$

17. Let $u = 1 - 2y$. Then $du = -2\,dy$, so $\int (1-2y)^{1.3}\,dy = \int u^{1.3}\left(-\frac{1}{2}\right)du$

$$= -\tfrac{1}{2}\frac{u^{2.3}}{2.3} + C = -\frac{(1-2y)^{2.3}}{4.6} + C$$

19. Let $u = 2\theta$. Then $du = 2\,d\theta$, so $\int \cos 2\theta\,d\theta = \int \cos u \frac{1}{2}\,du = \frac{1}{2}\sin u + C = \frac{1}{2}\sin 2\theta + C$

21. Let $u = x + 2$. Then $du = dx$, so $\displaystyle\int \frac{x}{\sqrt[4]{x+2}}\,dx = \int \frac{u-2}{\sqrt[4]{u}}\,du = \int (u^{3/4} - 2u^{-1/4})\,du$

$$= (4/7)u^{7/4} - 2(4/3)u^{3/4} + C = (4/7)(x+2)^{7/4} - (8/3)(x+2)^{3/4} + C$$

23. Let $u = t^2$. Then $du = 2t\,dt$, so $\int t\sin(t^2)\,dt = \int \sin u (1/2)\,du = -(1/2)\cos u + C$

$$= -(1/2)\cos(t^2) + C$$

25. Let $u = 1 - x^2$. Then $x^2 = 1 - u$ and $2x\,dx = -du$, so $\int x^3(1-x^2)^{3/2}\,dx$

$$= \int (1-x^2)^{3/2}x^2\cdot x\,dx = \int u^{3/2}(1-u)(-\tfrac{1}{2})\,du = \tfrac{1}{2}\int (u^{5/2} - u^{3/2})\,du$$

$$= \tfrac{1}{2}\left[\tfrac{2}{7}u^{7/2} - \tfrac{2}{5}u^{5/2}\right] + C = \tfrac{1}{7}(1-x^2)^{7/2} - \tfrac{1}{5}(1-x^2)^{5/2} + C$$

27. Let $u = \sin x$. Then $du = \cos x\,dx$, so $\int \sin^3 x\cos x\,dx = \int u^3\,du = \frac{u^4}{4} + C = \frac{1}{4}\sin^4 x + C$

29. Let $u = 2x - 1$. Then $du = 2\,dx$, so $\displaystyle\int \frac{dx}{2x-1} = \int \frac{(1/2)\,du}{u} = (1/2)\ln|u| + C$

$$= (1/2)\ln|2x-1| + C.$$

31. Let $u = \ln x$. Then $du = \frac{dx}{x}$, so $\displaystyle\int \frac{(\ln x)^2}{x}\,dx = \int u^2\,du = u^3/3 + C = (\ln x)^3/3 + C$

33. Let $u = 1 + e^x$. Then $du = e^x\,dx$, so $\int e^x(1+e^x)^{10}\,dx = \int u^{10}\,du = u^{11}/11 + C$
    $= (1+e^x)^{11}/11 + C$

35. Let $u = 1 + \sec x$. Then $du = \sec x \tan x\,dx$, so $\int \sec x \tan x\sqrt{1+\sec x}\,dx = \int u^{1/2}\,du$
    $= (2/3)u^{3/2} + C = (2/3)(1+\sec x)^{3/2} + C$.

37. Let $u = ax^2 + 2bx + c$. Then $du = 2(ax+b)\,dx$, so $\displaystyle\int \frac{(ax+b)\,dx}{\sqrt{ax^2+2bx+c}} = \int \frac{(1/2)\,du}{\sqrt{u}}$
    $= \frac{1}{2}\int u^{-1/2}\,du = u^{1/2} + C = \sqrt{ax^2+2bx+c} + C$

39. Let $u = 2x + 3$. Then $du = 2\,dx$, so $\int \sin(2x+3)\,dx = \int \sin u\,(1/2)\,du$
    $= -(1/2)\cos u + C = -(1/2)\cos(2x+3) + C$

41. Let $u = 3x$. Then $du = 3\,dx$, so $\int (\sin 3\alpha - \sin 3x)\,dx = \int (\sin 3\alpha - \sin u)(1/3)\,du$
    $= (1/3)\big[(\sin 3\alpha)u + \cos u\big] + C = (\sin 3\alpha)x + (1/3)\cos 3x + C$

43. Let $u = b + cx^{a+1}$. Then $du = (a+1)cx^a\,dx$, so $\int x^a\sqrt{b+cx^{a+1}}\,dx = \int u^{1/2}\frac{1}{(a+1)c}\,du$
    $= \frac{1}{(a+1)c}\frac{2}{3}u^{3/2} + C = \frac{2}{3c(a+1)}(b+cx^{a+1})^{3/2} + C$

45. Let $u = \ln x$. Then $du = \frac{dx}{x}$, so $\displaystyle\int \frac{dx}{x\ln x} = \int \frac{du}{u} = \ln|u| + C = \ln|\ln x| + C$

47. $\displaystyle\int \frac{e^x + 1}{e^x}\,dx = \int (1 + e^{-x})\,dx = x - e^{-x} + C$ [*Substitute* $u = -x$.]

49. Let $u = x^2 + 2x$. Then $du = 2(x+1)\,dx$, so $\displaystyle\int \frac{x+1}{x^2+2x}\,dx = \int \frac{(1/2)\,du}{u} = \frac{1}{2}\ln|u| + C$
    $= \frac{1}{2}\ln|x^2 + 2x| + C$

51. Let $u = e^x$. Then $du = e^x\,dx$, so $\displaystyle\int \frac{e^x}{e^{2x}+1}\,dx = \int \frac{du}{u^2+1} = \tan^{-1}u + C = \tan^{-1}(e^x) + C$

53. $\displaystyle\int \frac{1+x}{1+x^2}\,dx = \int \frac{1}{1+x^2}\,dx + \int \frac{x}{1+x^2}\,dx = \tan^{-1}x + \frac{1}{2}\int \frac{2x\,dx}{1+x^2}$
    $= \tan^{-1}x + \frac{1}{2}\ln(1+x^2) + C$ [*The second integral is* $\int du/u$ *where* $u = 1 + x^2$.]

55. Let $u = x^2$. Then $du = 2x\,dx$, so $\displaystyle\int \frac{x}{1+x^4}\,dx = \int \frac{(1/2)\,du}{1+u^2} = (1/2)\tan^{-1}u + C$
    $= (1/2)\tan^{-1}(x^2) + C$

57. Let $u = 2x - 1$. Then $du = 2\,dx$, so $\displaystyle\int_0^1 (2x-1)^{100}\,dx = \int_{-1}^{1} u^{100}\frac{1}{2}\,du = \int_0^1 u^{100}\,du$
    [*since this is an even function*] $= \left[\frac{u^{101}}{101}\right]_0^1 = \frac{1}{101}$

59. Let $u = x^4 + x$. Then $du = (4x^3+1)\,dx$, so $\displaystyle\int_0^1 (x^4+x)^5(4x^3+1)\,dx = \int_0^2 u^5\,du = \left[\frac{u^6}{6}\right]_0^2$
    $= \frac{2^6}{6} = \frac{32}{3}$

61. Let $u = x - 1$. Then $du = dx$, so $\int_1^2 x\sqrt{x-1}\,dx = \int_0^1 (u+1)\sqrt{u}\,du = \int_0^1 (u^{3/2} + u^{1/2})\,du$

$= \left[\frac{2}{5}u^{5/2} + \frac{2}{3}u^{3/2}\right]_0^1 = \frac{2}{5} + \frac{2}{3} = \frac{16}{15}$

63. Let $u = \pi t$. Then $du = \pi\,dt$, so $\int_0^1 \cos \pi t\,dt = \int_0^\pi \cos u\left(\frac{1}{\pi}du\right) = \frac{1}{\pi}\sin u\big]_0^\pi = \frac{1}{\pi}(0 - 0) = 0$

65. Let $u = 1 + \frac{1}{x}$. Then $du = -\frac{dx}{x^2}$, so $\int_1^4 \frac{1}{x^2}\sqrt{1 + \frac{1}{x}}\,dx = \int_2^{5/4} u^{1/2}(-du) = \int_{5/4}^2 u^{1/2}\,du$

$= \frac{2}{3}u^{3/2}\big]_{5/4}^2 = \frac{2}{3}\left[2\sqrt{2} - \frac{5\sqrt{5}}{8}\right] = \frac{4\sqrt{2}}{3} - \frac{5\sqrt{5}}{12}$

67. Let $u = \cos \theta$. Then $du = -\sin \theta\,d\theta$, so $\int_0^{\pi/3} \frac{\sin \theta}{\cos^2 \theta}\,d\theta = \int_1^{1/2} \frac{-du}{u^2} = \int_{1/2}^1 u^{-2}\,du$

$= \left[-\frac{1}{u}\right]_{1/2}^1 = -1 + 2 = 1$

69. Let $u = 1 + 2x$. Then $du = 2\,dx$, so $\int_0^{13} \frac{dx}{\sqrt[3]{(1+2x)^2}} = \int_1^{27} u^{-2/3}\frac{1}{2}\,du = \frac{1}{2}3u^{1/3}\big]_1^{27}$

$= \frac{3}{2}(3 - 1) = 3$

71. Let $u = a^2 - x^2$. Then $du = -2x\,dx$, so $\int_0^a x\sqrt{a^2 - x^2}\,dx = \int_{a^2}^0 u^{1/2}\left(-\frac{1}{2}du\right)$

$= \frac{1}{2}\int_0^{a^2} u^{1/2}\,du = \frac{1}{2}\cdot\frac{2}{3}u^{3/2}\big]_0^{a^2} = \frac{a^3}{3}$

73. $\int_{-a}^a x\sqrt{x^2 + a^2}\,dx = 0$ since $f(x) = x\sqrt{x^2 + a^2}$ is an odd function.

75. Let $u = 2x + 3$. Then $du = 2\,dx$, so $\int_0^3 \frac{dx}{2x+3} = \int_3^9 \frac{(1/2)\,du}{u} = \frac{1}{2}\ln u\big]_3^9 = \frac{1}{2}(\ln 9 - \ln 3)$

$= \frac{1}{2}(\ln 3^2 - \ln 3) = \frac{1}{2}(2\ln 3 - \ln 3) = \frac{1}{2}\ln 3$

77. Let $u = \ln x$. Then $du = \frac{dx}{x}$, so $\int_e^{e^4} \frac{dx}{x\sqrt{\ln x}} = \int_1^4 u^{-1/2}\,du = 2u^{1/2}\big]_1^4 = 2\cdot 2 - 2\cdot 1 = 2$

79. Let $u = x + 1$. Then $du = dx$, so area $= \int_0^3 \sqrt{x+1}\,dx = \int_1^4 u^{1/2}\,du = \frac{2}{3}u^{3/2}\big]_1^4$

$= \frac{2}{3}(8 - 1) = \frac{14}{3}$

81. Let $u = x/2$. Then $du = dx/2$, so area $= \int_0^{\pi/3} \sin(x/2)\,dx = \int_0^{\pi/6} \sin u \cdot 2\,du$

$= 2(-\cos u)\big]_0^{\pi/6} = -\sqrt{3} - (-2) = 2 - \sqrt{3}$

83. Let $u = x + 1$. Then $du = dx$, so area $= \int_0^{10} \frac{dx}{(x+1)^2} = \int_1^{11} u^{-2}\,du = \left[-\frac{1}{u}\right]_1^{11}$

$= -\frac{1}{11} + 1 = \frac{10}{11}$

85. Let $u = -2x$. Then $du = -2\,dx$, so area $= \int_0^1 2e^{-2x}\,dx = \int_0^{-2} e^u(-du) = \int_{-2}^0 e^u\,du$

$= e^u\big]_{-2}^0 = 1 - e^{-2}$

87. The volume of inhaled air in the lungs at time t is $V(t) = \int_0^t f(u)\,du$

$= \int_0^t (1/2)\sin(2\pi u/5)\,du = \int_0^{2\pi t/5} \frac{1}{2}\sin v \frac{5}{2\pi}\,dv$, $[v = 2\pi u/5 \Rightarrow dv = (2\pi/5)\,du]$

$$= \tfrac{5}{4\pi}(-\cos v)]_0^{2\pi t/5} = \tfrac{5}{4\pi}\left[-\cos(2\pi t/5) + 1\right] = \tfrac{5}{4\pi}[1 - \cos(2\pi t/5)] \text{ liters}$$

89. Let $u = -x$. Then $du = -dx$. When $x = a$, $u = -a$; when $x = b$, $u = -b$. So

$$\int_a^b f(-x)\,dx = \int_{-a}^{-b} f(u)(-du) = \int_{-b}^{-a} f(u)\,du = \int_{-b}^{-a} f(x)\,dx$$

91. Let $u = 1 - x$. Then $du = -dx$. When $x = 1$, $u = 0$ and when $x = 0$, $u = 1$. So

$$\int_0^1 x^a(1-x)^b dx = -\int_1^0 (1-u)^a u^b du = \int_0^1 u^b(1-u)^a du = \int_0^1 x^b(1-x)^a\,dx.$$

93. (a) $\frac{d}{dx}(\ln|\sin x| + C) = \frac{1}{\sin x}\cos x = \cot x$

(b) Let $u = \sin x$. Then $du = \cos x\,dx$, so $\int \cot x\,dx = \int \frac{\cos x}{\sin x}\,dx$

$$= \int \frac{du}{u} = \ln|u| + C = \ln|\sin x| + C$$

EXERCISES 5.7

1. (a)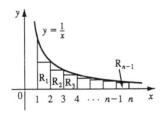

We interpret $\ln 1.5$ as the area under the curve $y = \frac{1}{x}$ from $x = 1$ to $x = 1.5$. The area of the square BCDE is $\frac{1}{2} \times \frac{2}{3} = \frac{1}{3}$. The area of the trapezoid ABCD is $\frac{1}{2} \times \frac{1}{2}(1 + \frac{2}{3}) = \frac{5}{12}$. Thus, by comparing areas, one observes that $\frac{1}{3} < \ln 1.5 < \frac{5}{12}$

(b) With $f(t) = 1/t$, $n = 10$, and $\Delta x = 0.05$, we have

$$\ln 1.5 = \int_1^{1.5} (1/t)\,dt \approx (0.05)[f(1.025) + f(1.075) + \cdots + f(1.475)]$$

$$= (0.05)\left[\tfrac{1}{1.025} + \tfrac{1}{1.075} + \cdots + \tfrac{1}{1.475}\right] \approx 0.4054.$$

3.

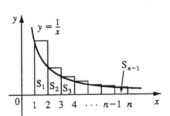

The area of R_i is $\frac{1}{i+1}$ and so $\frac{1}{2} + \frac{1}{3} + \cdots + \frac{1}{n} < \int_1^n \frac{1}{t}\,dt = \ln n$.

The area of S_i is $\frac{1}{i}$ and so $1 + \frac{1}{2} + \cdots + \frac{1}{n-1} > \int_1^n \frac{1}{t}\,dt = \ln n$.

5. If $f(x) = \ln(x^r)$, then $f'(x) = (1/x^r)(rx^{r-1}) = r/x$. But if $g(x) = r \ln x$, then

$g'(x) = r/x$. So f and g must differ by a constant : $\ln (x^r) = r \ln x + C$.

Put $x = 1$: $\ln (1^r) = r \ln 1 + C \Rightarrow C = 0$, so $\ln (x^r) = r \ln x$.

7. From (5.41c) and (5.49) we have $\ln [(e^x)^r] = r \ln (e^x) = rx = \ln (e^{rx})$

and, since ln is one-to-one, it follows that $(e^x)^r = e^{rx}$

9. Using (5.53), (5.41), and (5.51), we have $(ab)^x = e^{x \ln ab}$

$= e^{x (\ln a + \ln b)} = e^{x \ln a + x \ln b} = e^{x \ln a} e^{x \ln b} = a^x b^x$.

11. Let $\log_a x = r$ and $\log_a y = s$. Then $a^r = x$ and $a^s = y$.

(a) $xy = a^r a^s = a^{r+s} \Rightarrow \log_a (xy) = r + s = \log_a x + \log_a y$

(b) $\frac{x}{y} = \frac{a^r}{a^s} = a^{r-s} \Rightarrow \log_a \left[\frac{x}{y}\right] = r - s = \log_a x - \log_a y$

(c) $x^y = (a^r)^y = a^{ry} \Rightarrow \log_a (x^y) = ry = y \log_a x$

REVIEW EXERCISES FOR CHAPTER 5

1. True by Theorem 5.2 (b).

3. True by repeated application of Theorem 2.12(b).

5. True by Property 5.13.2.

7. True by Property 5.14.7.

9. True. The integrand is an odd function that is continuous on $[-1, 1]$, so the result follows from Property 5.35(b).

11. False. The function $f(x) = 1/x^4$ is not bounded on the interval $[-2, 1]$. It has an infinite discontinuity at $x = 0$, so it is not integrable on the interval. (If the integral were to exist, a positive value would be expected by Property 5.14.6)

13. $\displaystyle\sum_{i=2}^{5} \frac{1}{(-10)^i} = \frac{1}{(-10)^2} + \frac{1}{(-10)^3} + \frac{1}{(-10)^4} + \frac{1}{(-10)^5} = .01 - .001 + .0001 - .00001$

$= 0.00909$

15. $\displaystyle\sum_{k=1}^{n} (2 + k^3) = \sum_{k=1}^{n} 2 + \sum_{k=1}^{n} k^3 = 2n + n^2(n+1)^2/4$

17. $\displaystyle\sum_{i=1}^{n} f(x_i^*) \Delta x_i = \sum_{i=1}^{4} f\left(\frac{i-1}{2}\right) \cdot \frac{1}{2} = \frac{1}{2}\left[f(0) + f(1/2) + f(1) + f(3/2)\right]$.

Since $f(x) = 2 + (x-2)^2$, we have $f(0) = 6$, $f(1/2) = 4.25$, $f(1) = 3$, and $f(3/2) = 2.25$.

Thus $\sum_{i=1}^{n} f(x_i^*)\,\Delta x_i = \frac{1}{2}(15.5) = 7.75$

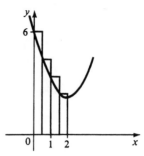

19. By 5.12, $\int_2^4 (3-4x)\,dx = \lim_{n\to\infty} \frac{2}{n}\sum_{i=1}^{n}\left[3-4\left(2+\frac{2i}{n}\right)\right] = \lim_{n\to\infty}\frac{2}{n}\sum_{i=1}^{n}\left[-5-\frac{8}{n}i\right]$

$= \lim_{n\to\infty}\frac{2}{n}\left[-5n - \frac{8}{n}\frac{n(n+1)}{2}\right] = \lim_{n\to\infty}\left[-10 - 8\cdot1\left(1+\frac{1}{n}\right)\right] = -10 - 8 = -18$

21. By 5.12, $\int_0^5 (x^3 - 2x^2)\,dx = \lim_{n\to\infty}\frac{5}{n}\sum_{i=1}^{n}\left[\left(\frac{5i}{n}\right)^3 - 2\left(\frac{5i}{n}\right)^2\right]$

$= \lim_{n\to\infty}\left[\frac{625}{n^4}\sum_{i=1}^{n}i^3 - \frac{250}{n^3}\sum_{i=1}^{n}i^2\right] = \lim_{n\to\infty}\left[\frac{625}{n^4}\frac{n^2(n+1)^2}{4} - \frac{250}{n^3}\frac{n(n+1)(2n+1)}{6}\right]$

$= \lim_{n\to\infty}\left[\frac{625}{4}\left(1+\frac{1}{n}\right)^2 - \frac{125}{3}\left(1+\frac{1}{n}\right)\left(2+\frac{1}{n}\right)\right] = \frac{625}{4} - \frac{250}{3} = \frac{875}{12}$

23. $\int_0^5 (x^3 - 2x^2)\,dx = \frac{x^4}{4} - \frac{2}{3}x^3]_0^5 = \frac{625}{4} - \frac{250}{3} = \frac{875}{12}$

NOTE: We evaluated this integral in #21.

25. $\int_0^1 (1-x^9)\,dx = \left[x - \frac{x^{10}}{10}\right]_0^1 = 1 - \frac{1}{10} = \frac{9}{10}$

27. $\int_1^8 \sqrt[3]{x}\,(x-1)\,dx = \int_1^8 (x^{4/3} - x^{1/3})\,dx = \frac{3}{7}x^{7/3} - \frac{3}{4}x^{4/3}]_1^8 = \left(\frac{3}{7}\cdot128 - \frac{3}{4}\cdot16\right) - \left(\frac{3}{7} - \frac{3}{4}\right)$

$= \frac{1209}{28}$

29. Let $u = 1 + 2x^3$. Then $du = 6x^2\,dx$, so $\int_0^2 x^2(1+2x^3)^3\,dx = \int_1^{17} u^3\frac{1}{6}\,du = \left[\frac{u^4}{24}\right]_1^{17}$

$= \frac{17^4 - 1}{24} = 3480.$

31. Let $u = 2x + 3$. Then $du = 2\,dx$, so $\int_3^{11}\frac{dx}{\sqrt{2x+3}} = \int_9^{25} u^{-1/2}\frac{1}{2}\,du = u^{1/2}]_9^{25} = 5 - 3 = 2.$

33. $\int_{-2}^{-1}\frac{dx}{(2x+3)^4}$ does not exist since the integrand has an infinite discontinuity at $-\frac{3}{2}$.

35. Let $u = 2 + x^5$. Then $du = 5x^4\,dx$, so $\int\frac{x^4\,dx}{(2+x^5)^6} = \int u^{-6}\frac{1}{5}\,du = \frac{1}{5}\frac{u^{-5}}{-5} + C = \frac{-1}{25u^5} + C$

$= \frac{-1}{25(2+x^5)^5} + C$

37. Let $u = \pi x$. Then $du = \pi\,dx$, so $\int \sin\pi x\,dx = \int \sin u\frac{1}{\pi}\,du = \frac{1}{\pi}(-\cos u) + C$

$= -\frac{1}{\pi}\cos\pi x + C$

39. Let $u = 1/t$. Then $du = -dt/t^2$, so $\int \cos(1/t)\,t^{-2}\,dt = \int \cos u\,(-du) = -\sin u + C$

$= -\sin(1/t) + C$

41. Let $u = \cos x$. Then $du = -\sin x\, dx$, so $\int \sin x \sec^2 (\cos x)\, dx = \int \sec^2 u\, (-du)$

    $= -\tan u + C = -\tan (\cos x) + C$

43. Let $u = \sqrt{x}$. Then $du = \frac{dx}{2\sqrt{x}}$ so $\int \frac{e^{\sqrt{x}}}{\sqrt{x}}\, dx = \int e^u 2\, du = 2e^u + C = 2e^{\sqrt{x}} + C$

45. Let $u = \ln(\cos x)$. Then $du = \frac{-\sin x}{\cos x}\, dx = -\tan x\, dx$, so $\int \tan x \ln(\cos x)\, dx = -\int u\, du$

    $= -u^2/2 + C = -\frac{1}{2}\big[\ln(\cos x)\big]^2 + C$

47. Let $u = 1 + x^4$. Then $du = 4x^3\, dx$, so $\int \frac{x^3\, dx}{1 + x^4} = \int \frac{(1/4)\, du}{u} = \frac{1}{4} \ln|u| + C$

    $= \frac{1}{4} \ln(1 + x^4) + C$

49. $\int_0^{2\pi} |\sin x|\, dx = \int_0^\pi \sin x\, dx - \int_\pi^{2\pi} \sin x\, dx = 2\int_0^\pi \sin x\, dx = -2\cos x\big]_0^\pi = -2\big[(-1) - 1\big] = 4$

51. $F(x) = \int_1^x \sqrt{1 + t^2 + t^4}\, dt \Rightarrow F'(x) = \sqrt{1 + x^2 + x^4}$

53. $g(x) = \int_0^{x^3} \frac{t\, dt}{\sqrt{1 + t^3}}$. Let $y = g(x)$ and $u = x^3$. Then $g'(x) = \frac{dy}{dx} = \frac{dy}{du}\frac{du}{dx} = \frac{u}{\sqrt{1 + u^3}} 3x^2$

    $= \frac{x^3}{\sqrt{1 + x^9}} 3x^2 = \frac{3x^5}{\sqrt{1 + x^9}}$

55. $y = \int_{\sqrt{x}}^x \frac{\cos\theta}{\theta}\, d\theta = \int_1^x \frac{\cos\theta}{\theta}\, d\theta + \int_{\sqrt{x}}^1 \frac{\cos\theta}{\theta}\, d\theta = \int_1^x \frac{\cos\theta}{\theta}\, d\theta - \int_1^{\sqrt{x}} \frac{\cos\theta}{\theta}\, d\theta \Rightarrow$

    $y' = \frac{\cos x}{x} - \frac{\cos\sqrt{x}}{\sqrt{x}}\frac{1}{2\sqrt{x}} = \frac{2\cos x - \cos\sqrt{x}}{2x}$

57. $f'(x) = \frac{d}{dx} \int_1^{\sqrt{x}} \frac{e^s}{s}\, ds = \frac{e^{\sqrt{x}}}{\sqrt{x}} \frac{d}{dx} \sqrt{x} = \frac{e^{\sqrt{x}}}{\sqrt{x}} \frac{1}{2\sqrt{x}} = \frac{e^{\sqrt{x}}}{2x}$

59. If $1 \le x \le 3$, then $2 \le \sqrt{x^2 + 3} \le 2\sqrt{3}$, so $2(3 - 1) \le \int_1^3 \sqrt{x^2 + 3}\, dx \le 2\sqrt{3}(3 - 1)$;

    that is, $4 \le \int_1^3 \sqrt{x^2 + 3}\, dx \le 4\sqrt{3}$

61. $|\cos x| \le 1 \Rightarrow \cos^8 x \le \cos^6 x \Rightarrow \int_0^\pi \cos^8 x\, dx \le \int_0^\pi \cos^6 x\, dx$

63. $0 \le x \le 1 \Rightarrow 0 \le \cos x \le 1 \Rightarrow x^2 \cos x \le x^2 \Rightarrow \int_0^1 x^2 \cos x\, dx \le \int_0^1 x^2\, dx = \frac{1}{3}x^3]_0^1 = \frac{1}{3}$

    [*Property 7*]

65. $\cos x \le 1 \Rightarrow e^x \cos x \le e^x \Rightarrow \int_0^1 e^x \cos x\, dx \le \int_0^1 e^x\, dx = [e^x]_0^1 = e - 1$

67. The greatest integer function is constant on any interval $[i - 1, i)$ and is equal to $i - 1$, where i is any positive integer. Now if we divide the interval of integration into intervals where $[\![x]\!]$ is constant, then the integration can be carried out:

    $\int_0^n [\![x]\!]\, dx = \sum_{i=1}^n \int_{i-1}^i [\![x]\!]\, dx = \sum_{i=1}^n \int_{i-1}^i (i - 1)\, dx = \sum_{i=1}^n (i - 1)$

    $= 0 + 1 + 2 + \cdots + (n - 1) = \sum_{i=1}^{n-1} i = \frac{(n - 1)[(n - 1) + 1]}{2} \ [using\ 5.3(c)] = \frac{n(n - 1)}{2}$

69. $y = \sqrt{16 - x^2}$ is a semi-circle with a radius of 4. So $\int_{-4}^{4} \sqrt{16 - x^2}\,dx$ represents the area

between the semicircle $y = \sqrt{16 - x^2}$ and the x$-$axis and is equal to $\frac{1}{2}\pi(4)^2 = 8\pi$.

71.   Following the hint, we have $\Delta x_i = x_i - x_{i-1} = \dfrac{i^2}{n^2} - \dfrac{(i-1)^2}{n^2} = \dfrac{2i-1}{n^2}$. So

$$\int_0^1 \sqrt{x}\,dx = \lim_{n\to\infty} \sum_{i=1}^{n} \Delta x_i\, f(x_i^*) = \lim_{n\to\infty} \sum_{i=1}^{n} \left(\frac{2i-1}{n^2}\right)\sqrt{i^2/n^2} = \lim_{n\to\infty} \sum_{i=1}^{n} \frac{2i^2 - i}{n^3}$$

$$= \lim_{n\to\infty} \frac{1}{n^3}\left[2\sum_{i=1}^{n} i^2 - \sum_{i=1}^{n} i\right] = \lim_{n\to\infty} \frac{1}{n^3}\left[\frac{n(n+1)(2n+1)}{3} - \frac{n(n+1)}{2}\right]$$

$$(by\ Theorem\ 5.3(c), (d)) = \lim_{n\to\infty}\left[\frac{(1 + \frac{1}{n})(2 + \frac{1}{n})}{3} - \frac{1 + \frac{1}{n}}{2n}\right] = \frac{2}{3} - 0 = \frac{2}{3}.$$

73.   Let $u = f(x)$ so $du = f'(x)dx$. So $2\int_a^b f(x)f'(x)dx = 2\int_{f(a)}^{f(b)} u\,du = u^2\big|_{f(a)}^{f(b)} = [f(b)]^2 - [f(a)]^2$.

75.   Let $u = 1 - x$. Then $du = -\,dx$, so $\int_0^1 f(1-x)\,dx = \int_1^0 f(u)\,(-du) = \int_0^1 f(u)\,du$

$= \int_0^1 f(x)\,dx$

77.   The area of the triangle with vertices 0, P, and (cosh t, 0) is $\frac{1}{2}$ sinh t cosh t and the area under

the curve $x^2 - y^2 = 1$ from $x = 1$ to $x = \cosh t$ is $\displaystyle\int_1^{\cosh t} \sqrt{x^2 - 1}\ dx$. Therefore,

the area of the shaded region is $A(t) = \frac{1}{2}$ sinh t cosh t $- \displaystyle\int_1^{\cosh t} \sqrt{x^2 - 1}\ dx$. So, by Part 1 of

the Fundamental Theorem of Calculus, $A'(t) = \frac{1}{2}(\cosh^2 t + \sinh^2 t) - \sqrt{\cosh^2 t - 1}\ \sinh t$

$= \frac{1}{2}(\cosh^2 t + \sinh^2 t) - \sqrt{\sinh^2 t}\ \sinh t = \frac{1}{2}(\cosh^2 t + \sinh^2 t) - \sinh^2 t = \frac{1}{2}(\cosh^2 t - \sinh^2 t)$

$= \frac{1}{2}(1) = \frac{1}{2}$. Thus, $A(t) = \frac{1}{2}t + C$ since $A'(t) = \frac{1}{2}$. To calculate C, we let $t = 0$. Thus,

$A(0) = \frac{1}{2}$ sinh 0 cosh 0 $- \displaystyle\int_1^{\cosh 0} \sqrt{x^2 - 1}\ dx = \frac{1}{2}(0) + C \Rightarrow C = 0$. Thus, $A(t) = \frac{1}{2}t$.

79. $\displaystyle\lim_{x\to -1} F(x) = \lim_{x\to -1} \frac{b^{x+1} - a^{x+1}}{x + 1} \overset{H}{=} \lim_{x\to -1} \frac{b^{x+1}\ \ln b - a^{x+1}\ \ln a}{1} = \ln b - \ln a = F(-1)$,

so F is continuous at -1.

APPLICATIONS PLUS (page 368)

1. (a) $I = \dfrac{k\cos\theta}{d^2} = \dfrac{k\dfrac{h}{d}}{d^2} = k\dfrac{h}{d^3} = k\dfrac{h}{\left(\sqrt{1600+h^2}\right)^3} = k\dfrac{h}{(1600+h^2)^{3/2}}$

$\dfrac{dI}{dh} = k\dfrac{(16000+h^2)^{3/2} - kh(3/2)(1600+h^2)^{1/2}\cdot 2h}{(1600+h^2)^3}$

$= \dfrac{k(1600+h^2)^{1/2}\left[1600+h^2-3h^2\right]}{(1600+h^2)^3} = \dfrac{k(1600-2h^2)}{(1600+h^2)^{5/2}}$

Set $\dfrac{dI}{dh} = 0$: $1600 - 2h^2 = 0 \Rightarrow h^2 = 800 \Rightarrow h = \sqrt{800} = 20\sqrt{2}$.

By the First Derivative Test, I has a relative maximum at $h = 20\sqrt{2} \approx 28$ ft.

(b)

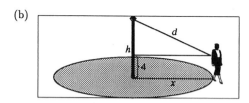

$\dfrac{dx}{dt} = 4$ ft/s

$I = \dfrac{k\cos\theta}{d^2} = \dfrac{k\left(\dfrac{h-4}{d}\right)}{d^2} = \dfrac{k(h-4)}{d^3} = \dfrac{k(h-4)}{\left([h-4]^2+x^2\right)^{3/2}} = k(h-4)\left([h-4]^2+x^2\right)^{-3/2}$

$\dfrac{dI}{dt} = k(h-4)(-3/2)\left([h-4]^2+x^2\right)^{-5/2}\cdot 2x\cdot\dfrac{dx}{dt}$

$= k(h-4)(-3x)\left([h-4]^2+x^2\right)^{-5/2}\cdot 4 = \dfrac{-12xk(h-4)}{[(h-4)^2+x^2]^{5/2}}$

$\dfrac{dI}{dt}\bigg|_{x=40} = -\dfrac{480k(h-4)}{[(h-4)^2+1600]^{5/2}}$

3. (a) First note that 90 mi/h $= 90\times\dfrac{5280}{3600}$ ft/s $= 132$ ft/s. Then $a(t) = 4$ ft/s^2 $\Rightarrow$
$v(t) = 4t = 132$ when $t = \dfrac{132}{4} = 33$ s. It takes 33 s to reach 132 ft/s.
Therefore, taking $s(0) = 0$, we have $s(t) = 2t^2$, $0 \le t \le 33$. So $s(33) = 2178$ ft.
For $33 \le t \le 933$ we have $v(t) = 132$ ft/s $\Rightarrow s(t) = 132(t-33)+C$ and $s(33) = 2178 \Rightarrow$
$C = 2178$, so $s(t) = 132(t-33)+2178$, $33 \le t \le 933$.
Therefore $s(933) = 132(933)+2178 = 120{,}978$ ft $= 22.9125$ mi.

(b) As in part (a), the train accelerates for 33 s and travels 2178 ft while doing so. Similarly,
it decelerates for 33 s and travels 2178 ft at the end of its trip. During the remaining
$900 - 66 = 834$ s it travels at 132 ft/s, so the distance traveled is $132\cdot 834 = 110{,}088$ ft.
Thus the total distance is $2178 + 110{,}088 + 2178 = 114{,}444$ ft $= 21.675$ mi.

5. (a) Let $F(t) = \int_0^t f(s)\,ds$. Then, by Part 1 of the Fundamental Theorem of Calculus,

$F'(t) = f(t) =$ rate of depreciation, so $F(t)$ represents the loss in value over the interval

$[0, t]$.

(b) $C(t) = [A + F(t)]/t$ represents the average expenditure over the interval $[0, t]$.

The company wants to minimize average expenditure.

(c) $C(t) = \frac{1}{t}\left(A + \int_0^t f(s)\,ds\right)$. Using Part 1 of the Fundamental Theorem of Calculus, we

have $C'(t) = -\frac{1}{t^2}\left(A + \int_0^t f(s)\,ds\right) + \frac{1}{t}f(t) = 0$ when $t\,f(t) = A + \int_0^t f(s)\,ds \Rightarrow$

$f(t) = \frac{1}{t}\left[A + \int_0^t f(s)\,ds\right] = C(t)$.

7. (a) $P = \dfrac{\text{area under } y = L\sin\theta}{\text{area of rectangle}} = \dfrac{\int_0^\pi L\sin\theta\,d\theta}{\pi L} = \dfrac{-L\cos\theta\,\Big|_0^\pi}{\pi L} = \dfrac{-(-1)+1}{\pi} = \dfrac{2}{\pi}$

(b) $P = \dfrac{\text{area under } y = \frac{L}{2}\sin\theta}{\text{area of rectangle}} = \dfrac{\int_0^\pi \frac{L}{2}\sin\theta\,d\theta}{\pi L} = \dfrac{1}{2\pi}\int_0^\pi \sin\theta\,d\theta = \dfrac{1}{2\pi}\left(-\cos\theta\,\Big|_0^\pi\right) = \dfrac{1}{2\pi}\cdot 2 = \dfrac{1}{\pi}$

(c) $P = \dfrac{\text{area under } y = \frac{L}{5}\sin\theta}{\text{area of rectangle}} = \dfrac{\int_0^\pi \frac{L}{5}\sin\theta\,d\theta}{\pi L} = \dfrac{1}{5\pi}\int_0^\pi \sin\theta\,d\theta = \dfrac{1}{5\pi}\cdot 2 = \dfrac{2}{5\pi}$

9. (a) coefficient of inequality $= \dfrac{\text{area between Lorenz curve and straight line}}{\text{area under straight line}}$

$= \dfrac{\int_0^1 [x - L(x)]\,dx}{\int_0^1 x\,dx} = \dfrac{\int_0^1 [x - L(x)]\,dx}{\frac{x^2}{2}\Big|_0^1} = \dfrac{\int_0^1 [x - L(x)]\,dx}{\frac{1}{2}} = 2\int_0^1 [x - L(x)]\,dx$

(b) $L(x) = \frac{5}{12}x^2 + \frac{7}{12}x \Rightarrow L(\frac{1}{2}) = \frac{5}{48} + \frac{7}{24} = \frac{19}{48} = .39583$, so the bottom 50% of the households

receive about 40% of the income.

coeff. of inequality $= 2\int_0^1\left[x - \frac{5}{12}x^2 - \frac{7}{12}x\right]dx = 2\int_0^1 \frac{5}{12}(x - x^2)\,dx = \frac{5}{6}\left(\frac{x^2}{2} - \frac{x^3}{3}\right)\Big|_0^1 = \frac{5}{36}$

(c) coeff. of inequality $= 2\int_0^1 [x - L(x)]\,dx = 2\int_0^1\left(x - \frac{5x^3}{4 + x^2}\right)dx$

$= 2\int_0^1\left[x - \left(5x - \frac{20x}{x^2 + 4}\right)\right]dx = 2\int_0^1\left(-4x + \frac{20x}{x^2 + 4}\right)dx = 2\left[-2x^2 + 10\ln(x^2 + 4)\right]_0^1$

$$= 2[-2 + 10\ln 5 - 10\ln 4] = -4 + 20\ln \tfrac{5}{4} \approx 0.46$$

11. (a) $y = (\tan\theta)x - \dfrac{g}{2v^2\cos^2\theta}x^2$

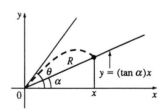

The parabola intersects the line when

$$(\tan\alpha)\,x = (\tan\theta)\,x - \frac{g}{2v^2\cos^2\theta}x^2 \;\Rightarrow\; x = \frac{(\tan\theta - \tan\alpha)2v^2\cos^2\theta}{g} \;\Rightarrow$$

$$R(\theta) = \frac{x}{\cos\alpha} = \frac{(\tan\theta - \tan\alpha)2v^2\cos^2\theta}{g\cos\alpha} = \left(\frac{\sin\theta}{\cos\theta} - \frac{\sin\alpha}{\cos\alpha}\right)\frac{2v^2\cos^2\theta}{g\cos\alpha}$$

$$= \left(\frac{\sin\theta}{\cos\theta} - \frac{\sin\alpha}{\cos\alpha}\right)(\cos\theta\cos\alpha)\frac{2v^2\cos\theta}{g\cos^2\alpha} = (\sin\theta\cos\alpha - \sin\alpha\cos\theta)\frac{2v^2\cos\theta}{g\cos^2\alpha}$$

$$= \sin(\theta - \alpha)\frac{2v^2\cos\theta}{g\cos^2\alpha}.$$

(b) $R'(\theta) = \dfrac{2v^2}{g\cos^2\alpha}\big[\cos\theta\cdot\cos(\theta-\alpha) + \sin(\theta-\alpha)(-\sin\theta)\big] = \dfrac{2v^2}{g\cos^2\alpha}\cos[\theta + (\theta - \alpha)]$

$$= \frac{2v^2}{g\cos^2\alpha}\cos(2\theta - \alpha) = 0 \text{ when } \cos(2\theta - \alpha) = 0 \;\Rightarrow\; 2\theta - \alpha = \pi/2 \;\Rightarrow\; \theta = \frac{\pi/2 + \alpha}{2}$$

$= \pi/4 + \alpha/2$. The First Derivative Test shows that this gives a maximum value for $R(\theta)$.

[This could be done without calculus by applying Formula 18a in Appendix B to $R(\theta)$.]

(c)

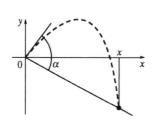

Replacing α by $-\alpha$ in part (a), we get $R(\theta) = \dfrac{2v^2\cos\theta\sin(\theta + \alpha)}{g\cos^2\alpha}$.

Proceeding as in part (b), or simply by replacing α by $-\alpha$ in the result of part (b), we see that $R(\theta)$ is a maximum when $\theta = \dfrac{\pi}{4} - \dfrac{\alpha}{2}$.

CHAPTER SIX

EXERCISES 6.1

1. $A = \int_{-1}^{1} \left[(x^2 + 3) - x \right] dx$

    $= \int_{-1}^{1} (x^2 - x + 3) dx = \left[\frac{x^3}{3} - \frac{x^2}{2} + 3x \right]_{-1}^{1}$

    $= (\frac{1}{3} - \frac{1}{2} + 3) - (-\frac{1}{3} - \frac{1}{2} - 3) = \frac{20}{3}$

    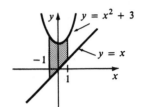

3. $A = \int_{-1}^{2} \left[y^2 - (y - 5) \right] dy$

    $= [\frac{y^3}{3} - \frac{y^2}{2} + 5y]_{-1}^{2}$

    $= (\frac{8}{3} - 2 + 10) - (-\frac{1}{3} - \frac{1}{2} - 5)$

    $= 16.5$

    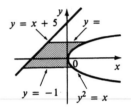

5. $A = \int_{0}^{6} \left[2x - \left(x^2 - 4x \right) \right] dx$

    $= \int_{0}^{6} (6x - x^2) dx = \left[3x^2 - \frac{x^3}{3} \right]_{0}^{6}$

    $= 108 - 72 = 36$

    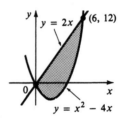

7. $A = \int_{0}^{1} (x - x^2) dx = \left[\frac{x^2}{2} - \frac{x^3}{3} \right]_{0}^{1}$

    $= \frac{1}{2} - \frac{1}{3} = \frac{1}{6}$

    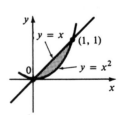

9. $A = \int_{0}^{1} (\sqrt{x} - x^2) dx = \left[\frac{2}{3}x^{3/2} - \frac{x^3}{3} \right]_{0}^{1}$

    $= \frac{2}{3} - \frac{1}{3} = \frac{1}{3}$

    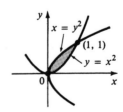

11. $A = \int_0^4 \left(\sqrt{x} - \frac{x}{2}\right) dx = \left[\frac{2}{3}x^{3/2} - \frac{x^2}{4}\right]_0^4$

$= \left(\frac{16}{3} - 4\right) - 0 = \frac{4}{3}$

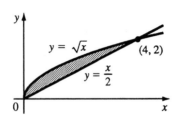

13. $A = \int_{-1}^1 \left[(x^2 + 3) - 4x^2\right] dx$

$= 2\int_0^1 (3 - 3x^2) dx = [2(3x - x^3)]_0^1$

$= 2(3 - 1) - 0 = 4$

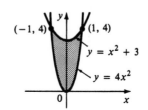

15. $A = \int_0^3 \left[(2x + 5) - (x^2 + 2)\right] dx + \int_3^6 \left[(x^2 + 2) - (2x + 5)\right] dx$

$= \int_0^3 (-x^2 + 2x + 3) dx + \int_3^6 (x^2 - 2x - 3) dx$

$= \left[-\frac{x^3}{3} + x^2 + 3x\right]_0^3 + \left[\frac{x^3}{3} - x^2 - 3x\right]_3^6$

$= (-9 + 9 + 9) - 0 + (72 - 36 - 18) - (9 - 9 - 9)$

$= 36$

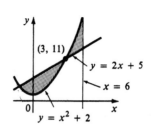

17. $A = \int_{-2}^{-1} \left[(x^2 + 1) - (3 - x^2)\right] dx + \int_{-1}^1 \left[(3 - x^2) - (x^2 + 1)\right] dx$

$+ \int_1^2 \left[(x^2 + 1) - (3 - x^2)\right] dx = \int_{-2}^{-1} (2x^2 - 2) dx + \int_{-1}^1 (2 - 2x^2) dx + \int_1^2 (2x^2 - 2) dx$

$= 2\int_0^1 (2 - 2x^2) dx + 2\int_1^2 (2x^2 - 2) dx$

$[by\ symmetry] = 2\left[2x - \frac{2}{3}x^3\right]_0^1 + 2\left[\frac{2}{3}x^3 - 2x\right]_1^2$

$= 2\left(2 - \frac{2}{3}\right) + 2\left(\frac{16}{3} - 4\right) - 2\left(\frac{2}{3} - 2\right) = 8$

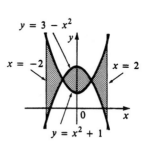

19. $A = \int_{-1}^3 (2y + 3 - y^2) dy$

$= \left[y^2 + 3y - \frac{y^3}{3}\right]_{-1}^3$

$= (9 + 9 - 9) - (1 - 3 + \frac{1}{3}) = \frac{32}{3}$

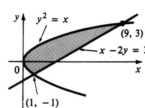

21. $A = \int_{-1}^{1} \left[(1-y^2) - (y^2-1)\right] dy$

$= \int_{-1}^{1} 2(1-y^2) dy = 4 \int_0^1 (1-y^2) dy$

$= 4 \left[y - \frac{y^3}{3}\right]_0^1 = 4\left(1 - \frac{1}{3}\right) = \frac{8}{3}$

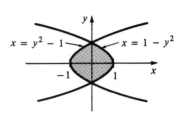

23. $A = \int_{-2}^0 \left[x^3 - (2x-x^2)\right] dx + \int_0^1 \left[(2x-x^2) - x^3\right] dx$

$= \left[\frac{x^4}{4} + \frac{x^3}{3} - x^2\right]_{-2}^0 + \left[-\frac{x^4}{4} - \frac{x^3}{3} + x^2\right]_0^1$

$= 0 - (4 - \frac{8}{3} - 4) + (-\frac{1}{4} - \frac{1}{3} + 1) - 0 = \frac{37}{12}$

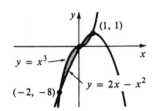

25. $A = \int_{-\pi/4}^0 (\sin x - x) dx + \int_0^{\pi/2} (x - \sin x) dx$

$= \left[-\cos x - \frac{x^2}{2}\right]_{-\pi/4}^0 + \left[\frac{x^2}{2} + \cos x\right]_0^{\pi/2}$

$= -1 - (-1/\sqrt{2} - \pi^2/32) + \pi^2/8 - 1$

$= 5\pi^2/32 + 1/\sqrt{2} - 2$

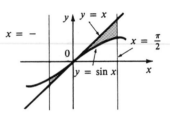

27. Notice that $\cos x = \sin 2x = 2\sin x \cos x \Leftrightarrow 2\sin x = 1$ or $\cos x = 0 \Leftrightarrow x = \pi/6$ or $\pi/2$.

$A = \int_0^{\pi/6} (\cos x - \sin 2x) dx + \int_{\pi/6}^{\pi/2} (\sin 2x - \cos x) dx$

$= \left[\sin x + \frac{1}{2}\cos 2x\right]_0^{\pi/6} + \left[-\frac{1}{2}\cos 2x - \sin x\right]_{\pi/6}^{\pi/2}$

$= \frac{1}{2} + \frac{1}{2}\cdot\frac{1}{2} - (0 + \frac{1}{2}\cdot 1) + (\frac{1}{2} - 1) - (-\frac{1}{2}\cdot\frac{1}{2} - \frac{1}{2}) = \frac{1}{2}$

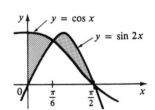

29. $\cos x = \sin 2x = 2\sin x \cos x \Leftrightarrow \cos x = 0$ or $\sin x = 1/2 \Leftrightarrow x = \pi/2$ or $5\pi/6$.

$A = \int_{\pi/2}^{5\pi/6} (\cos x - \sin 2x) dx + \int_{5\pi/6}^{\pi} (\sin 2x - \cos x) dx$

$= \left[(\sin x + \frac{1}{2}\cos 2x)\right]_{\pi/2}^{5\pi/6} - \left[(\sin x + \frac{1}{2}\cos 2x)\right]_{5\pi/6}^{\pi}$

$= (\frac{1}{2} + \frac{1}{2}\cdot\frac{1}{2}) - (1 - \frac{1}{2}) - (0 + \frac{1}{2}) + (\frac{1}{2} + \frac{1}{2}\cdot\frac{1}{2}) = \frac{1}{2}$

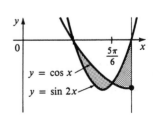

31. $A = \int_{-4}^{0} \left[-x - \left[(x+1)^2 - 7 \right] \right] dx + \int_{0}^{2} \left[x - \left[(x+1)^2 - 7 \right] \right] dx$

$= \int_{-4}^{0} (-x^2 - 3x + 6) \, dx + \int_{0}^{2} (-x^2 - x + 6) \, dx$

$= \left[-\frac{x^3}{3} - \frac{3x^2}{2} + 6x \right]_{-4}^{0} + \left[-\frac{x^3}{3} - \frac{x^2}{2} + 6x \right]_{0}^{2}$

$= 0 - (\frac{64}{3} - 24 - 24) + (-\frac{8}{3} - 2 + 12) - 0 = 34$

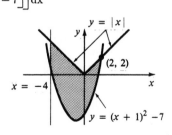

33. $A = \int_{0}^{3} \left[\frac{x}{3} - (-x) \right] dx + \int_{3}^{6} \left[\left(8 - \frac{7}{3}x \right) - (-x) \right] dx = \int_{0}^{3} \frac{4}{3} x \, dx + \int_{3}^{6} \left(-\frac{4}{3} x + 8 \right) dx$

$= \left[\frac{2}{3} x^2 \right]_{0}^{3} + \left[-\frac{2}{3} x^2 + 8x \right]_{3}^{6} = (6 - 0) + (24 - 18) = 12$

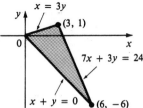

35. $A = \int_{1}^{2} \left(\frac{1}{x} - \frac{1}{x^2} \right) dx = \left[\ln x + \frac{1}{x} \right]_{1}^{2}$

$= \left(\ln 2 + \frac{1}{2} \right) - \left(\ln 1 + 1 \right) = \ln 2 - \frac{1}{2}$

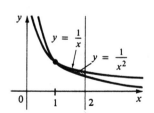

37. $A = 2 \int_{0}^{1} \left(\frac{2}{x^2 + 1} - x^2 \right) dx = \left[4 \tan^{-1} x - \frac{2}{3} x^3 \right]_{0}^{1}$

$= 4 \cdot \frac{\pi}{4} - \frac{2}{3} = \pi - \frac{2}{3}$

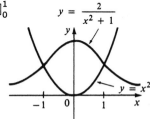

39. $A = \int_{0}^{1} (e^{3x} - e^x) \, dx = \frac{1}{3} e^{3x} - e^x \big]_{0}^{1}$

$= \left(\frac{e^3}{3} - e \right) - \left(\frac{1}{3} - 1 \right) = \frac{e^3}{3} - e + \frac{2}{3}$

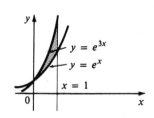

41.   (a) $A = \int_{-4}^{-1} \left[(2x+4) + \sqrt{-4x} \right] dx + \int_{-1}^{0} 2\sqrt{-4x}\, dx$

$$= [x^2 + 4x]_{-4}^{-1} + 2\int_{-4}^{-1} \sqrt{-x}\, dx + 4\int_{-1}^{0} \sqrt{-x}\, dx$$

$$= (-3-0) + 2\int_{1}^{4} \sqrt{u}\, du + 4\int_{0}^{1} \sqrt{u}\, du \quad [u = -x]$$

$$= -3 + \left[\tfrac{4}{3} u^{3/2} \right]_{1}^{4} + \left[\tfrac{8}{3} u^{3/2} \right]_{0}^{1} = -3 + \tfrac{28}{3} + \tfrac{8}{3} = 9$$

  (b) $A = \int_{-4}^{2} \left[-\tfrac{y^2}{4} - \left(\tfrac{y}{2} - 2 \right) \right] dy = [-\tfrac{y^3}{12} - \tfrac{y^2}{4} + 2y]_{-4}^{2}$

$$= \left(-\tfrac{2}{3} - 1 + 4 \right) - \left(\tfrac{16}{3} - 4 - 8 \right) = 9$$

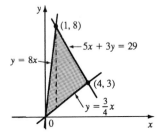

43. $A = \int_{0}^{1} \left(8x - \tfrac{3}{4} x \right) dx + \int_{1}^{4} \left[\left(-\tfrac{5}{3} x + \tfrac{29}{3} \right) - \tfrac{3}{4} x \right] dx = \tfrac{29}{4} \int_{0}^{1} x\, dx + \int_{1}^{4} \left(-\tfrac{29}{12} x + \tfrac{29}{3} \right) dx$

$$= \tfrac{29}{4} \left[\tfrac{x^2}{2} \right]_{0}^{1} - \tfrac{29}{12} \left[\tfrac{x^2}{2} - 4x \right]_{1}^{4} = \tfrac{29}{8} - \tfrac{29}{12} \left(-8 - \tfrac{1}{2} + 4 \right) = 14.5$$

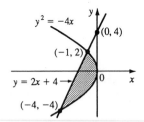

45. $\int_{0}^{2} |x^2 - x^3|\, dx = \int_{0}^{1} (x^2 - x^3)\, dx + \int_{1}^{2} (x^3 - x^2)\, dx = \left[\tfrac{x^3}{3} - \tfrac{x^4}{4} \right]_{0}^{1} + \left[\tfrac{x^4}{4} - \tfrac{x^3}{3} \right]_{1}^{2}$

$$= \tfrac{1}{3} - \tfrac{1}{4} + \left(4 - \tfrac{8}{3} \right) - \left(\tfrac{1}{4} - \tfrac{1}{3} \right) = 1.5$$

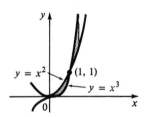

47. $\int_{-1}^{2} x^3 \, dx = \left[\frac{x^4}{4}\right]_{-1}^{2}$

$= 4 - \frac{1}{4} = \frac{15}{4} = 3.75$

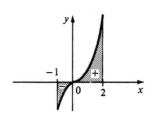

49. $\int_{0}^{\pi} \cos x \, dx = [\sin x]_{0}^{\pi} = \sin \pi - \sin 0 = 0$

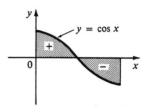

51. Let $f(x) = \sqrt{1+x^3} - (1-x)$, $\Delta x = \frac{2-0}{4} = \frac{1}{2}$.

$A = \int_{0}^{2} \left[\sqrt{1+x^3} - (1-x)\right] dx \approx \frac{1}{2}[f(1/4) + f(3/4) + f(5/4) + f(7/4)]$

$= \frac{1}{2}\left[\left(\frac{\sqrt{65}}{8} - \frac{3}{4}\right) + \left(\frac{\sqrt{91}}{8} - \frac{1}{4}\right) + \left(\frac{3\sqrt{21}}{8} + \frac{1}{4}\right) + \left(\frac{\sqrt{407}}{8} + \frac{3}{4}\right)\right]$

$= \frac{1}{16}(\sqrt{65} + \sqrt{91} + 3\sqrt{21} + \sqrt{407}) \approx 3.22$.

53. By the symmetry of the problem, we consider only the 1st quadrant where $y = x^2 \Rightarrow x = \sqrt{y}$.

We are looking for a number b such that

$\int_{0}^{4} x \, dy = 2\int_{0}^{b} x \, dy \Rightarrow \int_{0}^{4} \sqrt{y} \, dy = 2\int_{0}^{b} \sqrt{y} \, dy \Rightarrow \frac{2}{3}\left[y^{3/2}\right]_{0}^{4} = \frac{4}{3}\left[y^{3/2}\right]_{0}^{b}$

$\Rightarrow \frac{2}{3}(8 - 0) = \frac{4}{3}(b^{3/2} - 0) \Rightarrow b^{3/2} = 4 \Rightarrow b = 4^{2/3}$.

EXERCISES 6.2

1. $V = \int_{0}^{1} \pi(x^2)^2 \, dx = \pi \int_{0}^{1} x^4 \, dx = \pi\left[\frac{x^5}{5}\right]_{0}^{1} = \frac{\pi}{5}$

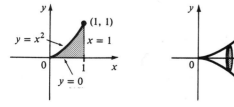

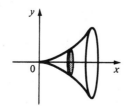

3. $V = \int_0^1 \pi(-x+1)^2\,dx = \pi\int_0^1 (x^2 - 2x + 1)\,dx = \pi\left[\dfrac{x^3}{3} - x^2 + x\right]_0^1 = \pi\left(\dfrac{1}{3} - 1 + 1\right) = \dfrac{\pi}{3}$

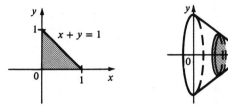
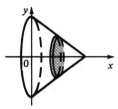

5. $V = \int_0^4 \pi(\sqrt{y})^2\,dy = \pi\int_0^4 y\,dy = \pi\left[\dfrac{y^2}{2}\right]_0^4 = 8\pi$

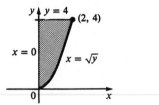
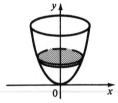

7. $V = \pi\int_0^1 \left[(\sqrt{x})^2 - (x^2)^2\right]dx = \pi\int_0^1 (x - x^4)\,dx = \pi\left[\dfrac{x^2}{2} - \dfrac{x^5}{5}\right]_0^1 = \pi\left(\dfrac{1}{2} - \dfrac{1}{5}\right) = \dfrac{3\pi}{10}$

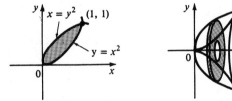
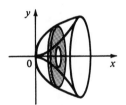

9. $V = \pi\int_0^2 \left[(2y)^2 - (y^2)^2\right]dy = \pi\int_0^2 (4y^2 - y^4)\,dy = \pi\left[\dfrac{4}{3}y^3 - \dfrac{y^5}{5}\right]_0^2 = \pi\left(\dfrac{32}{3} - \dfrac{32}{5}\right) = \dfrac{64\pi}{15}$

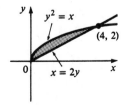
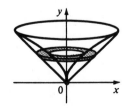

11. $V = \pi \int_{-1}^{1} \left[(2-x^4)^2 - 1^2\right] dx = 2\pi \int_{0}^{1} (3 - 4x^4 + x^8) \, dx = 2\pi \left[3x - \frac{4}{5}x^5 + \frac{x^9}{9}\right]_0^1$

$= 2\pi(3 - \frac{4}{5} + \frac{1}{9}) = \frac{208\pi}{45}$

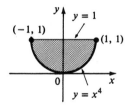

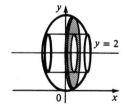

13. $V = \pi \int_{0}^{8} \left(\frac{x}{4}\right)^2 dx = \frac{\pi}{16}\left[\frac{x^3}{3}\right]_0^8 = \frac{32\pi}{3}$

15. $V = \pi \int_{0}^{2} (8 - 4y)^2 \, dy = \pi\left[64y - 32y^2 + \frac{16}{3}y^3\right]_0^2 = \pi\left(128 - 128 + \frac{128}{3}\right) = \frac{128\pi}{3}$

17. $V = \pi \int_{0}^{8} \left[(\sqrt[3]{x})^2 - \left(\frac{x}{4}\right)^2\right] = \pi \int_{0}^{8} \left(x^{2/3} - \frac{x^2}{16}\right) dx = \pi\left[\frac{3}{5}x^{5/3} - \frac{x^3}{48}\right]_0^8 = \pi\left(\frac{96}{5} - \frac{32}{3}\right)$

$= \frac{128\pi}{15}$

19. $V = \pi \int_{0}^{8} \left[\left(2 - \frac{x}{4}\right)^2 - (2 - \sqrt[3]{x})^2\right] dx = \pi \int_{0}^{8} \left(-x + \frac{x^2}{16} + 4x^{1/3} - x^{2/3}\right) dx$

$= \pi\left[-\frac{x^2}{2} + \frac{x^3}{48} + 3x^{4/3} - \frac{3}{5}x^{5/3}\right]_0^8 = \pi\left(-32 + \frac{32}{3} + 48 - \frac{96}{5}\right) = \frac{112\pi}{15}$

21. $V = \pi \int_{0}^{8} (2^2 - x^{2/3}) \, dx = \pi\left[4x - \frac{3}{5}x^{5/3}\right]_0^8 = \pi\left(32 - \frac{96}{5}\right) = \frac{64\pi}{5}$

23. $V = \pi \int_{0}^{8} (2 - \sqrt[3]{x})^2 \, dx = \pi \int_{0}^{8} (4 - 4x^{1/3} + x^{2/3}) \, dx = \pi\left[4x - 3x^{4/3} + \frac{3}{5}x^{5/3}\right]_0^8$

$= \pi(32 - 48 + \frac{96}{5}) = \frac{16\pi}{5}$

25. $V = \pi \int_{0}^{2} (x^2 - 1)^2 \, dx = \pi \int_{0}^{2} (x^4 - 2x^2 + 1) \, dx = \pi\left[\frac{x^5}{5} - \frac{2}{3}x^3 + x\right]_0^2 = \pi\left(\frac{32}{5} - \frac{16}{3} + 2\right)$

$= \frac{46\pi}{15}$

27. $V = \pi \int_{-1}^{1} (\sec^2 x - 1^2) \, dx = \pi\left[\tan x - x\right]_{-1}^{1} = \pi\left[(\tan 1 - 1) - (-\tan 1 + 1)\right]$

$= 2\pi(\tan 1 - 1)$

29. $V = \pi \int_{-3}^{-2} (-x - 2)^2 \, dx + \pi \int_{-2}^{0} (x + 2)^2 \, dx = \pi \int_{-3}^{0} (x + 2)^2 \, dx = \left[\frac{\pi}{3}(x + 2)^3\right]_{-3}^{0}$

$= \frac{\pi}{3}\left[8 - (-1)\right] = 3\pi$

31. $V = \int_{0}^{1} \pi(e^x)^2 \, dx = \int_{0}^{1} \pi e^{2x} \, dx = \frac{1}{2}\left[\pi e^{2x}\right]_0^1 = \frac{\pi}{2}(e^2 - 1)$

33. $V = \pi \int_{0}^{\pi/4} \left[1^2 - \tan^2 x\right] dx$

35. $x - 1 = (x - 4)^2 + 1 \Leftrightarrow x^2 - 9x + 18 = 0 \Leftrightarrow x = 3 \text{ or } 6$, so

$V = \pi \int_3^6 \left[[6 - (x - 4)^2]^2 - (8 - x)^2 \right] dx = \pi \int_3^6 (x^4 - 16x^3 + 83x^2 - 144x + 36) \, dx$

37. $V = \pi \int_0^{\pi/2} \left[(1 + \cos x)^2 - 1^2 \right] dx = \pi \int_0^{\pi/2} (2 \cos x + \cos^2 x) \, dx$

39. $V = \pi \int_0^1 3^2 \, dx + \pi \int_1^4 1^2 \, dx + \pi \int_4^5 3^2 \, dx$

$= 9\pi + 3\pi + 9\pi = 21\pi$

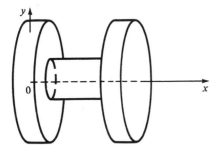

41. The solid is obtained by rotating the region under the curve $y = \tan x$, from $x = 0$ to $x = \frac{\pi}{4}$, about the x–axis.

43. The solid is obtained by rotating the region between the curves $x = y$ and $x = \sqrt{y}$ about the y–axis.

45. The solid is obtained by rotating the region between the curves $y = 5 - 2x^2$ and $y = 5 - 2x$ about the x–axis.

OR: The solid is obtained by rotating the region bounded by the curves $y = 2x$ and $y = 2x^2$ about the line $y = 5$.

47. $V = \pi \int_0^h \left(-\frac{r}{h} y + r \right)^2 dy$

$= \pi \int_0^h \left[\frac{r^2}{h^2} y^2 - \frac{2r^2}{h} y + r^2 \right] dy$

$= \pi \left[\frac{r^2}{3h^2} y^3 - \frac{r^2}{h} y^2 + r^2 y \right]_0^h = \frac{1}{3} \pi r^2 h$

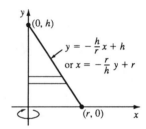

$y = -\frac{h}{r} x + h$

or $x = -\frac{r}{h} y + r$

49. $V = \pi \int_{r-h}^{r} (r^2 - y^2) \, dy = \pi \left[r^2 y - \dfrac{y^3}{3} \right]_{r-h}^{r}$

$= \pi \left[\left(r^3 - \dfrac{r^3}{3} \right) - \left(r^2(r-h) - \dfrac{(r-h)^3}{3} \right) \right]$

$= \pi h^2(r - h/3)$

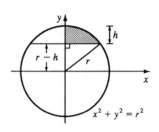

$x^2 + y^2 = r^2$

51. A typical cross section at height y above the base has dimensions $b\left(1 - \dfrac{y}{h}\right)$ and

$2b\left(1 - \dfrac{y}{h}\right)$, so $V = \int_0^h A(y) \, dy = \int_0^h 2b^2 \left(1 - \dfrac{y}{h}\right)^2 dy = 2b^2 \int_0^h \left(1 - \dfrac{2y}{h} + \dfrac{y^2}{h^2}\right) dy$

$= 2b^2 \left[y - \dfrac{y^2}{h} + \dfrac{y^3}{3h^2} \right]_0^h = 2b^2 \left[h - h + \dfrac{h}{3} \right] = \dfrac{2}{3} b^2 h$

$[= \dfrac{1}{3} Bh$, where B is the area of the base].

53. Area of cross-section at height z is

$A(z) = \dfrac{1}{2} \cdot 3\left(1 - \dfrac{z}{5}\right) \cdot 4\left(1 - \dfrac{z}{5}\right) = 6\left(1 - \dfrac{z}{5}\right)^2,$

so $V = \int_0^5 A(z) \, dz = 6 \int_0^5 \left(1 - \dfrac{z}{5}\right)^2 dz$

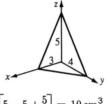

$= 6 \int_0^5 \left(1 - \dfrac{2z}{5} + \dfrac{z^2}{25}\right) dz = 6\left[z - \dfrac{z^2}{5} + \dfrac{z^3}{75} \right]_0^5 = 6\left[5 - 5 + \dfrac{5}{3} \right] = 10 \, \text{cm}^3$

55. $V = \int_{-2}^{2} A(x) \, dx = 2 \int_0^2 A(x) \, dx$

$= 2 \int_0^2 \dfrac{1}{2} (\sqrt{2} y)^2 \, dx = 2 \int_0^2 y^2 \, dx$

$= \dfrac{1}{2} \int_0^2 (36 - 9x^2) \, dx = \dfrac{9}{2} \int_0^2 (4 - x^2) \, dx$

$= \dfrac{9}{2} \left[4x - \dfrac{x^3}{3} \right]_0^2 = \dfrac{9}{2}\left(8 - \dfrac{8}{3}\right) = 24$

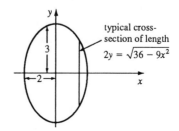

typical cross-
section of length

$2y = \sqrt{36 - 9x^2}$

57. The cross-section of the base corresponding
to the coordinate y has length $2x = 2\sqrt{y}$, so

$V = \int_0^1 A(y) \, dy = \int_0^1 (2x)^2 \, dy = \int_0^1 4x^2 \, dy$

$= \int_0^1 4y \, dy = 2y^2]_0^1 = 2$

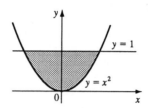

$y = 1$

$y = x^2$

59. Assume that the base of each isosceles triangle lies in the base of S. Then

$$V = \int_0^2 A(x)\,dx = \int_0^2 \tfrac{1}{2}y^2\,dx = \tfrac{1}{2}\int_0^2 \left(1 - \tfrac{x}{2}\right)^2 dx = \tfrac{1}{2}\left[\tfrac{2}{3}\left(\tfrac{x}{2} - 1\right)^3\right]_0^2 = \tfrac{1}{3}$$

61. (a) $V = \int_{-r}^{r} \pi\left[(R + \sqrt{r^2 - y^2})^2 - (R - \sqrt{r^2 - y^2})^2\right]dy = 2\pi\int_0^r 4R\sqrt{r^2 - y^2}\,dy$

$$= 8\pi R\int_0^r \sqrt{r^2 - y^2}\,dy$$

(b) Observe that the integral represents a quarter of the area of a circle with radius r,

so $8\pi R\int_0^r \sqrt{r^2 - y^2}\,dy = 8\pi R\,\tfrac{1}{4}(\pi r^2) = 2\pi^2 r^2 R$

63. The cross-sections perpendicular to the y-axis in Fig. 6.26 are rectangles. The rectangle

corresponding to the coordinate y has a base of length $2\sqrt{16 - y^2}$ in the xy-plane and a height of

$y/\sqrt{3}$. [Since $\angle BAC = 30°$, $|BC| = |AB|/\sqrt{3}$.] Thus $A(y) = (2/\sqrt{3})y\sqrt{16 - y^2}$ and

$$V = \int_0^4 A(y)\,dy = \int_0^4 A(y)\,dy = \tfrac{2}{\sqrt{3}}\int_0^4 \sqrt{16 - y^2}\,y\,dy = \tfrac{2}{\sqrt{3}}\int_{16}^0 u^{1/2}\left(-\tfrac{1}{2}du\right) \quad [u = 16 - y^2,$$

$$du = -2y\,dy] \quad = \tfrac{1}{\sqrt{3}}\int_0^{16} u^{1/2}\,du = \tfrac{1}{\sqrt{3}}\tfrac{2}{3}\left[u^{3/2}\right]_0^{16} = \tfrac{2}{3\sqrt{3}}(64) = \tfrac{128}{3\sqrt{3}}$$

65. Take the cylinder to be $x^2 + y^2 = R^2$ and drill out

the interior of the cylinder $y^2 + z^2 = r^2$. Taking

cross-sections perpendicular to the y-axis, we compute

$A(y) = (2\sqrt{R^2 - y^2})(2\sqrt{r^2 - y^2})$. [The cross-sections

are rectangles with length $2\sqrt{R^2 - y^2}$ in the x-direction

and height $2\sqrt{r^2 - y^2}$ in the z-direction.] Thus the volume cut

out is $V = 2\int_0^r A(y)\,dy = 8\int_0^r \sqrt{R^2 - y^2}\sqrt{r^2 - y^2}\,dy$.

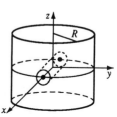

67. (a) Volume $(S_1) = \int_0^h A(z)\,dz = $ Volume (S_2) since the cross-sectional area $A(z)$ at

each height z is the same for both solids.

(b) By Cavalieri's Principle, the volume of the cylinder in the figure is the same as that of a

right circular cylinder with radius r and height h, that is, $\pi r^2 h$.

EXERCISES 6.3

1. $V = \int_1^2 2\pi x \cdot x^2 \, dx = 2\pi \int_1^2 x^3 \, dx = 2\pi \left[\frac{x^4}{4} \right]_1^2 = 2\pi \left(\frac{15}{4} \right) = \frac{15\pi}{2}$

3. $V = \int_0^4 2\pi x \sqrt{4 + x^2} \, dx = \pi \int_0^4 \sqrt{x^2 + 4} \, 2x \, dx = \left[\pi \frac{2}{3} (x^2 + 4)^{3/2} \right]_0^4 = \left(\frac{2\pi}{3} \right) (20\sqrt{20} - 8)$

     $= \frac{16\pi}{3} (5\sqrt{5} - 1)$

5. $V = \int_0^2 2\pi x (4 - x^2) \, dx = 2\pi \int_0^2 (4x - x^3) \, dx = 2\pi \left[2x^2 - x^4/4 \right]_0^2 = 2\pi (8 - 4) = 8\pi$

     $\left[\textit{Note: If we integrated from } -2 \textit{ to } 2, \textit{ we would be generating the volume twice.} \right]$

7. $V = \int_0^1 2\pi x (x^2 - x^3) \, dx = 2\pi \int_0^1 (x^3 - x^4) \, dx = 2\pi \left[\frac{x^4}{4} - \frac{x^5}{5} \right]_0^1 = 2\pi \left(\frac{1}{4} - \frac{1}{5} \right) = \frac{\pi}{10}$

9.   The two curves intersect at $(2, 2)$ and $(4, 2)$. For $2 < x < 4$, $-x^2 + 6x - 6 > x^2 - 6x + 10$.

     [To see this, just notice that

     $(x^2 - 6x + 10) - (-x^2 + 6x - 6) = 2x^2 - 12x + 16 = 2(x - 2)(x - 4) < 0$ for $2 < x < 4$.]

     Thus $V = \int_2^4 2\pi x \left[(-x^2 + 6x - 6) - (x^2 - 6x + 10) \right] dx = 2\pi \int_2^4 x(-2x^2 + 12x - 16) \, dx$

     $= 4\pi \int_2^4 (-x^3 + 6x^2 - 8x) \, dx = 4\pi \left[-\frac{x^4}{4} + 2x^3 - 4x^2 \right]_2^4$

     $= 4\pi \left[(-64 + 128 - 64) - (-4 + 16 - 16) \right] = 16\pi$

11. $V = \int_0^{16} 2\pi y \sqrt[4]{y} \, dy = 2\pi \int_0^{16} y^{5/4} \, dy = 2\pi \left[\frac{4}{9} y^{9/4} \right]_0^{16} = \frac{8\pi}{9} (512 - 0) = \frac{4096\pi}{9}$

13. $V = \int_0^9 2\pi y \cdot 2\sqrt{y} \, dy = 4\pi \int_0^9 y^{3/2} \, dy = 4\pi \left[\frac{2}{5} y^{5/2} \right]_0^9 = \frac{8\pi}{5} (243 - 0) = \frac{1944\pi}{5}$

15. $V = \int_0^1 2\pi y \left[(2 - y) - y^2 \right] dy = 2\pi \left[y^2 - \frac{y^3}{3} - \frac{y^4}{4} \right]_0^1 = 2\pi \left(1 - \frac{1}{3} - \frac{1}{4} \right) = \frac{5\pi}{6}$

17. $V = \int_1^4 2\pi x \sqrt{x} \, dx = 2\pi \int_1^4 x^{3/2} \, dx = 2\pi \left[\frac{2}{5} x^{5/2} \right]_1^4 = \frac{4\pi}{5} (32 - 1) = \frac{124\pi}{5}$

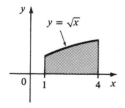
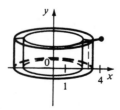

19. $V = \int_1^2 2\pi(x-1)x^2 \, dx = 2\pi \left[\frac{x^4}{4} - \frac{x^3}{3}\right]_1^2 = 2\pi\left[\left(4 - \frac{8}{3}\right) - \left(\frac{1}{4} - \frac{1}{3}\right)\right] = \frac{17\pi}{6}$

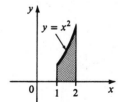
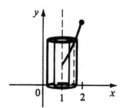

21. $V = \int_0^2 2\pi(3-y)(5-x) \, dy = \int_0^2 2\pi(3-y)(5-y^2-1) \, dy = \int_0^2 2\pi(12 - 4y - 3y^2 + y^3) \, dy$

$= 2\pi\left[12y - 2y^2 - y^3 + \frac{1}{4}y^4\right]_0^2 = 2\pi(24 - 8 - 8 + 4) = 24\pi$

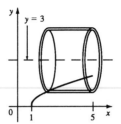
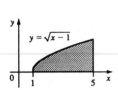

23. $V = \int_{2\pi}^{3\pi} 2\pi x \sin x \, dx$

25. $V = \int_0^{\pi/4} 2\pi y \cos y \, dy$

27. $V = \int_0^1 2\pi(x+1)(\sin\frac{\pi x}{2} - x^4) \, dx$

29. $V = \int_1^e 2\pi x \ln x \, dx$

31. $V = \int_1^\pi 2\pi y \cdot \ln y \, dy$

33. $V = \int_1^2 2\pi(y-1)\ln y \, dy = 2\pi \int_1^2 (y\ln y - \ln y) \, dy$

35. The solid is obtained by rotating the region bounded by the curve $y = \cos x$ and $y = 0$, from $x = 0$ to $x = \frac{\pi}{2}$, about the y–axis.

37. The solid is obtained by rotating the region (in the first quadrant) bounded by the curves $y = x^2$ and $y = x^6$ about the y–axis.

39. Use disks: $V = \int_{-2}^1 \pi(x^2 + x - 2)^2 \, dx = \pi \int_{-2}^1 (x^4 + 2x^3 - 3x^2 - 4x + 4) \, dx$

$= \pi\left[\frac{x^5}{5} + \frac{x^4}{2} - x^3 - 2x^2 + 4x\right]_{-2}^1 = \pi\left[\left(\frac{1}{5} + \frac{1}{2} - 1 - 2 + 4\right) - \left(-\frac{32}{5} + 8 + 8 - 8 - 8\right)\right]$

$$= \pi\left(\frac{33}{5} + \frac{3}{2}\right) = \frac{81\pi}{10}$$

41. Use disks: $V = \pi \int_{-1}^{1} (1 - y^2)^2 \, dy = 2\pi \int_{0}^{1} (y^4 - 2y^2 + 1) \, dy = 2\pi\left[\frac{y^5}{5} - \frac{2y^3}{3} + y\right]_{0}^{1}$

$$= 2\pi\left(\frac{1}{5} - \frac{2}{3} + 1\right) = \frac{16\pi}{15}$$

43. Use disks: $V = \pi \int_{0}^{2} \left[\sqrt{1 - (y-1)^2}\right]^2 \, dy = \pi \int_{0}^{2} (2y - y^2) \, dy = \pi\left[y^2 - \frac{y^3}{3}\right]_{0}^{2} = \pi\left(4 - \frac{8}{3}\right)$

$$= \frac{4\pi}{3}$$

45. $V = 2\int_{0}^{r} 2\pi x\sqrt{r^2 - x^2} \, dx = -2\pi \int_{0}^{r} (r^2 - x^2)^{1/2}(-2x) \, dx = \left[-2\pi\frac{2}{3}(r^2 - x^2)^{3/2}\right]_{0}^{r}$

$$= -\left(\frac{4\pi}{3}\right)(0 - r^3) = \frac{4}{3}\pi r^3$$

47. $V = 2\pi \int_{0}^{r} x\left(-\frac{h}{r}x + h\right) dx = 2\pi h \int_{0}^{r} \left(-\frac{x^2}{r} + x\right) dx$

$$= 2\pi h\left[-\frac{x^3}{3r} + \frac{x^2}{2}\right]_{0}^{r} = 2\pi h\left(\frac{r^2}{6}\right)$$

$$= \frac{\pi r^2 h}{3}$$

49. If $a < b \le 0$, then a typical cylindrical shell has radius $-x$ and height $f(x)$, so

$$V = \int_{a}^{b} 2\pi(-x)f(x) \, dx = -\int_{a}^{b} 2\pi x f(x) \, dx.$$

51. $\Delta x = \frac{\pi/4 - 0}{4} = \frac{\pi}{16}$.

$V = \int_{0}^{\pi/4} 2\pi x \tan x \, dx \approx 2\pi \frac{\pi}{16}\left(\frac{\pi}{32}\tan\frac{\pi}{32} + \frac{3\pi}{32}\tan\frac{3\pi}{32} + \frac{5\pi}{32}\tan\frac{5\pi}{32} + \frac{7\pi}{32}\tan\frac{7\pi}{32}\right) \approx 1.142$

EXERCISES 6.4

1. By (6.13), $W = Fd = (900)(8) = 7200\,\text{J}$.

3. By (6.15), $W = \int_{a}^{b} f(x) \, dx = \int_{0}^{10} (5x^2 + 1) \, dx = \left[\frac{5}{3}x^3 + x\right]_{0}^{10} = \frac{5000}{3} + 10 = \frac{5030}{3}\,\text{ft-lb}$.

5. $10 = f(x) = kx = \frac{k}{3}$ [4 inches $= \frac{1}{3}$ foot], so $k = 30$ [The units for k are pounds per foot.] Now 6

   inches $= \frac{1}{2}$ foot, so $W = \int_{0}^{1/2} 30x \, dx = \left[15x^2\right]_{0}^{1/2} = \frac{15}{4}\,\text{ft-lb}$.

7. If $\int_{0}^{.12} kx \, dx = 2\,\text{J}$, then $2 = \left[\frac{1}{2}kx^2\right]_{0}^{.12} = \frac{1}{2}k(.0144) = .0072k$ and $k = \frac{2}{.0072}$

   $= \frac{2500}{9} \approx 277.78$. Thus the work needed to stretch the spring from 35 cm to 40 cm is

   $\int_{.05}^{.10} \frac{2500}{9}x \, dx = \left[\frac{1250}{9}x^2\right]_{1/20}^{1/10} = \frac{1250}{9}\left(\frac{1}{100} - \frac{1}{400}\right) = \frac{25}{24} \approx 1.04\,\text{J}$.

9. $f(x) = kx$, so $30 = \frac{2500}{9}x$ and $x = \frac{270}{2500}$ m $= 10.8$ cm.

11. First notice that the exact height of the building is immaterial. The portion of the rope from x ft to $x + \Delta x$ ft below the top of the building weighs $\frac{1}{2}\Delta x$ lb and must be lifted x ft (approximately), so its contribution to the total work is $\frac{1}{2}x\Delta x$ ft-lb. The total work is
$$W = \int_0^{50} \frac{1}{2}x\,dx = \left[\frac{1}{4}x^2\right]_0^{50} = \frac{2500}{4} = 625 \text{ ft-lb.}$$

13. The work needed to lift the cable is $\int_0^{500} 2x\,dx = x^2\big]_0^{500} = 250{,}000$ ft-lb. The work needed to lift the coal is $800 \text{ lb}\cdot 500 \text{ ft} = 400{,}000$ ft-lb. Thus the total work required is
$$250{,}000 + 400{,}000 = 650{,}000 \text{ ft-lb.}$$

15. A "slice" of water Δx m thick and lying at a depth of x m $(0 \le x \le \frac{1}{2})$ has volume $2\Delta x$ m^3, a mass of $2000\Delta x$ kg, weighs $(9.8)(2000\Delta x) = 19{,}600\,\Delta x$ N, and requires $19{,}600x\Delta x$ J of work for its removal. Thus $W = \int_0^{1/2} 19{,}600x\,dx = 9800x^2\big]_0^{1/2}$
$= 2450$ J. [This answer is approximate because the value $g = 9.8$ m/s^2 is approximate. Thus we should really write $W \approx 2.45 \times 10^3$ J]

17. A "slice" of water Δx m thick and lying x ft above the bottom has volume $8x\Delta x$ m^3 and weighs $\approx (9.8 \times 10^3)(8x\Delta x)$ N. It must be lifted $5 - x$ m by the pump, so the work needed is $\approx (9.8 \times 10^3)(5 - x)(8x\Delta x)$ J. The total work required is
$$W \approx \int_0^3 (9.8 \times 10^3)(5-x)8x\,dx = (9.8 \times 10^3)\int_0^3 (40x - 8x^2)\,dx = (9.8 \times 10^3)\left[20x^2 - \frac{8}{3}x^3\right]_0^3 = (9$$
$.8 \times 10^3)(180 - 72) = (9.8 \times 10^3)(108) = 1058.4 \times 10^3 \approx 1.06 \times 10^6$ J.

19. Measure depth x downward from the flat top of the tank, so that $0 \le x \le 2$ ft. Then
$\Delta W = (62.5)(2\sqrt{4-x^2})(8\Delta x)(x+1)$ ft-lb, so $W \approx (62.5)(16)\int_0^2 (x+1)\sqrt{4-x^2}\,dx$
$$= 1000\left[\int_0^2 x\sqrt{4-x^2}\,dx + \int_0^2 \sqrt{4-x^2}\,dx\right] = 1000\left[\int_0^4 u^{1/2}\frac{1}{2}\,du + \frac{1}{4}\pi(2^2)\right]$$
$$\left[u = 4 - x^2,\ du = -2x\,dx\right] = 1000\left[\left[\frac{1}{2}\cdot\frac{2}{3}u^{3/2}\right]_0^4 + \pi\right] = 1000\left(\frac{8}{3} + \pi\right) \approx 5.8 \times 10^3 \text{ ft-lb.}$$
Note: The second integral was computed by noticing that it represents the area of a quarter-circle of radius 2.

21. The only change needed is to replace 1000 kg/m^3 by 680 kg/m^3, so we multiply the answer to #17 by .680 to get $\approx .720 \times 10^6 = 7.20 \times 10^5$ J.

23. $V = \pi r^2 x$, so V is a function of x and P can also be regarded as a function of x. If $V_1 = \pi r^2 x_1$ and $V_2 = \pi r^2 x_2$, then $W = \int_{x_1}^{x_2} F(x)\,dx = \int_{x_1}^{x_2} \pi r^2 P(V(x))\,dx$
$$= \int_{x_1}^{x_2} P(V(x))\,dV(x)\ \left[V(x) = \pi r^2 x,\text{ so } dV(x) = \pi r^2\,dx\right] = \int_{V_1}^{V_2} P(V)\,dV \text{ by}$$
the Substitution Rule.

25. $W = \int_a^b F(r)\,dr = \int_a^b G\frac{m_1 m_2}{r^2}\,dr = Gm_1 m_2\left[\frac{-1}{r}\right]_a^b = Gm_1 m_2\left(\frac{1}{a} - \frac{1}{b}\right)$

EXERCISES 6.5

1. $f_{ave} = \frac{1}{3-0} \int_0^3 (1-2x)\,dx = \frac{1}{3}(x-x^2)\big]_0^3 = \frac{1}{3}(3-9) = -2$

3. $f_{ave} = \frac{1}{2-(-2)} \int_{-2}^2 (x^2+2x-5)\,dx = \frac{1}{4}\left[\frac{x^3}{3}+x^2-5x\right]_{-2}^2$

    $= \frac{1}{4}\left[\left(\frac{8}{3}+4-10\right)-\left(-\frac{8}{3}+4+10\right)\right] = -\frac{11}{3}$

5. $f_{ave} = \frac{1}{1-(-1)} \int_{-1}^1 x^4\,dx = \frac{1}{2}2\int_0^1 x^4\,dx = \left[\frac{x^5}{5}\right]_0^1 = \frac{1}{5}$

7. $f_{ave} = \frac{1}{(\pi/4)-(-\pi/2)} \int_{-\pi/2}^{\pi/4} \sin^2 x \cos x\,dx = \frac{4}{3\pi} \int_{-\pi/2}^{\pi/4} \sin^2 x \cos x\,dx = \frac{4}{3\pi} \int_{-1}^{1/\sqrt{2}} u^2\,du$

    $[u = \sin x \Rightarrow du = \cos x\,dx] \quad = \frac{4}{3\pi}\left[\frac{u^3}{3}\right]_{-1}^{1/\sqrt{2}} = \frac{4}{9\pi}\left(\frac{1}{2\sqrt{2}}+1\right) = \frac{4}{9\pi}\left(\frac{\sqrt{2}}{4}+1\right) = \frac{\sqrt{2}+4}{9\pi}$

9. (a) $f_{ave} = \frac{1}{3-0} \int_0^3 2x\,dx = \frac{1}{3}x^2\big]_0^3 = \frac{1}{3}(9-0) = 3$

    (b) $f_{ave} = f(c)$ when $3 = 2c$, that is when $c = \frac{3}{2}$

    (c)
    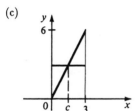

11. (a) $f_{ave} = \frac{1}{2-0} \int_0^2 (4-x^2)\,dx$

    $= \frac{1}{2}\left[4x - \frac{x^3}{3}\right]_0^2 = \frac{1}{2}\left[\left(8-\frac{8}{3}\right)-0\right] = \frac{8}{3}$

    (b) $f_{ave} = f(c) \Leftrightarrow \frac{8}{3} = 4-c^2 \Leftrightarrow c^2 = \frac{4}{3}$

    $\Leftrightarrow c = \frac{2}{\sqrt{3}}$

    (c)
    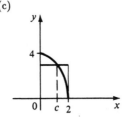

13. $T_{ave} = \frac{1}{12} \int_0^{12} \left[50 + 14\sin\frac{\pi t}{12}\right]dt = \frac{1}{12}\left[50t - 14\frac{12}{\pi}\cos\frac{\pi t}{12}\right]_0^{12}$

    $= \frac{1}{12}\left[50\cdot 12 + 14\frac{12}{\pi} + 14\frac{12}{\pi}\right] = (50 + \frac{28}{\pi})°F \approx 59°F$

15. $\rho_{ave} = \frac{1}{8} \int_0^8 \frac{12}{\sqrt{x+1}}\,dx = \frac{3}{2}\int_0^8 (x+1)^{-1/2}\,dx = 3\sqrt{x+1}\big]_0^8 = 9-3 = 6\,kg/m$

17. $V_{ave} = \frac{1}{5} \int_0^5 V(t)\,dt = \frac{1}{5}\int_0^5 \frac{5}{4\pi}\left[1-\cos(2\pi t/5)\right]dt = \frac{1}{4\pi}\int_0^5 \left[1-\cos(2\pi t/5)\right]dt$

$= \frac{1}{4\pi}\left[t - \frac{5}{2\pi}\sin\left(2\pi t/5\right)\right]_0^5 = \frac{1}{4\pi}\left[(5-0)-0\right] = \frac{5}{4\pi} \approx 0.4\,\text{L}$

19. Let $F(x) = \int_a^x f(t)\,dt$ for x in [a, b]. Then F is continuous on [a, b] and differentiable on (a, b), so by the Mean Value Theorem there is a number c in (a, b) such that $F(b) - F(a) = F'(c)(b-a)$. But $F'(x) = f(x)$ by the Fundamental Theorem of Calculus. Therefore $\int_a^b f(t)\,dt - 0 = f(c)(b-a)$.

REVIEW EXERCISES FOR CHAPTER 6

1. $A = \int_1^3 \left[0 - (x^2 - 4x + 3)\right]dx = \int_1^3 (-x^2 + 4x - 3)\,dx = -\frac{x^3}{3} + 2x^2 - 3x\Big]_1^3$

 $= (-9 + 18 - 9) - (-1/3 + 2 - 3) = 4/3$

3. $A = \int_0^6 \left[(12x - 2x^2) - (x^2 - 6x)\right]dx = \int_0^6 (18x - 3x^2)\,dx = 9x^2 - x^3\big]_0^6$

 $= 9\cdot 36 - 216 = 108$

5. By symmetry, $A = 2\int_0^1 (x^{1/3} - x^3)\,dx = 2\left[\frac{3}{4}x^{4/3} - \frac{x^4}{4}\right]_0^1 = 2\left[\frac{3}{4} - \frac{1}{4}\right] = 1$

7. $A = \int_0^\pi |\sin x - (-\cos x)|\,dx = \int_0^{3\pi/4} (\sin x + \cos x)\,dx - \int_{3\pi/4}^\pi (\sin x + \cos x)\,dx$

 $= \left[\sin x - \cos x\right]_0^{3\pi/4} - \left[-\cos x + \sin x\right]_{3\pi/4}^\pi$

 $= (1/\sqrt{2} + 1/\sqrt{2}) - (0-1) - (1+0) + (1/\sqrt{2} + 1/\sqrt{2}) = \sqrt{2} + 1 - 1 + \sqrt{2} = 2\sqrt{2}$

9. $V = \int_1^3 \pi(\sqrt{x-1})^2\,dx = \pi\int_1^3 (x-1)\,dx = \pi\left[\frac{x^2}{2} - x\right]_1^3 = \pi\left[\left(\frac{9}{2} - 3\right) - \left(\frac{1}{2} - 1\right)\right] = 2\pi$

11. $V = \int_1^3 2\pi y(-y^2 + 4y - 3)\,dy = 2\pi\int_1^3 (-y^3 + 4y^2 - 3y)\,dy = 2\pi\left[-\frac{y^4}{4} + \frac{4}{3}y^3 - \frac{3}{2}y^2\right]_1^3$

 $= 2\pi\left[\left(-\frac{81}{4} + 36 - \frac{27}{2}\right) - \left(-\frac{1}{4} + \frac{4}{3} - \frac{3}{2}\right)\right] = \frac{16\pi}{3}$

13. $V = \int_a^{a+h} 2\pi x \cdot 2\sqrt{x^2 - a^2}\,dx = 2\pi\int_0^{2ah+h^2} u^{1/2}\,du \ \left[u = x^2 - a^2,\ du = 2x\,dx\right]$

 $= 2\pi\left[\frac{2}{3}u^{3/2}\right]_0^{2ah+h^2} = \frac{4\pi}{3}(2ah + h^2)^{3/2}$

15. $V = \int_0^1 \pi\left[(1-x^3)^2 - (1-x^2)^2\right]dx$

17. (a) $V = \int_0^1 \pi(x^2 - x^4)\,dx = \pi\left[\frac{1}{3}x^3 - \frac{1}{5}x^5\right]_0^1 = \pi\left[\frac{1}{3} - \frac{1}{5}\right] = \frac{2\pi}{15}$

 OR: $V = \int_0^1 2\pi y(\sqrt{y} - y)\,dy = 2\pi\left[\frac{2}{5}y^{5/2} - \frac{y^3}{3}\right]_0^1 = \frac{2\pi}{15}.$

 (b) $V = \int_0^1 \pi(\sqrt{y}^2 - y^2)\,dy = \pi\left[\frac{1}{2}y^2 - \frac{1}{3}y^3\right]_0^1 = \pi\left[\frac{1}{2} - \frac{1}{3}\right] = \frac{\pi}{6}$

OR: $V = \int_0^1 2\pi x(x - x^2)\, dx = 2\pi\left[\frac{x^3}{3} - \frac{x^4}{4}\right]_0^1 = \frac{\pi}{6}$.

(c) $\quad V = \int_0^1 \pi\left[(2 - x^2)^2 - (2 - x)^2\right] dx = \int_0^1 \pi(x^4 - 5x^2 + 4x)\, dx = \pi\left[\frac{1}{5}x^5 - \frac{5}{3}x^3 + 2x^2\right]_0^1$

$\quad = \pi\left[\frac{1}{5} - \frac{5}{3} + 2\right] = \frac{8\pi}{15}$.

OR: $V = \int_0^1 2\pi(2 - y)(\sqrt{y} - y)\, dy = 2\pi \int_0^1 (y^2 - y^{3/2} - 2y + 2y^{1/2})\, dy$

$\quad = 2\pi\left[\frac{y^3}{3} - \frac{2}{5}y^{5/2} - y^2 + \frac{4}{3}y^{3/2}\right]_0^1 = \frac{8\pi}{15}$

19. The solid is obtained by rotating the region under the curve $y = \sin x$, above $y = 0$, from $x = 0$ to $x = \pi$, about the x–axis.

21. The solid is obtained by rotating the region in the first quadrant bounded by the curve $x = 4 - y^2$ and the coordinate axes about the x–axis.

23. Take the base to be the disk $x^2 + y^2 \leq 9$. Then $V = \int_{-3}^3 A(x)\, dx$, where $A(x_0)$ is the area of the isosceles right triangle whose hypotenuse lies along the line $x = x_0$ in the xy-plane. $A(x) = \frac{1}{4}(2\sqrt{9 - x^2})^2 = 9 - x^2$, so $V = 2\int_0^3 A(x)\, dx = 2\int_0^3 (9 - x^2)\, dx$
$= 2\left[9x - \frac{x^3}{3}\right]_0^3 = 2(27 - 9) = 36$.

25. Equilateral triangles with sides measuring $\frac{x}{4}$ meters have height $\frac{x}{4}\sin 60° = \frac{\sqrt{3}\,x}{8}$.
Therefore, $A(x) = \frac{1}{2}\cdot\frac{x}{4}\cdot\frac{\sqrt{3}\,x}{8} = \frac{\sqrt{3}}{64}x^2$.
$V = \int_0^{20} A(x)\, dx = \frac{\sqrt{3}}{64}\int_0^{20} x^2\, dx = \frac{\sqrt{3}}{64}\left[\frac{1}{3}x^3\right]_0^{20} = \frac{1000\sqrt{3}}{24} = \frac{125\sqrt{3}}{3}\, m^3$.

27. $30\,N = f(x) = kx = k(.03\,m)$, so $k = \frac{30}{.03} = 1000\frac{N}{m}$. $W = \int_0^{.08} kx\, dx = 1000\int_0^{.08} x\, dx$
$= 500\left[x^2\right]_0^{.08} = 500(.08)^2 = 3.2\,J$

29. $W = \int_0^4 \pi(2\sqrt{y})^2\, 62.5(4 - y)\, dy$
$= 250\pi \int_0^4 y(4 - y)\, dy = 250\pi\left[2y^2 - \frac{y^3}{3}\right]_0^4$
$= 250\pi\left(32 - \frac{64}{3}\right) = 8000\pi/3$ ft-lb

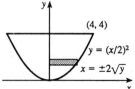

31. $\lim_{h \to 0} f_{ave} = \lim_{h \to 0} \frac{1}{h}\int_x^{x+h} f(t)\, dt = \lim_{h \to 0} \frac{F(x + h) - F(x)}{h}$, where $F(x) = \int_a^x f(t)\, dt$.
But we recognize this limit as being $F'(x)$ by the definition of derivative. Therefore
$\lim_{h \to 0} f_{ave} = F'(x) = f(x)$ by Part 1 of the Fundamental Theorem.

PROBLEMS PLUS (page 404)

1. Differentiating both sides of the equation $x \sin \pi x = \int_0^{x^2} f(t)\,dt$ (using part 1 of the Fundamental Theorem for the right side) gives $\sin \pi x + \pi x \cos \pi x = 2x\,f(x^2)$. Putting $x = 2$, we obtain $\sin 2\pi + 2\pi \cos 2\pi = 4\,f(4)$, so $f(4) = \dfrac{0 + 2\pi \cdot 1}{4} = \dfrac{\pi}{2}$.

3. For $1 \le x \le 2$, we have $x^4 \le 2^4 = 16$, so $1 + x^4 \le 17$ and $\dfrac{1}{1+x^4} \ge \dfrac{1}{17}$.

   Thus $\displaystyle\int_1^2 \dfrac{1}{1+x^4}\,dx \ge \int_1^2 \dfrac{1}{17}\,dx = \dfrac{1}{17}$. Also $1 + x^4 > x^4$ for $1 \le x \le 2$, so $\dfrac{1}{1+x^4} < \dfrac{1}{x^4}$ and

   $\displaystyle\int_1^2 \dfrac{1}{1+x^4}\,dx < \int_1^2 x^{-4}\,dx = \left[\dfrac{x^{-3}}{-3}\right]_1^2 = -\dfrac{1}{24} + \dfrac{1}{3} = \dfrac{7}{24}$. Thus we have the estimate

   $\dfrac{1}{17} \le \displaystyle\int_1^2 \dfrac{1}{1+x^4}\,dx \le \dfrac{7}{24}$.

5. Consider the statement that $\dfrac{d^n}{dx^n}(e^{ax}\sin bx) = r^n e^{ax}\sin(bx + n\theta)$. For $n = 1$,

   $\dfrac{d}{dx}(e^{ax}\sin bx) = ae^{ax}\sin bx + be^{ax}\cos bx$, and

   $re^{ax}\sin(bx + \theta) = re^{ax}[\sin bx \cos\theta + \cos bx \sin\theta] = re^{ax}(\tfrac{a}{r}\sin bx + \tfrac{b}{r}\cos bx)$

   $= ae^{ax}\sin bx + be^{ax}\cos bx$, since $\tan\theta = \tfrac{b}{a} \Rightarrow \sin\theta = \tfrac{b}{r}$ and $\cos\theta = \tfrac{a}{r}$. So the statement is

   true for $n = 1$. Assume it is true for $n = k$. Then

   $\dfrac{d^{k+1}}{dx^{k+1}}(e^{ax}\sin bx) = \dfrac{d}{dx}[r^k e^{ax}\sin(bx + k\theta)] = r^k ae^{ax}\sin(bx + k\theta) + r^k e^{ax}b\cos(bx + k\theta)$

   $= r^k e^{ax}[a\sin(bx + k\theta) + b\cos(bx + k\theta)]$. But

   $\sin[bx + (k+1)\theta] = \sin[(bx + k\theta) + \theta] = \sin(bx + k\theta)\cos\theta + \sin\theta\cos(bx + k\theta)$

   $= \tfrac{a}{r}\sin(bx + k\theta) + \tfrac{b}{r}\cos(bx + k\theta)$. Hence

   $a\sin(bx + k\theta) + b\cos(bx + k\theta) = r\sin[bx + (k+1)\theta]$. So that

   $\dfrac{d^{k+1}}{dx^{k+1}}(e^{ax}\sin bx) = r^k e^{ax}[a\sin(bx + k\theta) + b\sin(bx + k\theta)]$

   $= r^k e^{ax}[r\sin(bx + (k+1)\theta)] = r^{k+1}e^{ax}[\sin(bx + (k+1)\theta)]$. Therefore the statement is true

   for all n by mathematical induction.

7. The volume generated from $x = 0$ to $x = b$ is $\displaystyle\int_0^b \pi[f(x)]^2\,dx$. Hence we are given that

   $b^2 = \displaystyle\int_0^b \pi[f(x)]^2\,dx$ for all $b > 0$. Differentiating both sides of this equation using the

   Fundamental Theorem of Calculus gives $2b = \pi[f(b)]^2 \Rightarrow f(b) = \sqrt{2b/\pi}$, since f is positive.

   Therefore $f(x) = \sqrt{2x/\pi}$.

9. Let the line through A and B have equation $y = mx + b$. Now $x^2 = mx + b$ gives

$x^2 - mx - b = 0$ and hence $x = \dfrac{m \pm \sqrt{m^2 + 4b}}{2}$. So A has x–coordinate $x_1 = \dfrac{m - \sqrt{m^2 + 4b}}{2}$

and B has x–coordinate $x_2 = \dfrac{m + \sqrt{m^2 + 4b}}{2}$. So the parabolic segment has area

$$\int_{x_1}^{x_2} [(mx + b) - x^2]\,dx = \left[\frac{mx^2}{2} + bx - \frac{x^3}{3}\right]_{x_1}^{x_2}$$

$$= \tfrac{m}{2}(x_2^2 - x_1^2) + b(x_2 - x_1) - \tfrac{1}{3}(x_2^3 - x_1^3)$$

$$= (x_2 - x_1)[\tfrac{m}{2}(x_2 + x_1) + b - \tfrac{1}{3}(x_2^2 + x_1 x_2 + x_1^2)]$$

$$= \sqrt{m^2 + 4b}\left[\frac{m^2}{2} + b - \tfrac{1}{3}(mx_2 + b - b + mx_1 + b)\right]$$

$$= \sqrt{m^2 + 4b}\left[\frac{m^2}{2} + b - \tfrac{1}{3}(m^2 + b)\right] = \frac{(m^2 + 4b)^{3/2}}{6}.$$

Now since the line through C has slope m, we see that if C has x–coordinate c, then $2c = m$

[since $(x^2)' = 2x$] and hence $c = \tfrac{m}{2}$. Hence C has coordinates $(m/2, m^2/4)$. The line through

AC has slope $\dfrac{m^2/4 - x_1^2}{m/2 - x_1} = \tfrac{m}{2} + x_1$ and has equation $y - x_1^2 = (\tfrac{m}{2} + x_1)(x - x_1)$ or

$y = (\tfrac{m}{2} + x_1)x - \tfrac{m}{2}x_1$. Similarly, the equation of the line through BC is $y = (\tfrac{m}{2} + x_2)x - \tfrac{m}{2}x_2$.

So the area of the triangular region is

$$\int_{x_1}^{m/2} [(mx + b) - ((\tfrac{m}{2} + x_1)x - \tfrac{m}{2}x_1)]\,dx + \int_{m/2}^{x_2} [(mx + b) - ((\tfrac{m}{2} + x_2)x - \tfrac{m}{2}x_2)]\,dx$$

$$= \int_{x_1}^{m/2} [(\tfrac{m}{2} - x_1)x + (b + \tfrac{m}{2}x_1)]\,dx + \int_{m/2}^{x_2} [(\tfrac{m}{2} - x_2)x + (b + \tfrac{m}{2}x_2)]\,dx$$

$$= \left[\frac{(m/2 - x_1)x^2}{2} + (b + \tfrac{m}{2}x_1)x\right]_{x_1}^{m/2} + \left[\frac{(m/2 - x_2)x^2}{2} + (b + \tfrac{m}{2}x_2)\right]_{m/2}^{x_2}'$$

$$= \frac{m^3}{16} - \frac{m^2}{8}x_1 + \frac{bm}{2} + \frac{m^2}{4}x_1 - \frac{m}{4}x_1^2 + \frac{x_1^3}{2} - bx_1 - \frac{m}{2}x_1^2 + \frac{m}{4}x_2^2 - \frac{x_2^3}{2} + bx_2 +$$

$$\frac{m}{2}x_2^2 - \frac{m^3}{16} + \frac{m^2}{8}x_2 - \frac{bm}{2} - \frac{m^2}{4}x_2$$

$$= (b - \tfrac{m^2}{8})(x_2 - x_1) + \tfrac{3}{4}m(x_2^2 - x_1^2) - \tfrac{1}{2}(x_2^3 - x_1^3)$$

$$= (x_2 - x_1)[(b - \tfrac{m^2}{8}) + \tfrac{3}{4}m(x_2 + x_1) - \tfrac{1}{2}(x_2^2 + x_2 x_1 + x_1^2)]$$

$$= \sqrt{m^2 + 4b}\left[(b - \tfrac{m^2}{8}) + \tfrac{3}{4}m^2 - \tfrac{1}{2}(m^2 + b)\right] = \sqrt{m^2 + 4b}\left(\frac{m^2 + 4b}{8}\right) = \frac{(m^2 + 4b)^{3/2}}{8}.$$

The result follows since $\tfrac{4}{3}\left(\dfrac{(m^2 + 4b)^{3/2}}{8}\right) = \dfrac{(m^2 + 4b)^{3/2}}{6}$.

ALTERNATE SOLUTION: Let $A = (a, a^2)$, $B = (b, b^2)$. Then $m_{AB} = (b^2 - a^2)/(b - a)$

$= a + b$, so the equation of AB is $y - a^2 = (a + b)(x - a)$, or $y = (a + b)x - ab$, and the area

of the parabolic segment is $\int_a^b [(a + b)x - ab - x^2] dx = [(a + b)\frac{1}{2}x^2 - abx - \frac{1}{3}x^3]_a^b$

$= \frac{1}{2}(a + b)(b^2 - a^2) - ab(b - a) - \frac{1}{3}(b^3 - a^3) = \frac{1}{6}(b - a)^3$. At C, $y' = 2x = b + a$, so the x-

coordinate of C is $\frac{1}{2}(a + b)$. If we calculate the area of triangle ABC as in Problems Plus 21

after Chapter 4 (by subtracting areas of trapezoids) we find that the area is

$\frac{1}{2}(b - a)[\frac{1}{2}(a + b) - a][b - \frac{1}{2}(a + b)] = \frac{1}{2}(b - a)\frac{1}{2}(b - a)\frac{1}{2}(b - a) = \frac{1}{8}(b - a)^3$.

This is $\frac{3}{4}$ of the area of the parabolic segment calculated above.

11. By l'Hospital's Rule, and the Fundamental Theorem,

$$\lim_{x \to 0} \frac{\int_0^x (1 - \tan 2t)^{1/t} dt}{x}$$

$$= \lim_{x \to 0} \frac{(1 - \tan 2x)^{1/x}}{1} = e^{\displaystyle\lim_{x \to 0} \frac{\ln(1 - \tan 2x)}{x}} \overset{H}{=} e^{\displaystyle\lim_{x \to 0} \frac{-2\sec^2 2x}{1 - \tan 2x}} = e^{\frac{-2 \cdot 1^2}{1 - 0}} = e^{-2}.$$

13. We first show that $\dfrac{x}{1 + x^2} < \tan^{-1}x$ for $x > 0$. Let $f(x) = \tan^{-1}x - \dfrac{x}{1 + x^2}$. Then

$f'(x) = \dfrac{1}{1 + x^2} - \dfrac{1(1 + x^2) - x(2x)}{(1 + x^2)^2} = \dfrac{(1 + x^2) - (1 - x^2)}{(1 + x^2)^2} = \dfrac{2x^2}{(1 + x^2)^2} > 0$ for $x > 0$. So f(x) is

increasing on $[0, \infty)$. Hence $0 < x \Rightarrow 0 = f(0) < f(x) = \tan^{-1}x - \dfrac{x}{1 + x^2}$. So

$\dfrac{x}{1 + x^2} < \tan^{-1}x$ for $0 < x$. We next show that $\tan^{-1}x < x$ for $x > 0$. Let

$h(x) = x - \tan^{-1}x$. So $h'(x) = 1 - \dfrac{1}{1 + x^2} = \dfrac{x^2}{1 + x^2} > 0$. So h(x) is increasing on $[0, \infty)$. So

for $0 < x$, $0 = h(0) < h(x) = x - \tan^{-1}x$. Hence $\tan^{-1}x < x$ for $x > 0$.

15. $f(x) = \displaystyle\int_0^{g(x)} \frac{1}{\sqrt{1 + t^3}} dt$, where $g(x) = \displaystyle\int_0^{\cos x} [1 + \sin(t^2)] dt$. Using both part 1 of the

Fundamental Theorem and the chain rule twice we have that

$f'(x) = \dfrac{1}{\sqrt{1 + [g(x)]^3}} g'(x) = \dfrac{1}{\sqrt{1 + [g(x)]^3}}[1 + \sin(\cos^2 x)](-\sin x)$.

Now $g(\frac{\pi}{2}) = \displaystyle\int_0^0 [1 + \sin(t^2)] dt = 0$, so $f'(\frac{\pi}{2}) = \dfrac{1}{\sqrt{1 + 0}}[1 + \sin 0](-1) = 1 \cdot 1 \cdot (-1) = -1$.

17. By the Fundamental Theorem of Calculus, $f'(x) = \sqrt{1 + x^3} > 0$ for $x > -1$. So f is increasing

on $[-1, \infty)$ and hence is one-to-one. Note that $f(1) = 0$, so $f^{-1}(1) = 0 \Rightarrow$

$(f^{-1})'(0) = \dfrac{1}{f'(1)} = \dfrac{1}{\sqrt{2}}$.

19. Consider the function $f(x) = 2 + x - x^2 = (-x+2)(x+1)$.

$f(x) = 0 \Leftrightarrow x = 2$ or $x = -1$, and $f(x) \geq 0$ for $x \in [-1, 2]$ and $f(x) < 0$ everywhere else.

The integral $\int_a^b (2 + x - x^2)\, dx$ is a maximum on the interval where the integrand is positive, which is $[-1, 2]$. (Any larger interval gives a smaller integral since $f(x) < 0$ outside $[-1, 2]$. Any smaller interval also gives a smaller integral since $f(x) \geq 0$ in $[-1, 2]$.)

21. (a) Let $f(x) = x - \ln x - 1$, so $f'(x) = 1 - \frac{1}{x} = \frac{x-1}{x}$. Since $x > 0$, $f'(x) < 0$ for $0 < x < 1$ and $f'(x) > 0$ for $x > 1$. So there is an absolute minimum at $x = 1$ with $f(1) = 0$. So for $x > 0$, $x \neq 1$, $x - \ln x - 1 = f(x) > f(1) = 0$, and hence $\ln x < x - 1$.

(b) Here let $f(x) = \ln x - \frac{x-1}{x} = \ln x - 1 + \frac{1}{x}$. So $f'(x) = \frac{1}{x} - \frac{1}{x^2} = \frac{x-1}{x^2}$. As in (a), we see that there is an absolute minimum value at $x = 1$ and that $f(1) = 0$. So for $x > 0$, $x \neq 1$, $\ln x - \frac{x-1}{x} = f(x) > f(1) = 0$ and hence $\frac{x-1}{x} < \ln x$.

(c) Let $a > b > 0$, so $\frac{a}{b} > 1$. Letting $x = \frac{a}{b}$ in the inequalities in (a) and (b) gives
$$\frac{a-b}{a} = \frac{a/b-1}{a/b} < \ln\frac{a}{b} < \frac{a}{b} - 1 = \frac{a-b}{b}.$$ Noting that $\ln\frac{a}{b} = \ln a - \ln b$, the result follows after dividing through by $a - b$.

23. Note that $\frac{d}{dx}\left(\int_0^x \left(\int_0^u f(t)\, dt\right) du\right) = \int_0^x f(t)\, dt$ by Part 1 of the Fundamental Theorem, while
$$\frac{d}{dx}\left(\int_0^x f(u)(x-u)\, du\right) = \frac{d}{dx}\left(x\int_0^x f(u)\, du\right) - \frac{d}{dx}\left(\int_0^x f(u)\, u\, du\right) = \int_0^x f(u)\, du + x f(x) - f(x) x$$
$$= \int_0^x f(u)\, du.$$ Hence $\int_0^x f(u)(x-u)\, du = \int_0^x \left(\int_0^u f(t)\, dt\right) du + C$. Setting $x = 0$ gives

$C = 0$.

25. Note that $f(0) = 0$, so for $x \neq 0$, $\left|\frac{f(x) - f(0)}{x - 0}\right| = \left|\frac{f(x)}{x}\right| = \frac{|f(x)|}{|x|} \leq \frac{|\sin x|}{|x|} = \frac{\sin x}{x}$.

Therefore $\left|f'(0)\right| = \left|\lim_{x \to 0} \frac{f(x) - f(0)}{x - 0}\right| = \lim_{x \to 0}\left|\frac{f(x) - f(0)}{x - 0}\right| \leq \lim_{x \to 0} \frac{\sin x}{x} = 1$. But

$f'(x) = a_1 \cos x + 2a_2 \cos 2x + \cdots + na_n \cos nx$, so $\left|f'(0)\right| = \left|a_1 + 2a_2 + \cdots + na_n\right| \leq 1$.

ANOTHER SOLUTION: We are given that $\left|\sum_{k=1}^n a_k \sin kx\right| \leq |\sin x|$. So for x close to 0, and x ≠ 0, we have $\left|\sum_{k=1}^{n} a_k \frac{\sin kx}{\sin x}\right| \le 1 \Rightarrow \lim_{x \to 0} \left|\sum_{k=1}^{n} a_k \frac{\sin kx}{\sin x}\right| \le 1 \Rightarrow$

$\left|\sum_{k=1}^{n} a_k \lim_{x \to 0} \frac{\sin kx}{\sin x}\right| \le 1.$ But $\lim_{x \to 0} \frac{\sin kx}{\sin x} \overset{H}{=} \lim_{x \to 0} \frac{k \cos kx}{\cos x} = k$, so $\left|\sum_{k=1}^{n} k a_k\right| \le 1.$

27. Differentiating the equation $\int_0^x f(t)\,dt = [f(x)]^2$ using Part 1 of the Fundamental

Theorem gives $f(x) = 2 f(x) f'(x) \Rightarrow f(x)\,[2 f'(x) - 1] = 0$, so $f(x) = 0$ or $f'(x) = \frac{1}{2}$.

$f'(x) = \frac{1}{2} \Rightarrow f(x) = \frac{1}{2}x + C$. To find C we substitute into the original equation to get

$\int_0^x (\frac{1}{2}t + C)\,dt = (\frac{1}{2}x + C)^2 \Leftrightarrow \frac{x^2}{4} + Cx = \frac{x^2}{4} + Cx + C^2$. It follows that $C = 0$ so

$f(x) = \frac{1}{2}x$. Therefore $f(x) = 0$ or $f(x) = \frac{1}{2}x$.

CHAPTER SEVEN

EXERCISES 7.1

1. Let $u = x$, $dv = e^{2x}dx \Rightarrow du = dx$, $v = \frac{1}{2}e^{2x}$. Then by (7.2),
$\int xe^{2x}dx = \frac{1}{2}xe^{2x} - \int \frac{1}{2}e^{2x}dx = \frac{1}{2}xe^{2x} - \frac{1}{4}e^{2x} + C$.

3. Let $u = x$, $dv = \sin 4x\, dx \Rightarrow du = dx$, $v = -\frac{1}{4}\cos 4x$. Then
$\int x\sin 4x\, dx = -\frac{1}{4}x\cos 4x - \int -\frac{1}{4}\cos 4x\, dx = -\frac{1}{4}x\cos 4x + \frac{1}{16}\sin 4x + C$.

5. Let $u = x^2$, $dv = \cos 3x\, dx \Rightarrow du = 2x\, dx$, $v = \frac{1}{3}\sin 3x$. Then $I = \int x^2\cos 3x\, dx$
$= \frac{1}{3}x^2\sin 3x - \frac{2}{3}\int x\sin 3x\, dx$ by (7.2). Next let $U = x$, $dV = \sin 3x\, dx \Rightarrow dU = dx$,
$V = -\frac{1}{3}\cos 3x$ to get $\int x\sin 3x\, dx = -\frac{1}{3}x\cos 3x + \frac{1}{3}\int \cos 3x\, dx$
$= -\frac{1}{3}x\cos 3x + \frac{1}{9}\sin 3x + C_1$. Substituting for $\int x\sin 3x\, dx$, we get
$I = \frac{1}{3}x^2\sin 3x - \frac{2}{3}\left[-\frac{1}{3}x\cos 3x + \frac{1}{9}\sin 3x + C_1\right]$
$= \frac{1}{3}x^2\sin 3x + \frac{2}{9}x\cos 3x - \frac{2}{27}\sin 3x + C$, where $C = -\frac{2}{3}C_1$.

7. Let $u = (\ln x)^2$, $dv = dx \Rightarrow du = 2\ln x \cdot \frac{1}{x}dx$, $v = x$. Then
$I = \int (\ln x)^2\, dx = x(\ln x)^2 - 2\int \ln x\, dx$. Taking $U = \ln x$, $dV = dx \Rightarrow dU = \frac{1}{x}dx$,
$V = x$, we find that $\int \ln x\, dx = x\ln x - \int x \cdot \frac{1}{x}dx = x\ln x - x + C_1$. Thus
$I = x(\ln x)^2 - 2x\ln x + 2x + C$, where $C = -2C_1$.

9. $I = \int \theta\sin\theta\cos\theta\, d\theta = \frac{1}{4}\int 2\theta\sin 2\theta\, d\theta = \frac{1}{8}\int t\sin t\, dt$ $[t = 2\theta \Rightarrow dt = d\theta/2]$.
Let $u = t$, $dv = \sin t\, dt \Rightarrow du = dt$, $v = -\cos t$. Then $I = \frac{1}{8}(-t\cos t + \int \cos t\, dt)$
$= \frac{1}{8}(-t\cos t + \sin t) + C = \frac{1}{8}(\sin 2\theta - 2\theta\cos 2\theta) + C$.

11. Let $u = \ln t$, $dv = t^2 \Rightarrow du = \frac{1}{t}dt$, $v = \frac{1}{3}t^3$. Then
$\int t^2\ln t\, dt = \frac{1}{3}t^3\ln t - \int \frac{1}{3}t^3 \cdot \frac{1}{t}dt = \frac{1}{3}t^3\ln t - \frac{1}{9}t^3 + C = \frac{1}{9}t^3(3\ln t - 1) + C$.

13. First let $u = \sin 3\theta$, $dv = e^{2\theta}d\theta \Rightarrow du = 3\cos 3\theta\, d\theta$, $v = \frac{1}{2}e^{2\theta}$. Then
$I = \int e^{2\theta}\sin 3\theta\, d\theta = \frac{1}{2}e^{2\theta}\sin 3\theta - \frac{3}{2}\int e^{2\theta}\cos 3\theta\, d\theta$. Next let $U = \cos 3\theta$,
$dU = -3\sin 3\theta\, d\theta$, $dV = e^{2\theta}d\theta$, $v = \frac{1}{2}e^{2\theta}$ to get
$\int e^{2\theta}\cos 3\theta\, d\theta = \frac{1}{2}e^{2\theta}\cos 3\theta + \frac{3}{2}\int e^{2\theta}\sin 3\theta\, d\theta$. Substituting in the previous formula
gives $I = \frac{1}{2}e^{2\theta}\sin 3\theta - \frac{3}{4}e^{2\theta}\cos 3\theta - \frac{9}{4}\int e^{2\theta}\sin 3\theta\, d\theta$ or $\frac{13}{4}\int e^{2\theta}\sin 3\theta\, d\theta$
$= \frac{1}{2}e^{2\theta}\sin 3\theta - \frac{3}{4}e^{2\theta}\cos 3\theta + C_1$. Hence $\int e^{2\theta}\sin 3\theta\, d\theta = \frac{1}{13}e^{2\theta}(2\sin 3\theta - 3\cos 3\theta) + C$,
where $C = \frac{4}{13}C_1$.

15. Let $u = y$, $dv = \sinh y\, dy \Rightarrow du = dy$, $v = \cosh y$. Then
$\int y\sinh y\, dy = y\cosh y - \int \cosh y\, dy = y\cosh y - \sinh y + C$.

17. Let $u = t$, $dv = e^{-t}dt \Rightarrow du = dt$, $v = -e^{-t}$. Then (7.6) says

237

$$\int_0^1 te^{-t}dt = \left[-te^{-t}\right]_0^1 + \int_0^1 e^{-t}dt = -\tfrac{1}{e} + \left[-e^{-t}\right]_0^1 = -\tfrac{1}{e} - \tfrac{1}{e} + 1 = 1 - \tfrac{2}{e}.$$

19. Let $u = x$, $dv = \cos 2x\,dx \Rightarrow du = dx$, $v = \tfrac{1}{2}\sin 2x\,dx$. Then $\displaystyle\int_0^{\pi/2} x\cos 2x\,dx$

$$= \left[\tfrac{1}{2}x\sin 2x\right]_0^{\pi/2} - \tfrac{1}{2}\int_0^{\pi/2}\sin 2x\,dx = 0 + \left[\tfrac{1}{4}\cos 2x\right]_0^{\pi/2} = \tfrac{1}{4}(-1-1) = -\tfrac{1}{2}.$$

21. Let $u = \cos^{-1}x$, $dv = dx \Rightarrow du = -\dfrac{dx}{\sqrt{1-x^2}}$, $v = x$. Then

$$I = \int_0^{1/2}\cos^{-1}x\,dx = \left[x\cos^{-1}x\right]_0^{1/2} + \int_0^{1/2}\frac{x\,dx}{\sqrt{1-x^2}} = \tfrac{1}{2}\tfrac{\pi}{3} + \int_1^{3/4} t^{-1/2}\left[-\tfrac{1}{2}dt\right]$$

   where $t = 1-x^2 \Rightarrow dt = -2x\,dx$.

   Thus $I = \tfrac{\pi}{6} + \tfrac{1}{2}\displaystyle\int_{3/4}^1 t^{-1/2}dt = \left[\sqrt{t}\right]_{3/4}^1 = \tfrac{\pi}{6} + 1 - \dfrac{\sqrt{3}}{2} = \dfrac{\pi + 6 - 3\sqrt{3}}{6}.$

23. Let $u = \sin 3x$, $dv = \cos 5x\,dx \Rightarrow du = 3\cos 3x\,dx$, $v = \tfrac{1}{5}\sin 5x$. Then

   $I = \int \sin 3x\cos 5x\,dx = \tfrac{1}{5}\sin 3x\sin 5x - \tfrac{3}{5}\int \sin 5x\cos 3x\,dx$. Now use parts with

   $U = \cos 3x$, $dV = \sin 5x\,dx \Rightarrow dU = -3\sin 3x\,dx$, $V = -\tfrac{1}{5}\cos 5x$. Thus $\int \sin 5x\cos 3x\,dx$

   $= -\tfrac{1}{5}\cos 3x\cos 5x - \tfrac{3}{5}\int \sin 3x\cos 5x\,dx \Rightarrow I = \tfrac{1}{5}\sin 3x\sin 5x - \tfrac{3}{5}\left(-\tfrac{1}{5}\cos 3x\cos 5x - \tfrac{3}{5}I\right).$

   Solving for I, we get $I = \tfrac{1}{16}(5\sin 3x\sin 5x + 3\cos 3x\cos 5x) + C$.

   [Another method: Write $\sin 3x\cos 5x = \tfrac{1}{2}(\sin 8x - \sin 2x)$ as in Section 7.2. This gives

   $I = -\tfrac{1}{16}\cos 8x + \tfrac{1}{4}\cos 2x + C$.]

25. Let $u = \ln(\sin x)$, $dv = \cos x\,dx \Rightarrow du = \dfrac{\cos x}{\sin x}dx$, $v = \sin x$. Then

   $I = \int \cos x\ln(\sin x)\,dx = \sin x\ln(\sin x) - \int \cos x\,dx = \sin x\ln(\sin x) - \sin x + C$.

   [Another method: Substitute $t = \sin x$, so $dt = \cos x\,dx$. Then

   $I = \int \ln t\,dt = t\ln t - t + C$ [see Example 2] and so $I = \sin x(\ln\sin x - 1) + C$.]

27. Let $u = 2x + 3$, $dv = e^x dx \Rightarrow du = 2\,dx$, $v = e^x$. Then

   $\int (2x+3)e^x\,dx = (2x+3)e^x - \int e^x \cdot 2\,dx = (2x+3)e^x - 2e^x + C = (2x+1)e^x + C$.

29. Let $w = \ln x \Rightarrow dw = \dfrac{dx}{x}$. Then $x = e^w$ and $dx = e^w dw$, so

   $\int \cos(\ln x)\,dx = \int e^w \cos w\,dw = \tfrac{1}{2}e^w(\sin w + \cos w) + C$ [by the method of Example 4]

   $= \tfrac{1}{2}x[\sin(\ln x) + \cos(\ln x)] + C$.

31. $I = \displaystyle\int_1^4 \ln\sqrt{x}\,dx = \tfrac{1}{2}\int_1^4 \ln x\,dx = \tfrac{1}{2}[x\ln x - x]_1^4$ as in Example 2. So

   $I = \tfrac{1}{2}[(4\ln 4 - 4) - (0-1)] = 4\ln 2 - \tfrac{3}{2}$.

33. Let $w = \sqrt{x}$, so that $x = w^2$ and $dx = 2w\,dw$. Then use $u = 2w$, $dv = \sin w\,dw$. Thus

$\int \sin \sqrt{x} \, dx = \int 2w \sin w \, dw = -2w \cos w + \int 2 \cos w \, dw = -2w \cos w + 2 \sin w + C$

$= -2\sqrt{x} \cos \sqrt{x} + 2 \sin \sqrt{x} + C.$

35. $\int x^5 e^{x^2} dx = \int (x^2)^2 e^{x^2} x \, dx = \int t^2 e^t \frac{1}{2} dt$ [where $t = x^2 \Rightarrow dt/2 = x \, dx$]

$= \frac{1}{2}(t^2 - 2t + 2)e^t + C$ [by Example 3] $= \frac{1}{2}(x^4 - 2x^2 + 2)e^{x^2} + C.$

37. (a) Take n = 2 in Example 6 to get

$\int \sin^2 x \, dx = -\frac{1}{2} \cos x \sin x + \frac{1}{2} \int 1 \, dx = \frac{x}{2} - \frac{\sin 2x}{4} + C.$

(b) $\int \sin^4 x \, dx = -\frac{1}{4} \cos x \sin^3 x + \frac{3}{4} \int \sin^2 x \, dx = -\frac{1}{4} \cos x \sin^3 x + \frac{3}{8} x - \frac{3}{16} \sin 2x + C.$

39. (a) $\int_0^{\pi/2} \sin^n x \, dx$

$= \left[-\frac{1}{n} \cos x \sin^{n-1} x \right]_0^{\pi/2} + \frac{n-1}{n} \int_0^{\pi/2} \sin^{n-2} x \, dx = \frac{n-1}{n} \int_0^{\pi/2} \sin^{n-2} x \, dx.$

(b) $\int_0^{\pi/2} \sin^3 x \, dx = \frac{2}{3} \int_0^{\pi/2} \sin x \, dx = \left[-\frac{2}{3} \cos x \right]_0^{\pi/2} = \frac{2}{3}.$

$\int_0^{\pi/2} \sin^5 x \, dx = \frac{4}{5} \int_0^{\pi/2} \sin^3 x \, dx = \frac{4}{5} \cdot \frac{2}{3} = \frac{8}{15}.$

(c) Let $m = \frac{n-1}{2}$, so that $n = 2m + 1$, $m \geq 1$. The formula holds for $m = 1$

(i.e., n = 3) by (b). Assume it holds for some value of $m \geq 1$. Then

$\int_0^{\pi/2} \sin^{2m+1} x \, dx = \frac{2 \cdot 4 \cdot 6 \cdots (2m)}{3 \cdot 5 \cdot 7 \cdots (2m+1)}.$ By the reduction formula in Example 6,

$\int_0^{\pi/2} \sin^{2m+3} x \, dx = \frac{2m+2}{2m+3} \int_0^{\pi/2} \sin^{2m+1} x \, dx = \frac{2 \cdot 4 \cdot 6 \cdots [2(m+1)]}{2 \cdot 4 \cdot 6 \cdots [2(m+1)+1]}$ as

desired. By induction, the formula holds for all $m \geq 1$.

41. Let $u = (\ln x)^n$, $dv = dx \Rightarrow du = n(\ln x)^{n-1} \frac{1}{x} dx$, $v = x$. Then

$\int (\ln x)^n dx = x(\ln x)^n - n \int (\ln x)^{n-1} dx$ by (7.2).

43. Let $u = (x^2 + a^2)^n$, $dv = dx \Rightarrow du = n(x^2 + a^2)^{n-1} 2x \, dx$, $v = x$. Then

$\int (x^2 + a^2)^n dx = x(x^2 + a^2)^n - 2n \int x^2 (x^2 + a^2)^{n-1} dx$

$= x(x^2 + a^2)^n - 2n \left[\int (x^2 + a^2)^n dx - a^2 \int (x^2 + a^2)^{n-1} dx \right]$ [since $x^2 = (x^2 + a^2) - a^2$]

$\Rightarrow (2n + 1) \int (x^2 + a^2)^n dx = x(x^2 + a^2)^n + 2na^2 \int (x^2 + a^2)^{n-1} dx$ and $\int (x^2 + a^2)^n dx$

$= \frac{x(x^2 + a^2)^n}{2n+1} + \frac{2na^2}{2n+1} \int (x^2 + a^2)^{n-1} dx$ (provided $2n + 1 \neq 0$).

45. Take n = 3 in #41 to get

$\int (\ln x)^3 dx = x(\ln x)^3 - 3 \int (\ln x)^2 dx = x(\ln x)^3 - 3x(\ln x)^2 + 6x \ln x - 6x + C$ [by #7].

47. Let $u = \sin^{-1}x$, $dv = dx \Rightarrow du = \dfrac{dx}{\sqrt{1-x^2}}$, $v = x$. Then

$$\text{area} = \int_0^{1/2} \sin^{-1}x\,dx = \left[x\sin^{-1}x\right]_0^{1/2} - \int_0^{1/2} \frac{x}{\sqrt{1-x^2}}\,dx$$

$$= \tfrac{1}{2}\left(\tfrac{\pi}{6}\right) + \left[\sqrt{1-x^2}\right]_0^{1/2} = \frac{\pi}{12} + \frac{\sqrt{3}}{2} - 1 = \frac{\pi + 6\sqrt{3} - 12}{12}.$$

49. $\text{Area} = \displaystyle\int_1^e \ln x\,dx = \left[x\ln x - x\right]_1^e$ [by *Example 2*]

$$= (e\ln e - e) - (1\ln 1 - 1) = 0 - (-1) = 1.$$

51. $\text{Volume} = \int_{2\pi}^{3\pi} 2\pi x \sin x\,dx$. Let $u = x$, $dv = \sin x\,dx \Rightarrow du = dx$, $v = -\cos x \Rightarrow$
$V = 2\pi\left[-x\cos x + \sin x\right]_{2\pi}^{3\pi} = 2\pi\left[(3\pi + 0) - (-2\pi + 0)\right] = 2\pi(5\pi) = 10\pi^2.$

53. $\text{Volume} = \int_{-1}^0 2\pi(1-x)e^{-x}\,dx$. Let $u = 1-x$, $dv = e^{-x}\,dx \Rightarrow du = -dx$, $v = -e^{-x}$
$\Rightarrow V = 2\pi\left[xe^{-x}\right]_{-1}^0 = 2\pi(0 + e) = 2\pi e.$

55. Since $v(t) > 0$ for all t, the desired distance $s(t) = \displaystyle\int_0^t v(w)\,dw = \int_0^t w^2 e^{-w}\,dw.$

Let $u = w^2$, $dv = e^{-w}\,dw \Rightarrow du = 2w\,dw$, $v = -e^{-w}$. Then

$$s(t) = \left[-w^2 e^{-w}\right]_0^t + 2\int_0^t we^{-w}\,dw. \text{ Now let } U = w, dV = e^{-w}\,dw \Rightarrow dU = dw,$$

$V = -e^{-w}$. Then $s(t) = -t^2 e^{-t} + 2\left[\left[-we^{-w}\right]_0^t + \int_0^t e^{-w}\,dw\right]$

$$= -t^2 e^{-t} - 2te^{-t} - 2e^{-t} + 2 = 2 - e^{-t}(t^2 + 2t + 2) \text{ meters.}$$

57. Take $g(x) = x$ in (7.1).

59. By #58,

$$\int_1^e \ln x\,dx = e\ln e - 1\ln 1 - \int_{\ln 1}^{\ln e} e^y\,dy = e - \int_0^1 e^y\,dy = e - \left[e^y\right]_0^1 = e - (e-1) = 1.$$

61. Using the formula for volumes of rotation (5.7), and the figure, we see that

$$\text{Volume} = \int_0^d \pi b^2\,dy - \int_0^c \pi a^2\,dy - \int_c^d \pi[g(y)]^2\,dy = \pi b^2 d - \pi a^2 c - \int_c^d \pi[g(y)]^2\,dy.$$
Let $y = f(x)$, which gives $dy = f'(x)\,dx$ and $g(y) = x$, so that
$V = \pi b^2 d - \pi a^2 c - \pi\int_a^b x^2 f'(x)\,dx.$ Now integrate by parts with
$u = x^2$, and $dv = f'(x)\,dx \Rightarrow du = 2x\,dx$, $v = f(x)$, and
$$\int_a^b x^2 f'(x)\,dx = \left[x^2 f(x)\right]_a^b - \int_a^b 2xf(x)\,dx = b^2 f(b) - a^2 f(a) - \int_a^b 2xf(x)\,dx, \text{ but } f(a) = c$$

and $f(b) = d \Rightarrow V = \pi b^2 d - \pi a^2 c - \pi \left[b^2 d - a^2 c - \int_a^b 2xf(x)\,dx \right] = \int_a^b 2\pi xf(x)\,dx.$

EXERCISES 7.2

1. $\displaystyle\int_0^{\pi/2} \sin^2 3x\,dx = \int_0^{\pi/2} \tfrac{1}{2}(1 - \cos 6x)\,dx = \left[\tfrac{x}{2} - \tfrac{\sin 6x}{12}\right]_0^{\pi/2} = \tfrac{\pi}{4}$

3. $\displaystyle\int \cos^4 x\,dx = \int \left[\tfrac{1}{2}(1 + \cos 2x)\right]^2 dx = \tfrac{1}{4}\int (1 + 2\cos 2x + \cos^2 2x)\,dx = \tfrac{x}{4} + \tfrac{\sin 2x}{4}$

   $+ \tfrac{1}{4}\int \tfrac{1}{2}(1 + \cos 4x)\,dx = \tfrac{1}{4}\left[x + \sin 2x + \tfrac{x}{2} + \tfrac{\sin 4x}{8}\right] + C = \tfrac{3}{8}x + \tfrac{1}{4}\sin 2x + \tfrac{1}{32}\sin 4x + C$

5. Let $u = \cos x \Rightarrow du = -\sin x\,dx.$ Then $\int \sin^3 x\cos^4 x\,dx = \int \cos^4 x(1 - \cos^2 x)\sin x\,dx$

   $= \int u^4(1 - u^2)(-du) = \int (u^6 - u^4)\,du = \tfrac{1}{7}u^7 - \tfrac{1}{5}u^5 + C = \tfrac{1}{7}\cos^7 x - \tfrac{1}{5}\cos^5 x + C.$

7. $\displaystyle\int_0^{\pi/4} \sin^4 x\cos^2 x\,dx = \int_0^{\pi/4} \sin^2 x(\sin x\cos x)^2\,dx = \int_0^{\pi/4} \tfrac{1}{2}(1 - \cos 2x)(\tfrac{1}{2}\sin 2x)^2\,dx$

   $= \tfrac{1}{8}\int_0^{\pi/4} (1 - \cos 2x)\sin^2 2x\,dx = \tfrac{1}{8}\int_0^{\pi/4} \sin^2 2x\,dx - \tfrac{1}{8}\int_0^{\pi/4} \sin^2 2x\cos 2x\,dx$

   $= \tfrac{1}{16}\int_0^{\pi/4} (1 - \cos 4x)\,dx - \tfrac{1}{16}\left[\tfrac{\sin^3 2x}{3}\right]_0^{\pi/4}$

   $= \tfrac{1}{16}\left[x - \tfrac{\sin 4x}{4} - \tfrac{\sin^3 2x}{3}\right]_0^{\pi/4} = \tfrac{1}{16}\left(\tfrac{\pi}{4} - 0 - \tfrac{1}{3}\right) = \tfrac{3\pi - 4}{192}$

9. $\int (1 - \sin 2x)^2\,dx = \int (1 - 2\sin 2x + \sin^2 2x)\,dx = \int \left[1 - 2\sin 2x + \tfrac{1}{2}(1 - \cos 4x)\right]dx$

   $= \int \left[\tfrac{3}{2} - 2\sin 2x - \tfrac{1}{2}\cos 4x\right]dx = \tfrac{3}{2}x + \cos 2x - \tfrac{1}{8}\sin 4x + C$

11. Let $u = \sin x \Rightarrow du = \cos x\,dx.$ Then $\int \cos^5 x\sin^5 x\,dx = \int u^5(1 - u^2)^2\,du$

    $= \int u^5(1 - 2u^2 + u^4)\,du = \int (u^5 - 2u^7 + u^9)\,du = \tfrac{u^{10}}{10} - \tfrac{u^8}{4} + \tfrac{u^6}{6} + C$

    $= \tfrac{1}{10}\sin^{10} x - \tfrac{1}{4}\sin^8 x + \tfrac{1}{6}\sin^6 x + C$

    OR: Let $v = \cos x,\ dv = -\sin x\,dx.$ Then $\int \cos^5 x\sin^5 x\,dx = \int v^5(1 - v^2)^2(-dv)$

    $= \int (-v^5 + 2v^7 - v^9)\,dv = -\tfrac{v^{10}}{10} + \tfrac{v^8}{4} - \tfrac{v^6}{6} + C = -\tfrac{1}{10}\cos^{10} x + \tfrac{1}{4}\cos^8 x - \tfrac{1}{6}\cos^6 x + C.$

13. $\int \cos^6 x\,dx = \int \left[\tfrac{1}{2}(1 + \cos 2x)\right]^3 dx = \tfrac{1}{8}\int \left[1 + 3\cos 2x + 3\cos^2 2x + \cos^3 2x\right]dx$

    $= \tfrac{1}{8}\int \left[1 + 3\cos 2x + \tfrac{3}{2}(1 + \cos 4x) + (1 - \sin^2 2x)\cos 2x\right]dx$

    $= \tfrac{1}{8}\int \left[\tfrac{5}{2} + 4\cos 2x + \tfrac{3}{2}\cos 4x - \sin^2 2x\cos 2x\right]dx$

$$= \tfrac{1}{8}\Big[\tfrac{5}{2}x + 2\sin 2x + \tfrac{3}{8}\sin 4x - \tfrac{1}{6}\sin^3 2x\Big] + C.$$

15. Let $u = \cos x \Rightarrow du = -\sin x\,dx$. Then $\int \sin^5 x\,dx = \int (1 - \cos^2 x)^2 \sin x\,dx$

$$= \int (1 - u^2)^2(-du) = \int (-1 + 2u^2 - u^4)\,du = -\frac{u^5}{5} + \frac{2u^3}{3} - u + C$$

$$= -\frac{\cos^5 x}{5} + \frac{2\cos^3 x}{3} - \cos x + C.$$

17. Let $u = \cos x$, $du = -\sin x\,dx$. Then $\int \sin^3 x\,\sqrt{\cos x}\,dx = \int (1 - \cos^2 x)\,\sqrt{\cos x}\,\sin x\,dx$

$$= \int (1 - u^2)u^{1/2}(-du) = \int (u^{5/2} - u^{1/2})\,du = \tfrac{2}{7}u^{7/2} - \tfrac{2}{3}u^{3/2} + C$$

$$= \tfrac{2}{7}(\cos x)^{7/2} - \tfrac{2}{3}(\cos x)^{3/2} + C = \Big[\tfrac{2}{7}\cos^3 x - \tfrac{2}{3}\cos x\Big]\sqrt{\cos x} + C.$$

19. Let $u = \sqrt{x}$ so that $x = u^2$ and $dx = 2u\,du$. Then $\displaystyle\int \frac{\cos^2\sqrt{x}}{\sqrt{x}}\,dx = \int \frac{\cos^2 u}{u}\,2u\,du$

$$= \int 2\cos^2 u\,du = \int (1 + \cos 2u)\,du = u + \tfrac{1}{2}\sin 2u + C = \sqrt{x} + \tfrac{1}{2}\sin(2\sqrt{x}) + C.$$

21. Let $u = \cos x \Rightarrow du = -\sin x\,dx$. Then $\int \cos^2 x\tan^3 x\,dx = \displaystyle\int \frac{\sin^3 x}{\cos x}\,dx$

$$= \int \frac{(1 - u^2)(-du)}{u} = \int \Big[\tfrac{-1}{u} + u\Big]du = -\ln|u| + \tfrac{1}{2}u^2 + C = \tfrac{1}{2}\cos^2 x - \ln|\cos x| + C.$$

23. $\displaystyle\int \frac{1 - \sin x}{\cos x}\,dx = \int (\sec x - \tan x)\,dx = \ln|\sec x + \tan x| - \ln|\sec x| + C$ [by *Example* 8]

$$= \ln\big|(\sec x + \tan x)\cos x\big| + C = \ln|1 + \sin x| + C = \ln(1 + \sin x) + C \text{ since } 1 + \sin x \geq 0.$$

OR: $\displaystyle\int \frac{1 - \sin x}{\cos x}\,dx = \int \frac{1 - \sin x}{\cos x}\cdot\frac{1 + \sin x}{1 + \sin x}\,dx = \int \frac{(1 - \sin^2 x)\,dx}{\cos x(1 + \sin x)} = \int \frac{\cos x\,dx}{1 + \sin x} = \int \frac{dw}{w}$

[*where* $w = 1 + \sin x$, $dw = \cos x\,dx$] $= \ln|w| + C = \ln|1 + \sin x| + C = \ln(1 + \sin x) + C.$

25. $\int \tan^2 x\,dx = \int (\sec^2 x - 1)\,dx = \tan x - x + C.$

27. $\int \sec^4 x\,dx = \int (\tan^2 x + 1)\sec^2 x\,dx = \int \tan^2 x\sec^2 x\,dx + \int \sec^2 x\,dx$

$$= \tfrac{1}{3}\tan^3 x + \tan x + C.$$

29. Let $u = \tan x \Rightarrow du = \sec^2 x\,dx$. Then $\displaystyle\int_0^{\pi/4} \tan^4 x\sec^2 x\,dx = \int_0^1 u^4\,du = \Big[\tfrac{1}{5}u^5\Big]_0^1 = \tfrac{1}{5}.$

31. Let $u = \sec x \Rightarrow du = \sec x\tan x\,dx$. Then $\int \tan x\sec^3 x\,dx = \int \sec^2 x\sec x\tan x\,dx$

$$= \int u^2\,du = \tfrac{1}{3}u^3 + C = \tfrac{1}{3}\sec^3 x + C.$$

33. $\int \tan^5 x\,dx = \int (\sec^2 x - 1)^2\tan x\,dx = \int \sec^4 x\tan x\,dx - 2\int \sec^2 x\tan x\,dx + \int \tan x\,dx$

$$= \int \sec^3 x\sec x\tan x\,dx - 2\int \tan x\sec^2 x\,dx + \int \tan x\,dx$$

$$= \tfrac{1}{4}\sec^4 x - \tan^2 x + \ln|\sec x| + C \quad [\text{OR: } \tfrac{1}{4}\sec^4 x - \sec^2 x + \ln|\sec x| + C].$$

35. Let $u = \sec x \Rightarrow du = \sec x\tan x\,dx$. Then

$$\int_0^{\pi/3} \tan^5 x \sec x \, dx = \int_0^{\pi/3} (\sec^2 x - 1)^2 \sec x \tan x \, dx = \int_1^2 (u^2 - 1)^2 \, du$$

$$= \int_1^2 (u^4 - 2u^2 + 1) \, du = \left[\tfrac{1}{5}u^5 - \tfrac{2}{3}u^3 + u\right]_1^2 = \left[\tfrac{32}{5} - \tfrac{16}{3} + 2\right] - \left[\tfrac{1}{5} - \tfrac{2}{3} + 1\right] = \tfrac{38}{15}.$$

37. Let $u = \sec x \Rightarrow du = \sec x \tan x \, dx$. Then $\int \tan x \sec^6 x \, dx = \int \sec^5 x \sec x \tan x \, dx$

$= \int u^5 \, du = \frac{u^6}{6} + C = \frac{1}{6}\sec^6 x + C.$

[OR: Let $u = \tan x$, $du = \sec^2 x$. Then $\int \tan x \sec^6 x \, dx = \int \tan x \, (1 + \tan^2 x)^2 \sec^2 x \, dx$

$= \int \tan x \sec^2 x \, dx + 2\int \tan^3 x \sec^2 x \, dx + \int \tan^5 x \sec^2 x \, dx$

$= \frac{\tan^2 x}{2} + \frac{\tan^4 x}{2} + \frac{\tan^6 x}{6} + C.]$

39. Let $u = \tan x \Rightarrow du = \sec^2 x \, dx$. Then $\displaystyle\int \frac{\sec^2 x}{\cot x} \, dx = \int \tan x \sec^2 x \, dx$

$= \int u \, du = \frac{u^2}{2} + C = \frac{1}{2}\tan^2 x + C.$

41. $\displaystyle\int_{\pi/6}^{\pi/2} \cot^2 x \, dx = \int_{\pi/6}^{\pi/2} (\csc^2 x - 1) \, dx = [-\cot x - x]_{\pi/6}^{\pi/2}$

$= (0 - \tfrac{\pi}{2}) - (-\sqrt{3} - \tfrac{\pi}{6}) = \sqrt{3} - \tfrac{\pi}{3}.$

43. Let $u = \cot x \Rightarrow du = -\csc^2 x \, dx$. Then $\int \cot^4 x \csc^4 x \, dx = \int u^4(u^2 + 1)(-du)$

$= -\int (u^6 + u^4) \, du = -\tfrac{1}{7}u^7 - \tfrac{1}{5}u^5 + C = -\tfrac{1}{7}\cot^7 x - \tfrac{1}{5}\cot^5 x + C.$

45. $I = \int \csc x \, dx = \displaystyle\int \frac{\csc x(\csc x - \cot x)}{\csc x - \cot x} \, dx = \int \frac{-\csc x \cot x + \csc^2 x}{\csc x - \cot x} \, dx.$

Let $u = \csc x - \cot x \Rightarrow du = (-\csc x \cot x + \csc^2 x) \, dx.$

Then $I = \int \frac{1}{u} \, du = \ln|u| = \ln|\csc x - \cot x| + C.$

47. $\displaystyle\int \frac{\cos^2 x}{\sin x} \, dx = \int \frac{1 - \sin^2 x}{\sin x} \, dx = \int (\csc x - \sin x) \, dx = \ln|\csc x - \cot x| + \cos x + C$ by #45.

49. $\int \sin 5x \sin 2x \, dx = \int \tfrac{1}{2}[\cos(5x - 2x) - \cos(5x + 2x)] \, dx$

$= \tfrac{1}{2}\int (\cos 3x - \cos 7x) \, dx = \tfrac{1}{6}\sin 3x - \tfrac{1}{14}\sin 7x + C.$

51. $\int \cos 3x \cos 4x \, dx = \int \tfrac{1}{2}[\cos(3x - 4x) + \cos(3x + 4x)] \, dx$

$= \tfrac{1}{2}\int (\cos x + \cos 7x) \, dx = \tfrac{1}{2}\sin x + \tfrac{1}{14}\sin 7x + C.$

53. $\int \sin x \cos 5x \, dx = \int \tfrac{1}{2}[\sin(x - 5x) + \sin(x + 5x)] \, dx$

$= \tfrac{1}{2}\int (-\sin 4x + \sin 6x) \, dx = \tfrac{1}{8}\cos 4x - \tfrac{1}{12}\cos 6x + C.$

55. $\displaystyle\int \frac{1 - \tan^2 x}{\sec^2 x} \, dx = \int (\cos^2 x - \sin^2 x) \, dx = \int \cos 2x \, dx = \tfrac{1}{2}\sin 2x + C.$

57. $f_{ave} = \frac{1}{2\pi} \displaystyle\int_{-\pi}^{\pi} \sin^2 x \cos^3 x \, dx = \frac{1}{2\pi} \int_{-\pi}^{\pi} \sin^2 x(1 - \sin^2 x) \cos x \, dx = \frac{1}{2\pi} \int_0^0 u^2(1 - u^2) \, du$

[*where* $u = \sin x$] $= 0.$

59. For $0 < x < \frac{\pi}{2}$, we have $0 < \sin x < 1$, so $\sin^3 x < \sin x$. Hence

the area $= \displaystyle\int_0^{\pi/2} (\sin x - \sin^3 x)\,dx = \int_0^{\pi/2} \sin x(1 - \sin^2 x)\,dx = \int_0^{\pi/2} \cos^2 x \sin x\,dx.$

Now let $u = \cos x \Rightarrow du = -\sin x\,dx$. Then area $= \displaystyle\int_1^0 u^2(-du) = \int_0^1 u^2\,du = \left[\frac{u^3}{3}\right]_0^1 = \frac{1}{3}.$

61. $V = \int_{\pi/2}^{\pi} \pi \sin^2 x\,dx = \pi\int_{\pi/2}^{\pi} \frac{1}{2}(1 - \cos 2x)\,dx = \pi\left[\frac{x}{2} - \frac{1}{4}\sin 2x\right]_{\pi/2}^{\pi} = \pi\left(\frac{\pi}{2} - 0 - \frac{\pi}{4} + 0\right) = \frac{\pi^2}{4}$

63. Volume $= \pi\int_0^{\pi/2}\left[(1 + \cos x)^2 - 1^2\right]dx = \pi\int_0^{\pi/2}(2\cos x + \cos^2 x)\,dx$

$= \pi\left[2\sin x + \frac{x}{2} + \frac{\sin 2x}{4}\right]_0^{\pi/2} = \pi\left(2 + \frac{\pi}{4}\right) = 2\pi + \frac{\pi^2}{4}.$

65. $s = f(t) = \displaystyle\int_0^t \sin \omega u \cos^2 \omega u\,du$. Let $y = \cos \omega u \Rightarrow dy = -\omega \sin \omega u\,du$.

Then $s = \frac{-1}{\omega}\displaystyle\int_1^{\cos \omega t} y^2\,dy = \left[\frac{-1}{\omega}\frac{1}{3}y^3\right]_1^{\cos \omega t} = \frac{1}{3\omega}(1 - \cos^3 \omega t).$

67. $\displaystyle\int_{-\pi}^{\pi} \sin mx \sin nx\,dx = \int_{-\pi}^{\pi} \frac{1}{2}[\cos(m-n)x - \cos(m+n)x]\,dx.$ If $m \neq n$,

this $= \frac{1}{2}\left[\frac{1}{m-n}\sin(m-n)x - \frac{1}{m+n}\sin(m+n)x\right]_{-\pi}^{\pi} = 0.$ If $m = n$,

we get $\displaystyle\int_{-\pi}^{\pi} \frac{1}{2}[1 - \cos(m+n)x]\,dx = \left[\frac{x}{2}\right]_{-\pi}^{\pi} - \left[\frac{1}{2(m+n)}\sin(m+n)x\right]_{-\pi}^{\pi} = \pi - 0 = \pi.$

69. $\frac{1}{\pi}\displaystyle\int_{-\pi}^{\pi} f(x)\sin mx\,dx = \sum_{n=1}^{N}\frac{a_n}{\pi}\int_{-\pi}^{\pi} \sin mx \sin nx\,dx.$ By #67, every term is zero except

the mth one, and that term is $\frac{a_m}{\pi}\cdot\pi = a_m.$

EXERCISES 7.3

1. Let $x = \sin \theta$, where $-\frac{\pi}{2} \leq \theta \leq \frac{\pi}{2}$. Then $dx = \cos \theta\,d\theta$ and $\sqrt{1 - x^2} = |\cos \theta| = \cos \theta$

(since $\cos \theta > 0$ for θ in $[-\frac{\pi}{2}, \frac{\pi}{2}]$). Thus $\displaystyle\int_{1/2}^{\sqrt{3}/2} \frac{dx}{x^2\sqrt{1 - x^2}} = \int_{\pi/6}^{\pi/3} \frac{\cos \theta\,d\theta}{\sin^2 \theta \cos \theta}$

$= \displaystyle\int_{\pi/6}^{\pi/3} \csc^2 \theta\,d\theta = [-\cot \theta]_{\pi/6}^{\pi/3} = -\frac{1}{\sqrt{3}} - (-\sqrt{3}) = \frac{3}{\sqrt{3}} - \frac{1}{\sqrt{3}} = \frac{2}{\sqrt{3}}.$

3. Let $u = 1 - x^2$. Then $du = -2x\,dx$, so $\displaystyle\int \frac{x}{\sqrt{1 - x^2}}\,dx = -\frac{1}{2}\int \frac{du}{\sqrt{u}}$

$= -\sqrt{u} + C = -\sqrt{1 - x^2} + C.$

5. Let $2x = \sin\theta$, where $-\frac{\pi}{2} \le \theta \le \frac{\pi}{2}$. Then $x = \frac{1}{2}\sin\theta$, $dx = \frac{1}{2}\cos\theta\,d\theta$, and

$$\sqrt{1-4x^2} = \sqrt{1-(2x)^2} = \cos\theta. \quad \int \sqrt{1-4x^2}\,dx = \int \cos\theta\left(\frac{1}{2}\cos\theta\right)d\theta$$

$$= \frac{1}{4}\int (1+\cos 2\theta)\,d\theta = \frac{1}{4}\left(\theta + \frac{1}{2}\sin 2\theta\right) + C$$

$$= \frac{1}{4}(\theta + \sin\theta\cos\theta) + C$$

$$= \frac{1}{4}\left[\sin^{-1}(2x) + 2x\sqrt{1-4x^2}\right] + C.$$

7. Let $x = 3\tan\theta$, where $-\frac{\pi}{2} < \theta < \frac{\pi}{2}$. Then $dx = 3\sec^2\theta\,d\theta$ and $\sqrt{9+x^2} = 3\sec\theta$.

$$\int_0^3 \frac{dx}{\sqrt{9+x^2}} = \int_0^{\pi/4} \frac{3\sec^2\theta\,d\theta}{3\sec\theta} = \int_0^{\pi/4} \sec\theta\,d\theta = \left[\ln|\sec\theta + \tan\theta|\right]_0^{\pi/4}$$

$$= \ln(\sqrt{2}+1) - \ln 1 = \ln(\sqrt{2}+1).$$

9. Let $x = 4\sec\theta$, where $0 \le \theta < \frac{\pi}{2}$ or $\pi \le \theta < \frac{3\pi}{2}$. Then $dx = 4\sec\theta\tan\theta\,d\theta$ and

$$\sqrt{x^2-16} = 4|\tan\theta| = 4\tan\theta. \text{ Thus } \int \frac{dx}{x^3\sqrt{x^2-16}} = \int \frac{4\sec\theta\tan\theta\,d\theta}{64\sec^3\theta\,4\tan\theta}$$

$$= \frac{1}{64}\int \cos^2\theta\,d\theta = \frac{1}{128}\int (1+\cos 2\theta)\,d\theta = \frac{1}{128}\left(\theta + \frac{1}{2}\sin 2\theta\right) + C = \frac{1}{128}(\theta + \sin\theta\cos\theta) + C$$

$$= \frac{1}{128}\left(\sec^{-1}\frac{x}{4} + \frac{4\sqrt{x^2-16}}{x^2}\right) + C \text{ by the diagrams for } 0 \le \theta < \frac{\pi}{2} \text{ and } \pi \le \theta < \frac{3\pi}{2} \text{ (where}$$

the labels of the legs in the second diagram indicate the x- and y-coordinates of P rather than the lengths of those sides). Henceforth we omit the second diagram from our solutions.

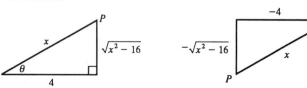

11. $9x^2 - 4 = (3x)^2 - 4$, so let $3x = 2\sec\theta$, where $0 \le \theta < \frac{\pi}{2}$ or $\pi \le \theta < \frac{3\pi}{2}$. Then

$$dx = \frac{2}{3}\sec\theta\tan\theta\,d\theta \text{ and } \sqrt{9x^2-4} = 2\tan\theta. \quad \int \frac{\sqrt{9x^2-4}}{x}\,dx = \int \frac{2\tan\theta}{(2/3)\sec\theta}\frac{2}{3}\sec\theta\tan\theta\,d\theta$$

$$= 2\int \tan^2\theta\,d\theta = 2\int (\sec^2\theta - 1)\,d\theta$$

$$= 2(\tan\theta - \theta) + C$$

$$= \sqrt{9x^2-4} - 2\sec^{-1}\left(\frac{3x}{2}\right) + C.$$

13. Let $x = a\sin\theta$, where $-\frac{\pi}{2} \le \theta \le \frac{\pi}{2}$. Then $dx = a\cos\theta\,d\theta$ and

$$\int \frac{x^2\,dx}{(a^2-x^2)^{3/2}} = \int \frac{a^2\sin^2\theta\,a\cos\theta\,d\theta}{a^3\cos^3\theta} = \int \tan^2\theta\,d\theta$$

$$= \int (\sec^2 \theta - 1)\, d\theta = \tan\theta - \theta + C$$

$$= \frac{x}{\sqrt{a^2 - x^2}} - \sin^{-1}\left(\tfrac{x}{a}\right) + C.$$

15. Let $x = \sqrt{3}\tan\theta$, where $-\frac{\pi}{2} < \theta < \frac{\pi}{2}$. Then

$$\int \frac{dx}{x\sqrt{x^2+3}} = \int \frac{\sqrt{3}\sec^2\theta\, d\theta}{\sqrt{3}\tan\theta\,\sqrt{3}\sec\theta} = \frac{1}{\sqrt{3}}\int \csc\theta\, d\theta$$

$$= \frac{1}{\sqrt{3}}\ln|\csc\theta - \cot\theta| + C = \frac{1}{\sqrt{3}}\ln\left|\frac{\sqrt{x^2+3}-\sqrt{3}}{x}\right| + C.$$

17. Let $u = 4 - 9x^2 \Rightarrow du = -18x\, dx$. Then $x^2 = \frac{1}{9}(4-u)$ and

$$\int_0^{2/3} x^3\sqrt{4-9x^2}\, dx = \int_4^0 \tfrac{1}{9}(4-u)u^{1/2}\left(-\tfrac{1}{18}\right)du = \frac{1}{162}\int_0^4 \left(4u^{1/2} - u^{3/2}\right)du$$

$$= \frac{1}{162}\left[\tfrac{8}{3}u^{3/2} - \tfrac{2}{5}u^{5/2}\right]_0^4 = \frac{1}{162}\left[\tfrac{64}{3} - \tfrac{64}{5}\right] = \frac{64}{1215}.$$

OR: Let $3x = 2\sin\theta$, where $-\frac{\pi}{2} \le \theta \le \frac{\pi}{2}$.

19. Let $u = 1 + x^2$, $du = 2x\, dx$.

Then $\int 5x\sqrt{1+x^2}\, dx = \frac{5}{2}\int u^{1/2}\, du = \frac{5}{3}u^{3/2} + C = \frac{5}{3}(1+x^2)^{3/2} + C.$

21. Let $x = \sqrt{2}\sec\theta$, where $0 \le \theta < \frac{\pi}{2}$ or $\pi \le \theta < \frac{3\pi}{2}$.

$$\text{Then } \int \frac{dx}{x^4\sqrt{x^2-2}} = \int \frac{\sqrt{2}\sec\theta\tan\theta\, d\theta}{4\sec^4\theta\,\sqrt{2}\tan\theta} = \frac{1}{4}\int \cos^3\theta\, d\theta$$

$$= \frac{1}{4}\int (1-\sin^2\theta)\cos\theta\, d\theta = \frac{1}{4}\left[\sin\theta - \tfrac{1}{3}\sin^3\theta\right] + C$$

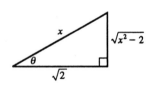

[*substitute* $u = \sin\theta$]

$$= \frac{1}{4}\left[\frac{\sqrt{x^2-2}}{x} - \frac{(x^2-2)^{3/2}}{3x^3}\right] + C.$$

23. $2x - x^2 = -(x^2 - 2x + 1) + 1 = 1 - (x-1)^2$. Let $u = x - 1$. Then $du = dx$ and

$$\int \sqrt{2x-x^2}\, dx = \int \sqrt{1-u^2}\, du = \int \cos^2\theta\, d\theta \ [\text{where } u = \sin\theta,\ -\tfrac{\pi}{2} \le \theta \le \tfrac{\pi}{2}]$$

$$= \frac{1}{2}\int (1+\cos 2\theta)\, d\theta = \frac{1}{2}(\theta + \tfrac{1}{2}\sin 2\theta) + C$$

$$= \frac{1}{2}(\sin^{-1}u + u\sqrt{1-u^2}) + C = \frac{1}{2}(\sin^{-1}(x-1) + (x-1)\sqrt{2x-x^2}) + C.$$

25. $9x^2 + 6x - 8 = (3x+1)^2 - 9$, so let $u = 3x + 1$, $du = 3\, dx$. Then

$$\int \frac{dx}{\sqrt{9x^2 + 6x - 8}} = \int \frac{(1/3)\,du}{\sqrt{u^2 - 9}}. \quad \text{Now let } u = 3\sec\theta, \text{ where } 0 \le \theta < \frac{\pi}{2} \text{ or } \pi \le \theta < \frac{3\pi}{2}.$$

Then $du = 3\sec\theta \tan\theta\,d\theta$ and $\sqrt{u^2 - 9} = 3\tan\theta$, so $\displaystyle\int \frac{(1/3)\,du}{\sqrt{u^2 - 9}} = \int \frac{\sec\theta \tan\theta\,d\theta}{3\tan\theta}$

$$= \tfrac{1}{3}\int \sec\theta\,d\theta = \tfrac{1}{3}\ln|\sec\theta + \tan\theta| + C_1 = \tfrac{1}{3}\ln\left|\frac{u + \sqrt{u^2 - 9}}{3}\right| + C_1$$

$$= \tfrac{1}{3}\ln\left|u + \sqrt{u^2 - 9}\right| + C = \tfrac{1}{3}\ln\left|3x + 1 + \sqrt{9x^2 + 6x - 8}\right| + C.$$

27. $x^2 + 2x + 2 = (x+1)^2 + 1$. Let $u = x + 1$, $du = dx$.

Then $\displaystyle\int \frac{dx}{(x^2 + 2x + 2)^2} = \int \frac{du}{(u^2 + 1)^2} = \int \frac{\sec^2\theta\,d\theta}{\sec^4\theta}$ $\quad [u = \tan\theta,\ du = \sec^2\theta\,d\theta,\ and$

$u^2 + 1 = \sec^2\theta] = \int \cos^2\theta\,d\theta = \tfrac{1}{2}(\theta + \sin\theta\cos\theta) + C$ [as in #23]

$$= \tfrac{1}{2}\left[\tan^{-1}u + \frac{u}{1 + u^2}\right] + C = \tfrac{1}{2}\left[\tan^{-1}(x+1) + \frac{x+1}{x^2 + 2x + 2}\right] + C.$$

29. Let $u = e^t \Rightarrow du = e^t\,dt$. Then $\int e^t\sqrt{9 - e^{2t}}\,dt = \int \sqrt{9 - u^2}\,du = \int (3\cos\theta)3\cos\theta\,d\theta$

[where $u = 3\sin\theta$, $-\frac{\pi}{2} \le \theta \le \frac{\pi}{2}$] $= 9\int \cos^2\theta\,d\theta = \tfrac{9}{2}(\theta + \sin\theta\cos\theta) + C$

[as in #23] $= \tfrac{9}{2}\left[\sin^{-1}\left(\tfrac{u}{3}\right) + \tfrac{u}{3}\cdot\frac{\sqrt{9 - u^2}}{3}\right] + C = \tfrac{9}{2}\sin^{-1}(e^t/3) + \tfrac{1}{2}e^t\sqrt{9 - e^{2t}} + C.$

31. (a) Let $x = a\tan\theta$, where $-\frac{\pi}{2} < \theta < \frac{\pi}{2}$. Then $\sqrt{x^2 + a^2} = a\sec\theta$ and

$$\int \frac{dx}{\sqrt{x^2 + a^2}} = \int \frac{a\sec^2\theta\,d\theta}{a\sec\theta} = \int \sec\theta\,d\theta = \ln|\sec\theta + \tan\theta| + C_1$$

$$= \ln\left|\frac{\sqrt{x^2 + a^2}}{a} + \frac{x}{a}\right| + C_1 = \ln\left(x + \sqrt{x^2 + a^2}\right) + C, \text{ where } C = C_1 - \ln|a|.$$

(b) Let $x = a\sinh t$, so that $dx = a\cosh t\,dt$ and $\sqrt{x^2 + a^2} = a\cosh t$.

Then $\displaystyle\int \frac{dx}{\sqrt{x^2 + a^2}} = \int \frac{a\cosh t\,dt}{a\cosh t} = t + C = \sinh^{-1}(x/a) + C.$

33. Area of $\triangle POQ = \tfrac{1}{2}(r\cos\theta)(r\sin\theta) = \tfrac{1}{2}r^2\sin\theta\cos\theta$. Area of region

$$PQR = \int_{r\cos\theta}^{r} \sqrt{r^2 - x^2}\,dx. \text{ Let } x = r\cos u \Rightarrow dx = -r\sin u\,du \text{ for } \theta \le u \le \tfrac{\pi}{2}. \text{ Then}$$

we obtain $\int \sqrt{r^2 - x^2}\,dx = \int r\sin u\,(-r\sin u)\,du = -r^2\int \sin^2 u\,du$

$= -\frac{1}{2}r^2(u - \sin u \cos u) + C = -\frac{1}{2}r^2 \cos^{-1}(x/r) + \frac{1}{2}x\sqrt{r^2 - x^2} + C.$ Area of region

$\text{PQR} = \frac{1}{2}\left[-r^2\cos^{-1}(x/r) + x\sqrt{r^2 - x^2}\right]_{r\cos\theta}^{r} = \frac{1}{2}\left[0 - (-r^2\theta + r\cos\theta\ r\sin\theta)\right]$

$= \frac{1}{2}r^2\theta - \frac{1}{2}r^2\sin\theta\cos\theta,$ so area of sector POR = area of $\triangle\text{POQ}$ + area of region

$\text{PQR} = \frac{1}{2}r^2\theta.$

35.   We use cylindrical shells and assume that $R > r$.

$x^2 = r^2 - (y - R)^2 \Rightarrow x = \pm\sqrt{r^2 - (y - R)^2},$ so $g(y) = 2\sqrt{r^2 - (y - R)^2}$ in Formula 5.11.

$\text{Volume} = \int_{R-r}^{R+r} 2\pi y \cdot 2\sqrt{r^2 - (y - R)^2}\,dy = \int_{-r}^{r} 4\pi(u + R)\sqrt{r^2 - u^2}\,du\ [where\ u = y - R.]$

$= 4\pi\int_{-r}^{r} u\sqrt{r^2 - u^2}\,du + 4\pi R\int_{-r}^{r}\sqrt{r^2 - u^2}\,du\ \ [Let\ u = r\sin\theta,\ du = r\cos\theta\,d\theta\ in\ 2nd\ integral]$

$= 4\pi\left[-\frac{1}{3}(r^2 - u^2)^{3/2}\right]_{-r}^{r} + 4\pi R\int_{-\pi/2}^{\pi/2} r^2\cos^2\theta\,d\theta = -\frac{4\pi}{3}(0 - 0) + 4\pi Rr^2\int_{-\pi/2}^{\pi/2}\cos^2 d\theta$

$= 2\pi Rr^2\int_{-\pi/2}^{\pi/2}(1 + \cos 2\theta)\,d\theta = 2\pi Rr^2\left[\theta + \frac{1}{2}\sin 2\theta\right]_{-\pi/2}^{\pi/2} = 2\pi^2 Rr^2.$

<u>Another</u> <u>Method</u> : Use washers instead of shells, so $V = 8\pi R\int_0^r\sqrt{r^2 - y^2}\,dy$ as in

Exercise 5.2.59(a), but evaluate the integral using $y = r\sin\theta$.

EXERCISES 7.4

1. $\dfrac{1}{(x - 1)(x + 2)} = \dfrac{A}{x - 1} + \dfrac{B}{x + 2}.$

3. $\dfrac{x + 1}{x(x + 2)} = \dfrac{A}{x} + \dfrac{B}{x + 2}.$

5. $\dfrac{x^2 + 3x - 4}{(2x - 1)^2(2x + 3)} = \dfrac{A}{2x - 1} + \dfrac{B}{(2x - 1)^2} + \dfrac{C}{2x + 3}.$

7. $\dfrac{1}{x^4 - x^3} = \dfrac{1}{x^3(x - 1)} = \dfrac{A}{x} + \dfrac{B}{x^2} + \dfrac{C}{x^3} + \dfrac{D}{x - 1}.$

9. $\dfrac{x^2 + 1}{x^2 - 1} = 1 + \dfrac{2}{(x - 1)(x + 1)} = 1 + \dfrac{A}{x - 1} + \dfrac{B}{x + 1}.$

11. $\dfrac{x^2 - 2}{x(x^2 + 2)} = \dfrac{A}{x} + \dfrac{Bx + C}{x^2 + 2}.$

13. $\dfrac{x^4+x^2+1}{(x^2+1)(x^2+4)^2} = \dfrac{Ax+B}{x^2+1} + \dfrac{Cx+D}{x^2+4} + \dfrac{Ex+F}{(x^2+4)^2}.$

15. $\dfrac{x^4}{(x^2+9)^3} = \dfrac{Ax+B}{x^2+9} + \dfrac{Cx+D}{(x^2+9)^2} + \dfrac{Ex+F}{(x^2+9)^3}.$

17. $\dfrac{x^3+x^2+1}{x^4+x^3+2x^2} = \dfrac{x^3+x^2+1}{x^2(x^2+x+2)} = \dfrac{A}{x} + \dfrac{B}{x^2} + \dfrac{Cx+D}{x^2+x+2}.$

19. $\displaystyle\int \dfrac{x^2}{x+1}\,dx = \int\left(x-1+\dfrac{1}{x+1}\right)dx = \tfrac{1}{2}x^2 - x + \ln|x+1| + C.$

21. $\dfrac{4x-1}{(x-1)(x+2)} = \dfrac{A}{x-1} + \dfrac{B}{x+2} \Rightarrow 4x-1 = A(x+2) + B(x-1).$ Take $x = 1$ to get

$3 = 3A$, then $x = -2$ to get $-9 = -3B \Rightarrow A = 1,\ B = 3.$

Now $\displaystyle\int_2^4 \dfrac{4x-1}{(x-1)(x+2)}\,dx = \int_2^4\left[\dfrac{1}{x-1} + \dfrac{3}{x+2}\right]dx = \Big[\ln(x-1) + 3\ln(x+2)\Big]_2^4$

$= \ln 3 + 3\ln 6 - \ln 1 - 3\ln 4 = 4\ln 3 - 3\ln 2 = \ln(81/8).$

23. $\displaystyle\int \dfrac{6x-5}{2x+3}\,dx = \int\left[3 - \dfrac{14}{2x+3}\right]dx = 3x - 7\ln|2x+3| + C.$

25. $\dfrac{x^2+1}{x^2-x} = 1 + \dfrac{x+1}{x(x-1)} = 1 - \dfrac{1}{x} + \dfrac{2}{x-1},$

so $\displaystyle\int \dfrac{x^2+1}{x^2-x}\,dx = x - \ln|x| + 2\ln|x-1| + C = x + \ln\big[(x-1)^2/|x|\big] + C.$

27. $\dfrac{2x+3}{(x+1)^2} = \dfrac{A}{x+1} + \dfrac{B}{(x+1)^2} \Rightarrow 2x+3 = A(x+1) + B.$ Take $x = -1$ to get $B = 1$, and

equate coefficients of x to get $A = 2$. Now $\displaystyle\int_0^1 \dfrac{2x+3}{(x+1)^2}\,dx = \int_0^1\left[\dfrac{2}{x+1} + \dfrac{1}{(x+1)^2}\right]dx$

$= \Big[2\ln(x+1) - \dfrac{1}{x+1}\Big]_0^1 = 2\ln 2 - \tfrac{1}{2} - (2\ln 1 - 1) = 2\ln 2 + \tfrac{1}{2}.$

29. $\dfrac{1}{x(x+1)(2x+3)} = \dfrac{A}{x} + \dfrac{B}{x+1} + \dfrac{C}{2x+3}$

$\Rightarrow 1 = A(x+1)(2x+3) + B(x)(2x+3) + C(x)(x+1).$ Set $x = 0$ to get $A = 1/3.$

Take $x = -1$ to get $B = -1$, and finally setting $x = -3/2$ gives $C = 4/3.$

Now $\displaystyle\int \dfrac{dx}{x(x+1)(2x+3)} = \int\left[\dfrac{1/3}{x} - \dfrac{1}{x+1} + \dfrac{4/3}{2x+3}\right]dx$

$= \tfrac{1}{3}\ln|x| - \ln|x+1| + \tfrac{2}{3}\ln|2x+3| + C.$

31. $\dfrac{6x^2+5x-3}{x^3+2x^2-3x} = \dfrac{A}{x} + \dfrac{B}{x+3} + \dfrac{C}{x-1}$

$\Rightarrow 6x^2+5x-3 = A(x+3)(x-1) + B(x)(x-1) + C(x)(x+3).$ Setting $x = 0$ gives

$A = 1$. Take $x = -3$ to get $B = 3$ and finally set $x = 1$ to get $C = 2$

$$\Rightarrow \int_2^3 \frac{6x^2 + 5x - 3}{x^3 + 2x^2 - 3x}\, dx = \int_2^3 \left[\frac{1}{x} + \frac{3}{x+3} + \frac{2}{x-1}\right] dx = \Big[\ln x + 3\ln(x+3) + 2\ln(x-1)\Big]_2^3$$

$$= (\ln 3 + 3\ln 6 + 2\ln 2) - (\ln 2 + 3\ln 5) = 4\ln 6 - 3\ln 5.$$

33. $\dfrac{1}{(x-1)^2(x+4)} = \dfrac{A}{x-1} + \dfrac{B}{(x-1)^2} + \dfrac{C}{x+4} \Rightarrow 1 = A(x-1)(x+4) + B(x+4) + C(x-1)^2$

Set $x = 1$ to get $B = 1/5$ and take $x = -4$ to get $C = 1/25$.

Now equating the coefficients of x^2, we get $0 = Ax^2 + Cx^2$ or $A = -C = -\frac{1}{25}$

$$\Rightarrow \int \frac{dx}{(x-1)^2(x+4)} = \int \left[\frac{-1/25}{x-1} + \frac{1/5}{(x-1)^2} + \frac{1/25}{x+4}\right] dx$$

$$= \tfrac{-1}{25}\ln|x-1| - \tfrac{1}{5}\tfrac{1}{x-1} + \tfrac{1}{25}\ln|x+4| + C = \tfrac{1}{25}\left[\ln\left|\tfrac{x+4}{x-1}\right| - \tfrac{5}{x-1}\right] + C.$$

35. $\dfrac{5x^2 + 3x - 2}{x^3 + 2x^2} = \dfrac{5x^2 + 3x - 2}{x^2(x+2)} = \dfrac{A}{x} + \dfrac{B}{x^2} + \dfrac{C}{x+2}$. Multiply by $x^2(x+2)$ to get

$5x^2 + 3x - 2 = A(x)(x+2) + B(x+2) + Cx^2$. Setting $x = -2$ gives $C = 3$, and take

$x = 0$ to get $B = -1$. Equating the coefficients of x^2 gives $5x^2 = Ax^2 + Cx^2$ or

$A = 2$. So $\int \dfrac{5x^2 + 3x - 2}{x^3 + 2x^2}\, dx = \int \left[\dfrac{2}{x} - \dfrac{1}{x^2} + \dfrac{3}{x+2}\right] dx = 2\ln|x| + \tfrac{1}{x} + 3\ln|x+2| + C.$

37. Let $u = x^3 + 3x^2 + 4$. Then $du = 3(x^2 + 2x)\, dx \Rightarrow$

$$\int \frac{x^2 + 2x}{x^3 + 3x^2 + 4}\, dx = \tfrac{1}{3} \int \frac{du}{u} = \tfrac{1}{3}\ln|x^3 + 3x^2 + 4| + C.$$

39. $\dfrac{x^2}{(x+1)^3} = \dfrac{A}{x+1} + \dfrac{B}{(x+1)^2} + \dfrac{C}{(x+1)^3}$. Multiply by $(x+1)^3$ to get

$x^2 = A(x+1)^2 + B(x+1) + C$. Setting $x = -1$ gives $C = 1$. Equating the

coefficients of x^2 gives $A = 1$, and set $x = 0$ to get $B = -2$. Now

$$\int \frac{x^2\, dx}{(x+1)^3} = \int \left[\frac{1}{x+1} - \frac{2}{(x+1)^2} + \frac{1}{(x+1)^3}\right] dx = \ln|x+1| + \frac{2}{x+1} - \frac{1}{2(x+1)^2} + C.$$

41. $\dfrac{1}{x^4 - x^2} = \dfrac{1}{x^2(x-1)(x+1)} = \dfrac{A}{x} + \dfrac{B}{x^2} + \dfrac{C}{x-1} + \dfrac{D}{x+1}$. Multiply by $x^2(x-1)(x+1)$ to

get $1 = A(x)(x-1)(x+1) + B(x-1)(x+1) + Cx^2(x+1) + Dx^2(x-1)$. Setting $x = 1$

gives $C = 1/2$. Take $x = -1$ to get $D = -1/2$. Equating the coefficients of x^3 gives

$0 = A + C + D = A$. And finally set $x = 0$ to get $B = -1$.

Now $\int \frac{dx}{x^4 - x^2} = \int \left[\frac{-1}{x^2} + \frac{1/2}{x-1} - \frac{1/2}{x+1}\right] dx = \frac{1}{x} + \frac{1}{2}\ln\left|\frac{x-1}{x+1}\right| + C.$

43. $\frac{x^3}{x^2+1} = \frac{(x^3+x)-x}{x^2+1} = x - \frac{x}{x^2+1},$ so $\int_0^1 \frac{x^3}{x^2+1}\, dx$

$= \int_0^1 x\, dx - \int_0^1 \frac{x\, dx}{x^2+1} = \left[\frac{x^2}{2}\right]_0^1 - \frac{1}{2}\int_1^2 \frac{1}{u}\, du \quad [\text{where } u = x^2 + 1,\ du = 2x\, dx]$

$= \frac{1}{2} - \left[\frac{1}{2}\ln u\right]_1^2 = \frac{1}{2} - \frac{1}{2}\ln 2 = \frac{1 - \ln 2}{2}.$

45. Complete the square: $x^2 + x + 1 = (x + 1/2)^2 + 3/4$ and let $u = x + 1/2$. Then

$\int_0^1 \frac{x}{x^2+x+1}\, dx = \int_{1/2}^{3/2} \frac{u - 1/2}{u^2 + 3/4}\, du = \int_{1/2}^{3/2} \frac{u}{u^2+3/4}\, du - \frac{1}{2}\int_{1/2}^{3/2} \frac{1}{u^2+3/4}\, du$

$= \frac{1}{2}\ln(u^2 + 3/4) - \frac{1}{2}\frac{1}{\sqrt{3}/2}\left[\tan^{-1}\left(\frac{2u}{\sqrt{3}}\right)\right]_{1/2}^{3/2} = \frac{1}{2}\ln 3 - \frac{1}{\sqrt{3}}\left(\frac{\pi}{3} - \frac{\pi}{6}\right) = \ln\sqrt{3} - \frac{\pi}{6\sqrt{3}}.$

47. $\frac{3x^2 - 4x + 5}{(x-1)(x^2+1)} = \frac{A}{x-1} + \frac{Bx+C}{x^2+1} \Rightarrow 3x^2 - 4x + 5 = A(x^2+1) + (Bx+C)(x-1).$

Take $x = 1$ to get $4 = 2A$ or $A = 2$. Now

$(Bx + C)(x-1) = 3x^2 - 4x + 5 - 2(x^2+1) = x^2 - 4x + 3.$ Equating coefficients of x^2 and then comparing the constant terms, we get $B = 1$ and $C = -3$. Hence

$\int \frac{3x^2 - 4x + 5}{(x-1)(x^2+1)}\, dx = \int \left[\frac{2}{x-1} + \frac{x-3}{x^2+1}\right] dx = 2\ln|x-1| + \int \frac{x\, dx}{x^2+1} - 3\int \frac{dx}{x^2+1}$

$= 2\ln|x-1| + \frac{1}{2}\ln(x^2+1) - 3\tan^{-1}x + C$

$= \ln(x-1)^2 + \ln\sqrt{x^2+1} - 3\tan^{-1}x + C.$

49. $\frac{x^2 + 7x - 6}{(x+1)(x^2 - 4x + 7)} = \frac{A}{x+1} + \frac{Bx+C}{x^2 - 4x + 7} \Rightarrow$

$x^2 + 7x - 6 = A(x^2 - 4x + 7) + (Bx + C)(x+1).$ Take $x = -1$ to get $-12 = 12A$ or

$A = -1$. Then $2x^2 + 3x + 1 = (Bx + C)(x+1)$, so $B = 2$ and $C = 1$. Now

$\int \frac{x^2 + 7x - 6}{(x+1)(x^2 - 4x + 7)}\, dx = \int \left[\frac{-1}{x+1} + \frac{2x+1}{x^2 - 4x + 7}\right] dx$

$= -\ln|x+1| + \int \frac{2x-4}{x^2 - 4x + 7}\, dx + 5\int \frac{dx}{(x-2)^2 + 3}$

$= -\ln|x+1| + \ln(x^2 - 4x + 7) + \frac{5}{\sqrt{3}}\tan^{-1}\left(\frac{x-2}{\sqrt{3}}\right) + C.$

51. $\dfrac{1}{x^3-1} = \dfrac{1}{(x-1)(x^2+x+1)} = \dfrac{A}{x-1} + \dfrac{Bx+C}{x^2+x+1} \Rightarrow$

$1 = A(x^2+x+1) + (Bx+C)(x-1)$. Take $x=1$ to get $A=1/3$. Then equate

coefficients of x^2 and 1 to get $0 = 1/3+B$ and $1 = 1/3-C$; i.e., $B = -1/3$ and

$C = -2/3$. Then $\displaystyle\int \dfrac{dx}{x^3-1} = \int \dfrac{1/3}{x-1}\,dx + \int \dfrac{(-1/3)x-2/3}{x^2+x+1}\,dx$

$= \frac{1}{3}\ln|x-1| - \frac{1}{3}\displaystyle\int \dfrac{x+2}{x^2+x+1}\,dx = \frac{1}{3}\ln|x-1| - \frac{1}{3}\int \dfrac{x+1/2}{x^2+x+1}\,dx - \frac{1}{3}\int \dfrac{(3/2)\,dx}{(x+1/2)^2+3/4}$

$= \frac{1}{3}\ln|x-1| - \frac{1}{6}\ln(x^2+x+1) - \frac{1}{2}(2/\sqrt{3})\tan^{-1}[(x+\frac{1}{2})/(\sqrt{3}/2)] + C$

$= \frac{1}{3}\ln|x-1| - \frac{1}{6}\ln(x^2+x+1) - (1/\sqrt{3})\tan^{-1}[(2x+1)/\sqrt{3}] + C.$

53. $\dfrac{x^2-2x-1}{(x-1)^2(x^2+1)} = \dfrac{A}{x-1} + \dfrac{B}{(x-1)^2} + \dfrac{Cx+D}{x^2+1} \Rightarrow$

$x^2-2x-1 = A(x-1)(x^2+1) + B(x^2+1) + (Cx+D)(x-1)^2.$ Setting $x=1$ gives

$B = -1$. Equating the coefficients of x^3 gives $A = -C$. Equating the constant terms

gives $-1 = -A-1+D$, so $D = A$, and set $x = 2$ to get $-1 = 5A - 5 - 2A + A$ or

$A = 1$. We have $\displaystyle\int \dfrac{x^2-2x-1}{(x-1)^2(x^2+1)}\,dx = \int\left[\dfrac{1}{x-1} - \dfrac{1}{(x-1)^2} - \dfrac{x-1}{x^2+1}\right]dx$

$= \ln|x-1| + 1/(x-1) - \frac{1}{2}\ln(x^2+1) + \tan^{-1}x + C.$

55. $\dfrac{3x^3-x^2+6x-4}{(x^2+1)(x^2+2)} = \dfrac{Ax+B}{x^2+1} + \dfrac{Cx+D}{x^2+2} \Rightarrow$

$3x^3-x^2+6x-4 = (Ax+B)(x^2+2) + (Cx+D)(x^2+1).$ Equating the coefficients

gives $A+C = 3$, $B+D = -1$, $2A+C = 6$, and $2B+D = -4 \Rightarrow A = 3$, $C = 0$,

$B = -3$, and $D = 2$. Now $\displaystyle\int \dfrac{3x^3-x^2+6x-4}{(x^2+1)(x^2+2)}\,dx = 3\int \dfrac{x-1}{x^2+1}\,dx + 2\int \dfrac{dx}{x^2+2}$

$= \frac{3}{2}\ln(x^2+1) - 3\tan^{-1}x + \sqrt{2}\tan^{-1}(x/\sqrt{2}) + C.$

57. $\displaystyle\int \dfrac{x-3}{(x^2+2x+4)^2}\,dx = \int \dfrac{x-3}{[(x+1)^2+3]^2}\,dx = \int \dfrac{u-4}{(u^2+3)^2}\,du \quad [u = x+1]$

$= \displaystyle\int \dfrac{u\,du}{(u^2+3)^2} - 4\int \dfrac{du}{(u^2+3)^2} = \frac{1}{2}\int \dfrac{dv}{v^2} - 4\int \dfrac{\sqrt{3}\sec^2\theta\,d\theta}{9\sec^4\theta}$

$[v = u^2+3 \text{ in the } 1^{st} \text{ integral; } u = \sqrt{3}\tan\theta \text{ in the } 2^{nd}]$

$= \dfrac{-1}{(2v)} - \dfrac{4\sqrt{3}}{9}\displaystyle\int \cos^2\theta\,d\theta = \dfrac{-1}{2(u^2+3)} - \dfrac{2\sqrt{3}}{9}(\theta + \sin\theta\cos\theta) + C$

$$= \frac{-1}{2(x^2 + 2x + 4)} - \frac{2\sqrt{3}}{9}\left[\tan^{-1}[(x+1)/\sqrt{3}] + \frac{\sqrt{3}(x+1)}{x^2 + 2x + 4}\right] + C$$

$$= \frac{-1}{2(x^2 + 2x + 4)} - \frac{2\sqrt{3}}{9}\left[\tan^{-1}[(x+1)/\sqrt{3}]\right] - \frac{2(x+1)}{3(x^2 + 2x + 4)} + C.$$

59. $\dfrac{x^4 + 1}{x(x^2 + 1)^2} = \dfrac{A}{x} + \dfrac{Bx + C}{x^2 + 1} + \dfrac{Dx + E}{(x^2 + 1)^2}$

$\Rightarrow x^4 + 1 = A(x^2 + 1)^2 + (Bx + C)(x)(x^2 + 1) + (Dx + E)(x)$. Setting $x = 0$ gives

$A = 1$, and equating the coefficients of x^4 gives $1 = A + B$, so $B = 0$. Now

$$\frac{C}{x^2 + 1} + \frac{Dx + E}{(x^2 + 1)^2} = \frac{x^4 + 1}{x(x^2 + 1)^2} - \frac{1}{x} = \frac{1}{x}\left[\frac{x^4 + 1 - (x^4 + 2x^2 + 1)}{(x^2 + 1)^2}\right] = \frac{-2x}{(x^2 + 1)^2},$$

so we can take $C = 0$, $D = -2$, and $E = 0$.

Hence $\displaystyle\int \frac{x^4 + 1}{x(x^2 + 1)^2}\,dx = \int\left[\frac{1}{x} - \frac{2x}{(x^2 + 1)^2}\right]dx = \ln|x| + \frac{1}{x^2 + 1} + C.$

61. Let $u = x^2 + 4$.

Then $\displaystyle\int \frac{8x\,dx}{(x^2 + 4)^3} = 4\int \frac{2x\,dx}{(x^2 + 4)^3} = 4\int u^{-3}\,du = -2u^{-2} + C = -2/(x^2 + 4)^2 + C.$

63. Let $u = \sin^2 x - 3\sin x + 2$. Then $du = (2\sin x\cos x - 3\cos x)\,dx$

so $\displaystyle\int \frac{(2\sin x - 3)\cos x}{\sin^2 x - 3\sin x + 2}\,dx = \int \frac{du}{u} = \ln|u| + C = \ln\left|\sin^2 x - 3\sin x + 2\right| + C.$

65. If $|x| < a$, then $\displaystyle\int \frac{dx}{a^2 - x^2} = \int \frac{a\,\text{sech}^2 u\,du}{a^2\,\text{sech}^2 u}\ [x = a\tanh u] = \frac{1}{a}u + C = \frac{1}{a}\tanh^{-1}(x/a) + C.$

If $|x| > a$, then $\displaystyle\int \frac{dx}{a^2 - x^2} = \int \frac{-a\,\text{csch}^2 u\,du}{-a^2\,\text{csch}^2 u}\ [x = a\coth u] = \frac{1}{a}u + C = \frac{1}{a}\coth^{-1}(x/a) + C.$

67. $\displaystyle\int \frac{dx}{x^2 - 2x} = \int \frac{dx}{(x-1)^2 - 1} = \int \frac{du}{u^2 - 1}\ [u = x - 1] = \frac{1}{2}\ln\left|\frac{u-1}{u+1}\right| + C\ [by\ (7.20)]$

$= \frac{1}{2}\ln|(x-2)/x| + C.$

69. $\displaystyle\int \frac{x\,dx}{x^2 + x - 1} = \frac{1}{2}\int \frac{(2x+1)\,dx}{x^2 + x - 1} - \frac{1}{2}\int \frac{dx}{(x + 1/2)^2 - 5/4}$

$= \frac{1}{2}\ln\left|x^2 + x - 1\right| - \frac{1}{2}\int \frac{du}{u^2 - (\sqrt{5}/2)^2}\ [u = x + 1/2]$

$= \frac{1}{2}\ln\left|x^2 + x - 1\right| - \left(\frac{1}{2\sqrt{5}}\right)\ln\left|(u - \sqrt{5}/2)/(u + \sqrt{5}/2)\right| + C$

$= \frac{1}{2}\ln\left|x^2 + x - 1\right| - \left(\frac{1}{2\sqrt{5}}\right)\ln\left|(2x + 1 - \sqrt{5})/(2x + 1 + \sqrt{5})\right| + C.$

71. $\frac{x+1}{x-1} = 1 + \frac{2}{x-1} > 0$ for $2 \leq x \leq 3$, so the area $= \int_2^3 \left[1 + \frac{2}{x-1} \right] dx$

$= \left[x + 2\ln|x-1| \right]_2^3 = (3 + 2\ln 2) - (2 + 2\ln 1) = 1 + 2\ln 2.$

73. There are only finitely many values of x where $Q(x) = 0$ (assuming that Q is not the zero polynomial). At all other values of x, $F(x)/Q(x) = G(x)/Q(x)$, so $F(x) = G(x)$. In other words, the values of F and G agree at all except perhaps finitely many values of x. By continuity of F and G, the polynomials F and G must agree at those values of x too. (If a is a value of x such that $Q(a) = 0$, then $Q(x) \neq 0$ for all x sufficiently near a. Thus $F(a) = \lim\limits_{x \to a} F(x)$ [by continuity of F] $= \lim\limits_{x \to a} G(x)$ [since $F(x) = G(x)$ wherever $Q(x) \neq 0$] $= G(a)$ [by continuity of G].)

EXERCISES 7.5

1. Let $u = \sqrt{x}$. Then $x = u^2$, $dx = 2u\,du \Rightarrow \int_0^1 \frac{dx}{1 + \sqrt{x}} = \int_0^1 \frac{2u\,du}{1 + u} = 2\int_0^1 \left[1 - \frac{1}{1+u} \right] du$

$= 2\left[u - \ln(1+u) \right]_0^1 = 2(1 - \ln 2).$

3. Let $u = \sqrt{x}$. Then $x = u^2$, $dx = 2u\,du \Rightarrow \int \frac{\sqrt{x}\,dx}{x+1} = \int \frac{u \cdot 2u\,du}{u^2 + 1} = 2\int \left[1 - \frac{1}{u^2+1} \right] du$

$= 2(u - \tan^{-1} u) + C = 2(\sqrt{x} - \tan^{-1}\sqrt{x}) + C.$

5. Let $u = \sqrt[3]{x}$. Then $x = u^3$, $dx = 3u^2\,du \Rightarrow \int \frac{dx}{x - \sqrt[3]{x}} = \int \frac{3u^2\,du}{u^3 - u}$

$= 3\int \frac{u\,du}{u^2 - 1} = \frac{3}{2}\ln|u^2 - 1| + C = \frac{3}{2}\ln|x^{2/3} - 1| + C.$

7. Let $u = \sqrt{x-1}$. Then $x = u^2 + 1$, $dx = 2u\,du \Rightarrow \int_5^{10} \frac{x^2\,dx}{\sqrt{x-1}} = \int_2^3 \frac{(u^2+1)^2\,2u\,du}{u}$

$= 2\int_2^3 (u^4 + 2u^2 + 1)\,du = 2\left[\frac{1}{5}u^5 + \frac{2}{3}u^3 + u \right]_2^3$

$= 2\left(\frac{243}{5} + 18 + 3 \right) - 2\left(\frac{32}{5} + \frac{16}{3} + 2 \right) = \frac{1676}{15}.$

9. Let $u = \sqrt{x}$. Then $x = u^2$, $dx = 2u\,du \Rightarrow \int \frac{dx}{\sqrt{1 + \sqrt{x}}} = \int \frac{2u\,du}{\sqrt{1 + u}}$

$= 2\int \frac{(v^2 - 1)2v\,dv}{v} \quad [v = \sqrt{1+u},\ u = v^2 - 1,\ du = 2v\,dv]$

$= 4\int (v^2 - 1)\,dv = \frac{4}{3}v^3 - 4v + C = \frac{4}{3}(1 + \sqrt{x})^{3/2} - 4\sqrt{1 + \sqrt{x}} + C.$

11. Let $u = \sqrt{x}$. Then $x = u^2$, $dx = 2u\,du \Rightarrow \int \dfrac{\sqrt{x}+1}{\sqrt{x}-1}dx = \int \dfrac{u+1}{u-1}2u\,du = 2\int \dfrac{u^2+u}{u-1}du$

$= 2\int \left[u + 2 + \dfrac{2}{u-1}\right]du = u^2 + 4u + 4\ln|u-1| + C = x + 4\sqrt{x} + 4\ln|\sqrt{x}-1| + C.$

13. Let $u = \sqrt[3]{x^2+1}$. Then $x^2 = u^3 - 1$, $2x\,dx = 3u^2\,du \Rightarrow \int \dfrac{x^3\,dx}{\sqrt[3]{x^2+1}}$

$= \int \dfrac{(u^3-1)(3/2)u^2\,du}{u} = (3/2)\int (u^4 - u)\,du = (3/10)u^5 - (3/4)u^2 + C$

$= (3/10)(x^2+1)^{5/3} - (3/4)(x^2+1)^{2/3} + C.$

15. Let $u = \sqrt[6]{x}$. Then $x = u^6$, $dx = 6u^5\,du \Rightarrow \int \dfrac{dx}{\sqrt{x}+\sqrt[3]{x}} = \int \dfrac{6u^5\,du}{u^3+u^2} = 6\int \dfrac{u^3\,du}{u+1}$

$= 6\int \left[u^2 - u + 1 - \dfrac{1}{u+1}\right]du = 2u^3 - 3u^2 + 6u - 6\ln|u+1| + C$

$= 2\sqrt{x} - 3\sqrt[3]{x} + 6\sqrt[6]{x} - 6\ln\left(\sqrt[6]{x}+1\right) + C.$

17. Let $u = \sqrt[4]{x}$. Then $x = u^4$, $dx = 4u^3\,du \Rightarrow \int \dfrac{dx}{\sqrt{x}+\sqrt[4]{x}} = \int \dfrac{4u^3\,du}{u^2+u} = 4\int \dfrac{u^2\,du}{u+1}$

$= 4\int \left[u - 1 + \dfrac{1}{u+1}\right]du = 2u^2 - 4u + 4\ln|u+1| + C = 2\sqrt{x} - 4\sqrt[4]{x} + 4\ln\left(\sqrt[4]{x}+1\right) + C.$

19. Let $u = \sqrt{x}$. Then $x = u^2$, $dx = 2u\,du \Rightarrow \int \dfrac{dx}{(bx+c)\sqrt{x}} = \int \dfrac{2u\,du}{(bu^2+c)u} = \dfrac{2}{b}\int \dfrac{du}{u^2 + (c/b)}$

$= \dfrac{2}{b}\sqrt{\dfrac{b}{c}}\tan^{-1}\left(\sqrt{\dfrac{b}{c}}u\right) + C = (2/\sqrt{bc})\tan^{-1}\left(\sqrt{bx/c}\right) + C.$

21. Let $u = \sqrt[3]{x-1}$. Then $x = u^3 + 1$, $dx = 3u^2\,du \Rightarrow \int \dfrac{\sqrt[3]{x-1}}{x}dx = \int \dfrac{u \cdot 3u^2\,du}{u^3+1}$

$= 3\int \left[1 - \dfrac{1}{u^3+1}\right]du = 3\int \left[1 - \dfrac{1/3}{u+1} + \dfrac{(1/3)u - 2/3}{u^2-u+1}\right]du = 3u - \ln|u+1|$

$+ \int \dfrac{u-2}{u^2-u+1}du = 3u - \ln|u+1| + \dfrac{1}{2}\int \dfrac{2u-1}{u^2-u+1}du - \dfrac{3}{2}\int \dfrac{du}{(u-1/2)^2+3/4}$

$= 3u - \ln|u+1| + \dfrac{1}{2}\ln(u^2-u+1) - \dfrac{3}{2}\dfrac{2}{\sqrt{3}}\tan^{-1}[(2/\sqrt{3})(u-1/2)] + C$

$= 3\sqrt[3]{x-1} - \ln|\sqrt[3]{x-1}+1| + \dfrac{1}{2}\ln[(x-1)^{2/3} - (x-1)^{1/3} + 1]$

$- \sqrt{3}\tan^{-1}[(2\sqrt[3]{x-1}-1)/\sqrt{3}] + C.$

23. Let $u = \sqrt{x}$. Then $x = u^2$, $dx = 2u\,du \Rightarrow \int \sqrt{\dfrac{1-x}{x}}dx = \int \dfrac{\sqrt{1-u^2}}{u}2u\,du = 2\int \sqrt{1-u^2}\,du$

$= 2\int \cos^2\theta\,d\theta \quad [u = \sin\theta] = \theta + \sin\theta\cos\theta + C = \sin^{-1}\sqrt{x} + \sqrt{x(1-x)} + C.$

[OR: Let $u = \sqrt{(1-x)/x}$. This gives $I = \sqrt{x(1-x)} - \tan^{-1}\sqrt{(1-x)/x} + C$.]

25. Let $u = \sin x$. Then $du = \cos x \, dx \Rightarrow \int \frac{\cos x \, dx}{\sin^2 x + \sin x} = \int \frac{du}{u^2 + u} = \int \frac{du}{u(u+1)}$

$= \int \left[\frac{1}{u} - \frac{1}{u+1} \right] du = \ln\left| \frac{u}{u+1} \right| + C = \ln\left| \frac{\sin x}{1 + \sin x} \right| + C$.

27. Let $u = e^x$. Then $x = \ln u$, $dx = du/u \Rightarrow \int \frac{e^{2x} \, dx}{e^{2x} + 3e^x + 2} = \int \frac{u^2(du/u)}{u^2 + 3u + 2}$

$= \int \frac{u \, du}{(u+1)(u+2)} = \int \left[\frac{-1}{u+1} + \frac{2}{u+2} \right] du = 2\ln|u+2| - \ln|u+1| + C$

$= \ln\left[(e^x + 2)^2/(e^x + 1) \right] + C$.

29. Let $u = e^x$. Then $x = \ln u$, $dx = du/u \Rightarrow \int \sqrt{1 - e^x}\,dx = \int \sqrt{1-u}\,(du/u) = \int \frac{\sqrt{1-u}\,du}{u}$

$= \int \frac{v(-2v)\,dv}{1 - v^2} \quad [v = \sqrt{1-u},\ u = 1 - v^2,\ du = -2v\,dv] = 2\int \left[1 + \frac{1}{v^2 - 1} \right] dv$

$= 2\left[v + \frac{1}{2}\ln\left| \frac{v-1}{v+1} \right| \right] + C = 2\sqrt{1 - e^x} + \ln\left[(1 - \sqrt{1 - e^x})/(1 + \sqrt{1 - e^x}) \right] + C$.

31. Let $t = \tan(x/2)$. Then $dx = 2\,dt/(1 + t^2)$, $\cos x = (1 - t^2)/(1 + t^2) \Rightarrow$

$\int \frac{dx}{1 - \cos x} = \int \frac{2\,dt/(1 + t^2)}{2t^2/(1 + t^2)} = \int t^{-2}\,dt = -1/t + C = -\cot\left(\frac{x}{2}\right) + C$.

OR: $\int \frac{dx}{1 - \cos x} = \int \frac{1 + \cos x}{1 - \cos^2 x}\,dx = \int \frac{1 + \cos x}{\sin^2 x}\,dx$

$= \int (\csc^2 x + \cot x \csc x)\,dx = -\cot x - \csc x + C$.

33. Let $t = \tan(x/2)$. Then, by (7.26), $\displaystyle\int_0^{\pi/2} \frac{dx}{\sin x + \cos x} = \int_0^1 \frac{2\,dt}{2t + 1 - t^2}$

$= -2\int_0^1 \frac{dt}{t^2 - 2t - 1} = -2\int_0^1 \frac{dt}{(t-1)^2 - 2} = -\frac{1}{\sqrt{2}}\left[\ln\left| \frac{t - 1 - \sqrt{2}}{t - 1 + \sqrt{2}} \right| \right]_0^1$

$= -(1/\sqrt{2})\left[\ln 1 - \ln[(\sqrt{2} + 1)/(\sqrt{2} - 1)] \right] = (1/\sqrt{2})\ln(\sqrt{2} + 1)^2 = \sqrt{2}\ln(\sqrt{2} + 1)$

or $-\sqrt{2}\ln(\sqrt{2} - 1)$ [since $\sqrt{2} + 1 = 1/(\sqrt{2} - 1)$]

or $(1/\sqrt{2})\ln(3 + 2\sqrt{2})$ [since $(\sqrt{2} + 1)^2 = 3 + 2\sqrt{2}$].

35. Let $t = \tan(x/2)$. Then, by (7.26), $\displaystyle\int \frac{dx}{3\sin x + 4\cos x} = \int \frac{2\,dt}{6t + 4(1 - t^2)}$

$$= \int \frac{-dt}{2t^2 - 3t - 2} = -\int \left[\frac{-2/5}{2t+1} + \frac{1/5}{t-2}\right] dt = \frac{1}{5}\ln\left|\frac{2t+1}{t-2}\right| + C$$

$$= \frac{1}{5}\ln\left|\frac{2\tan(x/2)+1}{\tan(x/2)-2}\right| + C.$$

37. Let $t = \tan(x/2)$. Then $\int \frac{dx}{2\sin x + \sin 2x} = \frac{1}{2}\int \frac{dx}{\sin x + \sin x \cos x}$

$$= \frac{1}{2}\int \frac{2\,dt/(1+t^2)}{2t/(1+t^2) + 2t(1-t^2)/(1+t^2)^2} = \frac{1}{2}\int \frac{(1+t^2)\,dt}{t(1+t^2) + t(1-t^2)}$$

$$= \frac{1}{4}\int \frac{(1+t^2)\,dt}{t} = \frac{1}{4}\int \left[\frac{1}{t} + t\right] dt = \frac{1}{4}\ln|t| + \frac{1}{8}t^2 + C$$

$$= (1/4)\ln|\tan(x/2)| + (1/8)\tan^2(x/2) + C.$$

39. Let $t = \tan(x/2)$. Then $\int \frac{dx}{a\sin x + b\cos x} = \int \frac{2\,dt}{a(2t) + b(1-t^2)}$

$$= \frac{-2}{b}\int \frac{dt}{t^2 - 2(a/b)t - 1} = \frac{-2}{b}\int \frac{dt}{(t-a/b)^2 - (1 + a^2/b^2)}$$

$$= \frac{-1}{b}\frac{b}{\sqrt{a^2+b^2}}\ln\left|\frac{t - a/b - \sqrt{a^2+b^2}/b}{t - a/b + \sqrt{a^2+b^2}/b}\right| + C$$

$$= (1/\sqrt{a^2+b^2})\ln\left|(b\tan(\tfrac{x}{2}) - a + \sqrt{a^2+b^2})/(b\tan(\tfrac{x}{2}) - a - \sqrt{a^2+b^2})\right| + C.$$

41. (a) Let $t = \tan(x/2)$. Then $\int \sec x\,dx = \int \frac{dx}{\cos x} = \int \frac{2\,dt}{1-t^2} = \int \left[\frac{1}{1-t} + \frac{1}{1+t}\right] dt$

$$= \ln|1+t| - \ln|1-t| + C = \ln|(1+t)/(1-t)| + C$$

$$= \ln\left|\frac{1 + \tan(x/2)}{1 - \tan(x/2)}\right| + C.$$

(b) $\tan(\pi/4 + x/2) = \frac{\tan(\pi/4) + \tan(x/2)}{1 - \tan(\pi/4)\tan(x/2)} = \frac{1 + \tan(x/2)}{1 - \tan(x/2)}.$ Substituting in the

formula from part (a), we get $\int \sec x\,dx = \ln|\tan(\pi/4 + x/2)| + C.$

43. According to (7.10), $\int \sec x\,dx = \ln|\sec x + \tan x| + C.$ Now

$$\frac{1 + \tan\left(\tfrac{x}{2}\right)}{1 - \tan\left(\tfrac{x}{2}\right)} = \frac{1 + \sin\left(\tfrac{x}{2}\right)/\cos\left(\tfrac{x}{2}\right)}{1 - \sin\left(\tfrac{x}{2}\right)/\cos\left(\tfrac{x}{2}\right)} = \frac{\cos\left(\tfrac{x}{2}\right) + \sin\left(\tfrac{x}{2}\right)}{\cos\left(\tfrac{x}{2}\right) - \sin\left(\tfrac{x}{2}\right)}$$

$$= \frac{\left(\cos\left(\tfrac{x}{2}\right) + \sin\left(\tfrac{x}{2}\right)\right)^2}{\left(\cos\left(\tfrac{x}{2}\right) - \sin\left(\tfrac{x}{2}\right)\right)\left(\cos\left(\tfrac{x}{2}\right) + \sin\left(\tfrac{x}{2}\right)\right)} = \frac{1 + 2\cos\left(\tfrac{x}{2}\right)\sin\left(\tfrac{x}{2}\right)}{\cos^2\left(\tfrac{x}{2}\right) - \sin^2\left(\tfrac{x}{2}\right)}$$

$$= \tfrac{1+\sin x}{\cos x} \ [using \ (15a) \ and \ (15b) \ from \ Appendix \ B] = \sec x + \tan x,$$

so $\ln \left| \dfrac{1+\tan(\frac{x}{2})}{1-\tan(\frac{x}{2})} \right| = \ln|\sec x + \tan x|$. Thus the formula in Exercise 41(a) agrees with (7.10).

EXERCISES 7.6

1. $\displaystyle \int \frac{2x+5}{x-3}\,dx = \int \frac{(2x-6)+11}{x-3}\,dx = \int \left[2+\frac{11}{x-3}\right]dx = 2x + 11\ln|x-3| + C.$

3. $\int \sin^2 x \cos^3 x\,dx = \int \sin^2 x\,(1-\sin^2 x)\cos x\,dx = \int u^2(1-u^2)\,du \quad [u = \sin x]$
   $= \int (u^2 - u^4)\,du = u^3/3 - u^5/5 + C = \tfrac{1}{3}\sin^3 x - \tfrac{1}{5}\sin^5 x + C.$

5. Let $u = 1-x^2$. Then $du = -2x\,dx \Rightarrow \displaystyle\int_0^{1/2} \frac{x\,dx}{\sqrt{1-x^2}} = -\int_1^{3/4} \frac{du}{2\sqrt{u}}$
   $= \displaystyle\int_{3/4}^1 \frac{du}{2\sqrt{u}} = \left[\sqrt{u}\right]_{3/4}^1 = 1 - \sqrt{3}/2.$

7. Let $u = \sqrt{x-2}$. Then $x = u^2 + 2$, $dx = 2u\,du \Rightarrow \displaystyle\int \frac{\sqrt{x-2}}{x+2}\,dx = \int \frac{u\cdot 2u\,du}{u^2+4}$
   $= 2\displaystyle\int \left[1 - \frac{4}{u^2+4}\right]du = 2u - \tfrac{8}{2}\tan^{-1}\left(\tfrac{u}{2}\right) + C = 2\sqrt{x-2} - 4\tan^{-1}\left(\sqrt{x-2}/2\right) + C.$

9. Use integration by parts: $u = \ln(1+x^2)$, $dv = dx \Rightarrow du = \dfrac{2x}{1+x^2}\,dx$, $v = x$, so
   $\int \ln(1+x^2)\,dx = x\ln(1+x^2) - \displaystyle\int x\cdot \frac{2x\,dx}{1+x^2} = x\ln(1+x^2) - 2\int \left[1 - \frac{1}{1+x^2}\right]dx$
   $= x\ln(1+x^2) - 2x + 2\tan^{-1}x + C.$

11. Let $u = 1+\sqrt{x}$. Then $x = (u-1)^2$, $dx = 2(u-1)\,du \Rightarrow \displaystyle\int_0^1 (1+\sqrt{x})^8\,dx$
    $= \displaystyle\int_1^2 u^8 \cdot 2(u-1)\,du = 2\int_1^2 (u^9 - u^8)\,du = \left[u^{10}/5 - 2u^9/9\right]_1^2$
    $= \dfrac{1024}{5} - \dfrac{1024}{9} - \dfrac{1}{5} + \dfrac{2}{9} = \dfrac{4097}{45}.$

13. $\displaystyle\int \frac{x\,dx}{x^2-2x+2} = \tfrac{1}{2}\int \frac{(2x-2)\,dx}{x^2-2x+2} + \int \frac{dx}{(x-1)^2+1} = \tfrac{1}{2}\ln(x^2-2x+2) + \tan^{-1}(x-1) + C.$

15. Let $u = \sqrt{9-x^2}$. Then $u^2 = 9-x^2$, $u\,du = -x\,dx \Rightarrow \displaystyle\int \frac{\sqrt{9-x^2}}{x}\,dx = \int \frac{\sqrt{9-x^2}}{x^2}\,x\,dx$
    $= \displaystyle\int \frac{u}{9-u^2}(-u)\,du = \int \left[1 - \frac{9}{9-u^2}\right]du = u + 9\int \frac{du}{u^2-9} = u + \frac{9}{2\cdot 3}\ln\left|\frac{u-3}{u+3}\right| + C$

$$= \sqrt{9-x^2} + \tfrac{3}{2}\ln\left|\frac{\sqrt{9-x^2}-3}{\sqrt{9-x^2}+3}\right| + C = \sqrt{9-x^2} + \tfrac{3}{2}\ln\frac{(\sqrt{9-x^2}-3)^2}{x^2} + C$$

$$= \sqrt{9-x^2} + 3\ln\left|(3-\sqrt{9-x^2})/x\right| + C. \quad \text{OR:} \quad \text{Put } x = 3\sin\theta.$$

17. Integrate by parts: $u = x^2$, $dv = \cosh x\, dx \Rightarrow du = 2x\, dx$, $v = \sinh x$, so

$I = \int x^2 \cosh x\, dx = x^2 \sinh x - \int 2x \sinh x\, dx$. Now let $U = x$, $dV = \sinh x\, dx$

$\Rightarrow dU = dx$, $V = \cosh x$. So $I = x^2 \sinh x - 2\left(x\cosh x - \int \cosh x\, dx\right)$

$= x^2 \sinh x - 2\left[x\cosh x - \sinh x\right] = (x^2 + 2)\sinh x - 2x\cosh x + C.$

19. Let $u = \sin x$. Then $\displaystyle\int \frac{\cos x\, dx}{1 + \sin^2 x} = \int \frac{du}{1 + u^2} = \tan^{-1} u + C = \tan^{-1}(\sin x) + C.$

21. $\displaystyle\int_0^1 \cos\pi x \tan\pi x\, dx = \int_0^1 \sin\pi x\, dx = \tfrac{-1}{\pi}\int_0^1 -\pi\sin\pi x\, dx = \tfrac{-1}{\pi}\Big[\cos\pi x\Big]_0^1 = \tfrac{-1}{\pi}(-1-1) = \tfrac{2}{\pi}.$

23. Integrate by parts twice, first with $u = e^{3x}$, $dv = \cos 5x\, dx$: $\int e^{3x} \cos 5x\, dx$

$= \tfrac{1}{5}e^{3x}\sin 5x - \int \tfrac{3}{5}e^{3x}\sin 5x\, dx = \tfrac{1}{5}e^{3x}\sin 5x + \tfrac{3}{25}e^{3x}\cos 5x - \tfrac{9}{25}\int e^{3x}\cos 5x\, dx$, so

$\tfrac{34}{25}\int e^{3x}\cos 5x\, dx = \tfrac{1}{25}e^{3x}(5\sin 5x + 3\cos 5x) + C_1$ and $\int e^{3x}\cos 5x\, dx$

$= \tfrac{1}{34}e^{3x}(5\sin 5x + 3\cos 5x) + C.$

25. $\displaystyle\int \frac{dx}{x^3 + x^2 + x + 1} = \int \frac{dx}{(x+1)(x^2+1)} = \int\left[\frac{1/2}{x+1} - \frac{(1/2)x - 1/2}{x^2+1}\right]dx$

$= \tfrac{1}{2}\int\left[\frac{1}{x+1} - \frac{x}{x^2+1} + \frac{1}{x^2+1}\right]dx = \tfrac{1}{2}\ln|x+1| - \tfrac{1}{4}\ln(x^2+1) + \tfrac{1}{2}\tan^{-1}x + C.$

27. Let $t = x^3$. Then $dt = 3x^2\, dx \Rightarrow I = \int x^5 e^{-x^3}\, dx = \tfrac{1}{3}\int t e^{-t}\, dt$. Now integrate by

parts with $u = t$, $dv = e^{-t}\, dt$: $I = \tfrac{-1}{3}te^{-t} + \tfrac{1}{3}\int e^{-t}\, dt = \tfrac{-1}{3}te^{-t} - \tfrac{1}{3}e^{-t} + C$

$= -\tfrac{1}{3}e^{-x^3}(x^3 + 1) + C.$

29. Let $u = 3x + 2$. Then $\displaystyle\int \frac{dx}{\sqrt{9x^2 + 12x - 5}} = \int \frac{dx}{\sqrt{(3x+2)^2 - 9}} = \tfrac{1}{3}\int \frac{du}{\sqrt{u^2 - 9}}$

$= \tfrac{1}{3}\cosh^{-1}(u/3) + C_1 = \tfrac{1}{3}\cosh^{-1}[(3x+2)/3] + C_1$

$= (1/3)\ln\left|3x + 2 + \sqrt{9x^2 + 12x - 5}\right| + C \text{ [by (6.81)]. [OR: Substitute } u = 3\sec\theta.\text{]}$

31. $\int x^{1/3}(1 - x^{1/2})\, dx = \int (x^{1/3} - x^{5/6})\, dx = \tfrac{3}{4}x^{4/3} - \tfrac{6}{11}x^{11/6} + C.$

33. Let $u = x^2 + 1$. Then $du = 2x\, dx$, so $\displaystyle\int \frac{x}{x^4 + 2x^2 + 10}\, dx = \int \frac{x}{(x^2+1)^2 + 9}\, dx = \tfrac{1}{2}\int \frac{du}{u^2 + 1}$

$= \tfrac{1}{2}\cdot\tfrac{1}{3}\tan^{-1}\left(\tfrac{u}{3}\right) + C = \tfrac{1}{6}\tan^{-1}\left(\tfrac{x^2+1}{3}\right) + C.$

35. $\int \sin^2 x \cos^4 x \, dx = \int (\sin x \cos x)^2 \cos^2 x \, dx = \int \frac{1}{4} \sin^2 2x \frac{1}{2} (1 + \cos 2x) \, dx$

$= \frac{1}{8} \int \sin^2 2x \, dx + \frac{1}{8} \int \sin^2 2x \cos 2x \, dx = \frac{1}{16} \int (1 - \cos 4x) \, dx + \frac{1}{16} \int \sin^2 2x \, (2 \cos 2x) \, dx$

$= x/16 - (1/64) \sin 4x + (1/48) \sin^3 2x + C.$

[OR: Write $\int \sin^2 x \cos^4 x \, dx = \frac{1}{8} \int (1 - \cos 2x)(1 + \cos 2x)^2 \, dx.$]

37. Let $u = 1 - x^2$. Then $du = -2x \, dx \Rightarrow \displaystyle\int \frac{x \, dx}{1 - x^2 + \sqrt{1 - x^2}} = \frac{-1}{2} \int \frac{du}{u + \sqrt{u}} = -\int \frac{v \, dv}{v^2 + v}$

$[v = \sqrt{u}, \; u = v^2, \; du = 2v \, dv] = -\displaystyle\int \frac{dv}{v + 1} = -\ln|v + 1| + C = -\ln \left(\sqrt{1 - x^2} + 1 \right) + C.$

39. Let $u = e^x$. Then $x = \ln u$, $dx = du/u \Rightarrow \displaystyle\int \frac{e^x \, dx}{e^{2x} - 1} = \int \frac{u(du/u)}{u^2 - 1}$

$= \displaystyle\int \frac{du}{u^2 - 1} = \frac{1}{2} \ln \left| \frac{u - 1}{u + 1} \right| + C = \frac{1}{2} \ln \left| (e^x - 1)/(e^x + 1) \right| + C.$

41. $\displaystyle\int_{-1}^{1} x^5 \cosh x \, dx = 0$ by Theorem 4.35 since $x^5 \cosh x$ is odd.

43. $\displaystyle\int_{-3}^{3} \left| x^3 + x^2 - 2x \right| dx = \int_{-3}^{3} \left| (x + 2)x(x - 1) \right| dx = -\int_{-3}^{-2} (x^3 + x^2 - 2x) \, dx$

$+ \displaystyle\int_{-2}^{0} (x^3 + x^2 - 2x) \, dx - \int_{0}^{1} (x^3 + x^2 - 2x) \, dx + \int_{1}^{3} (x^3 + x^2 - 2x) \, dx.$

Let $f(x) = x^4/4 + x^3/3 - x^2$. Then $f'(x) = x^3 + x^2 - 2x$,

so $\displaystyle\int_{-3}^{3} \left| x^3 + x^2 - 2x \right| dx = -f(-2) + f(-3) + f(0) - f(-2) - f(1) + f(0) + f(3) - f(1)$

$= f(-3) - 2f(-2) + 2f(0) - 2f(1) + f(3) = \frac{9}{4} - 2(-\frac{8}{3}) + 2 \cdot 0 - 2(-\frac{5}{12}) + \frac{81}{4} = \frac{86}{3}.$

45. Let $u = \ln(\sin x)$. Then $du = \cot x \, dx \Rightarrow$

$\int \cos x \ln (\sin x) \, dx = \int u \, du = u^2/2 + C = \frac{1}{2} [\ln (\sin x)]^2 + C.$

47. $\dfrac{x}{(x^2 + 1)(x^2 + 4)} = \dfrac{Ax + B}{x^2 + 1} + \dfrac{Cx + D}{x^2 + 4} \Rightarrow x = (Ax + B)(x^2 + 4) + (Cx + D)(x^2 + 1)$

$\Rightarrow 0 = A + C, \; 0 = B + D, \; 1 = 4A + C, \text{ and } 0 = 4B + D \Rightarrow A = -C = 1/3, \; B = D = 0.$

$\displaystyle\int \frac{x \, dx}{(x^2 + 1)(x^2 + 4)} = \frac{1}{3} \int \left[\frac{x}{x^2 + 1} - \frac{x}{x^2 + 4} \right] dx$

$= \frac{1}{6} \ln (x^2 + 1) - \frac{1}{6} \ln (x^2 + 4) + C = \frac{1}{6} \ln [(x^2 + 1)/(x^2 + 4)] + C.$

49. Let $u = \sqrt[3]{x + c}$. Then $x = u^3 - c \Rightarrow \displaystyle\int x \sqrt[3]{x + c} \, dx = \int (u^3 - c) u \cdot 3u^2 \, du$

$= 3 \int (u^6 - cu^3) \, du = \frac{3}{7} u^7 - \frac{3}{4} cu^4 + C = \frac{3}{7} (x + c)^{7/3} - \frac{3}{4} c(x + c)^{4/3} + C.$

51. Let $u = \sqrt{x + 1}$. Then $x = u^2 - 1 \Rightarrow \displaystyle\int \frac{dx}{x + 4 + 4\sqrt{x + 1}} = \int \frac{2u \, du}{u^2 + 3 + 4u}$

$= \displaystyle\int \left[\frac{-1}{u + 1} + \frac{3}{u + 3} \right] du = 3 \ln |u + 3| - \ln |u + 1| + C$

$= 3 \ln (\sqrt{x + 1} + 3) - \ln (\sqrt{x + 1} + 1) + C.$

53. Use parts twice. First let $u = x^2 + 4x - 3$, $dv = \sin 2x \, dx \Rightarrow du = (2x + 4) \, dx$,

$v = -\frac{1}{2}\cos 2x$. Then $I = \int (x^2 + 4x - 3)\sin 2x\, dx = (x^2 + 4x - 3)(-\frac{1}{2}\cos 2x)$

$+ \int (2x + 4)(\frac{1}{2}\cos 2x)\, dx$. Now let $U = 2x + 4$, $dV = \frac{1}{2}\cos 2x\, dx \Rightarrow dU = 2\,dx$,

$V = \frac{1}{4}\sin 2x$. Then $I = (x^2 + 4x - 3)(-\frac{1}{2}\cos 2x) + (2x + 4)(\frac{1}{4}\sin 2x) - \frac{1}{2}\int \sin 2x\, dx$

$= -\frac{1}{2}(x^2 + 4x - 3)\cos 2x + \frac{1}{2}(x + 2)\sin 2x + \frac{1}{4}\cos 2x + C$

$= (1/2)(x + 2)\sin 2x - (1/4)(2x^2 + 8x - 7)\cos 2x + C.$

55. Let $u = x^2$. Then $du = 2x\,dx \Rightarrow \displaystyle\int \frac{x\,dx}{\sqrt{16 - x^4}} = \frac{1}{2}\int \frac{du}{\sqrt{16 - u^2}}$

$= \frac{1}{2}\sin^{-1}\left(\frac{u}{4}\right) + C = \frac{1}{2}\sin^{-1}(x^2/4) + C.$

57. Let $u = \csc 2x$. Then $du = -2\cot 2x\csc 2x\, dx \Rightarrow \int \cot^3 2x\csc^3 2x\, dx$

$= \int \csc^2 2x(\csc^2 2x - 1)\cot 2x\csc 2x\, dx = \int u^2(u^2 - 1)(-\frac{1}{2}du) = -\frac{1}{2}\int (u^4 - u^2)\, du$

$= -\frac{1}{2}\left[\frac{1}{5}u^5 - \frac{1}{3}u^3\right] + C = \frac{1}{6}\csc^3 2x - \frac{1}{10}\csc^5 2x + C.$

59. Let $u = \arctan x$. Then $du = \dfrac{dx}{1 + x^2} \Rightarrow \displaystyle\int \frac{e^{\arctan x}}{1 + x^2}\, dx = \int e^u\, du$

$= e^u + C = e^{\arctan x} + C.$

61. Integrate by parts 3 times, first with $u = t^3$, $dv = e^{-2t}\, dt$: $\int t^3 e^{-2t}\, dt$

$= -\frac{1}{2}t^3 e^{-2t} + \frac{1}{2}\int 3t^2 e^{-2t}\, dt = -\frac{1}{2}t^3 e^{-2t} - \frac{3}{4}t^2 e^{-2t} + \frac{1}{2}\int 3te^{-2t}\, dt$

$= -e^{-2t}\left[\frac{1}{2}t^3 + \frac{3}{4}t^2\right] - \frac{3}{4}te^{-2t} + \frac{3}{4}\int e^{-2t}\, dt = -e^{-2t}\left[\frac{1}{2}t^3 + \frac{3}{4}t^2 + \frac{3}{4}t + \frac{3}{8}\right] + C$

$= -\frac{1}{8}e^{-2t}(4t^3 + 6t^2 + 6t + 3) + C.$

63. $\int \sin x\sin 2x\sin 3x\, dx = \int \sin x \cdot \frac{1}{2}[\cos(2x - 3x) - \cos(2x + 3x)]\, dx$

$= \frac{1}{2}\int (\sin x\cos x - \sin x\cos 5x)\, dx$

$= \frac{1}{4}\int \sin 2x\, dx - \frac{1}{2}\int \frac{1}{2}[\sin(x + 5x) + \sin(x - 5x)]\, dx$

$= -\frac{1}{8}\cos 2x - \frac{1}{4}\int (\sin 6x - \sin 4x)\, dx$

$= -\frac{1}{8}\cos 2x + \frac{1}{24}\cos 6x - \frac{1}{16}\cos 4x + C.$

65. As in Example 5, $\displaystyle\int \sqrt{\frac{1 + x}{1 - x}}\, dx = \int \frac{1 + x}{\sqrt{1 - x^2}}\, dx = \int \frac{dx}{\sqrt{1 - x^2}} + \int \frac{x\,dx}{\sqrt{1 - x^2}}$

$= \sin^{-1}x - \sqrt{1 - x^2} + C.$ [Another method: Substitute $u = \sqrt{(1 + x)/(1 - x)}$.]'

67. $\displaystyle\int \frac{x + a}{x^2 + a^2}\, dx = \frac{1}{2}\int \frac{2x\,dx}{x^2 + a^2} + a\int \frac{dx}{x^2 + a^2} = \frac{1}{2}\ln(x^2 + a^2) + a\cdot\frac{1}{a}\tan^{-1}\left(\frac{x}{a}\right) + C$

$= \ln\sqrt{x^2 + a^2} + \tan^{-1}(x/a) + C.$

69. Let $u = x^5$. Then $du = 5x^4\,dx \Rightarrow \displaystyle\int \frac{x^4\,dx}{x^{10} + 16} = \int \frac{(1/5)\,du}{u^2 + 16}$

$= \frac{1}{5} \cdot \frac{1}{4} \tan^{-1}\left(\frac{u}{4}\right) + C = \frac{1}{20} \tan^{-1}(x^5/4) + C.$

71. Integrate by parts with $u = x$, $dv = \sec x \tan x\, dx \Rightarrow du = dx$, $v = \sec x$:

$\int x \sec x \tan x\, dx = x \sec x - \int \sec x\, dx = x \sec x - \ln|\sec x + \tan x| + C.$

73. $\displaystyle \int \frac{dx}{\sqrt{x+1}+\sqrt{x}} = \int (\sqrt{x+1} - \sqrt{x})\, dx = \frac{2}{3}[(x+1)^{3/2} - x^{3/2}] + C.$

75. Let $u = \sqrt{x}$. Then $du = \frac{dx}{2\sqrt{x}} \Rightarrow \int \frac{\arctan \sqrt{x}}{\sqrt{x}}\, dx = \int \tan^{-1}u\; 2\, du = 2u \tan^{-1}u$

$\displaystyle - \int \frac{2u\, du}{1+u^2}$ [by parts] $= 2u \tan^{-1}u - \ln(1+u^2) + C = 2\sqrt{x}\tan^{-1}\sqrt{x} - \ln(1+x) + C.$

77. Let $u = e^x$. Then $x = \ln u$, $dx = du/u \Rightarrow \int \frac{dx}{e^{3x}-e^x} = \int \frac{du/u}{u^3-u} = \int \frac{du}{(u-1)u^2(u+1)}$

$\displaystyle = \int \left[\frac{1/2}{u-1} - \frac{1}{u^2} - \frac{1/2}{u+1}\right] du = \frac{1}{u} + \frac{1}{2}\ln|(u-1)/(u+1)| + C$

$= e^{-x} + (1/2)\ln|(e^x-1)/(e^x+1)| + C.$

79. Let $u = \sqrt{2x-25} \Rightarrow u^2 = 2x - 25 \Rightarrow 2u\, du = 2\, dx$

$\displaystyle \int \frac{dx}{x\sqrt{2x-25}} = \int \frac{u\, du}{\frac{u^2+25}{2}\cdot u} = 2\int \frac{du}{u^2+25} = \frac{2}{5}\tan^{-1}\left(\frac{u}{5}\right) + C = \frac{2}{5}\tan^{-1}\left(\frac{\sqrt{2x-25}}{5}\right) + C.$

EXERCISES 7.7

1. By Formula 99,

$\displaystyle \int e^{-3x}\cos 4x\, dx = \frac{e^{-3x}}{(-3)^2+4^2}(-3\cos 4x + 4\sin 4x) + C = \frac{e^{-3x}}{25}(-3\cos 4x + 4\sin 4x) + C.$

3. Let $u = 3x$. Then $du = 3\, dx$, so $\displaystyle \int \frac{\sqrt{9x^2-1}}{x^2}\, dx = \int \frac{\sqrt{u^2-1}}{u^2/9}\frac{du}{3} = 3\int \frac{\sqrt{u^2-1}}{u^2}\, du$

$\displaystyle = -\frac{3\sqrt{u^2-1}}{u} + 3\ln|u + \sqrt{u^2-1}| + C$ [by Formula 42]

$\displaystyle = -\frac{\sqrt{9x^2-1}}{x} + 3\ln|3x + \sqrt{9x^2-1}| + C.$

5. $\int x^2 e^{3x}\, dx = \frac{1}{3}x^2 e^{3x} - \frac{2}{3}\int xe^{3x}\, dx$ [Formula 97] $= \frac{1}{3}x^2 e^{3x} - \frac{2}{3}\left[\frac{1}{9}(3x-1)e^{3x}\right] + C$

[Formula 96] $= (1/27)(9x^2 - 6x + 2)e^{3x} + C.$

7. Let $u = x^2$. Then $du = 2x\, dx$, so $\int x \sin^{-1}(x^2)\, dx = \frac{1}{2}\int \sin^{-1}u\, du$

$= \frac{1}{2}\left(u \sin^{-1}u + \sqrt{1-u^2}\right) + C$ [Formula 87] $= \frac{1}{2}\left(x^2 \sin^{-1}(x^2) + \sqrt{1-x^4}\right) + C.$

9. Let $u = e^x$. Then $du = e^x dx$, so $\int e^x \operatorname{sech}(e^x) dx = \int \operatorname{sech} u \, du = \tan^{-1}|\sinh u| + C$

 [*Formula 107*] $= \tan^{-1}(\sinh(e^x)) + C$.

11. Let $u = x + 2$. Then $\int \sqrt{5 - 4x - x^2} \, dx = \int \sqrt{9 - (x+2)^2} \, dx = \int \sqrt{9 - u^2} \, du$

 $= \frac{u}{2}\sqrt{9 - u^2} + \frac{9}{2}\sin^{-1}(\frac{u}{3}) + C$ [*Formula 30*] $= \frac{x+2}{2}\sqrt{5 - 4x - x^2} + \frac{9}{2}\sin^{-1}\left(\frac{x+2}{3}\right) + C$.

13. $\int \sec^5 x \, dx = \frac{1}{4}\tan x \sec^3 x + \frac{3}{4}\int \sec^3 x \, dx$ [*Formula 77*]

 $= \frac{1}{4}\tan x \sec^3 x + \frac{3}{4}(\frac{1}{2}\tan x \sec x + \frac{1}{2}\int \sec x \, dx)$ [*Formula 77 again*]

 $= \frac{1}{4}\tan x \sec^3 x + \frac{3}{8}\tan x \sec x + \frac{3}{8}\ln|\sec x + \tan x| + C$ [*Formula 14*].

15. Let $u = \sin x$. Then $du = \cos x \, dx$, so $\int \sin^2 x \cos x \ln(\sin x) \, dx = \int u^2 \ln u \, du$

 $= \frac{u^3}{9}(3\ln u - 1) + C$ [*Formula 101*] $= (1/9)\sin^3 x [3\ln(\sin x) - 1] + C$.

17. $\int \sqrt{2 + 3\cos x} \tan x \, dx = -\int \frac{\sqrt{2 + 3\cos x}}{\cos x}(-\sin x \, dx) = -\int \frac{\sqrt{2 + 3u}}{u} \, du$ [$u = \cos x$]

 $= -2\sqrt{2 + 3u} - 2\int \frac{du}{u\sqrt{2 + 3u}}$ [*Formula 58*] $= -2\sqrt{2 + 3u} - 2 \cdot \frac{1}{\sqrt{2}}\ln\left|\frac{\sqrt{2 + 3u} - \sqrt{2}}{\sqrt{2 + 3u} + \sqrt{2}}\right| + C$

 [*Formula 57*] $= -2\sqrt{2 + 3\cos x} - \sqrt{2}\ln\left|\frac{\sqrt{2 + 3\cos x} - \sqrt{2}}{\sqrt{2 + 3\cos x} + \sqrt{2}}\right| + C$.

19. $\displaystyle\int_0^{\pi/2} \cos^5 x \, dx = \frac{1}{5}\left[\cos^4 x \sin x\right]_0^{\pi/2} + \frac{4}{5}\int_0^{\pi/2} \cos^3 x \, dx$ [*Formula 74*]

 $= 0 + \frac{4}{5}\left[\frac{1}{3}(2 + \cos^2 x)\sin x\right]_0^{\pi/2}$ [*Formula 68*] $= \frac{4}{15}(2 - 0) = \frac{8}{15}$.

21. Let $u = x^5$, $du = 5x^4 \, dx$. $\displaystyle\int \frac{x^4 \, dx}{\sqrt{x^{10} - 2}} = \frac{1}{5}\int \frac{du}{\sqrt{u^2 - 2}}$

 $= \frac{1}{5}\ln\left|u + \sqrt{u^2 - 2}\right| + C$ [*Formula 43*] $= \frac{1}{5}\ln\left|x^5 + \sqrt{x^{10} - 2}\right| + C$.

23. Let $u = 1 + e^x$, $du = e^x \, dx$. $\int e^x \ln(1 + e^x) \, dx = \int \ln u \, du = u \ln u - u + C$ [*Formula 100*]

 $= (1 + e^x)\ln(1 + e^x) - e^x - 1 + C = (1 + e^x)\ln(1 + e^x) - e^x + C_1$.

25. Let $u = e^x \Rightarrow \ln u = x \Rightarrow dx = \frac{du}{u}$. $\int \sqrt{e^{2x} - 1} \, dx = \int \frac{\sqrt{u^2 - 1}}{u} \, du$

 $= \sqrt{u^2 - 1} - \cos^{-1}(\frac{1}{u}) + C$ [*formula 41*] $= \sqrt{e^{2x} - 1} - \cos^{-1}\left(\frac{1}{e^x}\right) + C$.

 [OR: Let $u = \sqrt{e^{2x} - 1}$.]

27. Volume $= \displaystyle\int_0^1 \frac{2\pi x}{(1 + 5x)^2} \, dx = 2\pi\left[\frac{1}{25(1 + 5x)} + \frac{1}{25}\ln|1 + 5x|\right]_0^1$

 $= \frac{2\pi}{25}(\frac{1}{6} + \ln 6 - 1 - \ln 1) = \frac{2\pi}{25}(\ln 6 - \frac{5}{6})$.

29. (a) $\dfrac{d}{du}\left[\dfrac{1}{b^3}\left(a + bu - \dfrac{a^2}{a+bu} - 2a\ln|a+bu|\right) + C\right] = \dfrac{1}{b^3}\left[b + \dfrac{ba^2}{(a+bu)^2} - \dfrac{2ab}{(a+bu)}\right]$

$$= \dfrac{1}{b^3}\left[\dfrac{b(a+bu)^2 + ba^2 - (a+bu)2ab}{(a+bu)^2}\right] = \dfrac{1}{b^3}\left[\dfrac{b^3 u^2}{(a+bu)^2}\right] = \dfrac{u^2}{(a+bu)^2}$$

(b) Let $t = a + bu \Rightarrow dt = b\,du$. $\displaystyle\int \dfrac{u^2\,du}{(a+bu)^2} = \dfrac{1}{b^3}\int \dfrac{(t-a)^2}{t^2}\,dt$

$$= \dfrac{1}{b^3}\int\left(1 - \dfrac{2a}{t} + \dfrac{a^2}{t^2}\right)dt = \dfrac{1}{b^3}\left(t - 2a\ln|t| - \dfrac{a^2}{t}\right) + C$$

$$= \dfrac{1}{b^3}\left(a + bu - \dfrac{a^2}{a+bu} - 2a\ln|a+bu|\right) + C.$$

EXERCISES 7.8

1. $\displaystyle\int_0^1 x^3\,dx = \left[x^4/4\right]_0^1 = 0.25.\quad f(x) = x^3.$

$\underline{n = 4}$: $L_4 = \frac{1}{4}[0^3 + (1/4)^3 + (1/2)^3 + (3/4)^3] = .140625$

$R_4 = \frac{1}{4}[(1/4)^3 + (1/2)^3 + (3/4)^3 + 1^3] = .390625$

$T_4 = \frac{1}{4\cdot 2}[0^3 + 2(1/4)^3 + 2(1/2)^3 + 2(3/4)^3 + 1^3] = \frac{17}{64} = 0.265625$

$M_4 = \frac{1}{4}[(1/8)^3 + (3/8)^3 + (5/8)^3 + (7/8)^3] \approx .242188$

$E_L = \int_0^1 x^3\,dx - L_4 = \frac{1}{4} - .140625 = .109375,\ \ E_R = \frac{1}{4} - .390625 = -.140625,$

$E_T = \frac{1}{4} - .265625 = -.015625,\ \ E_M = \frac{1}{4} - .2421875 \approx .007813$

$\underline{n = 8}$: $L_8 = \frac{1}{8}\left[f(0) + f(1/8) + f(2/8) + \cdots + f(7/8)\right] \approx .191406$

$R_8 = \frac{1}{8}\left[f(1/8) + f(2/8) + \cdots + f(7/8) + f(1)\right] \approx .316406$

$T_8 = \frac{1}{8\cdot 2}\left[f(0) + 2\big(f(1/8) + f(2/8) + \cdots + f(7/8)\big) + f(1)\right] \approx .253906$

$M_8 = \frac{1}{8}\left[f(1/16) + f(3/16) + \cdots + f(13/16) + f(15/16)\right] = .248047$

$E_L \approx \frac{1}{4} - .191406 \approx .058594,\ \ E_R \approx \frac{1}{4} - .316406 \approx -.066406,$

$E_T \approx \frac{1}{4} - .253906 \approx -.003906,\ \ E_M \approx \frac{1}{4} - .248047 \approx .001953$

$\underline{n = 16}$: $L_{16} = \frac{1}{16}\left[f(0) + f(1/16) + f(2/16) + \cdots + f(15/16)\right] \approx .219727$

$R_{16} = \frac{1}{16}\left[f(1/16) + f(2/16) + \cdots + f(15/16) + f(1)\right] \approx .282227$

$T_{16} = \frac{1}{16\cdot 2}\left[f(0) + 2\big(f(1/16) + f(2/16) + \cdots + f(15/16)\big) + f(1)\right] \approx .250977$

$M_{16} = \frac{1}{16}\left[f(1/32) + f(3/32) + \cdots + f(31/32)\right] \approx .249512$

$E_L \approx \frac{1}{4} - .219727 \approx .030273,\ \ E_R \approx \frac{1}{4} - .282227 \approx -.032227,$

$$E_T \approx \tfrac{1}{4} - .250977 \approx -.000977, \quad E_M \approx \tfrac{1}{4} - .249512 \approx .000488$$

n	L_n	R_n	T_n	M_n
4	0.140625	0.390625	0.265625	0.242188
8	0.191406	0.316406	0.253906	0.248047
16	0.219727	0.282227	0.250977	0.249512

n	E_L	E_R	E_T	E_M
4	0.109375	−0.140625	−0.015625	0.007812
8	0.058594	−0.066406	−0.003906	0.001953
16	0.030273	−0.032227	−0.000977	0.000488

Observations:

1. E_L and E_R are always opposite in sign as are E_T and E_M.

2. As n is doubled, E_L and E_R are decreased by about a factor of 2, and E_T and E_M are decreased by a factor of about 4.

3. The Midpoint approximation is about twice as accurate as the Trapezoidal approximation.

4. All the approximations become more accurate as the value of n increases.

5. The Midpoint and Trapezoidal approximations are much more accurate than the endpoint approximations.

3. $\displaystyle\int_1^4 \sqrt{x}\,dx = \left[\tfrac{2}{3}x^{3/2}\right]_1^4 = \tfrac{2}{3}(8-1) = \tfrac{14}{3} \approx 4.666667$

$\underline{n = 6}$: $\Delta x = \dfrac{4-1}{6} = \dfrac{1}{2}$.

$T_6 = \dfrac{1}{2\cdot 2}\left[\sqrt{1} + 2\sqrt{1.5} + 2\sqrt{2} + 2\sqrt{2.5} + 2\sqrt{3} + 2\sqrt{3.5} + \sqrt{4}\right] \approx 4.661488$

$M_6 = \tfrac{1}{2}\left[\sqrt{1.25} + \sqrt{1.75} + \sqrt{2.25} + \sqrt{2.75} + \sqrt{3.25} + \sqrt{3.75}\right] \approx 4.669245$

$S_6 = \dfrac{1}{2\cdot 3}\left[\sqrt{1} + 4\sqrt{1.5} + 2\sqrt{2} + 4\sqrt{2.5} + 2\sqrt{3} + 4\sqrt{3.5} + \sqrt{4}\right] \approx 4.666563$

$E_T \approx \tfrac{14}{3} - 4.661488 \approx .005178, \quad E_M \approx \tfrac{14}{3} - 4.669245 \approx -.002578$

$E_S \approx \tfrac{14}{3} - 4.666563 \approx .000104$

$\underline{n = 12}$: $\Delta x = \dfrac{4-1}{12} = \dfrac{1}{4}$.

$T_{12} = \dfrac{1}{4\cdot 2}\left[f(1) + 2\big(f(1.25) + f(1.5) + \cdots + f(3.5) + f(3.75)\big) + f(4)\right] \approx 4.665367$

$M_{12} = \tfrac{1}{4}\left[f(1.125) + f(1.375) + f(1.625) + \cdots + f(3.875)\right] \approx 4.667316$

$S_{12} = \dfrac{1}{4\cdot 3}\left[f(1) + 4f(1.25) + 2f(1.5) + 4f(1.75) + \cdots + 4f(3.75) + f(4)\right] \approx 4.666659$

$E_T \approx \tfrac{14}{3} - 4.665367 \approx .001300, \quad E_M \approx \tfrac{14}{3} - 4.667316 \approx -.000649$

$$E_S \approx \tfrac{14}{3} - 4.666659 \approx .000007$$

NOTE: These errors were computed more precisely and then rounded to six places. They were not computed by comparing the rounded values of T_n, M_n, and S_n with the rounded value of the actual definite integral.

n	T_n	M_n	S_n
6	4.661488	4.669245	4.666563
12	4.665367	4.667316	4.666659

n	E_T	E_M	E_S
6	0.005178	−0.002578	0.000104
12	0.001300	−0.000649	0.000007

Observations:

1. E_T and E_M are opposite in sign and decrease by about a factor of 4 as n is doubled.

2. The Simpson's approximation is much more accurate than the Midpoint and Trapezoidal approximations, and seems to decrease by about a factor of 16 as n is doubled.

5. $\displaystyle\int_2^4 \frac{x\,dx}{x^2+1} = \left[\tfrac{1}{2}\ln(x^2+1)\right]_2^4 = \tfrac{1}{2}(\ln 17 - \ln 5) \approx 0.611888$

$\underline{n=4}$: $\Delta x = \dfrac{4-2}{4} = \dfrac{1}{2}$.

$T_4 = \dfrac{1}{2\cdot 2}\left[f(2) + 2f(2.5) + 2f(3) + 2f(3.5) + f(4)\right] \approx .613313$

$M_4 = \dfrac{1}{2}\left[f(2.25) + f(2.75) + f(3.25) + f(3.75)\right] \approx .611173$

$S_4 = \dfrac{1}{2\cdot 3}\left[f(2) + 4f(2.5) + 2f(3) + 4f(3.5) + f(4)\right] \approx .611868$

$E_T \approx .611888 - .613313 \approx -.001425$, $E_M \approx .611888 - .611173 \approx .000715$,

$E_S \approx .611888 - .611868 \approx .000020$

$\underline{n=8}$: $\Delta x = \dfrac{4-2}{8} = \dfrac{1}{4}$.

$T_8 = \dfrac{1}{4\cdot 2}\left[f(2) + 2\big(f(2.25) + f(2.5) + \cdots + f(3.5) + f(3.75)\big) + f(4)\right] \approx .612243$

$M_8 = \dfrac{1}{4}\left[f(2.125) + f(2.375) + f(2.625) + \cdots + f(3.875)\right] \approx .611710$

$S_8 = \dfrac{1}{4\cdot 3}\left[f(2) + 4f(2.25) + 2f(2.5) + 4f(2.75) + \cdots + 4f(3.75) + f(4)\right] \approx .611886$

$E_T \approx .611888 - .612243 \approx -.000355$, $E_M \approx .611888 - .611710 \approx .000178$

$E_S \approx .611888 - .611886 \approx .000002$

n	T_n	M_n	S_n
4	0.613313	0.611173	0.611868
8	0.612243	0.611710	0.611886

n	E_T	E_M	E_S
4	−0.001425	0.000715	0.000020
8	−0.000355	0.000178	0.000002

Observations:

1. E_T and E_M are opposite in sign and appear to decrease by a factor of 4 as n is doubled.

2. The Simpson's approximation is much more accurate than the Midpoint and Trapezoidal approximations, and decreases by a factor of about 16 (to the nearest power of 2) as n is doubled. (Though more precision is needed to see this.)

7. $f(x) = \sqrt{1+x^3}$, $\Delta x = \dfrac{1-(-1)}{8} = 1/4$.

   (a) $T_8 = (.25/2)[f(-1) + 2f(-3/4) + 2f(-1/2) + \cdots + 2f(1/2) + 2f(3/4) + f(1)]$
   ≈ 1.913972

   (b) $S_8 = (.25/3)[f(-1) + 4f(-3/4) + 2f(-1/2) + 4f(-1/4) + 2f(0) + 4f(1/4) + 2f(1/2)$
   $+ 4f(3/4) + f(1)] \approx 1.934766$

9. $f(x) = \dfrac{\sin x}{x}$, $\Delta x = (\pi - \pi/2)/6 = \pi/12$.

   (a) $T_6 = (\pi/24)[f(\pi/2) + 2f(7\pi/12) + 2f(2\pi/3) + 2f(3\pi/4) + 2f(5\pi/6)$
   $+ 2f(11\pi/12) + f(\pi)] \approx 0.481672$

   (b) $S_6 = (\pi/36)[f(\pi/2) + 4f(7\pi/12) + 2f(2\pi/3) + 4f(3\pi/4) + 2f(5\pi/6)$
   $+ 4f(11\pi/12) + f(\pi)] \approx 0.481172$

11. $f(x) = e^{-x^2}$, $\Delta x = (1-0)/10 = .1$

   (a) $T_{10} = (.1/2)[f(0) + 2f(.1) + 2f(.2) + \cdots + 2f(.8) + 2f(.9) + f(1)] \approx 0.746211$

   (b) $M_{10} = .1[f(.05) + f(.15) + f(.25) + \cdots + f(.75) + f(.85) + f(.95)] \approx .747131$

   (c) $S_{10} = (.1/3)[f(0) + 4f(.1) + 2f(.2) + 4f(.3) + 2f(.4) + 4f(.5) + 2f(.6) + 4f(.7) + 2f(.8)$
   $+ 4f(.9) + f(1)] \approx 0.746825$

13. $f(x) = \cos(e^x)$, $\Delta x = (1/2 - 0)/8 = 1/16$.

   (a) $T_8 = (1/32)[f(0) + 2f(1/16) + 2f(1/8) + \cdots + 2f(7/16) + f(1/2)] \approx 0.132465$

   (b) $M_8 = (1/16)[f(1/32) + f(3/32) + f(5/32) + \cdots + f(15/32)] \approx .132857$

   (c) $S_8 = (1/48)[f(0) + 4f(1/16) + 2f(1/8) + 4f(3/16) + 2f(1/4) + 4f(5/16) + 2f(3/8)$
   $+ 4f(7/16) + f(1/2)] \approx 0.132727$

15. $f(x) = x^5 e^x$, $\Delta x = (1-0)/10 = 1/10$.

(a) $T_{10} = (.1/2)\,[f(0) + 2f(.1) + 2f(.2) + \cdots + 2f(.9) + f(1)] \approx 0.409140$

(b) $M_{10} = .1[f(.05) + f(.15) + f(.25) + \cdots + f(.95)] \approx .388849$

(c) $S_{10} = (.1/3)\,[f(0) + 4f(.1) + 2f(.2) + 4f(.3) + 2f(.4) + 4f(.5) + 2f(.6) + 4f(.7) + 2f(.8)$
$+ 4f(.9) + f(1)] \approx 0.395802$

17. $f(x) = e^{1/x}$, $\Delta x = (2-1)/4 = 1/4$.

(a) $T_4 = \frac{1}{4 \cdot 2}[f(1) + 2f(1.25) + 2f(1.5) + 2f(1.75) + f(2)] \approx 2.031893$

(b) $M_4 = \frac{1}{4}[f(1.125) + f(1.375) + f(1.625) + f(1.875)] \approx 2.014207$

(c) $S_4 = \frac{1}{4 \cdot 3}[f(1) + 4f(1.25) + 2f(1.5) + 4f(1.75) + f(2)] \approx 2.020651$

19. $f(x) = \dfrac{1}{1+x^4}$, $\Delta x = (3-0)/6 = 1/2$.

(a) $T_6 = \frac{1}{2 \cdot 2}[f(0) + 2f(.5) + 2f(1) + 2f(1.5) + 2f(2) + 2f(2.5) + f(3)] \approx 1.098004$

(b) $M_6 = \frac{1}{2}[f(.25) + f(.75) + f(1.25) + f(1.75) + f(2.25) + f(2.75)] \approx 1.098709$

(c) $S_6 = \frac{1}{2 \cdot 3}[f(0) + 4f(.5) + 2f(1) + 4f(1.5) + 2f(2) + 4f(2.5) + f(3)] = 1.109031$

21. $f(x) = e^{-x^2}$, $\Delta x = (2-0)/10 = 1/5$.

(a) $T_{10} = \frac{1}{5 \cdot 2}[f(0) + 2\big(f(0.2) + f(.4) + \cdots + f(1.8)\big) + f(2)] \approx .881839$

$M_{10} = \frac{1}{5}[f(.1) + f(.3) + f(.5) + \cdots + f(1.7) + f(1.9)] \approx .882202$

(b) $f(x) = e^{-x^2}$, $f'(x) = -2xe^{-x^2}$, $f''(x) = (4x^2 - 2)e^{-x^2}$, $f'''(x) = 4x(3 - 2x^2)e^{-x^2}$.

$f'''(x) = 0 \Leftrightarrow x = 0$ or $x = \pm\sqrt{3/2}$. So to find the maximum value of $\left|f''(x)\right|$ on $[0, 2]$, we

need only consider its values at $x = 0$, $x = 2$, and $x = \sqrt{3/2}$. $\left|f''(0)\right| = 2$,

$\left|f''(2)\right| \approx 0.2564$ and $\left|f''(\sqrt{3/2})\right| = 4e^{-3/2} \approx .8925$. Thus, taking $M = 2$, $a = 0$, $b = 2$,

$n = 10$ in (7.32), we get $\left|E_T\right| \leq 2 \cdot 2^3/[12(10)^2] = 1/75 = 0.01\overline{3}$, and

$\left|E_M\right| \leq 2 \cdot 2^3/[24(10)^2] < .00\overline{6}$

23. (a) $T_{10} = \frac{1}{10 \cdot 2}[f(0) + 2\big(f(.1) + f(.2) + \cdots f(.9)\big) + f(1)] \approx 1.719713$

$S_{10} = \frac{1}{10 \cdot 3}[f(0) + 4f(.1) + 2f(.2) + 4f(.3) + \cdots 4f(.9) + f(1)] \approx 1.7182828$

Since $\int_0^1 e^x\,dx = \left[e^x\right]_0^1 = e - 1 \approx 1.71828183$, $E_T \approx -.00143166$ and $E_S \approx -.00000095$

(b) $f(x) = e^x \Rightarrow f''(x) = e^x \leq e$ for $0 \leq x \leq 1$. Taking $M = e$, $a = 0$, $b = 1$, $n = 10$ in

(7.32), we get $\left|E_T\right| \leq \dfrac{e(1)^3}{12(10)^2} \approx .002265 > .00143166$ [actual $\left|E_T\right|$ from (a)].

$f^{(4)}(x) = e^x < e$ for $0 \leq x \leq 1$. Using (7.34) we have

$\left|E_S\right| \leq \dfrac{e(1)^5}{180(10)^4} \approx .0000015 > .00000095$ [actual $\left|E_S\right|$ from (a)]. We see that the actual

errors are about 2/3 the size of the error estimates.

25. Take $M = 2$ (as in Exercise 21) in (7.32).

$$|E_T| \le \frac{M(b-a)^3}{12n^2} \le 10^{-5} \Leftrightarrow \frac{1}{6n^2} \le 10^{-5} \Leftrightarrow 6n^2 \ge 10^5 \Leftrightarrow n \ge 129.099\ldots \Leftrightarrow n \ge 130.$$

Take $n = 130$ in the trapezoidal method. For E_M, again take $M = 2$ in (7.32) to get

$$|E_M| \le \frac{2(1)^3}{24n^2} \le 10^{-5} \Leftrightarrow n^2 \ge \frac{2}{24(10^{-5})} \Leftrightarrow n \ge 91.3 \Leftrightarrow n \ge 92. \text{ Take } n = 92 \text{ for } M_n.$$

27. $\displaystyle\int_1^{3.2} y\,dx \approx \frac{2}{2}[4.9 + 2(5.4) + 2(5.8) + 2(6.2) + 2(6.7) + 2(7.0) + 2(7.3) + 2(7.5) + 2(8.0)$

$+ 2(8.2) + 2(8.3) + 8.3] = 15.4$

29. $\Delta t = 1 \text{ min} = \frac{1}{60} \text{ h}$, so distance $= \displaystyle\int_0^{1/6} v(t)\,dt \approx \frac{1/60}{3}[40 + 4(42) + 2(45) + 4(49)$

$+ 2(52) + 4(54) + 2(56) + 4(57) + 2(57) + 4(55) + 56] \approx 8.6 \text{ mi.}$

31. $\Delta x = (4 - 0)/4 = 1.$

(a) $T_4 = \frac{1}{2}[f(0) + 2f(1) + 2f(2) + 2f(3) + f(4)] \approx \frac{1}{2}[0 + 2(3) + 2(5) + 2(3) + 1] = 11.5$

(b) $M_4 = 1 \cdot [f(.5) + f(1.5) + f(2.5) + f(3.5)] \approx 1 + 4.5 + 4.5 + 2 = 12$

(c) $S_4 = \frac{1}{3}[f(0) + 4f(1) + 2f(2) + 4f(3) + f(4)] \approx \frac{1}{3}[0 + 4(3) + 2(5) + 4(3) + 1] = 11.\bar{6}$

33. Volume $= \pi\displaystyle\int_0^2 \left(\sqrt[3]{1+x^3}\right)^2 dx = \pi\int_0^2 (1+x^3)^{2/3}\,dx.$ $V \approx \pi \cdot S_{10}$ where $f(x) = (1+x^3)^{2/3}$

and $\Delta x = (2 - 0)/10 = 1/5.$ Therefore $V \approx \pi \cdot S_{10} = \pi\frac{1}{5\cdot 3}[f(0) + 4f(.2) + 2f(.4) + 4f(.6)$

$+ 2f(.8) + 4f(1) + 2f(1.2) + 4f(1.4) + 2f(1.6) + 4f(1.8) + f(2)] \approx 12.325078$

35. Since the Trapezoidal and Midpoint approximations on the interval $[a, b]$ are the sums of the Trapezoidal and Midpoint approximations on the subintervals $[x_{i-1}, x_i]$, $i = 1, 2, \ldots, n$, we can focus our attention on one such interval. The condition $f''(x) < 0$ for $a \le x \le b$ means that the graph of f is concave down as in Figure 7.10. In that figure, T_n is the area of the trapezoid AQRD, $\int_a^b f(x)\,dx$ is the area of the region AQPRD, and M_n is the area of the trapezoid ABCD, so $T_n < \int_a^b f(x)\,dx < M_n$. In general, the condition $f'' < 0$ implies that the graph of f on $[a, b]$ lies above the chord joining the points $(a, f(a))$ and $(b, f(b))$. Thus $\int_a^b f(x)\,dx > T_n$. Since M_n is the area under a tangent to the graph, and since $f'' < 0$ implies that the tangent lies above the graph, we also have $M_n > \int_a^b f(x)\,dx$. Thus $T_n < \int_a^b f(x)\,dx < M_n$.

EXERCISES 7.9

1. $\displaystyle\int_2^\infty \frac{dx}{\sqrt{x+3}} = \lim_{t\to\infty}\int_2^t \frac{dx}{\sqrt{x+3}} = \lim_{t\to\infty}\left[2\sqrt{x+3}\right]_2^t = \lim_{t\to\infty}(2\sqrt{t+3} - 2\sqrt{5}) = \infty.$ Divergent.

3. $\displaystyle\int_{-\infty}^1 \frac{dx}{(2x-3)^2} = \lim_{t\to-\infty}\frac{1}{2}\int_t^1 \frac{2\,dx}{(2x-3)^2} = \lim_{t\to-\infty}\frac{1}{2}\left[-\frac{1}{2x-3}\right]_t^1$

$$= \lim_{t \to -\infty} \left[\tfrac{1}{2} + \frac{1}{2(2t-3)}\right] = \tfrac{1}{2}.$$

5. $\displaystyle\int_{-\infty}^{\infty} x\,dx = \int_{-\infty}^{0} x\,dx + \int_{0}^{\infty} x\,dx.$ $\displaystyle\int_{-\infty}^{0} x\,dx = \lim_{t \to -\infty} \left[x^2/2\right]_t^0$

$= \displaystyle\lim_{t \to -\infty} (-t^2/2) = -\infty.$ Divergent.

7. $\displaystyle\int_{0}^{\infty} e^{-x}\,dx = \lim_{t \to \infty} \int_{0}^{t} e^{-x}\,dx = \lim_{t \to \infty} \left[-e^{-x}\right]_0^t = \lim_{t \to \infty}(-e^{-t} + 1) = 1.$

9. $\displaystyle\int_{-\infty}^{\infty} xe^{-x^2}\,dx = \int_{-\infty}^{0} xe^{-x^2}\,dx + \int_{0}^{\infty} xe^{-x^2}\,dx.$ $\displaystyle\int_{-\infty}^{0} xe^{-x^2}\,dx = \lim_{t \to -\infty} -\tfrac{1}{2}\left[e^{-x^2}\right]_0^t$

$= \displaystyle\lim_{t \to -\infty} -\tfrac{1}{2}\left(1 - e^{-t^2}\right) = -\tfrac{1}{2};$ $\displaystyle\int_{0}^{\infty} xe^{-x^2}\,dx = \lim_{t \to \infty} -\tfrac{1}{2}\left[e^{-x^2}\right]_t^0$

$= \displaystyle\lim_{t \to \infty} -\tfrac{1}{2}\left(e^{-t^2} - 1\right) = \tfrac{1}{2}.$ Therefore $\displaystyle\int_{-\infty}^{\infty} xe^{-x^2}\,dx = -\tfrac{1}{2} + \tfrac{1}{2} = 0.$

11. $\displaystyle\int_{0}^{\infty} \frac{dx}{(x+2)(x+3)} = \lim_{t \to \infty} \int_{0}^{t} \left[\frac{1}{x+2} - \frac{1}{x+3}\right]dx = \lim_{t \to \infty}\left[\ln\left(\frac{x+2}{x+3}\right)\right]_0^t$

$= \displaystyle\lim_{t \to \infty}\left[\ln\left(\frac{t+2}{t+3}\right) - \ln\tfrac{2}{3}\right] = \ln 1 - \ln\tfrac{2}{3} = -\ln\tfrac{2}{3}.$

13. $\displaystyle\int_{0}^{\infty} \cos x\,dx = \lim_{t \to \infty} [\sin x]_0^t = \lim_{t \to \infty} \sin t,$ which does not exist. Divergent.

15. $\displaystyle\int_{0}^{\infty} \frac{5\,dx}{2x+3} = \tfrac{5}{2} \lim_{t \to \infty} \int_{0}^{t} \frac{2\,dx}{2x+3} = \tfrac{5}{2} \lim_{t \to \infty}\left[\ln(2x+3)\right]_0^t$

$= \tfrac{5}{2} \displaystyle\lim_{t \to \infty}[\ln(2t+3) - \ln 3] = \infty.$ Divergent.

17. $\displaystyle\int_{-\infty}^{1} xe^{2x}\,dx = \lim_{t \to -\infty} \int_{t}^{1} xe^{2x}\,dx = \lim_{t \to -\infty}\left[\tfrac{1}{2}xe^{2x} - \tfrac{1}{4}e^{2x}\right]_t^1$ [by parts]

$= \displaystyle\lim_{t \to -\infty}\left[\tfrac{1}{2}e^2 - \tfrac{1}{4}e^2 - \tfrac{1}{2}te^{2t} + \tfrac{1}{4}e^{2t}\right] = \tfrac{1}{4}e^2 - 0 + 0 = \tfrac{1}{4}e^2$ since

$\displaystyle\lim_{t \to -\infty} te^{2t} = \lim_{t \to -\infty} \frac{t}{e^{-2t}} \overset{\text{H}}{=} \lim_{t \to -\infty} \frac{1}{-2e^{-2t}} = \lim_{t \to -\infty} -\tfrac{1}{2}e^{2t} = 0.$

19. $\displaystyle\int_{1}^{\infty} \frac{\ln x}{x}\,dx = \lim_{t \to \infty}\left[\frac{(\ln x)^2}{2}\right]_1^t = \lim_{t \to \infty} \frac{(\ln t)^2}{2} = \infty.$ Divergent.

21. $\displaystyle\int_{-\infty}^{\infty} \frac{x\,dx}{1+x^2} = \int_{-\infty}^{0} \frac{x\,dx}{1+x^2} + \int_{0}^{\infty} \frac{x\,dx}{1+x^2}$ and $\displaystyle\int_{-\infty}^{0} \frac{x\,dx}{1+x^2}$

$= \displaystyle\lim_{t \to -\infty}\left[\tfrac{1}{2}\ln(1+x^2)\right]_t^0 = \lim_{t \to -\infty}\left[0 - \tfrac{1}{2}\ln(1+t^2)\right] = -\infty.$ Divergent.

23. Integrate by parts with $u = \ln x$, $dv = dx/x^2 \Rightarrow du = dx/x$, $v = -1/x$.

$\displaystyle\int_{1}^{\infty} \frac{\ln x}{x^2}\,dx = \lim_{t \to \infty} \int_{1}^{t} \frac{\ln x}{x^2}\,dx = \lim_{t \to \infty}\left[-\frac{\ln x}{x} - \frac{1}{x}\right]_1^t = \lim_{t \to \infty}\left[-\frac{\ln t}{t} - \frac{1}{t} + 0 + 1\right]$

$= -0 - 0 + 0 + 1 = 1$ since $\displaystyle\lim_{t \to \infty} \frac{\ln t}{t} \overset{\text{H}}{=} \lim_{t \to \infty} \frac{1/t}{1} = 0.$

25. $\displaystyle\int_0^\infty \frac{dx}{2^x} = \lim_{t\to\infty} \int_0^t 2^{-x}\,dx = \lim_{t\to\infty}\left[\frac{-1}{\ln 2}2^{-x}\right]_0^t = \frac{1}{\ln 2}\lim_{t\to\infty}(1-2^{-t}) = \frac{1}{\ln 2}.$

27. $\displaystyle\int_0^3 \frac{dx}{\sqrt{x}} = \lim_{t\to 0^+}\int_t^3 \frac{dx}{\sqrt{x}} = \lim_{t\to 0^+}\left[2\sqrt{x}\right]_t^3 = \lim_{t\to 0^+}(2\sqrt3 - 2\sqrt t) = 2\sqrt 3.$

29. $\displaystyle\int_{-1}^0 \frac{dx}{x^2} = \lim_{t\to 0^-}\int_{-1}^t \frac{dx}{x^2} = \lim_{t\to 0^-}\left[\frac{-1}{x}\right]_{-1}^t = \lim_{t\to 0^-}\left[-\frac{1}{t} + \frac{1}{-1}\right] = \infty.$ Divergent.

31. $\displaystyle\int_{-2}^3 \frac{dx}{x^4} = \int_{-2}^0 \frac{dx}{x^4} + \int_0^3 \frac{dx}{x^4}$ and $\displaystyle\int_{-2}^0 \frac{dx}{x^4} = \lim_{t\to 0^-}\left[-\frac13 x^{-3}\right]_{-2}^t$

$= \displaystyle\lim_{t\to 0^-}\left[-\frac{1}{3t^3} - \frac{1}{24}\right] = \infty.$ Divergent.

33. $\displaystyle\int_4^5 \frac{dx}{(5-x)^{2/5}} = \lim_{t\to 5^-}\left[-\frac53(5-x)^{3/5}\right]_4^t = \lim_{t\to 5^-}\left[-\frac53(5-t)^{3/5} + \frac53\right] = 0 + \frac53 = \frac53.$

35. $\displaystyle\int_{\pi/4}^{\pi/2} \tan^2 x\,dx = \lim_{t\to\pi/2^-}\int_{\pi/4}^t (\sec^2 x - 1)\,dx = \lim_{t\to\pi/2^-}\left[\tan x - x\right]_{\pi/4}^t$

$= \frac{\pi}{4} - 1 + \displaystyle\lim_{t\to\pi/2^-}(\tan t - t) = \infty.$ Divergent.

37. $\displaystyle\int_0^2 \frac{x\,dx}{\sqrt{4-x^2}} = \lim_{t\to 2^-}\int_0^t \frac{x\,dx}{\sqrt{4-x^2}} = \lim_{t\to 2^-}\left[-\sqrt{4-x^2}\right]_0^t = \lim_{t\to 2^-}\left[2 - \sqrt{4-t^2}\right] = 2.$

39. $\displaystyle\int_0^\pi \sec x\,dx = \int_0^{\pi/2}\sec x\,dx + \int_{\pi/2}^\pi \sec x\,dx.$ $\displaystyle\int_0^{\pi/2}\sec x\,dx = \lim_{t\to\pi/2^-}\int_0^t \sec x\,dx$

$= \displaystyle\lim_{t\to\pi/2^-}\left[\ln|\sec x + \tan x|\right]_0^t = \lim_{t\to\pi/2^-}\ln|\sec t + \tan t| = \infty.$ Divergent.

41. $\displaystyle\int_{-2}^2 \frac{dx}{x^2-1} = \int_{-2}^{-1}\frac{dx}{x^2-1} + \int_{-1}^0 \frac{dx}{x^2-1} + \int_0^1 \frac{dx}{x^2-1} + \int_1^2 \frac{dx}{x^2-1}.$

$\displaystyle\int \frac{dx}{x^2-1} = \int\frac{dx}{(x-1)(x+1)} = \frac12\ln\left|\frac{x-1}{x+1}\right| + C,$ so $\displaystyle\int_0^1 \frac{dx}{x^2-1} = \lim_{t\to 1^-}\left[\frac12\ln\left|\frac{x-1}{x+1}\right|\right]_0^t$

$= \displaystyle\lim_{t\to 1^-}\frac12\ln\left|\frac{t-1}{t+1}\right| = -\infty.$ Divergent.

43. Let $u = \ln x$. Then $du = dx/x \Rightarrow \displaystyle\int_1^e \frac{dx}{x\sqrt[4]{\ln x}} = \lim_{t\to 1^+}\int_t^e \frac{dx}{x\sqrt[4]{\ln x}} = \lim_{t\to 1^+}\int_{\ln t}^1 \frac{du}{\sqrt[4]{u}}$

$= \displaystyle\lim_{t\to 1^+}\left[\frac43 u^{3/4}\right]_{\ln t}^1 = \lim_{t\to 1^+}\frac43\left[1 - (\ln t)^{3/4}\right] = \frac43.$

45. Integrate by parts with $u = \ln x,\ dv = x\,dx$: $\displaystyle\int_0^1 x\ln x\,dx = \lim_{t\to 0^+}\int_t^1 x\ln x\,dx$

$= \displaystyle\lim_{t\to 0^+}\left[\frac{x^2}{2}\ln x - \frac{x^2}{4}\right]_t^1 = -\frac14 - \lim_{t\to 0^+}\frac{t^2}{2}\ln t$

$= -\frac14 - \frac12\displaystyle\lim_{t\to 0^+}\frac{\ln t}{1/t^2} \overset{H}{=} -\frac14 - \frac12\lim_{t\to 0^+}\frac{1/t}{-2/t^3} = -\frac14 + \frac14\lim_{t\to 0^+}t^2 = -\frac14.$

47.

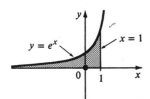

$$\text{Area} = \int_{-\infty}^{1} e^x \, dx = \lim_{t \to -\infty} \left[e^x \right]_t^1 = e - \lim_{t \to -\infty} e^t = e.$$

49.

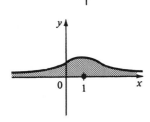

$$\text{Area} = \int_{-\infty}^{\infty} \frac{dx}{x^2 - 2x + 5}$$

$$= \int_{-\infty}^{0} \frac{dx}{(x-1)^2 + 4} + \int_{0}^{\infty} \frac{dx}{(x-1)^2 + 4}$$

$$= \lim_{t \to -\infty} \left[\tfrac{1}{2} \tan^{-1} \left(\tfrac{x-1}{2} \right) \right]_t^0 + \lim_{t \to \infty} \left[\tfrac{1}{2} \tan^{-1} \left(\tfrac{x-1}{2} \right) \right]_0^t$$

$$= \tfrac{1}{2} \tan^{-1} \left(-\tfrac{1}{2} \right) - \tfrac{1}{2} \left(-\tfrac{\pi}{2} \right) + \tfrac{1}{2} \left(\tfrac{\pi}{2} \right) - \tfrac{1}{2} \tan^{-1} \left(-\tfrac{1}{2} \right) = \tfrac{\pi}{2}.$$

51.

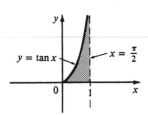

$$\text{Area} = \int_{0}^{\pi/2} \tan x \, dx = \lim_{t \to \pi/2^-} \left[\ln (\sec x) \right]_0^t = \infty,$$

so the area is infinite.

53. $\dfrac{\sin^2 x}{x^2} \le \dfrac{1}{x^2}$ on $[1, \infty)$. $\displaystyle\int_{1}^{\infty} \frac{dx}{x^2}$ is convergent by Example 4, so $\displaystyle\int_{1}^{\infty} \frac{\sin^2 x}{x^2} \, dx$ is convergent by the Comparison Theorem.

55. For $x \ge 1$, $x + e^{2x} > e^{2x} > 0 \Rightarrow \dfrac{1}{x + e^{2x}} \le \dfrac{1}{e^{2x}} = e^{-2x}$ on $[1, \infty)$.

$\displaystyle\int_{1}^{\infty} e^{-2x} \, dx = \lim_{t \to \infty} \left[-\tfrac{1}{2} e^{-2x} \right]_1^t = \lim_{t \to \infty} \left[-\tfrac{1}{2} e^{-2t} + \tfrac{1}{2} e^{-2} \right] = \tfrac{1}{2} e^{-2}$. Therefore $\displaystyle\int_{1}^{\infty} e^{-2x} \, dx$ is convergent and by the Comparison Theorem $\displaystyle\int_{1}^{\infty} \frac{dx}{x + e^{2x}}$ is also convergent.

57. $\dfrac{1}{x \sin x} \ge \dfrac{1}{x}$ on $(0, \pi/2]$ since $0 \le \sin x \le 1$. $\displaystyle\int_{0}^{\pi/2} \tfrac{1}{x} \, dx = \lim_{t \to 0^+} \int_{t}^{\pi/2} \tfrac{1}{x} \, dx$

$= \lim_{t \to 0^+} \left[\ln x \right]_t^{\pi/2}$. But $\ln t \to -\infty$ as $t \to 0^+$, so $\displaystyle\int_{0}^{\pi/2} \tfrac{1}{x} \, dx$ is divergent and by the

Comparison Theorem, $\displaystyle\int_{0}^{\pi/2} \frac{dx}{x \sin x}$ is also divergent.

59. $\displaystyle\int_{0}^{\infty} \frac{dx}{\sqrt{x}(1 + x)} = \int_{0}^{1} \frac{dx}{\sqrt{x}(1 + x)} + \int_{1}^{\infty} \frac{dx}{\sqrt{x}(1 + x)} = \lim_{t \to 0^+} \int_{t}^{1} \frac{dx}{\sqrt{x}(1 + x)}$

$+ \displaystyle\lim_{t \to \infty} \int_{1}^{t} \frac{dx}{\sqrt{x}(1 + x)} = \lim_{t \to 0^+} \int_{\sqrt{t}}^{1} \frac{2 \, du}{1 + u^2} + \lim_{t \to \infty} \int_{1}^{\sqrt{t}} \frac{2 \, du}{1 + u^2} \quad [u = \sqrt{x}, \, x = u^2]$

$$= \lim_{t \to 0^+} \left[2\tan^{-1}u\right]_{\sqrt{t}}^1 + \lim_{t \to \infty} \left[2\tan^{-1}u\right]_1^{\sqrt{t}} = \lim_{t \to 0^+} \left[2(\tfrac{\pi}{4}) - 2\tan^{-1}\sqrt{t}\right]$$

$$+ \lim_{t \to \infty} \left[2\tan^{-1}\sqrt{t} - 2(\tfrac{\pi}{4})\right] = \tfrac{\pi}{2} - 0 + 2(\tfrac{\pi}{2}) - \tfrac{\pi}{2} = \pi.$$

61. If $p = 1$, then $\int_0^1 \frac{dx}{x^p} = \lim_{t \to 0^+} [\ln x]_t^1 = \infty.$ Divergent. If $p \neq 1$, then $\int_0^1 \frac{dx}{x^p}$

$$= \lim_{t \to 0^+} \int_t^1 \frac{dx}{x^p} \text{ [Note that the integral is not improper if } p < 0.] = \lim_{t \to 0^+} \left[\frac{x^{-p+1}}{-p+1}\right]_t^1$$

$$= \lim_{t \to 0^+} \frac{1}{1-p}\left[1 - \frac{1}{t^{p-1}}\right]. \text{ If } p > 1, \text{ then } p - 1 > 0, \text{ so } 1/t^{p-1} \to \infty \text{ as } t \to 0^+ \text{ and the}$$

integral diverges. Finally, if $p < 1$, then $\int_0^1 \frac{dx}{x^p} = \frac{1}{1-p} \lim_{t \to 0^+} (1 - t^{1-p}) = \frac{1}{1-p}.$

Thus the integral converges $\Leftrightarrow p < 1$, and in that case its value is $1/(1-p)$.

63. First suppose $p = -1$. Then $\int_0^1 x^p \ln x \, dx = \int_0^1 \frac{\ln x}{x} \, dx = \lim_{t \to 0^+} \int_t^1 \frac{\ln x}{x} \, dx$

$$= \lim_{t \to 0^+} \left[\tfrac{1}{2}(\ln x)^2\right]_t^1 = -\tfrac{1}{2} \lim_{t \to 0^+} (\ln t)^2 = -\infty, \text{ so the integral diverges. Now}$$

suppose $p \neq -1$. Then integration by parts gives $\int x^p \ln x \, dx = \frac{x^{p+1}}{p+1} \ln x$

$$- \int \frac{x^p}{p+1} \, dx = \frac{x^{p+1}}{p+1} \ln x - \frac{x^{p+1}}{(p+1)^2} + C. \text{ If } p < -1, \text{ then } p + 1 < 0, \text{ so } \int_0^1 x^p \ln x \, dx$$

$$= \lim_{t \to 0^+} \left[\frac{x^{p+1}}{p+1} \ln x - \frac{x^{p+1}}{(p+1)^2}\right]_t^1 = \frac{-1}{(p+1)^2} - \frac{1}{p+1} \lim_{t \to 0^+} \left[t^{p+1}\left(\ln t - \frac{1}{p+1}\right)\right] = \infty. \text{ If}$$

$p > -1$, then $p + 1 > 0$ and $\int_0^1 x^p \ln x \, dx = \frac{-1}{(p+1)^2} - \frac{1}{p+1} \lim_{t \to 0^+} \frac{\ln t - \frac{1}{p+1}}{t^{-(p+1)}}$

$$\overset{\underline{H}}{=} \frac{-1}{(p+1)^2} - \frac{1}{p+1} \lim_{t \to 0^+} \frac{1/t}{-(p+1)t^{-(p+2)}} = \frac{-1}{(p+1)^2} + \frac{1}{(p+1)^2} \lim_{t \to 0^+} t^{p+1}$$

$$= \frac{-1}{(p+1)^2}. \text{ Thus the integral converges to } -1/(p+1)^2 \text{ if } p > -1 \text{ and diverges otherwise.}$$

65. (a) $\int_{-\infty}^{\infty} x \, dx = \int_{-\infty}^0 x \, dx + \int_0^{\infty} x \, dx.$ $\int_0^{\infty} x \, dx = \lim_{t \to \infty} \int_0^t x \, dx$

$$= \lim_{t \to \infty} \left[\frac{t^2}{2} - \frac{0^2}{2}\right] = \infty, \text{ so the integral is divergent.}$$

(b) $\int_{-t}^t x \, dx = \left[x^2/2\right]_{-t}^t = t^2/2 - t^2/2 = 0, \text{ so } \lim_{t \to \infty} \int_{-t}^t x \, dx = 0.$

Therefore $\int_{-\infty}^{\infty} x \, dx \neq \lim_{t \to \infty} \int_{-t}^t x \, dx.$

67. Volume $= \int_1^{\infty} \pi \left(\frac{1}{x}\right)^2 dx = \pi \lim_{t \to \infty} \int_1^t \frac{dx}{x^2} = \pi \lim_{t \to \infty} \left[-\frac{1}{x}\right]_1^t = \pi \lim_{t \to \infty} \left(1 - \frac{1}{t}\right) = \pi < \infty$

69. Work $= \int_R^{\infty} F \, dr = \lim_{t \to \infty} \int_R^t \frac{GmM}{r^2} \, dr = \lim_{t \to \infty} GmM\left(\frac{1}{R} - \frac{1}{t}\right) = \frac{GmM}{R}. \text{ The initial}$

273

kinetic energy provides the work, so $\frac{1}{2}mv_0^2 = \frac{GmM}{R} \Rightarrow v_0 = \sqrt{\frac{2GM}{R}}$.

71. (a) $F(s) = \int_0^\infty f(t)e^{-st}\,dt = \int_0^\infty e^{-st}\,dt = \lim_{n\to\infty}\left[-\frac{1}{s}e^{-st}\right]_0^n = \lim_{n\to\infty}\left(\frac{e^{-sn}}{-s} + \frac{1}{s}\right)$. This

   converges to $\frac{1}{s}$ only if $s > 0$. Therefore $F(s) = \frac{1}{s}$ and the domain of F is $\{s\,|\,s > 0\}$.

   (b) $F(s) = \int_0^\infty f(t)e^{-st}\,dt = \int_0^\infty e^t e^{-st}\,dt = \lim_{n\to\infty}\int_0^n e^{t(1-s)}\,dt$

   $= \lim_{n\to\infty}\left[\frac{1}{1-s}e^{t(1-s)}\right]_0^n = \lim_{n\to\infty}\left(\frac{e^{(1-s)n}}{1-s} - \frac{1}{1-s}\right)$. This converges only if

   $1 - s < 0 \Rightarrow s > 1$ in which case $F(s) = \frac{1}{s-1}$, and the domain is $\{s\,|\,s > 1\}$.

   (c) $F(s) = \int_0^\infty f(t)e^{-st}\,dt = \lim_{n\to\infty}\int_0^n te^{-st}\,dt$. Use integration by parts: let $u = t$,

   $dv = e^{-st}\,dt \Rightarrow du = dt, v = -\frac{1}{s}e^{-st}$. Then $F(s) = \lim_{n\to\infty}\left[-\frac{t}{s}e^{-st} - \frac{1}{s^2}e^{-st}\right]_0^n$

   $= \lim_{n\to\infty}\left(\frac{-n}{se^{sn}} - \frac{1}{s^2 e^{sn}} + 0 + \frac{1}{s^2}\right) = \frac{1}{s^2}$ only if $s > 0$. Therefore $F(s) = \frac{1}{s^2}$ and

   the domain of F is $\{s\,|\,s > 0\}$.

73. $G(s) = \int_0^\infty f'(t)e^{-st}\,dt$. Integrate by parts with $u = e^{-st}$, $dv = f'(t)\,dt \Rightarrow du = -se^{-st}$,

   $v = f(t)$. Therefore $G(s) = \lim_{n\to\infty}\left[f(t)e^{-st}\right]_0^n + s\int_0^\infty f(t)e^{-st}\,dt$

   $= \lim_{n\to\infty}f(n)e^{-sn} - f(0) + sF(s)$. But $0 \le f(t) \le Me^{at} \Rightarrow 0 \le f(t)e^{-st} \le Me^{at}e^{-st}$ and

   $\lim_{t\to\infty}Me^{t(a-s)} = 0$ for $s > a$. Therefore by the Squeeze Theorem,

   $\lim_{t\to\infty}f(t)e^{-st} = 0$ for $s > a \Rightarrow G(s) = 0 - f(0) + sF(s) = sF(s) - f(0)$ for $s > a$.

75. Use integration by parts: let $u = x$, $dv = xe^{-x^2}\,dx \Rightarrow du = dx, v = -\frac{1}{2}e^{-x^2}$

   $\int_0^\infty x^2 e^{-x^2}\,dx = \lim_{t\to\infty}\left[-\frac{x}{2}e^{-x^2}\right]_0^t + \frac{1}{2}\int_0^\infty e^{-x^2}\,dx = \lim_{t\to\infty}-\frac{t}{2e^{t^2}} + \frac{1}{2}\int_0^\infty e^{-x^2}\,dx$

   $= \frac{1}{2}\int_0^\infty e^{-x^2}\,dx$. (The limit is 0 by l'Hospital's Rule)

77. For the first part of the integral, let $x = 2\tan\theta \Rightarrow dx = 2\sec^2\theta\,d\theta$. $\int \frac{1}{\sqrt{x^2+4}}\,dx$

   $= \int \sec\theta = \ln|\sec\theta + \tan\theta|$. But $\tan\theta = \frac{x}{2}$, and $\sec\theta = \sqrt{1+\tan^2\theta} = \sqrt{1+\frac{x^2}{4}} = \frac{\sqrt{x^2+4}}{2}$

   So $\int_0^\infty \left(\frac{1}{\sqrt{x^2+4}} - \frac{C}{x+2}\right)dx = \lim_{t\to\infty}\left[\ln\left|\frac{\sqrt{x^2+4}}{2} + \frac{x}{2}\right| - C\ln|x+2|\right]_0^t$

   $= \lim_{t\to\infty}\ln\left(\frac{\sqrt{t^2+4}+t}{2(t+2)^C}\right) - (\ln 1 - C\ln 2) = \ln\left(\lim_{t\to\infty}\frac{t+\sqrt{t^2+4}}{(t+2)^C}\right) + \ln 2^{C-1}$.

By l'Hospital's Rule, $\lim\limits_{t\to\infty} \dfrac{t+\sqrt{t^2+4}}{(t+2)^C} = \lim\limits_{t\to\infty} \dfrac{1+t/\sqrt{t^2+4}}{C(t+2)^{C-1}} = \dfrac{2}{C\lim\limits_{t\to\infty}(t+2)^{C-1}}.$

If $C < 1$, we get $+\infty$ and the interval diverges. If $C = 1$, we get 2, so the original integral converges to $\ln 2 + \ln 2^0 = \ln 2$. If $C > 1$, we get 0, so the original integral diverges to $-\infty$.

REVIEW EXERCISES FOR CHAPTER 7

1. False. Since the numerator has a higher degree than the denominator,
$$\frac{x(x^2+4)}{x^2-4} = x + \frac{8x}{x^2-4} = x + \frac{A}{x+2} + \frac{B}{x-2}.$$

3. False. It can be put in the form $\frac{A}{x} + \frac{B}{x^2} + \frac{C}{x-4}$.

5. False. This is an improper integral, since the denominator vanishes at $x = 1$.

$$\int_0^4 \frac{x}{x^2-1}\,dx = \int_0^1 \frac{x}{x^2-1}\,dx + \int_1^4 \frac{x}{x^2-1}\,dx \text{ and } \int_0^1 \frac{x}{x^2-1} = \lim_{t\to1^-}\int_0^t \frac{x}{x^2-1}\,dx$$

$$= \lim_{t\to1^-}[\tfrac{1}{2}\ln|x^2-1|]_0^t = \lim_{t\to1^-}\tfrac{1}{2}\ln|t^2-1| = \infty.$$ Since this integral is infinite it diverges and so does the improper integral diverge.

7. False. See Section 7.9 #65.

9. $\displaystyle\int \frac{x-1}{x+1}\,dx = \int \left[1 - \frac{2}{x+1}\right]dx = x - 2\ln|x+1| + C.$

11. Let $u = \arctan x$. Then $du = dx/(1+x^2)$, so $\displaystyle\int \frac{(\arctan x)^5}{1+x^2}\,dx = \int u^5\,du$

$= \tfrac{1}{6}u^6 + C = \tfrac{1}{6}(\arctan x)^6 + C.$

13. Let $u = \sin x$. Then $\displaystyle\int \frac{\cos x\,dx}{e^{\sin x}} = \int e^{-u}\,du = -e^{-u} + C = \dfrac{-1}{e^{\sin x}} + C.$

15. Use integration by parts with $u = \ln x$, $dv = x^4\,dx \Rightarrow du = dx/x$, $v = x^5/5$:
$\displaystyle\int x^4 \ln x\,dx = \frac{x^5}{5}\ln x - \frac{1}{5}\int x^4\,dx = \frac{x^5}{5}\ln x - \frac{x^5}{25} + C = \frac{1}{25}x^5(5\ln x - 1) + C.$

17. Let $u = x^2$. Then $du = 2x\,dx$, so $\int x\sin(x^2)\,dx = (1/2)\int \sin u\,du$
$= -(1/2)\cos u + C = -(1/2)\cos(x^2) + C.$

19. $\displaystyle\int \frac{dx}{2x^2-5x+2} = \int \left[\frac{-2/3}{2x-1} + \frac{1/3}{x-2}\right]dx$
$= -\tfrac{1}{3}\ln|2x-1| + \tfrac{1}{3}\ln|x-2| + C = \tfrac{1}{3}\ln\left|\dfrac{x-2}{2x-1}\right| + C.$

21. Let $u = \sec x$. Then $du = \sec x \tan x \, dx$, so $\int \tan^7 x \sec^3 x \, dx$

$= \int \tan^6 x \sec^2 x \sec x \tan x \, dx = \int (u^2 - 1)^3 u^2 \, du = \int (u^8 - 3u^6 + 3u^4 - u^2) \, du$

$= u^9/9 - 3u^7/7 + 3u^5/5 - u^3/3 + C = \frac{1}{9}\sec^9 x - \frac{3}{7}\sec^7 x + \frac{3}{5}\sec^5 x - \frac{1}{3}\sec^3 x + C.$

23. Let $u = \sqrt{1 + 2x}$. Then $x = \frac{1}{2}(u^2 - 1)$, $dx = u \, du$, so $\displaystyle\int \frac{dx}{\sqrt{1+2x}+3} = \int \frac{u \, du}{u+3}$

$= \int \left[1 - \frac{3}{u+3}\right] du = u - 3\ln|u+3| + C = \sqrt{1+2x} - 3\ln(\sqrt{1+2x}+3) + C.$

25. $u = \sqrt{x} \Rightarrow du = \dfrac{dx}{2\sqrt{x}} \Rightarrow \displaystyle\int \frac{e^{\sqrt{x}}dx}{\sqrt{x}} = 2\int e^u \, du = 2e^u + C = 2e^{\sqrt{x}} + C.$

27. Let $u = x + 1$. Then $\int \ln(x^2 + 2x + 2) \, dx = \int \ln(u^2 + 1) \, du$

$= u\ln(u^2 + 1) - \displaystyle\int u \cdot \frac{2u \, du}{u^2 + 1}$ [integration by parts] $= u\ln(u^2 + 1) - 2\int \left[1 - \frac{1}{u^2 + 1}\right] du$

$= u\ln(u^2 + 1) - 2u + 2\tan^{-1}u + C$

$= (x+1)\ln(x^2 + 2x + 2) - 2(x+1) + 2\tan^{-1}(x+1) + C$

$= (x+1)\ln(x^2 + 2x + 2) - 2x + 2\tan^{-1}(x+1) + C_1.$

[OR: Integrate by parts with $u = \ln(x^2 + 2x + 2)$, $dv = dx$.]

29. $\displaystyle\int \frac{dx}{x^3 + x} = \int \left[\frac{1}{x} - \frac{x}{x^2 + 1}\right] dx = \ln|x| - \frac{1}{2}\ln(x^2 + 1) + C.$

31. $\int \cot^2 x \, dx = \int (\csc^2 x - 1) \, dx = -\cot x - x + C.$

33. Let $x = \sec\theta$. Then $\displaystyle\int \frac{dx}{(x^2 - 1)^{3/2}} = \int \frac{\sec\theta \tan\theta}{\tan^3 \theta} \, d\theta = \int \frac{\sec\theta}{\tan^2 \theta} \, d\theta = \int \frac{\cos\theta \, d\theta}{\sin^2 \theta}$

$= -\dfrac{1}{\sin\theta} + C = -\dfrac{x}{\sqrt{x^2 - 1}} + C.$

35. $\displaystyle\int \frac{2x^2 + 3x + 11}{x^3 + x^2 + 3x - 5} \, dx = \int \left[\frac{2}{x-1} - \frac{1}{x^2 + 2x + 5}\right] dx = 2\ln|x - 1| - \int \frac{dx}{(x+1)^2 + 4}$

$= 2\ln|x - 1| - \frac{1}{2}\tan^{-1}\left(\frac{x+1}{2}\right) + C.$

37. Let $u = \cot 4x$. Then $du = -4\csc^2 4x \, dx \Rightarrow \int \csc^4 4x \, dx = \int (\cot^2 4x + 1) \csc^2 4x \, dx$

$= \int (u^2 + 1)(-\frac{1}{4} \, du) = -\frac{1}{4}(u^3/3 + u) + C = -\frac{1}{12}(\cot^3 4x + 3\cot 4x) + C.$

39. Let $u = \ln x$. Then $\displaystyle\int \frac{\ln(\ln x)}{x} \, dx = \int \ln u \, du$. Now use parts with $w = \ln u$,

$dv = du \Rightarrow dw = \frac{1}{u} \, du$, $v = u \Rightarrow \int \ln u \, du = u\ln u - u + C = (\ln x)[\ln(\ln x) - 1] + C.$

41. Let $u = 2x + 1$. Then $du = 2 \, dx \Rightarrow \displaystyle\int \frac{dx}{\sqrt{4x^2 + 4x + 5}} = \int \frac{(1/2) \, du}{\sqrt{u^2 + 4}} = \frac{1}{2}\int \frac{2\sec^2\theta \, d\theta}{2\sec\theta}$

$[u = 2\tan\theta,\ du = 2\sec^2\theta \, d\theta] = \frac{1}{2}\displaystyle\int \sec\theta \, d\theta = \frac{1}{2}\ln|\sec\theta + \tan\theta| + C_1$

$$= \tfrac{1}{2}\ln\left|\frac{\sqrt{u^2+4}}{2}+\frac{u}{2}\right| + C_1 = \tfrac{1}{2}\ln\left(u+\sqrt{u^2+4}\right) + C = \tfrac{1}{2}\ln\left(2x+1+\sqrt{4x^2+4x+5}\right) + C.$$

43. $\int (\cos x + \sin x)^2 \cos 2x \, dx = \int (\cos^2 x + 2\sin x \cos x + \sin^2 x)\cos 2x \, dx = \int (1 + \sin 2x) \, \cos 2x \, dx.$

$= \int \cos 2x \, dx + \tfrac{1}{2}\int \sin 4x \, dx = \tfrac{1}{2}\sin 2x - \tfrac{1}{8}\cos 4x + C.$

[OR: $\int(\cos x + \sin x)^2 \cos 2x \, dx = \int(\cos x + \sin x)^2(\cos^2 x - \sin^2 x)\, dx$

$= \int(\cos x + \sin x)^3(\cos x - \sin x)\, dx = \tfrac{1}{4}(\cos x + \sin x)^4 + C_2.]$

45. $\displaystyle\int_0^{\pi/2} \cos^3 x \sin 2x \, dx = \int_0^{\pi/2} 2\cos^4 x \sin x \, dx = \left[-\tfrac{2}{5}\cos^5 x\right]_0^{\pi/2} = \tfrac{2}{5}.$

47. $\displaystyle\int_0^3 \frac{dx}{x^2 - x - 2} = \int_0^3 \frac{dx}{(x+1)(x-2)} = \int_0^2 \frac{dx}{(x+1)(x-2)} + \int_2^3 \frac{dx}{(x+1)(x-2)}$

$\displaystyle\int_2^3 \frac{dx}{x^2 - x - 2} = \lim_{t\to 2^+} \int_t^3 \left[\frac{-1/3}{x+1} + \frac{1/3}{x-2}\right] dx = \lim_{t\to 2^+} \left[\tfrac{1}{3}\ln\left|\frac{x-2}{x+1}\right|\right]_t^3$

$\displaystyle = \lim_{t\to 2^+} \left[\tfrac{1}{3}\ln\tfrac{1}{4} - \tfrac{1}{3}\ln\left|\frac{t-2}{t+1}\right|\right] = \infty.$ Divergent.

49. $\displaystyle\int_0^1 \frac{t^2 - 1}{t^2 + 1}\, dt = \int_0^1 \left[1 - \frac{2}{t^2 + 1}\right] dt = \left[t - 2\tan^{-1} t\right]_0^1 = \left(1 - 2\cdot\tfrac{\pi}{4}\right) - 0 = 1 - \tfrac{\pi}{2}.$

51. $\displaystyle\int_0^\infty \frac{dx}{(x+2)^4} = \lim_{t\to\infty} \left[\frac{-1}{3(x+2)^3}\right]_0^t = \lim_{t\to\infty}\left[\frac{1}{3\cdot 2^3} - \frac{1}{3(t+2)^3}\right] = \tfrac{1}{24}.$

53. Let $u = \ln x$. Then $\displaystyle\int_1^e \frac{dx}{x\sqrt{\ln x}} = \lim_{t\to 1^+} \int_t^e \frac{dx}{x\sqrt{\ln x}} = \lim_{t\to 1^+} \int_{\ln t}^1 \frac{du}{\sqrt{u}} = \lim_{t\to 1^+} \left[2\sqrt{u}\right]_{\ln t}^1$

$\displaystyle = \lim_{t\to 1^+} \left(2 - 2\sqrt{\ln t}\right) = 2.$

55. Let $u = \sqrt{x} + 2$. Then $x = (u - 2)^2$, $dx = 2(u - 2)\, du$, so $\displaystyle\int_1^4 \frac{\sqrt{x}\, dx}{\sqrt{x} + 2} = \int_3^4 \frac{2(u-2)^2\, du}{u}$

$\displaystyle = \int_3^4 \left[2u - 8 + \tfrac{8}{u}\right] du = \left[u^2 - 8u + 8\ln u\right]_3^4 = (16 - 32 + 8\ln 4) - (9 - 24 + 8\ln 3)$

$= -1 + 8\ln 4 - 8\ln 3 = 8\ln(4/3) - 1.$

57. Let $u = 2x + 1$. Then $\displaystyle\int_{-\infty}^\infty \frac{dx}{4x^2 + 4x + 5} = \int_{-\infty}^\infty \frac{(1/2)\, du}{u^2 + 4}$

$\displaystyle = \tfrac{1}{2}\int_{-\infty}^0 \frac{du}{u^2 + 4} + \tfrac{1}{2}\int_0^\infty \frac{du}{u^2 + 4}$

$\displaystyle = \tfrac{1}{2}\lim_{t\to -\infty} \left[\tfrac{1}{2}\tan^{-1}\left(\tfrac{u}{2}\right)\right]_t^0 + \tfrac{1}{2}\lim_{t\to\infty} \left[\tfrac{1}{2}\tan^{-1}\left(\tfrac{u}{2}\right)\right]_0^t$

$\displaystyle = \tfrac{1}{4}\left[0 - \tfrac{-\pi}{2}\right] + \tfrac{1}{4}\left[\tfrac{\pi}{2} - 0\right] = \tfrac{\pi}{4}.$

59. Let $x = \sec\theta$. Then $\displaystyle\int_1^2 \frac{\sqrt{x^2 - 1}}{x}\, dx = \int_0^{\pi/3} \frac{\tan\theta}{\sec\theta}\sec\theta\tan\theta\, d\theta = \int_0^{\pi/3} \tan^2\theta\, d\theta$

$\displaystyle = \int_0^{\pi/3} (\sec^2\theta - 1)\, d\theta = \left[\tan\theta - \theta\right]_0^{\pi/3} = \sqrt{3} - \tfrac{\pi}{3}.$

61. $\int_0^\infty e^{ax}\cos bx\,dx = \lim_{t\to\infty}\int_0^t e^{ax}\cos bx\,dx.$ Integrate by parts twice: $\int e^{ax}\cos bx\,dx$

$= \frac{1}{b}e^{ax}\sin bx - \frac{a}{b}\int e^{ax}\sin bx\,dx = \frac{1}{b}e^{ax}\sin bx + \frac{a}{b^2}e^{ax}\cos bx - \frac{a^2}{b^2}\int e^{ax}\cos bx\,dx,$ so

$\left(1+\frac{a^2}{b^2}\right)\int e^{ax}\cos bx\,dx = \frac{1}{b}e^{ax}\sin bx + \frac{a}{b^2}e^{ax}\cos bx + C_1.$

Thus $\int e^{ax}\cos bx\,dx = \frac{e^{ax}}{a^2+b^2}(b\sin bx + a\cos bx) + C.$ Now $\int_0^\infty e^{ax}\cos bx\,dx$

$= \lim_{t\to\infty}\left[\frac{e^{ax}}{a^2+b^2}(a\cos bx + b\sin bx)\right]_0^t = \lim_{t\to\infty}\frac{e^{at}}{a^2+b^2}(a\cos bt + b\sin bt) - \frac{a}{a^2+b^2}.$ If

$a \geq 0$, the limit does not exist and the integral is divergent. If $a < 0$, the limit is 0
(since $\left|e^{at}\cos bt\right| \leq e^{at}$ and $\left|e^{at}\sin bt\right| \leq e^{at}$), so the integral converges
to $-a/(a^2+b^2)$.

63. Let $u = e^x$. Then $du = e^x\,dx$, so $\int e^x\sqrt{1-e^{2x}}\,dx = \int\sqrt{1-u^2}\,du$

$= \frac{u}{2}\sqrt{1-u^2} + \frac{1}{2}\sin^{-1}u + C$ [*Formula 30*] $= \frac{1}{2}\left[e^x\sqrt{1-e^{2x}} + \sin^{-1}(e^x)\right] + C.$

65. Let $u = x + \frac{1}{2}$. Then $du = dx$, so $\int\sqrt{x^2+x+1}\,dx = \int\sqrt{(x+1/2)^2+3/4}\,dx$

$= \int\sqrt{u^2+(\sqrt{3}/2)^2}\,du = \frac{u}{2}\sqrt{u^2+3/4} + \frac{3}{8}\ln\left|u+\sqrt{u^2+3/4}\right| + C$ [*Formula 21*]

$= \frac{2x+1}{4}\sqrt{x^2+x+1} + \frac{3}{8}\ln\left|x+\frac{1}{2}+\sqrt{x^2+x+1}\right| + C.$

67. (a) $\frac{d}{du}\left(-\frac{1}{u}\sqrt{a^2-u^2} - \sin^{-1}\left(\frac{u}{a}\right) + C\right) = \frac{1}{u^2}\sqrt{a^2-u^2} + \frac{1}{\sqrt{a^2-u^2}} - \frac{1}{\sqrt{1-\frac{u^2}{a^2}}}\cdot\frac{1}{a}$

$= (a^2-u^2)^{-1/2}\left[\frac{1}{u^2}(a^2-u^2) + 1 - 1\right] = \frac{\sqrt{a^2-u^2}}{u^2}.$

(b) Let $u = a\sin\theta \Rightarrow du = a\cos\theta\,d\theta$, $a^2-u^2 = a^2(1-\sin^2\theta) = a^2\cos^2\theta$

$\int\frac{\sqrt{a^2-u^2}}{u^2}\,du = \int\frac{a^2\cos^2\theta}{a^2\sin^2\theta}\,d\theta = \int\frac{1-\sin^2\theta}{\sin^2\theta}\,d\theta$

$= \int(\csc^2\theta - 1)\,d\theta = -\cot\theta - \theta + C = -\frac{\sqrt{a^2-u^2}}{u} - \sin^{-1}\left(\frac{u}{a}\right) + C.$

69. $f(x) = \sqrt{1+x^4}$, $\Delta x = (b-a)/n = (1-0)/10 = 1/10.$

(a) $T_{10} = \frac{1}{2}[f(0) + 2f(.1) + 2f(.2) + \cdots + 2f(.8) + 2f(.9) + f(1)] \approx 1.090608$

(b) $M_{10} = .1[f(1/20) + f(3/20) + f(5/20) + \cdots + f(19/20)] \approx 1.088840$

(c) $S_{10} = \frac{1}{3}[f(0) + 4f(.1) + 2f(.2) + 4f(.3) + 2f(.4) + 4f(.5) + 2f(.6) + 4f(.7)$
$+ 2f(.8) + 4f(.9) + f(1)] \approx 1.089429$

71. $f(x) = (1+x^4)^{1/2}$, $f'(x) = (1/2)(1+x^4)^{-1/2}(4x^3) = 2x^3(1+x^4)^{-1/2},$

278

$f''(x) = (2x^6 + 6x^2)(1 + x^4)^{-3/2}$. Thus $|f''(x)| \le 8 \cdot 1^{-3/2} = 8$ on $[0, 1]$. By taking $M = 8$, we find that the error in #69 (a) is bounded by $M(b - a)^3/(12n^2) = 8/1200 = 1/150 < .0067$ and in (b) by $M(b - a)^3/(24n^2) = 1/300 = .00\overline{3}$.

73. $\dfrac{x^3}{x^5 + 2} \le \dfrac{x^3}{x^5} = \dfrac{1}{x^2}$ for x in $[1, \infty)$. $\displaystyle\int_1^\infty \dfrac{1}{x^2} dx$ is convergent by (7.36) with $p = 2 > 1$.

Therefore $\displaystyle\int_1^\infty \dfrac{x^3}{x^5 + 2} dx$ is convergent by the Comparison Theorem.

75. For x in $[0, \pi/2]$, $0 \le \cos^2 x \le \cos x$. For x in $[\pi/2, \pi]$, $\cos x \le 0 \le \cos^2 x$. Thus area

$= \displaystyle\int_0^{\pi/2} (\cos x - \cos^2 x) dx + \int_{\pi/2}^{\pi} (\cos^2 x - \cos x) dx$

$= \left[\sin x - \dfrac{x}{2} - \dfrac{1}{4}\sin 2x \right]_0^{\pi/2} + \left[\dfrac{x}{2} + \dfrac{1}{4}\sin 2x - \sin x \right]_{\pi/2}^{\pi}$

$= \left[\left(1 - \dfrac{\pi}{4}\right) - 0 \right] + \left[\dfrac{\pi}{2} - \left(\dfrac{\pi}{4} - 1\right) \right] = 2$.

77. For $n \ge 0$, $\displaystyle\int_0^\infty x^n dx = \lim_{t \to \infty} \left[x^{n+1}/(n+1) \right]_0^t = \infty$. For $n < 0$, $\displaystyle\int_0^\infty x^n dx$

$= \displaystyle\int_0^1 x^n dx + \int_1^\infty x^n dx$. Both integrals are improper. By (7.36) the second integral

diverges if $-1 \le n < 0$. By Ex. 61 of Section 7.9, the first integral diverges if $n \le -1$.

Thus $\displaystyle\int_0^\infty x^n dx$ is divergent for all values of n.

79. By the Fundamental Theorem of Calculus, $\displaystyle\int_0^\infty f'(x) dx = \lim_{t \to \infty} \int_0^t f'(x) dx$

$= \displaystyle\lim_{t \to \infty} [f(t) - f(0)] = \lim_{t \to \infty} f(t) - f(0) = 0 - f(0) = -f(0)$.

81. Let $u = \dfrac{1}{x} \Rightarrow x = 1/u \Rightarrow dx = -(1/u^2) du$. $\displaystyle\int_0^\infty \dfrac{\ln x}{1 + x^2} dx = \int_\infty^0 \dfrac{\ln\left(\frac{1}{u}\right)}{1 + \frac{1}{u^2}} \left(-\dfrac{du}{u^2}\right)$

$= \displaystyle\int_\infty^0 \dfrac{-\ln u}{u^2 + 1}(-du) = \int_\infty^0 \dfrac{\ln u}{1 + u^2} du = -\int_0^\infty \dfrac{\ln u}{1 + u^2} du$.

Therefore $\displaystyle\int_0^\infty \dfrac{\ln x}{1 + x^2} dx = -\int_0^\infty \dfrac{\ln x}{1 + x^2} dx = 0$.

Applications Plus (page 468)

1. (a) By Formula 6.7, $V = \int_0^h \pi [f(y)]^2 dy$.

    (b) $\frac{dV}{dt} = \frac{dV}{dh} \cdot \frac{dh}{dt} = \pi [f(h)]^2 \frac{dh}{dt}$

    (c) $kA\sqrt{h} = \pi [f(h)]^2 \frac{dh}{dt}$. Set $\frac{dh}{dt} = C$: $\pi [f(h)]^2 C = kA\sqrt{h} \Rightarrow [f(h)]^2 = \frac{kA}{\pi C}\sqrt{h} \Rightarrow$

    $f(h) = \sqrt{\frac{kA}{\pi C}} h^{1/4}$, i.e., $f(y) = \sqrt{\frac{kA}{\pi C}} y^{1/4}$.

    Advantage: The markings on the container will be equally spaced.

3. (a) The tangent to the curve $y = f(x)$ at $x = x_0$ has the equation $y - f(x_0) = f'(x_0)(x - x_0)$.

    The y-intercept of this tangent line is $f(x_0) - f'(x_0) x_0$. Thus L is the distance from the

    point $(0, f(x_0) - f'(x_0) x_0)$ to the point $(x_0, f'(x_0))$. That is, $L^2 = x_0^2 + [f'(x_0)]^2 x_0^2$, so

    $[f'(x_0)]^2 = \frac{L^2 - x_0^2}{x_0^2}$ and $f'(x_0) = -\frac{\sqrt{L^2 - x_0^2}}{x_0}$ for each $0 < x_0 < L$.

    (b) $\frac{dy}{dx} = -\frac{\sqrt{L^2 - x^2}}{x} \Rightarrow y = \int -\frac{\sqrt{L^2 - x^2}}{x} dx = \int \frac{-L\cos\theta\, L\cos\theta\, d\theta}{L\sin\theta}$ $\quad \begin{bmatrix} where \\ x = L\sin\theta \end{bmatrix}$

    $= L \int \frac{\sin^2\theta - 1}{\sin\theta} d\theta = L \int (\sin\theta - \csc\theta)\, d\theta = -L\cos\theta + L\ln|\csc\theta + \cot\theta| + C$

    $= -\sqrt{L^2 - x^2} + L\ln\left(\frac{L}{x} + \frac{\sqrt{L^2 - x^2}}{x}\right) + C$. When $x = L$, $0 = y = -0 + L\ln(1 + 0) + C$,

    so $C = 0$. Therefore $y = -\sqrt{L^2 - x^2} + L\ln\left(\frac{L + \sqrt{L^2 - x^2}}{x}\right)$.

5. (a) The volume above the surface is $\int_0^{L-h} A(y)dy = \int_{-h}^{L-h} A(y)dy - \int_{-h}^0 A(y)dy$. So the

    proportion of volume above the surface is $\dfrac{\int_0^{L-h} A(y)dy}{\int_{-h}^{L-h} A(y)dy} = \dfrac{\int_{-h}^{L-h} A(y)dy - \int_{-h}^0 A(y)dy}{\int_{-h}^{L-h} A(y)dy}$.

    Now by Archimedes' Principle, we have $\rho_f g \int_{-h}^0 A(y)dy = \rho_0 g \int_{-h}^{L-h} A(y)dy$, so

280

$$\int_{-h}^{0} A(y)\,dy = \frac{\rho_0 \displaystyle\int_{-h}^{L-h} A(y)\,dy}{\rho_f}$$

Therefore, $\dfrac{\displaystyle\int_{0}^{L-h} A(y)\,dy}{\displaystyle\int_{-h}^{L-h} A(y)\,dy} = \dfrac{\displaystyle\int_{-h}^{L-h} A(y)\,dy - \dfrac{\rho_0 \displaystyle\int_{-h}^{L-h} A(y)\,dy}{\rho_f}}{\displaystyle\int_{-h}^{L-h} A(y)\,dy} = \dfrac{\rho_f - \rho_0}{\rho_f}$

So the percentage of volume above the surface is $100\dfrac{\rho_f - \rho_0}{\rho_f}$.

(b) For an iceberg, the percentage of volume above the surface is $100\left(\dfrac{1030 - 917}{1030}\right)\% \approx 11\%$.

(c) The figure shows the instant when the height of the

exposed part of the ball is y. Using the formula in

Problem 2(a) with $r = 0.4$ and $h = 0.8 - y$, we see

that the volume of the submerged part of the

sphere is $\frac{1}{3}\pi(.8 - y)^2[1.2 - (.8 - y)]$

so its weight is $\dfrac{1000\pi}{3} g(.8 - y)^2(.4 + y)$.

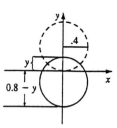

Let $s = .8 - y$. Then the work done to submerge the sphere is

$$W = \int_{0}^{.8} g\frac{1000\pi}{3} s^2(1.2 - s)\,ds = \frac{1000\pi}{3} g \int_{0}^{.8} (1.2\,s^2 - s^3)\,ds = \frac{1000\pi}{3} g\left(\frac{1.2 s^3}{3} - \frac{s^4}{4}\right)\Big|_{0}^{.8}$$

$$= g\frac{1000\pi}{3}\left(.4 s^3 - \frac{s^4}{4}\right)\Big|_{0}^{.8} = g\frac{1000\pi}{3}(.2048 - .1024) = 9.8\frac{1000\pi}{3}(.1024) \approx 1.05 \times 10^3 \text{ joules.}$$

Chapter 8

EXERCISES 8.1

1. $\frac{dy}{dx} = y^2 \Rightarrow \frac{dy}{y^2} = dx \; (y \neq 0) \Rightarrow \int \frac{dy}{y^2} = \int dx \Rightarrow -\frac{1}{y} = x + C \Rightarrow -y = \frac{1}{x+C} \Rightarrow$

   $y = \frac{-1}{x+C}$ and $y = 0$ is also a solution.

3. $yy' = x \Rightarrow \int y \, dy = \int x \, dx \Rightarrow \frac{y^2}{2} = \frac{x^2}{2} + C_1 \Rightarrow y^2 = x^2 + 2C_1 \Rightarrow x^2 - y^2 = C$

   $[C = -2C_1]$. This represents a family of hyperbolas.

5. $x^2 y' + y = 0 \Rightarrow \frac{dy}{dx} = \frac{-y}{x^2} \Rightarrow \int \frac{dy}{y} = \int \frac{-dx}{x^2} \; (y \neq 0) \Rightarrow \ln|y| = \frac{1}{x} + K \Rightarrow |y| = e^K e^{1/x} \Rightarrow$

   $y = Ce^{1/x}$, where now we allow C to be any constant.

7. $\frac{dy}{dx} = \frac{x\sqrt{x^2+1}}{ye^y} \Rightarrow \int ye^y \, dy = \int x\sqrt{x^2+1} \, dx \Rightarrow (y-1)e^y = \frac{1}{3}(x^2+1)^{3/2} + C.$

9. $\frac{du}{dt} = e^{u+2t} = e^u e^{2t} \Rightarrow \int e^{-u} \, du = \int e^{2t} \, dt \Rightarrow -e^{-u} = \frac{1}{2}e^{2t} + C_1 \Rightarrow e^{-u} = -\frac{1}{2}e^{2t} + C$

   [where $C = -C_1$ and the right-hand side is positive, since $e^{-u} > 0$] $\Rightarrow -u = \ln\left(C - \frac{1}{2}e^{2t}\right)$

   $\Rightarrow u = -\ln\left(C - \frac{1}{2}e^{2t}\right).$

11. $\frac{dy}{dx} = y^2 + 1$, $y(1) = 0$. $\int \frac{dy}{y^2+1} = \int dx \Rightarrow \tan^{-1}y = x + C$. $y = 0$ when $x = 1$,

    so $1 + C = \tan^{-1}0 = 0$ and $C = -1$. Thus $\tan^{-1}y = x - 1$ and $y = \tan(x-1)$.

13. $e^y y' = \frac{3x^2}{1+y}$, $y(2) = 0$. $\int e^y(1+y) \, dy = \int 3x^2 \, dx \Rightarrow ye^y = x^3 + C$. $y(2) = 0$

    so $0 = 2^3 + C$ and $C = -8$. Thus $ye^y = x^3 - 8$.

15. $\frac{dy}{dx} = \frac{\sin x}{\sin y}$, $y(0) = \frac{\pi}{2}$. $\int \sin y \, dy = \int \sin x \, dx \Rightarrow -\cos y = -\cos x + C$. $y(0) = \frac{\pi}{2}$

    so $-\cos \frac{\pi}{2} = -\cos 0 + C$ and $C = 1$. Thus $-\cos y = -\cos x + 1$ and $\cos y = \cos x - 1$.

17. $xe^{-t}\frac{dx}{dt} = t$, $x(0) = 1$. $\int x \, dx = \int te^t \, dt \Rightarrow \frac{x^2}{2} = (t-1)e^t + C$. $x(0) = 1$

    so $\frac{1}{2} = (0-1)e^0 + C$ and $C = \frac{3}{2}$. Thus $x^2 = 2(t-1)e^t + 3 \Rightarrow x = \sqrt{2(t-1)e^t + 3}$.

19. $\frac{du}{dt} = \frac{2t+1}{2(u-1)}$, $u(0) = -1$. $\int 2(u-1) \, du = \int (2t+1) \, dt \Rightarrow u^2 - 2u = t^2 + t + C$.

    $u(0) = -1$ so $(-1)^2 - 2(-1) = 0^2 + 0 + C$ and $C = 3$. Thus $u^2 - 2u = t^2 + t + 3$;

    the quadratic formula gives $u = 1 - \sqrt{t^2 + t + 4}$.

21. Let $y = f(x)$. Then $\frac{dy}{dx} = x^3 y$ and $y(0) = 1$. $\frac{dy}{y} = x^3\,dx$ $(y \neq 0)$, so $\int \frac{dy}{y} = \int x^3\,dx$ and

$\ln|y| = \frac{x^4}{4} + C$; $y(0) = 1 \Rightarrow C = 0$, so $\ln|y| = \frac{x^4}{4}$, $|y| = e^{x^4/4}$ and $y = f(x) = e^{x^4/4}$

[*since* $y(0) = 1$].

23. $\frac{dy}{dx} = 4x^3 y$, $y(0) = 7$. $\frac{dy}{y} = 4x^3\,dx$ $[y \neq 0] \Rightarrow \int \frac{dy}{y} = \int 4x^3\,dx \Rightarrow \ln|y| = x^4 + C$

$\Rightarrow y = Ae^{x^4}$; $y(0) = 7 \Rightarrow A = 7 \Rightarrow y = 7e^{x^4}$.

25. (a) Let $y(t)$ be the amount of salt (in kg) after t minutes. Then $y(0) = 15$. The amount of

liquid in the tank is $1000\,L$ at all times, so the concentration at time t (in minutes) is $y(t)/1000$

kg/L and $\frac{dy}{dt} = -\left(\frac{y(t)\,\text{kg}}{1000\,\text{L}}\right)\left(10\,\frac{\text{L}}{\text{min}}\right) = -\frac{y(t)}{100}\,\frac{\text{kg}}{\text{min}}$. $\int \frac{dy}{y} = -\frac{1}{100}\int dt \Rightarrow \ln y = -\frac{t}{100} + C$.

$y(0) = 15 \Rightarrow \ln 15 = C$, so $\ln y = \ln 15 - \frac{t}{100}$. It follows that $\ln\left(\frac{y}{15}\right) = -\frac{t}{100}$ and

$\frac{y}{15} = e^{-t/100}$, so $y = 15e^{-t/100}$ kg.

(b) After $20\,\text{min.}$, $y = 15e^{-20/100} = 15e^{-.2} \approx 12.3\,\text{kg}$.

27. $\frac{dx}{dt} = k(a-x)(b-x)$, $a \neq b$. $\int \frac{dx}{(a-x)(b-x)} = \int k\,dt \Rightarrow \frac{1}{b-a}\int \left(\frac{1}{a-x} - \frac{1}{b-x}\right) dx$

$= \int k\,dt \Rightarrow \frac{1}{b-a}(-\ln|a-x| + \ln|b-x|) = kt + C \Rightarrow \ln\left|\frac{b-x}{a-x}\right| = (b-a)(kt+C)$. Here

the concentrations $[A] = a - x$ and $[B] = b - x$ cannot be negative, so $\frac{b-x}{a-x} \geq 0$ and

$\left|\frac{b-x}{a-x}\right| = \frac{b-x}{a-x}$. We now have $\ln\left(\frac{b-x}{a-x}\right) = (b-a)(kt+C)$. Since $x(0) = 0$,

$\ln\left(\frac{b}{a}\right) = (b-a)C$. Hence $\ln\left(\frac{b-x}{a-x}\right) = (b-a)kt + \ln\left(\frac{b}{a}\right)$, $\frac{b-x}{a-x} = \frac{b}{a}e^{(b-a)kt}$, and

$x = \frac{b(e^{(b-a)kt} - 1)}{\frac{b}{a}e^{(b-a)kt} - 1} = \frac{ab(e^{(b-a)kt} - 1)}{be^{(b-a)kt} - a}\,\frac{\text{moles}}{\text{L}}$.

29. Let $P(t)$ be the world population in the year t. Then $\frac{dP}{dt} = .02P$, so $\int \frac{dP}{P} = \int .02\,dt$ and

$\ln P = .02t + C \Rightarrow P(t) = Ae^{.02t}$. $P(1986) = 5 \times 10^9 \Rightarrow P(t) = 5 \times 10^9 e^{.02(t-1986)}$.

(a) The population in 2000 will be approximately $P(2000) = 5e^{.28} \times 10^9 \approx 6.6$ billion $\Rightarrow$

number of $\text{ft}^2/\text{person} = 1.8 \times 10^{15}/6.6 \times 10^9 \approx 270{,}000$.

(b) The predicted population for the year 2100 is $P(2100) = 5e^{2.28} \times 10^9 \approx 49$ billion $\Rightarrow$

$1.8 \times 10^{15}/49 \times 10^9 \approx 37{,}000\ \text{ft}^2/\text{person}$.

(c) The prediction for 2500 is $P(2500) = 5e^{10.28} \times 10^9 \approx 146\,\text{trillion} \Rightarrow$

$1.8 \times 10^{15}/146 \times 10^{12} \approx 12\ \text{ft}^2/\text{person}$.

31. (a) Our assumption is that $\frac{dy}{dt} = ky(1-y)$, where y is the fraction of the population that has

heard the rumor.

(b) Take M $= 1$ in the solution of (8.51) to get $y = \dfrac{y_0}{y_0 + (1 - y_0)e^{-kt}}$.

(c) Let t be the number of hours since 8 A.M. Then $y_0 = y(0) = \dfrac{80}{1000} = .08$ and $y(4) = \frac{1}{2}$, so

$\frac{1}{2} = y(4) = \dfrac{.08}{.08 + .92e^{-4k}}$. Thus $.08 + .92e^{-4k} = .16$, $e^{-4k} = \frac{.08}{.92} = \frac{2}{23}$, and $e^{-k} = \left(\frac{2}{23}\right)^{1/4}$,

so $y = \dfrac{.08}{.08 + .92(2/23)^{t/4}} = \dfrac{2}{2 + 23(2/23)^{t/4}}$ and $\left(\frac{2}{23}\right)^{t/4} = \frac{2}{23}\frac{1-y}{y}$ or $\left(\frac{2}{23}\right)^{t/4-1} = \frac{1-y}{y}$.

It follows that $\frac{t}{4} - 1 = \ln\left(\frac{1-y}{y}\right)/\ln\left(\frac{2}{23}\right)$, so $t = 4\left[1 + \ln\left(\frac{1-y}{y}\right)/\ln\left(\frac{2}{23}\right)\right]$. When $y = .9$,

$\dfrac{(1-y)}{y} = \frac{1}{9}$, so $t = 4\left[1 - \dfrac{\ln 9}{\ln 23}\right] \approx 7.6$ h or 7 h 36 min. Thus 90% of the population will have

heard the rumor by 3:36 P.M.

33. y increases most rapidly when y' is maximal, that is, $y'' = 0$. But $y' = ky(M - y) \Rightarrow$

$y'' = ky'(M - y) + ky(-y') = ky'(M - 2y) = k^2y(M - y)(M - 2y)$. Since $0 < y < M$, we see

that $y'' = 0 \Leftrightarrow y = M/2$.

35. $RI + LI'(t) = V$, $I(0) = 0$. $LI' = V - RI \Rightarrow L\dfrac{dI}{dt} = V - RI \Rightarrow \displaystyle\int \dfrac{L\,dI}{V - RI} = \int dt$

$\Rightarrow -\frac{L}{R}\ln|V - RI| = t + C \Rightarrow V - RI = Ae^{-Rt/L} \Rightarrow I = \frac{V}{R} - \frac{A}{R}e^{-Rt/L}$.

$I(0) = 0 \Rightarrow 0 = \frac{V}{R} - \frac{A}{R}\cdot e^0 \Rightarrow A = V$. So, $I = \frac{V}{R}(1 - e^{-Rt/L})$.

EXERCISES 8.2

1. $L = \displaystyle\int_{-1}^{3}\sqrt{1 + \left(\frac{dy}{dx}\right)^2}\,dx = \int_{-1}^{3}\sqrt{1 + 2^2}\,dx = \sqrt{5}\,[3 - (-1)] = 4\sqrt{5}$.

The arc length can be calculated using the distance formula since the curve is a straight line, so

$L = $ [distance from $(-1, -1)$ to $(3, 7)$] $= \sqrt{(3 - (-1))^2 + (7 - (-1))^2} = \sqrt{80} = 4\sqrt{5}$.

3. $x^2 = 64y^3$, $y = \left(\frac{x}{8}\right)^{2/3} \Rightarrow \dfrac{dy}{dx} = \frac{1}{12}\left(\frac{x}{8}\right)^{-1/3} \Rightarrow 1 + \left(\frac{dy}{dx}\right)^2 = 1 + \frac{1}{144}\left(\frac{x}{8}\right)^{-2/3}$

$= 1 + \dfrac{1}{36x^{2/3}}$. So $L = \displaystyle\int_{8}^{64}\sqrt{1 + \dfrac{1}{(6\sqrt[3]{x})^2}}\,dx = \int_{2}^{4}\sqrt{1 + \dfrac{1}{(6u)^2}}\,3u^2\,du$

[Let $u = \sqrt[3]{x}$. Then $x = u^3$, $dx = 3u^2\,du$] $= \displaystyle\int_{2}^{4}\dfrac{\sqrt{(6u)^2 + 1}}{6u}\,3u^2\,du$

$$= \int_2^4 \tfrac{1}{2}\sqrt{(6u)^2 + 1}\, u\, du = \int_{12}^{24} \sqrt{v^2 + 1}\, \tfrac{v}{12}\, \tfrac{dv}{6} \quad [v = 6u \Rightarrow dv = 6\, du]$$

$$= \tfrac{1}{144} \int_{12}^{24} \sqrt{v^2 + 1}\, 2v\, dv = \tfrac{1}{144}\left[\tfrac{2}{3}(v^2 + 1)^{3/2}\right]_{12}^{24} = \frac{577^{3/2} - 145^{3/2}}{216}.$$

5. $y^2 = (x - 1)^3$, $y = (x - 1)^{3/2} \Rightarrow \frac{dy}{dx} = \frac{3}{2}(x - 1)^{1/2} \Rightarrow 1 + \left(\frac{dy}{dx}\right)^2 = 1 + \frac{9}{4}(x - 1)$.

So $L = \int_1^2 \sqrt{1 + \frac{9}{4}(x - 1)}\, dx = \int_1^2 \sqrt{\frac{9}{4}x - \frac{5}{4}}\, dx = \left[\frac{4}{9} \cdot \frac{2}{3}\left(\frac{9}{4}x - \frac{5}{4}\right)^{3/2}\right]_1^2 = \frac{13\sqrt{13} - 8}{27}.$

7. $12xy = 4y^4 + 3$, $x = \frac{y^3}{3} + \frac{y^{-1}}{4} \Rightarrow \frac{dx}{dy} = y^2 - \frac{y^{-2}}{4}$, so $\left(\frac{dx}{dy}\right)^2 = y^4 - \frac{1}{2} + \frac{y^{-4}}{16}$

$\Rightarrow 1 + \left(\frac{dx}{dy}\right)^2 = y^4 + \frac{1}{2} + \frac{y^{-4}}{16} \Rightarrow \sqrt{1 + \left(\frac{dx}{dy}\right)^2} = y^2 + \frac{y^{-2}}{4}.$

So $L = \int_1^2 \left(y^2 + \frac{y^{-2}}{4}\right) dy = \left[\frac{y^3}{3} - \frac{1}{4y}\right]_1^2 = \left(\frac{8}{3} - \frac{1}{8}\right) - \left(\frac{1}{3} - \frac{1}{4}\right) = \frac{59}{24}.$

9. $y = \frac{1}{3}(x^2 + 2)^{3/2} \Rightarrow \frac{dy}{dx} = \frac{1}{2}(x^2 + 2)^{1/2}(2x) = x\sqrt{x^2 + 2} \Rightarrow 1 + \left(\frac{dy}{dx}\right)^2 = 1 + x^2(x^2 + 2)$

$= (x^2 + 1)^2.$ So $L = \int_0^1 (x^2 + 1)\, dx = \left[\frac{x^3}{3} + x\right]_0^1 = \frac{4}{3}.$

11. $y = \frac{x^4}{4} + \frac{1}{8x^2} \Rightarrow \frac{dy}{dx} = x^3 - \frac{1}{4x^3} \Rightarrow 1 + \left(\frac{dy}{dx}\right)^2 = 1 + x^6 - \frac{1}{2} + \frac{1}{16x^6} = x^6 + \frac{1}{2} + \frac{1}{16x^6}.$

So $L = \int_1^3 \left(x^3 + \frac{1}{4}x^{-3}\right) dx = \left[\frac{x^4}{4} - \frac{x^{-2}}{8}\right]_1^3 = \left(\frac{81}{4} - \frac{1}{72}\right) - \left(\frac{1}{4} - \frac{1}{8}\right) = \frac{181}{9}.$

13. $y = \ln(\cos x) \Rightarrow y' = \frac{1}{\cos x}(-\sin x) = -\tan x \Rightarrow 1 + (y')^2 = 1 + \tan^2 x = \sec^2 x.$

So $L = \int_0^{\pi/4} \sec x\, dx = \ln(\sec x + \tan x)]_0^{\pi/4} = \ln(\sqrt{2} + 1).$

15. $y = \ln(1 - x^2) \Rightarrow \frac{dy}{dx} = \frac{-2x}{1 - x^2} \Rightarrow 1 + \left(\frac{dy}{dx}\right)^2 = 1 + \frac{4x^2}{(1 - x^2)^2} = \frac{(1 + x^2)^2}{(1 - x^2)^2}.$ So

$L = \int_0^{1/2} \frac{1 + x^2}{1 - x^2}\, dx = \int_0^{1/2}\left[-1 + \frac{2}{(1 - x)(1 + x)}\right] dx = \int_0^{1/2}\left[-1 + \frac{1}{1 + x} + \frac{1}{1 - x}\right] dx$

$= \left[-x + \ln(1 + x) - \ln(1 - x)\right]_0^{1/2} = -\frac{1}{2} + \ln\frac{3}{2} - \ln\frac{1}{2} - 0 = \ln 3 - \frac{1}{2}.$

17. $y = e^x \Rightarrow y' = e^x \Rightarrow 1 + (y')^2 = 1 + e^{2x}.$ So $L = \int_0^1 \sqrt{1 + e^{2x}}\, dx = \int_1^e \sqrt{1 + u^2}\, \frac{du}{u}$

$[u = e^x \Rightarrow x = \ln u, dx = \frac{du}{u}] = \int_1^e \frac{\sqrt{1 + u^2}}{u^2}\, u\, du = \int_{\sqrt{2}}^{\sqrt{1 + e^2}} \frac{v}{v^2 - 1}\, v\, dv$

$[v = \sqrt{1+u^2} \Rightarrow v^2 = 1+u^2, \; v\,dv = u\,du]$ $= \int_{\sqrt{2}}^{\sqrt{1+e^2}} \left(1 + \frac{1/2}{v-1} - \frac{1/2}{v+1} \right) dv$

$= \left[v + \tfrac{1}{2}\ln\frac{v-1}{v+1} \right]_{\sqrt{2}}^{\sqrt{1+e^2}} = \sqrt{1+e^2} - \sqrt{2} + \tfrac{1}{2}\ln\frac{\sqrt{1+e^2}-1}{\sqrt{1+e^2}+1} - \tfrac{1}{2}\ln\frac{\sqrt{2}-1}{\sqrt{2}+1}$

$= \sqrt{1+e^2} - \sqrt{2} + \ln(\sqrt{1+e^2}-1) - 1 - \ln(\sqrt{2}-1).$

[OR: Use Formula 23 for $\int \frac{\sqrt{1+u^2}}{u}\,du$, or substitute $u = \tan\theta$.]

19. $y = \cosh x \Rightarrow y' = \sinh x \Rightarrow 1 + (y')^2 = 1 + \sinh^2 x = \cosh^2 x.$
So $L = \int_0^1 \cosh x\,dx = [\sinh x]_0^1 = \sinh 1 = \tfrac{1}{2}(e - e^{-1}).$

21. $y = x^3 \Rightarrow y' = 3x^2 \Rightarrow 1 + (y')^2 = 1 + 9x^4.$ So $L = \int_0^1 \sqrt{1+9x^4}\,dx.$

23. $y = \sin x \Rightarrow y' = \cos x \Rightarrow 1 + (y')^2 = 1 + \cos^2 x.$ So $L = \int_0^\pi \sqrt{1+\cos^2 x}\,dx.$

25. $y = e^x \cos x \Rightarrow y' = e^x(\cos x - \sin x) \Rightarrow 1 + (y')^2 = 1 + e^{2x}(\cos^2 x - 2\cos x \sin x + \sin^2 x)$
$= 1 + e^{2x}(1 - \sin 2x).$ So $L = \int_0^{\pi/2} \sqrt{1+e^{2x}(1-\sin 2x)}\,dx.$

27. $y = 2x^{3/2} \Rightarrow y' = 3x^{1/2} \Rightarrow 1 + (y')^2 = 1 + 9x.$ The arc length function with starting
point $P_0(1,2)$ is $s(x) = \int_1^x \sqrt{1+9t}\,dt = \left[\tfrac{2}{27}(1+9t)^{3/2} \right]_1^x = \tfrac{2}{27}\left[(1+9x)^{3/2} - 10\sqrt{10} \right].$

29. $y = 4500 - \frac{x^2}{8000} \Rightarrow \frac{dy}{dx} = -\frac{x}{4000} \Rightarrow \left(\frac{dy}{dx}\right)^2 = \frac{x^2}{16{,}000{,}000}.$ When $y = 4500$ m, $x = 0$ m.

When $y = 0$ m, $x = 6000$ m. Therefore $L = \int_0^{6000} \sqrt{1+(x/4000)^2}\,dx$

$= \int_0^{3/2} \sqrt{1+u^2}\,4000\,du \; \left[u = \frac{x}{4000} \right] = 4000\left[\frac{u}{2}\sqrt{1+u^2} + \frac{1}{2}\ln(u + \sqrt{1+u^2}) \right]_0^{3/2}$ [Formula

21 or $u = \tan\theta$] $= 4000\left[\frac{3}{4}\sqrt{\frac{13}{4}} + \frac{1}{2}\ln\left(\frac{3}{2} + \sqrt{\frac{13}{4}}\right) \right] = 1500\sqrt{13} + 2000\ln\frac{3+\sqrt{13}}{2} \approx 7798\,\text{m}.$

31. $y^{2/3} = 1 - x^{2/3} \Rightarrow y = (1 - x^{2/3})^{3/2}$

$\Rightarrow \frac{dy}{dx} = \frac{3}{2}(1 - x^{2/3})^{1/2}\left(-\frac{2}{3}x^{-1/3}\right)$

$= -x^{-1/3}(1 - x^{2/3})^{1/2}$

$\Rightarrow \left(\frac{dy}{dx}\right)^2 = x^{-2/3}(1 - x^{2/3}) = x^{-2/3} - 1.$

Thus $L = 4\int_0^1 \sqrt{1 + (x^{-2/3} - 1)}\,dx = 4\int_0^1 x^{-1/3}\,dx$

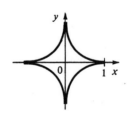

$$= 4 \lim_{t \to 0^+} \left[\tfrac{3}{2}x^{2/3}\right]_t^1 = 6.$$

33. $y = x^3 \Rightarrow 1 + (y')^2 = 1 + (3x^2)^2 = 1 + 9x^4$. So $L = \int_0^1 \sqrt{1 + 9x^4}\, dx$. Let $f(x) = \sqrt{1 + 9x^4}$.

Then by Simpson's rule with $n = 10$, $L \approx \dfrac{1/10}{3}\Big(f(0) + 4f(.1) + 2f(.2)$

$+ 4f(.3) + 2f(.4) + 4f(.5) + 2f(.6) + 4f(.7) + 2f(.8) + 4f(.9) + f(1)\Big) \approx 1.548$

35. $y = \sin x$, $1 + \left(\dfrac{dy}{dx}\right)^2 = 1 + \cos^2 x$, $L = \int_0^\pi \sqrt{1 + \cos^2 x}\, dx$. Let $g(x) = \sqrt{1 + \cos^2 x}$. Then

$L \approx \dfrac{\pi/10}{3}\Big(g(0) + 4g\big(\tfrac{\pi}{10}\big) + 2g\big(\tfrac{\pi}{5}\big) + 4g\big(\tfrac{3\pi}{10}\big) + 2g\big(\tfrac{2\pi}{5}\big) + 4g\big(\tfrac{\pi}{2}\big) + 2g\big(\tfrac{3\pi}{5}\big) + 4g\big(\tfrac{7\pi}{10}\big)$

$+ 2g\big(\tfrac{4\pi}{5}\big) + 4g\big(\tfrac{9\pi}{10}\big) + g(\pi)\Big) \approx 3.820$

37. $y = \int_1^x \sqrt{t^3 - 1}\, dt \Rightarrow \dfrac{dy}{dx} = \sqrt{x^3 - 1}$ [by the Fundamental Theorem of Calculus]

$\Rightarrow 1 + \left(\dfrac{dy}{dx}\right)^2 = x^3 \Rightarrow L = \int_1^4 x^{3/2}\, dx = \tfrac{2}{5}\left[x^{5/2}\right]_1^4 = \tfrac{2}{5}(32 - 1) = \tfrac{62}{5} = 12.4.$

EXERCISES 8.3

1. $y = \sqrt{x} \Rightarrow 1 + \left(\dfrac{dy}{dx}\right)^2 = 1 + \left(\dfrac{1}{2\sqrt{x}}\right)^2 = 1 + \dfrac{1}{4x}$. So $S = \displaystyle\int_4^9 2\pi y \sqrt{1 + \left(\dfrac{dy}{dx}\right)^2}\, dx$

$= \displaystyle\int_4^9 2\pi\sqrt{x}\sqrt{1 + \tfrac{1}{4x}}\, dx = 2\pi \int_4^9 \sqrt{x + \tfrac{1}{4}}\, dx = 2\pi\left[\tfrac{2}{3}\left(x + \tfrac{1}{4}\right)^{3/2}\right]_4^9 = \tfrac{4\pi}{3}\left[\tfrac{1}{8}(4x+1)^{3/2}\right]_4^9$

$= \tfrac{\pi}{6}(37\sqrt{37} - 17\sqrt{17}).$

3. $2y = x + 4$, $y = \tfrac{1}{2}x + 2 \Rightarrow y' = \tfrac{1}{2} \Rightarrow \sqrt{1 + (y')^2} = \dfrac{\sqrt{5}}{2}$. So $S = 2\pi \displaystyle\int_0^2 \tfrac{1}{2}(x + 4)\dfrac{\sqrt{5}}{2}\, dx$

$= \dfrac{\pi\sqrt{5}}{2}\left[\tfrac{x^2}{2} + 4x\right]_0^2 = 5\sqrt{5}\,\pi.$

5. $y = x^3 + \dfrac{1}{12x} \Rightarrow \dfrac{dy}{dx} = 3x^2 - \dfrac{1}{12x^2} \Rightarrow 1 + \left(\dfrac{dy}{dx}\right)^2 = 9x^4 + \tfrac{1}{2} + \dfrac{1}{144x^4} = \left(3x^2 + \dfrac{1}{12x^2}\right)^2$.

So $S = 2\pi \displaystyle\int_1^2 \left(x^3 + \dfrac{1}{12x}\right)\left(3x^2 + \dfrac{1}{12x^2}\right) dx = 2\pi \int_1^2 \left(3x^5 + \tfrac{x}{3} + \dfrac{1}{144x^3}\right) dx$

$= 2\pi\left[\tfrac{x^6}{2} + \tfrac{x^2}{6} - \dfrac{1}{288x^2}\right]_1^2 = 2\pi\left[\left(32 + \tfrac{2}{3} - \dfrac{1}{288 \cdot 4}\right) - \left(\tfrac{1}{2} + \tfrac{1}{6} - \dfrac{1}{288}\right)\right] = \dfrac{12\,289\pi}{192}.$

7. $y = \sin x \Rightarrow 1 + \left(\dfrac{dy}{dx}\right)^2 = 1 + \cos^2 x$. So $S = 2\pi \int_0^\pi \sin x\sqrt{1 + \cos^2 x}\, dx$

$= 2\pi \int_{-1}^1 \sqrt{1 + u^2}\, du$ [$u = -\cos x \Rightarrow du = \sin x\, dx$]

$= 4\pi \int_0^1 \sqrt{1 + u^2}\, du = 4\pi \int_0^{\pi/4} \sec^3\theta\, d\theta$ [$u = \tan\theta \Rightarrow du = \sec^2\theta\, d\theta$]

$= 2\pi\left[\sec\theta\tan\theta + \ln|\sec\theta + \tan\theta|\right]_0^{\pi/4} = 2\pi\left[\sqrt{2} + \ln(\sqrt{2} + 1)\right].$

9. $y = \cosh x \Rightarrow 1 + \left(\frac{dy}{dx}\right)^2 = 1 + \sinh^2 x = \cosh^2 x.$ So $S = 2\pi \int_0^1 \cosh x \cosh x\, dx$

$= 2\pi \int_0^1 \frac{1}{2}(1 + \cosh 2x)\, dx = \pi\left[x + \frac{1}{2}\sinh 2x\right]_0^1 = \pi\left(1 + \frac{1}{2}\sinh 2\right)$ or $\pi\left(1 + \frac{e^2 - e^{-2}}{4}\right).$

11. $x = \frac{y^4}{2} + \frac{1}{16y^2} \Rightarrow \frac{dx}{dy} = 2y^3 - \frac{1}{8y^3} \Rightarrow 1 + \left(\frac{dx}{dy}\right)^2 = 4y^6 + \frac{1}{2} + \frac{1}{64y^6} = \left(2y^3 + \frac{1}{8y^3}\right)^2.$

So $S = 2\pi \int_1^3 y\left[2y^3 + \frac{1}{8y^3}\right]dy = 2\pi \int_1^3 \left(2y^4 + \frac{1}{8}y^{-2}\right)dy = 2\pi\left[\frac{2}{5}y^5 - \frac{1}{8}y^{-1}\right]_1^3$

$= 2\pi\left[\frac{2}{5}(243) - \frac{1}{24} - \frac{2}{5} + \frac{1}{8}\right] = \frac{5813\pi}{30}.$

13. $x = 1 + 2y^2 \Rightarrow 1 + \left(\frac{dx}{dy}\right)^2 = 1 + (4y)^2 = 1 + 16y^2.$ So $S = 2\pi \int_1^2 y\sqrt{1 + 16y^2}\, dy$

$= \frac{\pi}{16}\int_1^2 (16y^2 + 1)^{1/2}32y\, dy = \frac{\pi}{16}\left[\frac{2}{3}(16y^2 + 1)^{3/2}\right]_1^2 = \frac{\pi}{24}(65\sqrt{65} - 17\sqrt{17}).$

15. $y = \sqrt[3]{x} \Rightarrow x = y^3 \Rightarrow 1 + \left(\frac{dx}{dy}\right)^2 = 1 + 9y^4.$ So $S = 2\pi \int_1^2 x\sqrt{1 + \left(\frac{dx}{dy}\right)^2}\, dy$

$= 2\pi \int_1^2 y^3\sqrt{1 + 9y^4}\, dy = \frac{2\pi}{36}\int_1^2 \sqrt{1 + 9y^4}\,36y^3\, dy = \frac{\pi}{18}\left[\frac{2}{3}(1 + 9y^4)^{3/2}\right]_1^2$

$= \pi(145\sqrt{145} - 10\sqrt{10})/27.$

17. $y^2 = x^3 \Rightarrow x = y^{2/3} \Rightarrow 1 + \left(\frac{dx}{dy}\right)^2 = 1 + \frac{4}{9}y^{-2/3}.$ So $S = 2\pi \int_1^8 y^{2/3}\sqrt{1 + \frac{4}{9}y^{-2/3}}\, dy$

$= 2\pi \int_1^2 u^2\sqrt{1 + \frac{4}{9u^2}}\,3u^2\, du\left[u = y^{1/3} \Rightarrow y = u^3,\ dy = 3u^2\, du\right] = 2\pi \int_1^2 u^3\sqrt{9u^2 + 4}\, du$

$= 2\pi \int_{13}^{40} \frac{1}{9}(v - 4)v^{1/2}\frac{1}{18}\, dv \left[v = 9u^2 + 4 \Rightarrow dv = 18u\, du,\ u^2 = \frac{1}{9}(v - 4)\right]$

$= \frac{\pi}{81}\int_{13}^{40} (v^{3/2} - 4v^{1/2})\, dv = \frac{\pi}{81}\left[\frac{2}{5}v^{5/2} - \frac{8}{3}v^{3/2}\right]_{13}^{40}$

$= \frac{\pi}{81}\left[\frac{3200}{5}\sqrt{40} - \frac{320}{3}\sqrt{40} - \frac{338}{5}\sqrt{13} + \frac{104}{3}\sqrt{13}\right] = \frac{\pi}{243}\left(3200\sqrt{10} - \frac{494}{5}\sqrt{13}\right).$

19. $x = e^{2y} \Rightarrow 1 + \left(\frac{dx}{dy}\right)^2 = 1 + 4e^{4y}.$ So $S = 2\pi \int_0^{1/2} e^{2y}\sqrt{1 + (2e^{2y})^2}\, dy$

$= 2\pi \int_2^{2e} \sqrt{1 + u^2}\,\frac{1}{4}\, du\ [u = 2e^{2y},\ du = 4e^{2y}\, dy] = \frac{\pi}{2}\int_2^{2e} \sqrt{1 + u^2}\, du$

$= \frac{\pi}{2}\left[\frac{u}{2}\sqrt{1 + u^2} + \frac{1}{2}\ln\left|u + \sqrt{1 + u^2}\right|\right]_2^{2e}\ [u = \tan\theta\ or\ Formula\ 21]$

$= \frac{\pi}{2}\left[e\sqrt{1 + 4e^2} + \frac{1}{2}\ln(2e + \sqrt{1 + 4e^2}) - \sqrt{5} - \frac{1}{2}\ln(2 + \sqrt{5})\right]$

$= \frac{\pi}{4}\left[2e\sqrt{1 + 4e^2} - 2\sqrt{5} + \ln\left((2e + \sqrt{1 + 4e^2})/(2 + \sqrt{5})\right)\right].$

21. $x = \frac{1}{2\sqrt{2}}(y^2 - \ln y) \Rightarrow \frac{dx}{dy} = \frac{1}{2\sqrt{2}}\left(2y - \frac{1}{y}\right) \Rightarrow 1 + \left(\frac{dx}{dy}\right)^2 = 1 + \frac{1}{8}\left(2y - \frac{1}{y}\right)^2$

$= 1 + \frac{1}{8}\left(4y^2 - 4 + \frac{1}{y^2}\right) = \frac{1}{8}\left(4y^2 + 4 + \frac{1}{y^2}\right) = \left[\frac{1}{2\sqrt{2}}\left(2y + \frac{1}{y}\right)\right]^2.$

So $S = 2\pi \int_1^2 \frac{1}{2\sqrt{2}}(y^2 - \ln y)\frac{1}{2\sqrt{2}}\left(2y + \frac{1}{y}\right)dy = \frac{\pi}{4}\int_1^2\left(2y^3 + y - 2y\ln y - \frac{\ln y}{y}\right)dy$

$= \frac{\pi}{4}\left[\frac{y^4}{2} + \frac{y^2}{2} - y^2\ln y + \frac{y^2}{2} - \frac{1}{2}(\ln y)^2\right]_1^2 = \frac{\pi}{8}\left[y^4 + 2y^2 - 2y^2\ln y - (\ln y)^2\right]_1^2$

$= \frac{\pi}{8}\left[16 + 8 - 8\ln 2 - (\ln 2)^2 - 1 - 2\right] = \frac{\pi}{8}\left[21 - 8\ln 2 - (\ln 2)^2\right].$

23. $S = 2\pi\int_0^1 x^4\sqrt{1 + (4x^3)^2}\,dx = 2\pi\int_0^1 x^4\sqrt{16x^6 + 1}\,dx \approx 2\pi\frac{1/10}{3}\Big(f(0) + 4f(.1) + 2f(.2)$

$+ 4f(.3) + 2f(.4) + 4f(.5) + 2f(.6) + 4f(.7) + 2f(.8) + 4f(.9) + f(1)\Big)$

$[where\ f(x) = x^4\sqrt{16x^6 + 1}] \approx 3.44$

25. The curve $8y^2 = x^2(1 - x^2)$ actually consists of two loops in the region described by the inequalities $|x| \le 1, |y| \le \sqrt{2}/8$. (The maximum value of $|y|$ is attained when $|x| = 1/\sqrt{2}$.) If we consider the loop in the region $x \ge 0$, the surface area S it generates when rotated about the x-axis is calculated as follows: $16y\frac{dy}{dx} = 2x - 4x^3$, so

$\left(\frac{dy}{dx}\right)^2 = \left(\frac{x - 2x^3}{8y}\right)^2 = \frac{x^2(1 - 2x^2)^2}{64y^2} = \frac{x^2(1 - 2x^2)^2}{8x^2(1 - x^2)} = \frac{(1 - 2x^2)^2}{8(1 - x^2)}$ for $x \ne 0, \pm 1$.

[The formula also holds for $x = 0$ by continuity.] $1 + \left(\frac{dy}{dx}\right)^2 = 1 + \frac{(1 - 2x^2)^2}{8(1 - x^2)}$

$= \frac{9 - 12x^2 + 4x^4}{8(1 - x^2)} = \frac{(3 - 2x^2)^2}{8(1 - x^2)}.$ So $S = 2\pi\int_0^1 \frac{\sqrt{x^2(1 - x^2)}}{2\sqrt{2}} \cdot \frac{3 - 2x^2}{2\sqrt{2}\sqrt{1 - x^2}}\,dx$

$= \frac{\pi}{4}\int_0^1 x(3 - 2x^2)\,dx = \frac{\pi}{4}\left[\frac{3x^2}{2} - \frac{x^4}{2}\right]_0^1 = \frac{\pi}{4}\left(\frac{3}{2} - \frac{1}{2}\right) = \frac{\pi}{4}.$

27. $S = 2\pi\int_1^\infty y\sqrt{1 + \left(\frac{dy}{dx}\right)^2}\,dx = 2\pi\int_1^\infty \frac{1}{x}\sqrt{1 + \frac{1}{x^4}}\,dx = 2\pi\int_1^\infty \frac{\sqrt{x^4 + 1}}{x^3}\,dx > 2\pi\int_1^\infty \frac{x^2}{x^3}\,dx$

$= 2\pi\int_1^\infty \frac{dx}{x} = 2\pi\lim_{t\to\infty}[\ln x]_1^t = 2\pi\lim_{t\to\infty}\ln t = \infty.$

29. $\frac{x^2}{a^2} + \frac{y^2}{b^2} = 1 \Rightarrow \frac{y(dy/dx)}{b^2} = \frac{-x}{a^2} \Rightarrow \frac{dy}{dx} = \frac{-b^2x}{a^2y} \Rightarrow 1 + \left(\frac{dy}{dx}\right)^2$

$= 1 + \frac{b^4x^2}{a^4y^2} = \frac{b^4x^2 + a^4y^2}{a^4y^2} = \frac{b^4x^2 + a^4b^2(1 - x^2/a^2)}{a^4b^2(1 - x^2/a^2)}$

$$= \frac{a^4 b^2 + b^4 x^2 - a^2 b^2 x^2}{a^4 b^2 - a^2 b^2 x^2} = \frac{a^4 + b^2 x^2 - a^2 x^2}{a^4 - a^2 x^2}$$

$$= \frac{a^4 - (a^2 - b^2)x^2}{a^2(a^2 - x^2)}. \quad \text{The ellipsoid's surface area is twice the area generated by rotating}$$

the first quadrant portion of the ellipse about the x-axis. Thus

$$S = 2 \int_0^a 2\pi y \sqrt{1 + \left(\frac{dy}{dx}\right)^2} \, dx = 4\pi \int_0^a \frac{b}{a} \sqrt{a^2 - x^2} \frac{\sqrt{a^4 - (a^2 - b^2)x^2}}{a\sqrt{a^2 - x^2}} \, dx$$

$$= \frac{4\pi b}{a^2} \int_0^a \sqrt{a^4 - (a^2 - b^2)x^2} \, dx = \frac{4\pi b}{a^2} \int_0^{a\sqrt{a^2 - b^2}} \sqrt{a^4 - u^2} \frac{du}{\sqrt{a^2 - b^2}}$$

$$\left[u = \sqrt{a^2 - b^2}\, x \right] \quad = \frac{4\pi b}{a^2 \sqrt{a^2 - b^2}} \left[\frac{u}{2} \sqrt{a^4 - u^2} + \frac{a^4}{2} \sin^{-1} \frac{u}{a^2} \right]_0^{a\sqrt{a^2 - b^2}} \quad [\text{Formula } 30]$$

$$= \frac{4\pi b}{a^2 \sqrt{a^2 - b^2}} \left[\frac{a\sqrt{a^2 - b^2}}{2} \sqrt{a^4 - a^2(a^2 - b^2)} + \frac{a^4}{2} \sin^{-1} \frac{\sqrt{a^2 - b^2}}{a} \right]$$

$$= 2\pi \left[b^2 + \frac{a^2 b \sin^{-1}(\sqrt{a^2 - b^2}/a)}{\sqrt{a^2 - b^2}} \right].$$

31. In the derivation of (8.22), we computed a typical contribution to the surface area to be $2\pi \frac{y_{i-1} + y_i}{2} |P_{i-1} P_i|$, the area of a frustum of a cone. When f(x) is not necessarily positive, the approximations $y_i = f(x_i) \approx f(x_i^*)$ and $y_{i-1} = f(x_{i-1}) \approx f(x_i^*)$ must be replaced by $y_i = |f(x_i)| \approx |f(x_i^*)|$ and $y_{i-1} = |f(x_{i-1})| \approx |f(x_i^*)|$. Thus $2\pi \frac{y_{i-1} + y_i}{2} |P_{i-1} P_i|$
$\approx 2\pi |f(x_i^*)| \sqrt{1 + \left[f'(x_i^*) \right]^2} \Delta x_i$. Continuing with the rest of the derivation as before, we obtain
$S = \int_a^b 2\pi |f(x)| \sqrt{1 + \left[f'(x) \right]^2} \, dx$.

33. For the upper semicircle, $f(x) = \sqrt{r^2 - x^2}$, $f'(x) = -x/\sqrt{r^2 - x^2}$. The surface area generated is
$$S_1 = \int_{-r}^r 2\pi(r - \sqrt{r^2 - x^2}) \sqrt{1 + x^2/(r^2 - x^2)} \, dx = 4\pi \int_0^r (r - \sqrt{r^2 - x^2})(r/\sqrt{r^2 - x^2}) \, dx$$
$$= 4\pi \int_0^r (r^2/\sqrt{r^2 - x^2} - r) \, dx. \quad \text{For the lower semicircle, } f(x) = -\sqrt{r^2 - x^2},$$
$f'(x) = x/\sqrt{r^2 - x^2}$ and the surface area generated is
$$S_2 = \int_{-r}^r 2\pi(r + \sqrt{r^2 - x^2}) \sqrt{1 + x^2/(r^2 - x^2)} \, dx$$
$$= 4\pi \int_0^r (r + \sqrt{r^2 - x^2})(r/\sqrt{r^2 - x^2}) \, dx = 4\pi \int_0^r (r^2/\sqrt{r^2 - x^2} + r) \, dx. \quad \text{Thus the total area is}$$
$$S = S_1 + S_2 = 8\pi \int_0^r (r^2/\sqrt{r^2 - x^2}) \, dx = 8\pi \left[r^2 \sin^{-1}\left(\frac{x}{r}\right) \right]_0^r = 8\pi r^2 \left(\frac{\pi}{2}\right) = 4\pi^2 r^2.$$

EXERCISES 8.4

1. $m_1 = 4, m_2 = 8$; $P_1(-1, 2), P_2(2, 4)$. $m = m_1 + m_2 = 12$. $M_x = 4 \cdot 2 + 8 \cdot 4 = 40$;

$M_y = 4 \cdot (-1) + 8 \cdot 2 = 12$; $\bar{x} = \dfrac{M_y}{m} = 1$ and $\bar{y} = \dfrac{M_x}{m} = \dfrac{10}{3}$, so the center of mass is

$(\bar{x}, \bar{y}) = \left(1, \dfrac{10}{3}\right)$.

3. $m = m_1 + m_2 + m_3 = 4 + 2 + 5 = 11$. $M_x = 4 \cdot (-2) + 2 \cdot 4 + 5 \cdot (-3) = -15$;

$M_y = 4 \cdot (-1) + 2 \cdot (-2) + 5 \cdot 5 = 17$, $(\bar{x}, \bar{y}) = \left(\dfrac{17}{11}, -\dfrac{15}{11}\right)$.

5. $A = \int_0^2 x^2 \, dx = \left[\dfrac{x^3}{3}\right]_0^2 = \dfrac{8}{3}$, $\bar{x} = \dfrac{1}{A}\int_0^2 x \cdot x^2 \, dx = \dfrac{3}{8}\left[\dfrac{x^4}{4}\right]_0^2 = \dfrac{3}{8} \cdot 4 = \dfrac{3}{2}$,

$\bar{y} = \dfrac{1}{A}\int_0^2 \dfrac{1}{2}(x^2)^2 \, dx = \dfrac{3}{8} \cdot \dfrac{1}{2}\left[\dfrac{x^5}{5}\right]_0^2 = \dfrac{3}{16} \cdot \dfrac{32}{5} = \dfrac{6}{5}$. Centroid $(\bar{x}, \bar{y}) = \left(\dfrac{3}{2}, \dfrac{6}{5}\right) = (1.5, 1.2)$.

7. $A = \int_{-1}^2 (3x + 5) \, dx = \left[\dfrac{3x^2}{2} + 5x\right]_{-1}^2 = (6 + 10) - \left(\dfrac{3}{2} - 5\right) = 16 + \dfrac{7}{2} = \dfrac{39}{2}$,

$\bar{x} = \dfrac{1}{A}\int_{-1}^2 x(3x + 5) \, dx = \dfrac{2}{39}\int_{-1}^2 (3x^2 + 5x) \, dx = \dfrac{2}{39}\left[x^3 + \dfrac{5x^2}{2}\right]_{-1}^2$

$= \dfrac{2}{39}\left[(8 + 10) - \left(-1 + \dfrac{5}{2}\right)\right] = \dfrac{2}{39}\left(\dfrac{36 - 3}{2}\right) = \dfrac{11}{13}$, $\bar{y} = \dfrac{1}{A}\int_{-1}^2 \dfrac{1}{2}(3x + 5)^2 \, dx$

$= \dfrac{1}{39}\int_{-1}^2 (9x^2 + 30x + 25) \, dx = \dfrac{1}{39}\left[3x^3 + 15x^2 + 25x\right]_{-1}^2$

$= \dfrac{1}{39}\left[(24 + 60 + 50) - (-3 + 15 - 25)\right] = \dfrac{147}{39} = \dfrac{49}{13}$. $(\bar{x}, \bar{y}) = \left(\dfrac{11}{13}, \dfrac{49}{13}\right)$.

9. $A = \int_0^1 \sqrt{x^2 + 1} \, dx = \left[\dfrac{x}{2}\sqrt{1 + x^2} + \dfrac{1}{2}\ln\left|x + \sqrt{1 + x^2}\right|\right]_0^1$ [Formula 21]

$= \left(\dfrac{1}{2}\sqrt{2} + \dfrac{1}{2}\ln(1 + \sqrt{2})\right) - \left(\dfrac{1}{2}\ln 1\right) = (\sqrt{2} + \ln(1 + \sqrt{2}))/2$, $\bar{x} = \dfrac{1}{A}\int_0^1 x\sqrt{x^2 + 1} \, dx$

$= \dfrac{1}{A}\left[\dfrac{1}{3}(x^2 + 1)^{3/2}\right]_0^1 = \dfrac{1}{A} \cdot \dfrac{1}{3}(2\sqrt{2} - 1) = \dfrac{2(2\sqrt{2} - 1)}{3(\sqrt{2} + \ln(1 + \sqrt{2}))}$, $\bar{y} = \dfrac{1}{A}\int_0^1 \dfrac{1}{2}(\sqrt{x^2 + 1})^2 \, dx$

$= \dfrac{1}{2A}\int_0^1 (x^2 + 1) \, dx = \dfrac{1}{2A}\left[\dfrac{x^3}{3} + x\right]_0^1 = \dfrac{1}{2A} \cdot \dfrac{4}{3} = \dfrac{2}{3A} = \dfrac{4}{3(\sqrt{2} + \ln(1 + \sqrt{2}))}$.

$(\bar{x}, \bar{y}) = \left(\dfrac{2(2\sqrt{2} - 1)}{3(\sqrt{2} + \ln(1 + \sqrt{2}))}, \dfrac{4}{3(\sqrt{2} + \ln(1 + \sqrt{2}))}\right)$.

11. By symmetry, $\bar{x} = 0$ and $A = 2\int_0^{\pi/4} \cos 2x \, dx = \sin 2x\Big]_0^{\pi/4} = 1$,

$\bar{y} = \dfrac{1}{A}\int_{-\pi/4}^{\pi/4} \dfrac{1}{2}\cos^2 2x \, dx = \int_0^{\pi/4} \cos^2 2x \, dx = \dfrac{1}{2}\int_0^{\pi/4} (1 + \cos 4x) \, dx = \dfrac{1}{2}\left[x + \dfrac{1}{4}\sin 4x\right]_0^{\pi/4}$

$= \dfrac{1}{2}\left(\dfrac{\pi}{4} + \dfrac{1}{4} \cdot 0\right) = \dfrac{\pi}{8}$. $(\bar{x}, \bar{y}) = \left(0, \dfrac{\pi}{8}\right)$.

13. $A = \int_0^1 e^x \, dx = [e^x]_0^1 = e - 1$, $\bar{x} = \dfrac{1}{A}\int_0^1 xe^x \, dx = \dfrac{1}{e - 1}\left[xe^x - e^x\right]_0^1$ [by integration by parts]

$= \frac{1}{e-1}[0-(-1)] = \frac{1}{e-1}$, $\bar{y} = \frac{1}{A}\int_0^1 \frac{1}{2}(e^x)^2\,dx = \frac{1}{e-1}\frac{1}{4}\left[e^{2x}\right]_0^1 = \frac{1}{4(e-1)}(e^2-1)$

$= \frac{e+1}{4}$. $(\bar{x},\bar{y}) = \left(\frac{1}{e-1}, \frac{e+1}{4}\right)$.

15. $A = \int_0^1 (\sqrt{x}-x)\,dx = \left[\frac{2}{3}x^{3/2} - \frac{x^2}{2}\right]_0^1 = \frac{2}{3} - \frac{1}{2} = \frac{1}{6}$, $\bar{x} = \frac{1}{A}\int_0^1 x(\sqrt{x}-x)\,dx$

$= 6\int_0^1 (x^{3/2}-x^2)\,dx = 6\left[\frac{2}{5}x^{5/2} - \frac{x^3}{3}\right]_0^1 = 6\left(\frac{2}{5}-\frac{1}{3}\right) = \frac{2}{5}$, $\bar{y} = \frac{1}{A}\int_0^1 \frac{1}{2}\left[(\sqrt{x})^2 - x^2\right]dx$

$= 3\int_0^1 (x-x^2)\,dx = 3\left[\frac{x^2}{2} - \frac{x^3}{3}\right]_0^1 = 3\left(\frac{1}{2}-\frac{1}{3}\right) = \frac{1}{2}$. $(\bar{x},\bar{y}) = \left(\frac{2}{5},\frac{1}{2}\right) = (0.4, 0.5)$.

17. $A = \int_0^{\pi/4} (\cos x - \sin x)\,dx = \sin x + \cos x]_0^{\pi/4} = \sqrt{2}-1$, $\bar{x} = \frac{1}{A}\int_0^{\pi/4} x(\cos x - \sin x)\,dx$

$= \frac{1}{A}\left[x(\sin x + \cos x) + \cos x - \sin x\right]_0^{\pi/4}$ [by integration by parts] $= \frac{1}{A}\left(\frac{\pi}{4}\sqrt{2}-1\right)$

$= \left(\frac{\pi}{4}\sqrt{2}-1\right)/(\sqrt{2}-1)$, $\bar{y} = \frac{1}{A}\int_0^{\pi/4} \frac{1}{2}(\cos^2 x - \sin^2 x)\,dx = \frac{1}{2A}\int_0^{\pi/4} \cos 2x\,dx$

$= \frac{1}{4A}\left[\sin 2x\right]_0^{\pi/4} = \frac{1}{4A} = 1/[4(\sqrt{2}-1)]$. $(\bar{x},\bar{y}) = \left((\pi\sqrt{2}-4)/[4(\sqrt{2}-1)], 1/[4(\sqrt{2}-1)]\right)$.

19. By symmetry, $M_y = 0$ and $\bar{x} = 0$. $A = \frac{1}{2}bh = \frac{1}{2}(2)(2) = 2$.

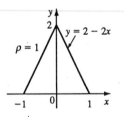

$M_x = 2\rho\int_0^1 \frac{1}{2}(2-2x)^2\,dx = 4\int_0^1 (1-x)^2\,dx$

$= 4\int_0^1 (1-2x+x^2)\,dx = 4\left[x-x^2+\frac{x^3}{3}\right]_0^1$

$= 4\left(1-1+\frac{1}{3}\right) = \frac{4}{3}$. $\bar{y} = \frac{1}{\rho A}M_x = \frac{2}{3}$. $(\bar{x},\bar{y}) = \left(0,\frac{2}{3}\right)$.

21. By symmetry, $M_y = 0$ and $\bar{x} = 0$. $A =$ area of triangle + area of square $= 1+4 = 5$, so

$m = \rho A = 4\cdot 5 = 20$. $M_x = \rho\cdot 2\int_0^1 \frac{1}{2}\left[(1-x)^2 - (-2)^2\right]dx = 4\int_0^1 (x^2-2x-3)\,dx$

$= 4\left[\frac{x^3}{3} - x^2 - 3x\right]_0^1 = 4\left(\frac{1}{3}-1-3\right) = 4(-11/3) = -44/3$. $\bar{y} = \frac{1}{m}M_x = \frac{1}{20}(-44/3)$

$= -11/15$. $(\bar{x},\bar{y}) = (0,-11/15)$.

23. Choose x- and y-axes so that the base (one side of the triangle) lies along the x-axis with the other vertex along the positive y-axis as shown. From geometry, we know the medians intersect at a point 2/3 of the way from each vertex (along the median) to the opposite side.

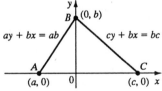

The median from B goes to the midpoint $((a+c)/2, 0)$ of side AC, so the point of intersection of the medians is $\left(\frac{2}{3}\cdot\frac{a+c}{2}, \frac{1}{3}\cdot b\right) = \left(\frac{a+c}{3}, \frac{b}{3}\right)$. This can also be verified by finding the equations of two medians, and solving them simultaneously to find their point of intersection. Now let us compute the location of the centroid of the triangle.

The area is $A = \frac{1}{2}(c-a)b$. $\bar{x} = \frac{1}{A}\left[\int_a^0 x \cdot \frac{b}{a}(a-x)\,dx + \int_0^c x \cdot \frac{b}{c}(c-x)\,dx\right]$

$= \frac{1}{A}\left[\frac{b}{a}\int_a^0 (ax-x^2)\,dx + \frac{b}{c}\int_0^c (cx-x^2)\,dx\right] = \frac{b}{Aa}\left[\frac{ax^2}{2}-\frac{x^3}{3}\right]_a^0 + \frac{b}{Ac}\left[\frac{cx^2}{2}-\frac{x^3}{3}\right]_0^c$

$= \frac{b}{Aa}\left[-\frac{a^3}{2}+\frac{a^3}{3}\right] + \frac{b}{Ac}\left[\frac{c^3}{2}-\frac{c^3}{3}\right] = \frac{2}{a(c-a)}\cdot\frac{-a^3}{6} + \frac{2}{c(c-a)}\cdot\frac{c^3}{6} = \frac{1}{3(c-a)}(c^2-a^2)$

$= \frac{a+c}{3}$ and $\bar{y} = \frac{1}{A}\left[\int_a^0 \frac{1}{2}\left(\frac{b}{a}(a-x)\right)^2 dx + \int_0^c \frac{1}{2}\left(\frac{b}{c}(c-x)\right)^2 dx\right]$

$= \frac{1}{A}\left[\frac{b^2}{2a^2}\int_a^0 (a^2-2ax+x^2)\,dx + \frac{b^2}{2c^2}\int_0^c (c^2-2cx+x^2)\,dx\right]$

$= \frac{1}{A}\left[\frac{b^2}{2a^2}\left[a^2x-ax^2+x^3/3\right]_a^0 + \frac{b^2}{2c^2}\left[c^2x-cx^2+x^3/3\right]_0^c\right]$

$= \frac{1}{A}\left[\frac{b^2}{2a^2}\left(-a^3+a^3-\frac{a^3}{3}\right) + \frac{b^2}{2c^2}\left(c^3-c^3+\frac{c^3}{3}\right)\right] = \frac{1}{A}\left(\frac{b^2}{6}(-a+c)\right)$

$= \frac{2}{(c-a)b}\cdot\frac{(c-a)b^2}{6} = \frac{b}{3}$. Thus $(\bar{x},\bar{y}) = \left(\frac{a+c}{3},\frac{b}{3}\right)$ as claimed.

Remarks. Actually the computation of $\bar{y}$ is all that is needed. By considering each side of the triangle in turn to be the base, we see that the centroid is $1/3$ of the way from each side to the opposite vertex and must therefore be the intersection of the medians. ,

The computation of $\bar{y}$ in this problem (and many others) can be simplified by using horizontal rather than vertical approximating rectangles. If the length of a thin rectangle at coordinate y is $l(y)$, then its area is $l(y)\Delta y$, its mass is $\rho l(y)\Delta y$,

and its moment about the x-axis is $\Delta M_x = \rho y l(y)\Delta y$.

Thus $M_x = \int \rho y l(y)\,dy$ and

$\bar{y} = \frac{\int \rho y l(y)\,dy}{\rho A} = \frac{1}{A}\int y l(y)\,dy$. In this problem

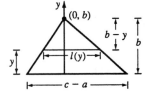

$l(y) = \frac{c-a}{b}(b-y)$ by similar triangles, so

$\bar{y} = \frac{1}{A}\int_0^b \frac{c-a}{b}y(b-y)\,dy = \frac{2}{b^2}\int_0^b (by-y^2)\,dy = \frac{2}{b^2}\left[\frac{by^2}{2}-\frac{y^3}{3}\right]_0^b = \frac{2}{b^2}\cdot\frac{b^3}{6} = \frac{b}{3}$.

Notice that only one integral is needed when this method is used.

Since the position of a centroid is independent of density, when it is constant, for convenience $\rho = 1$ will be assumed for Exercises 25-27.

25. Divide the lamina into two triangles and one rectangle with respective masses of 2, 2 and 4, so that the total mass is 8. Using the result of # 23, the triangles have centroids $\left(-1,\frac{2}{3}\right)$ and

$\left(1, \frac{2}{3}\right)$. The centroid of the rectangle (its center) is $\left(0, -\frac{1}{2}\right)$. So, using Formulas 8.31 and 8.33, we have

$$\bar{y} = \frac{\Sigma m_i y_i}{m} = \frac{2}{8}\left(\frac{2}{3}\right) + \frac{2}{8}\left(\frac{2}{3}\right) + \frac{4}{8}\left(-\frac{1}{2}\right) = \frac{1}{12}.$$

$\bar{x} = 0.$ [*since the lamina is symmetric about the line* $x = 0$]

Therefore $(\bar{x}, \bar{y}) = \left(0, \frac{1}{12}\right)$.

27. Consider first that the large rectangle is complete, so that its mass is $6 \times 3 = 18$. Its centroid is $\left(1, \frac{3}{2}\right)$. The mass removed from this object to create the one being studied is 3. The centroid of the cut–out piece is $\left(\frac{3}{2}, \frac{3}{2}\right)$. Therefore, for the true lamina, whose mass is 15,

$$\bar{x} = \frac{18}{15}(1) - \frac{3}{15}\left(\frac{3}{2}\right) = \frac{9}{10}$$

$$\bar{y} = \frac{3}{2} \text{ [\textit{since the lamina is symmetric about the line} } y = \frac{3}{2}]$$

Therefore $(\bar{x}, \bar{y}) = \left(\frac{9}{10}, \frac{3}{2}\right)$.

29. A cone of height h and radius r can be generated by rotating a right triangle about one of its legs as shown. By #23, $\bar{x} = \frac{r}{3}$, so the volume of the cone is $V = Ad = \frac{1}{2}rh \cdot 2\pi\frac{r}{3} = \frac{1}{3}\pi r^2 h$ by the Theorem of Pappus.

31. Suppose the region lies between two curves $y = f(x)$ and $y = g(x)$ where $f(x) \geq g(x)$, as illustrated in Fig. 8.21. Take a partition P by points x_i with $a = x_0 < x_1 < \cdots < x_n = b$ and choose x_i^* to be the midpoint of the ith subinterval; that is, $x_i^* = (x_{i-1} + x_i)/2$. Then the centroid of the ith approximating rectangle R_i is its center $C_i\left(x_i^*, \frac{1}{2}[f(x_i^*) + g(x_i^*)]\right)$. Its area is $[f(x_i^*) - g(x_i^*)]\Delta x_i$, so its mass is $\rho[f(x_i^*) - g(x_i^*)]\Delta x_i$. Thus $M_y(R_i) = \rho[f(x_i^*) - g(x_i^*)]\Delta x_i \cdot x_i^* = \rho x_i^*[f(x_i^*) - g(x_i^*)]\Delta x_i$ and $M_x(R_i) = \rho[f(x_i^*) - g(x_i^*)]\Delta x_i \cdot \frac{1}{2}[f(x_i^*) + g(x_i^*)] = \rho \cdot \frac{1}{2}[f(x_i^*)^2 - g(x_i^*)^2]\Delta x_i$.
Summing over i and taking the limit as $\|P\| \to 0$, we get

$$M_y = \lim_{\|P\| \to 0} \sum_i \rho x_i^*[f(x_i^*) - g(x_i^*)]\Delta x_i = \rho \int_a^b x[f(x) - g(x)]dx \text{ and}$$

$$M_x = \lim_{\|P\| \to 0} \sum_i \rho \cdot \frac{1}{2}[f(x_i^*)^2 - g(x_i^*)^2]\Delta x_i = \rho \int_a^b \frac{1}{2}[f(x)^2 - g(x)^2]dx. \text{ Thus}$$

$$\bar{x} = \frac{M_y}{m} = \frac{M_y}{\rho A} = \frac{1}{A}\int_a^b x[f(x) - g(x)]dx \text{ and } \bar{y} = \frac{M_x}{m} = \frac{M_x}{\rho A} = \frac{1}{A}\int_a^b \frac{1}{2}[f(x)^2 - g(x)^2]dx.$$

EXERCISES 8.5

1. (a) $P = \rho gd = (1000\,\text{kg/m}^3)(9.8\,\text{m/s}^2)(1\,\text{m}) = 9800\,\text{Pa} = 9.8\,\text{kPa}.$

    (b) $F = \rho gdA = PA = (9800\,\text{N/m}^2)(2\,\text{m}^2) = 1.96 \times 10^4\,\text{N}.$

    (c) $F = \int_0^1 \rho gx \cdot 1\,dx = 9800 \int_0^1 x\,dx = 4900x^2]_0^1 = 4.90 \times 10^3\,\text{N}.$

3. $F = \int_0^{10} \rho gx \cdot 2\sqrt{100 - x^2}\,dx = 9.8 \times 10^3 \int_0^{10} \sqrt{100 - x^2}\,2x\,dx$

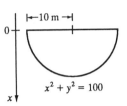

    $= 9.8 \times 10^3 \int_{100}^0 u^{1/2}(-du) \quad \left[u = 100 - x^2\right]$

    $= 9.8 \times 10^3 \int_0^{100} u^{1/2}\,du = 9.8 \times 10^3 \left[\tfrac{2}{3}u^{3/2}\right]_0^{100}$

    $= \tfrac{2}{3} \cdot 9.8 \times 10^6 \approx 6.5 \times 10^6\,\text{N}.$

5. $F = \int_{-r}^r \rho g(x + r) \cdot 2\sqrt{r^2 - x^2}\,dx = \rho g \int_{-r}^r \sqrt{r^2 - x^2}\,2x\,dx + 2\rho gr \int_{-r}^r \sqrt{r^2 - x^2}\,dx$

    $= \rho g \cdot 0 + 2\rho gr \cdot \tfrac{1}{2}\pi r^2$ [*The first integral is 0 because the integrand is an odd function*

    and the second integral is the area of a semicircular disk with radius r or use the trigonometric

    substitution $x = r\sin\theta$] $= \rho g\,\pi r^3 = 1000g\pi r^3\,\text{N}$ [*metric units assumed*].

7. $F = \int_0^6 \delta x \cdot \tfrac{2x}{3}\,dx = \left[\tfrac{2}{9}\delta x^3\right]_0^6$

    $= 48\delta \approx 48 \times 62.5 = 3000\,\text{lb}.$

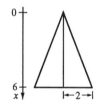

9. $F = \int_2^6 \delta(x - 2)\tfrac{2x}{3}\,dx = \tfrac{2}{3}\delta \int_2^6 (x^2 - 2x)\,dx = \tfrac{2}{3}\delta\left[\tfrac{x^3}{3} - x^2\right]_2^6 = \tfrac{2}{3}\delta\left[36 - \left(-\tfrac{4}{3}\right)\right] = \tfrac{224}{9}\delta$

    $\approx 1.56 \times 10^3\,\text{lb}.$

11. $F = \int_0^8 \delta x \cdot (12 + x)\,dx = \delta \int_0^8 (12x + x^2)\,dx$

    $= \delta\left[6x^2 + \tfrac{x^3}{3}\right]_0^8 = \delta\left(384 + \tfrac{512}{3}\right)$

    $= (62.5)\tfrac{1664}{3} \approx 3.47 \times 10^4\,\text{lb}.$

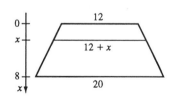

13. $F = \int_0^{4\sqrt{3}} \rho g (4\sqrt{3} - x) \frac{2x}{\sqrt{3}} \, dx = 8\rho g \int_0^{4\sqrt{3}} x \, dx - \frac{2\rho g}{\sqrt{3}} \int_0^{4\sqrt{3}} x^2 \, dx$

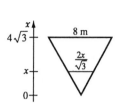

$= 4\rho g [x^2]_0^{4\sqrt{3}} - \frac{2\rho g}{3\sqrt{3}} [x^3]_0^{4\sqrt{3}} = 192\rho g - \frac{2\rho g}{3\sqrt{3}} 64 \cdot 3\sqrt{3}$

$= 192\rho g - 128\rho g = 64\rho g \approx 64(840)(9.8) \approx 5.27 \times 10^5 \, \text{N}.$

15. (a) $F = \rho g d A \approx (1000)(9.8)(.8)(.2)^2 \approx 314 \, \text{N}.$

(b) $F = \int_{.8}^1 \rho g x (.2) \, dx = .2\rho g \left[\frac{x^2}{2}\right]_{.8}^1 = (.2\rho g)(.18) = .036\rho g \approx (.036)(1000)(9.8) \approx 353 \, \text{N}.$

17. $F = \int_0^2 \rho g x \cdot 3 \cdot \sqrt{2} \, dx = 3\sqrt{2}\rho g \int_0^2 x \, dx$

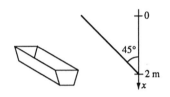

$= 3\sqrt{2}\rho g \left[\frac{x^2}{2}\right]_0^2 = 6\sqrt{2}\rho g \approx 8.32 \times 10^4 \, \text{N}.$

19. Assume that the pool is filled with water.

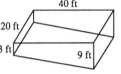

(a) $F = \int_0^3 \delta x 20 \, dx = 20\delta \left[\frac{x^2}{2}\right]_0^3 = 20\delta \cdot \frac{9}{2}$

$= 90\delta \approx 5625 \, \text{lb} \approx 5.63 \times 10^3 \, \text{lb}.$

(b) $F = \int_0^9 \delta x 20 \, dx = 20\delta \left[\frac{x^2}{2}\right]_0^9 = 810\delta \approx 50625 \, \text{lb} \approx 5.06 \times 10^4 \, \text{lb}.$

(c) $F = \int_0^3 \delta x 40 \, dx + \int_3^9 \delta x (40) \frac{9-x}{6} \, dx = 40\delta \left[\frac{x^2}{2}\right]_0^3 + \frac{20}{3}\delta \int_3^9 (9x - x^2) \, dx$

$= 180\delta + \frac{20}{3}\delta \left[\frac{9}{2}x^2 - \frac{x^3}{3}\right]_3^9 = 180\delta + \frac{20}{3}\delta \left[\left(\frac{729}{2} - 243\right) - \left(\frac{81}{2} - 9\right)\right] = 780\delta \approx 4.88 \times 10^4 \, \text{lb}.$

(d) $F = \int_3^9 \delta x 20 \frac{\sqrt{409}}{3} \, dx$

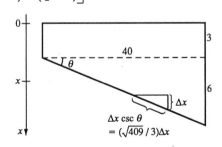

$= \frac{1}{3}(20\sqrt{409})\delta \left[\frac{x^2}{2}\right]_3^9$

$= \frac{1}{3} \cdot 10\sqrt{409}\delta(81 - 9)$

$\approx 3.03 \times 10^5 \, \text{lb}.$

21. $\bar{x} = \frac{1}{A} \int_a^b x w(x) \, dx$ (Formula 8.36) $\Rightarrow A\bar{x} = \int_a^b x w(x) \, dx \Rightarrow (\rho g \bar{x}) A = \int_a^b \rho g x w(x) \, dx = F$ by

Exercise 20.

EXERCISES 8.6

1. $C(2000) = C(0) + \int_0^{2000} C'(x)\,dx = 1{,}500{,}000 + \int_0^{2000} (0.006x^2 - 1.5x + 8)\,dx$

$= 1{,}500{,}000 + \left[0.002x^3 - 0.75x^2 + 8x\right]_0^{2000} = \$14{,}516{,}000.$

3. $C(5000) - C(3000) = \int_{3000}^{5000} (140 - 0.5x + 0.012x^2)\,dx = \left[140x - 0.25x^2 + 0.004x^3\right]_{3000}^{5000}$

$= 494{,}450{,}000 - 106{,}170{,}000 = \$388{,}280{,}000.$

5. $p(x) = 20 = \dfrac{1000}{x+20} \Rightarrow X + 20 = 50 \Rightarrow X = 30.$

Consumer's surplus $= \int_0^{30} [p(x) - 20]\,dx = \int_0^{30}\left(\dfrac{1000}{x+20} - 20\right)dx$

$= \left[1000\ln(x+20) - 20x\right]_0^{30} = 1000\ln\left(\dfrac{50}{20}\right) - 600 = 1000\ln\left(\dfrac{5}{2}\right) - 600 \approx \$316.29$

7. $P = p(x) = 10 = 5 + \frac{1}{10}\sqrt{x} \Rightarrow 50 = \sqrt{x} \Rightarrow x = 2500.$

Producer's surplus $= \int_0^{2500} [P - p(x)]\,dx = \int_0^{2500}\left(10 - 5 - \frac{1}{10}\sqrt{x}\right)dx$

$= \left[5x - \frac{1}{15}x^{3/2}\right]_0^{2500} \approx \$4166.67$

9. The demand function is linear, with slope $\dfrac{-10}{100}$ and $p(1000) = 450$. So, its equation is

$p - 450 = -\frac{1}{10}(x - 1000)$, or $p = -\frac{1}{10}x + 550.$

A selling price of $\$400 \Rightarrow 400 = -\frac{1}{10}x + 550 \Rightarrow x = 1500.$

Consumer's surplus $= \int_0^{1500}\left(550 - \frac{1}{10}x - 400\right)dx = \left[150x - \frac{1}{20}x^2\right]_0^{1500} = \$112{,}500.$

11. Pretend that it's five years later. Then the fund will start in five years and continue for 15

years, so the present value (five years from now) is

$\int_5^{20} 12{,}000e^{-.11t}\,dt = -\dfrac{12{,}000}{0.11}\left[e^{-.11t}\right]_5^{20} = \dfrac{12{,}000}{.11}(e^{-.55} - e^{-2.2}) \approx \$50{,}852.36.$

13. (a) $f(t) = A$, so Present Value $= \int_0^\infty Ae^{-rt}\,dt = \lim_{x\to\infty}\int_0^x Ae^{-rt}\,dt = \lim_{x\to\infty} -\frac{A}{r}\left[e^{-rt}\right]_0^x$

$= \lim_{x\to\infty} -\frac{A}{r}\left[e^{-rx} - 1\right] = \frac{A}{r}$ [since $r > 0$, $e^{-rx} \to 0$ as $x \to \infty$].

(b) $r = 0.1$, $A = 5000$, so Present Value $= \dfrac{5000}{0.1} = \$50{,}000$ [by (a)].

15. $f(8) - f(4) = \int_4^8 f'(t)\,dt = \int_4^8 \sqrt{t}\,dt = \frac{2}{3}t^{3/2}\Big]_4^8 = \frac{2}{3}(16\sqrt{2} - 8) = \dfrac{16(2\sqrt{2} - 1)}{3}$

$\approx \$9.75$ million.

17. $F = \dfrac{\pi PR^4}{8\eta\ell} = \dfrac{\pi(4000)(0.008)^4}{8(0.027)(2)} \approx 1.19 \times 10^{-4}$ cm^3/s.

19. $\int_0^{12} c(t)\,dt = \int_0^{12} \frac{1}{4}t(12-t)\,dt = \left[\frac{3}{2}t^2 - \frac{1}{12}t^3\right]_0^{12} = \frac{144}{2} = 72$ mg$\cdot$s/L

Therefore, $F = \frac{A}{72} = \frac{8}{72} = \frac{1}{9}$ L/s $= \frac{60}{9}$ L/min.

REVIEW EXERCISES FOR CHAPTER 8

1. $y^2\dfrac{dy}{dx} = x + \sin x \Rightarrow \int y^2\,dy = \int (x + \sin x)\,dx \Rightarrow \dfrac{y^3}{3} = \dfrac{x^2}{2} - \cos x + C \Rightarrow$

   $y^3 = \frac{3}{2}x^2 - 3\cos x + K$ [*where* $K = 3C$] $\Rightarrow y = \sqrt[3]{3x^2/2 - 3\cos x + K}.$

3. $y' = \dfrac{1}{x^2y - 2x^2 + y - 2} \Rightarrow \dfrac{dy}{dx} = \dfrac{1}{(x^2+1)(y-2)} \Rightarrow \int (y-2)\,dy = \int \dfrac{dx}{x^2+1}$

   $\Rightarrow \dfrac{y^2}{2} - 2y = \tan^{-1}x + K \Rightarrow y = 2 \pm \sqrt{2\tan^{-1}x + C}$, where $C = 4 + 2K$.

5. $xyy' = \ln x$, $y(1) = 2$. $y\,dy = \dfrac{\ln x}{x}\,dx \Rightarrow \int y\,dy = \int \dfrac{\ln x}{x}\,dx$ [*Make the substitution*

   $u = \ln x;$ *then* $du = \frac{1}{x}\,dx$] So, $\int y\,dy = \int u\,du \Rightarrow \frac{1}{2}y^2 = \frac{1}{2}u^2 + C \Rightarrow \frac{1}{2}y^2 = \frac{1}{2}(\ln x)^2 + C.$

   $y(1) = 2 \Rightarrow \frac{1}{2}2^2 = \frac{1}{2}(\ln 1)^2 + C = C \Leftrightarrow C = 2.$ Therefore, $\frac{1}{2}y^2 = \frac{1}{2}(\ln x)^2 + 2$, or

   $y = \sqrt{(\ln x)^2 + 4}$ (since $y(1) > 0$).

7. $3x = 2(y-1)^{3/2}$, $2 \le y \le 5$. $x = \frac{2}{3}(y-1)^{3/2}$, so $\dfrac{dx}{dy} = (y-1)^{1/2}$ and $1 + \left(\dfrac{dx}{dy}\right)^2$

   $= 1 + (y-1) = y \Rightarrow L = \int_2^5 \sqrt{1 + \left(\dfrac{dx}{dy}\right)^2}\,dy = \int_2^5 \sqrt{y}\,dy = \left[\frac{2}{3}y^{3/2}\right]_2^5 = \frac{2}{3}(5\sqrt{5} - 2\sqrt{2}).$

9. (a) $y = \frac{1}{6}x^3 + \frac{1}{2x}$, $1 \le x \le 2 \Rightarrow y' = \frac{1}{2}\left(x^2 - \frac{1}{x^2}\right) \Rightarrow (y')^2 = \frac{1}{4}\left(x^4 - 2 + \frac{1}{x^4}\right)$

   $\Rightarrow 1 + (y')^2 = \frac{1}{4}\left(x^4 + 2 + \frac{1}{x^4}\right) = \frac{1}{4}\left(x^2 + \frac{1}{x^2}\right)^2 \Rightarrow L = \int_1^2 \sqrt{1 + (y')^2}\,dy = \frac{1}{2}\int_1^2 \left(x^2 + \frac{1}{x^2}\right)dx$

   $= \frac{1}{2}\left[\frac{1}{3}x^3 - \frac{1}{x}\right]_1^2 = \frac{1}{2}\left(\frac{17}{6}\right) = \frac{17}{12}.$

   (b) $S = \int_1^2 2\pi y \sqrt{1 + \left(\dfrac{dy}{dx}\right)^2}\,dx = 2\pi \int_1^2 \left(\frac{x^3}{6} + \frac{1}{2x}\right)\frac{1}{2}\left(x^2 + \frac{1}{x^2}\right)dx$

   $= \pi \int_1^2 \left(\frac{x^5}{6} + \frac{2x}{3} + \frac{x^{-3}}{2}\right)dx = \pi\left[\frac{x^6}{36} + \frac{x^2}{3} - \frac{x^{-2}}{4}\right]_1^2$

   $= \pi\left[\left(\frac{64}{36} + \frac{4}{3} - \frac{1}{16}\right) - \left(\frac{1}{36} + \frac{1}{3} - \frac{1}{4}\right)\right] = \frac{47\pi}{16}.$

11. $y = \dfrac{1}{x^2}$, $1 \le x \le 2$. $\dfrac{dy}{dx} = \dfrac{-2}{x^3}$, so $1 + \left(\dfrac{dy}{dx}\right)^2 = 1 + \dfrac{4}{x^6}$. $L = \int_1^2 \sqrt{1 + \dfrac{4}{x^6}}\,dx$. By Simpson's

Rule with n = 10, $L \approx \dfrac{1/10}{3}\Big(f(1) + 4f(1.1) + 2f(1.2) + 4f(1.3) + 2f(1.4) + 4f(1.5)$

$+ 2f(1.6) + 4f(1.7) + 2f(1.8) + 4f(1.9) + f(2)\Big) \approx 1.297$ [where $f(x) = \sqrt{1 + \dfrac{4}{x^6}}$].

13. The loop lies between x = 0 and x = 3a and is symmetric about the x-axis. We can assume without loss of generality that a > 0. [We must have a ≠ 0; otherwise the curve consists of the parallel lines x = 0 and x = 3a, so there is no loop.] The upper half of the loop is given by
$y = [1/(3\sqrt{a})]\sqrt{x}(3a - x) = \sqrt{a}x^{1/2} - x^{3/2}/(3\sqrt{a})$, $0 \le x \le 3a$. The desired surface area is twice the area generated by the upper half of the loop; i.e.,

$$S = 2(2\pi)\int_0^{3a} x\sqrt{1 + \left(\frac{dy}{dx}\right)^2}\, dx. \quad \frac{dy}{dx} = \frac{\sqrt{a}}{2}x^{-1/2} - \frac{x^{1/2}}{2\sqrt{a}} \Rightarrow 1 + \left(\frac{dy}{dx}\right)^2 = \frac{a}{4x} + \frac{1}{2} + \frac{x}{4a}.$$

Therefore $S = 2(2\pi)\displaystyle\int_0^{3a} x\left(\frac{\sqrt{a}}{2}x^{-1/2} + \frac{x^{1/2}}{2\sqrt{a}}\right)dx = 2\pi\int_0^{3a}\left(\sqrt{a}x^{1/2} + \frac{x^{3/2}}{\sqrt{a}}\right)dx$

$= 2\pi\left[\dfrac{2\sqrt{a}}{3}x^{3/2} + \dfrac{2}{5\sqrt{a}}x^{5/2}\right]_0^{3a} = 2\pi\left[\dfrac{2\sqrt{a}}{3}3a\sqrt{3a} + \dfrac{2}{5\sqrt{a}}9a^2\sqrt{3a}\right] = \dfrac{56\sqrt{3}\pi a^2}{5}.$

15. $A = \int_{-2}^{1}\left[(4 - x^2) - (x + 2)\right]dx = \int_{-2}^{1}(2 - x - x^2)\,dx = \left[2x - \frac{1}{2}x^2 - \frac{1}{3}x^3\right]_{-2}^{1}$

$= \left(2 - \frac{1}{2} - \frac{1}{3}\right) - \left(-4 - 2 + \frac{8}{3}\right) = \frac{9}{2} \Rightarrow \bar{x} = \frac{1}{A}\int_{-2}^{1} x(2 - x - x^2)\,dx$

$= \frac{2}{9}\int_{-2}^{1}(2x - x^2 - x^3)\,dx = \frac{2}{9}\left[x^2 - \frac{1}{3}x^3 - \frac{1}{4}x^4\right]_{-2}^{1} = \frac{2}{9}\left[\left(1 - \frac{1}{3} - \frac{1}{4}\right) - \left(4 + \frac{8}{3} - 4\right)\right]$

$= -\frac{1}{2}$ and $\bar{y} = \frac{1}{A}\int_{-2}^{1}\frac{1}{2}\left[(4 - x^2)^2 - (x + 2)^2\right]dx = \frac{1}{9}\int_{-2}^{1}(x^4 - 9x^2 - 4x + 12)\,dx$

$= \frac{1}{9}\left[\frac{1}{5}x^5 - 3x^3 - 2x^2 + 12x\right]_{-2}^{1} = \frac{1}{9}\left[\left(\frac{1}{5} - 3 - 2 + 12\right) - \left(-\frac{32}{5} + 24 - 8 - 24\right)\right] = \frac{12}{5}.$

$(\bar{x}, \bar{y}) = \left(-\frac{1}{2}, \frac{12}{5}\right).$

17. The equation of the line passing through (0, 0) and (3, 2) is $y = \frac{2}{3}x$. $A = \frac{1}{2}3 \cdot 2 = 3.$

Therefore, $\bar{x} = \frac{1}{3}\int_0^3 x\frac{2}{3}x\,dx = \frac{2}{27}\left[x^3\right]_0^3 = 2$, and $\bar{y} = \frac{1}{3}\int_0^3\frac{1}{2}\left(\frac{2}{3}x\right)^2 dx = \frac{2}{81}\left[x^3\right]_0^3 = \frac{2}{3}.$

$(\bar{x}, \bar{y}) = \left(2, \frac{2}{3}\right).$ [Or, use Exercise 8.4.23.]

19. The centroid of this circle, (1, 0), travels a distance $2\pi(1)$ when the lamina is rotated about the y-axis. The area of the circle is $\pi(1)^2$. By the Theorem of Pappus

Volume $= A2\pi\bar{x} = \pi(1)^2 2\pi(1) = 2\pi^2.$

21. As in Example 1 of Section 8.5,

$F = \int_0^2 \rho g x(5 - x)\,dx = \rho g\left[\frac{5}{2}x^2 - \frac{1}{3}x^3\right]_0^2 = \rho g\frac{22}{3} = \frac{22}{3}\delta \approx \frac{22}{3}62.5 \approx 458$ lb.

23. $x = 100 \Rightarrow P = 2000 - 0.1(100) - 0.01(100)^2 = 1890$

Consumer's surplus $= \int_0^{100}\left[p(x) - P\right]dx = \int_0^{100}(2000 - 0.1x - 0.01x^2 - 1890)\,dx$

$$= \left[110x - 0.05x^2 - \frac{0.01}{3}x^3\right]_0^{100} = 11{,}000 - 500 - \frac{10{,}000}{3} \approx \$7166.67$$

25.　(a) $\frac{dL}{dt} \propto L_\infty - L \Rightarrow \frac{dL}{dt} = k(L_\infty - L) \Rightarrow \int \frac{dL}{L_\infty - L} = \int k\,dt$

$\Rightarrow -\ln|L_\infty - L| = kt + C \Rightarrow L_\infty - L = Ae^{-kt} \Rightarrow L = L_\infty - Ae^{-kt}$. At $t = 0$,

$L = L(0) = L_\infty - A \Rightarrow A = L_\infty - L(0) \Rightarrow L(t) = L_\infty - (L_\infty - L(0))e^{-kt}$.

(b) $L_\infty = 53$ cm, $L(0) = 10$ cm and $k = 0.2$.

So, $L(t) = 53 - (53 - 10)e^{-0.2t} = 53 - 43e^{-0.2t}$ cm.

27.　First note that, in this question, "weighs" is used in the informal sense, so what we really
require is Barbara's mass m in kg as a function of t. Barbara's net intake of calories per day at
time t (measured in days) is $c(t) = 1600 - 850 - 15\,m(t)$

$= 750 - 15\,m(t)$ where $m(t)$ is her mass at time t. We are given that $m(0) = 60\,\text{kg}$ and

$\frac{dm}{dt} = \frac{c(t)}{10{,}000}$, so $\frac{dm}{dt} = \frac{750 - 15m}{10{,}000} = \frac{150 - 3m}{2000} = \frac{-3(m - 50)}{2000}$ with $m(0) = 60$. From

$\int \frac{dm}{m - 50} = \int \frac{-3\,dt}{2000}$, we get $\ln|m - 50| = \frac{-3t}{2000} + C$. Since $m(0) = 60$, $C = \ln 10$. Now

$\ln(|m - 50|/10) = -3t/2000$, so $|m - 50| = 10e^{-3t/2000}$. The quantity $m - 50$ is continuous
and initially positive; the right-hand side is never zero. Thus $m - 50$ is positive for all t, and
$m(t) = 50 + 10e^{-3t/2000}$ kg. As $t \to \infty$, $m(t) \to 50\,\text{kg}$. Thus Barbara's mass gradually
settles down to 50 kg.

PROBLEMS PLUS (page 510)

1. **(a)** $\int_0^a f(x)\,dx = -\int_a^0 f(a-u)\,du \ \ [u = a-x] = \int_0^a f(a-u)\,du = \int_0^a f(a-x)\,dx.$

   (b) Let $I = \int_0^{\pi/2} \dfrac{\sin^n x}{\sin^n x + \cos^n x}\,dx = \int_0^{\pi/2} \dfrac{\sin^n\left(\frac{\pi}{2} - x\right)}{\sin^n\left(\frac{\pi}{2} - x\right) + \cos^n\left(\frac{\pi}{2} - x\right)}\,dx \ \ [by\ (a)]$

   $= \int_0^{\pi/2} \dfrac{\cos^n x}{\cos^n x + \sin^n x}\,dx.$

   Therefore, $2I = \int_0^{\pi/2} \dfrac{\sin^n x}{\sin^n x + \cos^n x}\,dx + \int_0^{\pi/2} \dfrac{\cos^n x}{\sin^n x + \cos^n x}\,dx$

   $= \int_0^{\pi/2} \dfrac{\sin^n x + \cos^n x}{\sin^n x + \cos^n x}\,dx = \int_0^{\pi/2}\,dx = \dfrac{\pi}{2} \Rightarrow I = \dfrac{\pi}{4}.$

3. $f'(x) = \lim\limits_{h \to 0} \dfrac{f(x+h) - f(x)}{h} = \lim\limits_{h \to 0} \dfrac{f(x)f(h) - f(x)}{h} \quad [since\ f(x+h) = f(x)f(h)]$

   $= \lim\limits_{h \to 0} \dfrac{f(x)[f(h) - 1]}{h} = f(x)\lim\limits_{h \to 0}\dfrac{f(h) - 1}{h} = f(x)\lim\limits_{h \to 0}\dfrac{f(h) - f(0)}{h - 0} = f(x)f'(0) = f(x).$

   Therefore, $f'(x) = f(x)$ for all x. In Leibniz notation,

   $\dfrac{dy}{dx} = y \Rightarrow \int \dfrac{dy}{y} = \int dx \Rightarrow \ln|y| = x + C \Rightarrow y = Ae^x$ (or use Theorem 6.59).

   $f(0) = 1 \Rightarrow A = 1 \Rightarrow f(x) = e^x.$

5. First we show that $x(1 - x) \le \frac{1}{4}$ for all x. Let $f(x) = x(1 - x) = x - x^2$. Then $f'(x) = 1 - 2x$
   $= 0$ when $x = \frac{1}{2}$ and $f'(x) > 0$ for $x < \frac{1}{2}$, $f'(x) < 0$ for $x > \frac{1}{2}$, so the absolute maximum of f is
   $f(\frac{1}{2}) = \frac{1}{4}$. Thus $x(1 - x) \le \frac{1}{4}$ for all x.

   Now suppose that the given assertion is false, i.e., $a(1-b) > \frac{1}{4}$ and $b(1-a) > \frac{1}{4}$.
   Multiply these inequalities: $a(1 - b)b(1 - a) > \frac{1}{16} \Rightarrow [a(1 - a)][b(1 - b)] > \frac{1}{16}$. But we know
   that $a(1 - a) \le \frac{1}{4}$ and $b(1 - b) \le \frac{1}{4} \Rightarrow [a(1 - a)][b(1 - b)] \le \frac{1}{16}$. Thus we have a
   contradiction, so the given assertion is proved.

7. Let $F(x) = \int_a^x f(t)\,dt - \int_x^b f(t)\,dt$. Then $F(a) = -\int_a^b f(t)\,dt$ and $F(b) = \int_a^b f(t)\,dt$. Also F is

   continuous by the Fundamental Theorem. So, by the Intermediate Value Theorem, there is a

   number x in $[a, b]$ such that $F(x) = 0$. For that number x, we have $\int_a^x f(t)\,dt = \int_x^b f(t)\,dt.$

9. Make the substitution $u = \sqrt{x} \Rightarrow 2\,du = \dfrac{dx}{\sqrt{x}}$, and $-u^2 = -x$. Then,

   $\int_1^\infty \dfrac{e^{-x}}{\sqrt{x}}\,dx = \lim\limits_{t \to \infty} \int_1^t \dfrac{e^{-x}}{\sqrt{x}}\,dx = \lim\limits_{t \to \infty} \int_1^{\sqrt{t}} e^{-u^2} \cdot 2\,du = 2\int_1^\infty e^{-u^2}\,du = 2\int_1^\infty e^{-x^2}\,dx$

11. First we find the domain explicitly. $5x^2 \geq x^4 + 4 \Leftrightarrow x^4 - 5x^2 + 4 \leq 0 \Leftrightarrow (x^2 - 4)(x^2 - 1) \leq 0$
$\Leftrightarrow (x-2)(x-1)(x+1)(x+2) \leq 0 \Leftrightarrow x \in [-2, -1]$ or $[1, 2]$. Therefore, the domain is
$\{x \mid 5x^2 \geq x^4 + 4\} = [-2, -1] \cup [1, 2]$. $f(x) = 3x - 2x^3 \Rightarrow f'(x) = 3 - 6x^2 \leq -3$. Since f is
decreasing on its domain, its maximum value is either $f(-2)$ or $f(1)$. $f(-2) = 10$ and $f(1) = 1$.
So the maximum value of f is 10.

13. To differentiate this function, use the Fundamental Theorem of Calculus.

(a) $y = \int_1^x \sqrt{te^t - 1}\,dt$, $x \geq 1 \Rightarrow y' = \sqrt{xe^x - 1}$, $x \geq 1 \Rightarrow y'' = \dfrac{e^x(1+x)}{2\sqrt{xe^x - 1}} \geq 0$ always, when

$x \geq 1$. Thus the function is concave upward on its entire domain.

(b) $y' = \sqrt{xe^x - 1} \Rightarrow 1 + (y')^2 = xe^x \Rightarrow L = \int_1^2 \sqrt{xe^x}\,dx = \left[xe^x\right]_1^2 - \int_1^2 e^x\,dx$ [after integrating

by parts] $= 2e^2 - e - \left[e^x\right]_1^2 = e^2$.

15. $0 < a < b$. $\int_0^1 [bx + a(1-x)]^t\,dx = \int_a^b \dfrac{u^t}{(b-a)}\,du$ $[u = bx + a(1-x)]$

$= \left[\dfrac{u^{t+1}}{(t+1)(b-a)}\right]_a^b = \dfrac{b^{t+1} - a^{t+1}}{(t+1)(b-a)}$.

Let $y = \lim\limits_{t \to 0}\left[\dfrac{b^{t+1} - a^{t+1}}{(t+1)(b-a)}\right]^{1/t}$. Then $\ln y = \lim\limits_{t \to 0} \dfrac{1}{t}\ln\left[\dfrac{b^{t+1} - a^{t+1}}{(t+1)(b-a)}\right]$.

This limit is of the form $0/0$, so we can apply l'Hospital's Rule to get

$\ln y = \lim\limits_{t \to 0}\left[\dfrac{b^{t+1}\ln b - a^{t+1}\ln a}{b^{t+1} - a^{t+1}} - \dfrac{1}{t+1}\right] = \dfrac{b\ln b - a\ln a}{b-a} - 1 = \dfrac{b}{b-a}\ln b - \dfrac{a}{b-a}\ln a - \ln e$

$= \ln\left(\dfrac{b^{b/(b-a)}}{ea^{a/(b-a)}}\right)$. Therefore, $y = e^{-1}\left(\dfrac{b^b}{a^a}\right)^{1/(b-a)}$

CHAPTER NINE

EXERCISES 9.1

1.   (a)

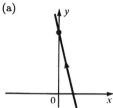

(b) $x = 1 - t, y = 2 + 3t$

$y = 2 + 3(1 - x) = 5 - 3x$, so $3x + y = 5$

3.   (a)

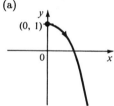

(0, 2)

(12, 12)

(b) $x = 3t^2, y = 2 + 5t, 0 \le t \le 2$

$x = 3\left[\dfrac{y - 2}{5}\right]^2 = \dfrac{3}{25}(y - 2)^2, 2 \le y \le 12$

5.   (a)

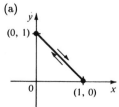

(0, 1)

(b) $x = \sqrt{t}, y = 1 - t$

$y = 1 - t = 1 - x^2, x \ge 0$

7.   (a)

(0, 1)

(0, −1)

(b) $x = \sin\theta, y = \cos\theta, 0 \le \theta \le \pi$

$x^2 + y^2 = \sin^2\theta + \cos^2\theta = 1, 0 \le x \le 1$

9.   (a)

(0, 1)

(1, 0)

(b) $x = \sin^2\theta, y = \cos^2\theta$

$x + y = \sin^2\theta + \cos^2\theta = 1, 0 \le x \le 1$

303

11. (a)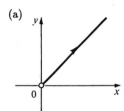

(b) $x = e^t,\ y = e^t$

$y = x,\ x > 0$

13. (a)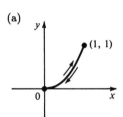

(b) $x = \cos^2 t,\ y = \cos^4 t$

$y = x^2,\ 0 \le x \le 1$

15. (a)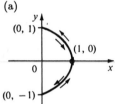

(b) $x = \cos^2 \theta,\ y = \sin \theta$

$x + y^2 = \cos^2 \theta + \sin^2 \theta = 1,\ -1 \le y \le 1$

17. (a)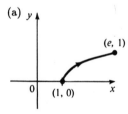

(b) $x = e^t,\ y = \sqrt{t}$

$x = e^{y^2},\ 0 \le y \le 1$

or: $y = \sqrt{\ln x},\ 1 \le x \le e$

19. (a)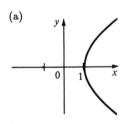

(b) $x = \cosh t,\ y = \sinh t$

$x^2 - y^2 = \cosh^2 t - \sinh^2 t = 1,\ x \ge 1$

21. $x^2 + y^2 = \cos^2 \pi t + \sin^2 \pi t = 1,\ 1 \le t \le 2$, so the particle moves counterclockwise along the circle $x^2 + y^2 = 1$ from $(-1, 0)$ to $(1, 0)$, along the lower half of the circle.

23. $x = 8t - 3,\ y = 2 - t,\ 0 \le t \le 1,\ \Rightarrow\ x = 8(2 - y) - 3 = 13 - 8y$, so the particle moves along the line $x + 8y = 13$ from $(-3, 2)$ to $(5, 1)$.

25. $(x/2)^2 + (y/3)^2 = \sin^2 t + \cos^2 t = 1$, so the particle moves once clockwise along the ellipse $\frac{x^2}{4} + \frac{y^2}{9} = 1$, starting and ending at $(0, 3)$.

27. $x = 3(t^2 - 3)$ and $y = t(t^2 - 3)$

t	-4	-3	-2	$-\sqrt{3}$	-1.5	-1	0	1	1.5	$\sqrt{3}$	2	3	4
x	39	18	3	0	-2.25	-6	-9	-6	-2.25	0	3	18	39
y	-52	-18	-2	0	1.125	2	0	-2	-1.125	0	2	18	52

Note that replacing t by $-t$ changes the sign of y but leaves x unchanged, so the curve is symmetric about the x-axis. Also $3y = tx$, so $9y^2 = t^2 x^2 = (\frac{1}{3}x + 3)x^2$ and $27y^2 = x^3 + 9x^2$. For x near 0, $27y^2 \approx 9x^2$, so $y \approx \pm x/\sqrt{3}$. For large x, $27y^2 \approx x^3$, so $y \approx \pm (x/3)^{3/2}$. The graph has a loop in it.

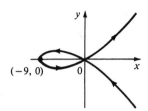

29. $x = \tan\theta + \sin\theta$, $y = \cos\theta$. Replacing θ by $-\theta$ changes the sign of x and leaves y unchanged, so the curve is symmetric about the y-axis. Clearly $|y| \leq 1$, but $x \to \pm\infty$ as $\theta \to \pm\pi/2$. The full curve is obtained from values of θ in $[-\pi, \pi]$ (but is not defined for $\theta = \pm\pi/2$). $y > 0$ for θ in $(-\pi/2, \pi/2)$; $y < 0$ for θ in $[-\pi, -\pi/2)$ and $(\pi/2, \pi]$. As $\theta \to (\pi/2)^-$, $x \to \infty$ and $y \to 0^+$. As $\theta \to (\pi/2)^+$, $x \to -\infty$ and $y \to 0^-$.

θ	$-\pi$	$-5\pi/6$	$-3\pi/4$	$-2\pi/3$	$-5\pi/9$	$-4\pi/9$	$-\pi/3$	$-\pi/4$	$-\pi/6$	0
x	0	.077	.293	.866	4.69	-6.66	-2.60	-1.71	-1.08	0
y	-1	$-.866$	$-.707$	$-.5$	$-.174$	.174	.5	.707	.866	1

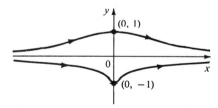

31. Clearly the curve passes through (x_1, y_1) when $t = 0$ and through (x_2, y_2) when $t = 1$. For $0 < t < 1$, x is strictly between x_1 and x_2 and y is strictly between y_1 and y_2. For every value of t, x and y satisfy the relation $y - y_1 = \frac{y_2 - y_1}{x_2 - x_1}(x - x_1)$, which is the equation of the straight line through (x_1, y_1) and (x_2, y_2). Finally, any point (x, y) on that line satisfies $\frac{y - y_1}{y_2 - y_1} = \frac{x - x_1}{x_2 - x_1}$; if we call that common value t, then the given parametric equations yield the point (x, y); and any (x, y) on the line between (x_1, y_1) and (x_2, y_2) yields a value of t in $[0, 1]$. This proves that the given parametric equations exactly specify the line segment from (x_1, y_1) to (x_2, y_2).

33. The case $\pi/2 < \theta < \pi$ is illustrated to the right.
    C has coordinates $(r\theta, r)$ as before, and Q has
    coordinates $(r\theta, r + r[\cos(\pi - \theta)])$
    $= (r\theta, r(1 - \cos\theta))$, so P has coordinates
    $(r\theta - r\sin(\pi - \theta), r(1 - \cos\theta))$
    $= (r(\theta - \sin\theta), r(1 - \cos\theta))$. Again we have the
    parametric equations $x = r(\theta - \sin\theta)$, $y = r(1 - \cos\theta)$.

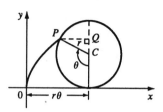

35. It is apparent that $x = |OQ|$ and $y = |QP| = |ST|$.
    From the diagram $x = |OQ| = a\cos\theta$ and
    $y = |ST| = b\sin\theta$. Thus the parametric equations
    are $x = a\cos\theta$ and $y = b\sin\theta$. To eliminate θ we
    rearrange:
    $\sin\theta = \dfrac{y}{b} \Rightarrow \sin^2\theta = \left(\dfrac{y}{b}\right)^2$ and $\cos\theta = \dfrac{x}{a}$
    $\Rightarrow \cos^2\theta = \left(\dfrac{x}{a}\right)^2$. Adding the two equations:
    $\sin^2\theta + \cos^2\theta = 1 = \dfrac{x^2}{a^2} + \dfrac{y^2}{b^2}$. Thus we have an ellipse.

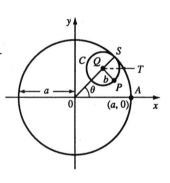

37. (a) The center Q of the smaller circle has coordinates
        $((a - b)\cos\theta, (a - b)\sin\theta)$. Arc PS on circle C has
        length $a\theta$ since it is equal in length to arc AS (the smal-
        ler circle rolls without slipping against the larger).
        Thus $\angle PQS = a\theta/b$ and $\angle PQT = (a\theta/b) - \theta$, so P
        has coordinates $x = (a - b)\cos\theta + b\cos(\angle PQT)$
        $= (a - b)\cos\theta + b\cos\left[\dfrac{a - b}{b}\theta\right]$, and
        $y = (a - b)\sin\theta - b\sin(\angle PQT)$
        $= (a - b)\sin\theta - b\sin\left[\dfrac{a - b}{b}\theta\right]$.

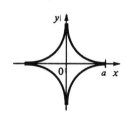

    (b) If $b = a/4$, then $a - b = \dfrac{3a}{4}$ and $\dfrac{a - b}{b} = 3$, so
        $x = \dfrac{3a}{4}\cos\theta + \dfrac{a}{4}\cos 3\theta =$
        $\dfrac{3a}{4}\cos\theta + \dfrac{a}{4}(4\cos^3\theta - 3\cos\theta) = a\cos^3\theta$,
        and $y = \dfrac{3a}{4}\sin\theta - \dfrac{a}{4}\sin 3\theta =$
        $\dfrac{3a}{4}\sin\theta - \dfrac{a}{4}(3\sin\theta - 4\sin^3\theta) = a\sin^3\theta$

        The curve is symmetric about the origin.

39. $C = (2a \cot \theta, 2a)$, so the x-coordinate of P is $x = 2a \cot \theta$. Let $B = (0, 2a)$. Then $\angle OAB$ is a right angle and $\angle OBA = \theta$, so $|OA| = 2a \sin \theta$ and $A = (2a \sin \theta \cos \theta, 2a \sin^2\theta)$. Thus the y-coordinate of P is $y = 2a \sin^2\theta$.

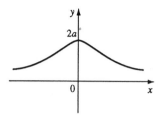

EXERCISES 9.2

1. $x = t^2 + t$, $y = t^2 - t$; $t = 0$. $\frac{dy}{dt} = 2t - 1$, $\frac{dx}{dt} = 2t + 1$, so $\frac{dy}{dx} = \frac{dy/dt}{dx/dt} = \frac{2t-1}{2t+1}$. When $t = 0$, $x = y = 0$ and $\frac{dy}{dx} = -1$. The tangent is $y - 0 = (-1)(x - 0)$ or $y = -x$.

3. $x = t^2 + t$, $y = \sqrt{t}$; $t = 4$. $\frac{dy}{dt} = \frac{1}{2\sqrt{t}}$, $\frac{dx}{dt} = 2t + 1$, so $\frac{dy}{dx} = \frac{dy/dt}{dx/dt} = \frac{1}{2\sqrt{t}(2t+1)}$. When $t = 4$, $(x, y) = (20, 2)$ and $\frac{dy}{dx} = \frac{1}{36}$, so the equation of the tangent is $y - 2 = \frac{1}{36}(x - 20)$ or $x - 36y + 52 = 0$.

5. $x = 2\sin\theta$, $y = 3\cos\theta$; $\theta = \pi/4$. $\frac{dx}{d\theta} = 2\cos\theta$, $\frac{dy}{d\theta} = -3\sin\theta$, $\frac{dy}{dx} = \frac{dy/d\theta}{dx/d\theta} = -\frac{3}{2}\tan\theta$. When $\theta = \pi/4$, $(x, y) = (\sqrt{2}, 3\sqrt{2}/2)$, and $\frac{dy}{dx} = -\frac{3}{2}$, so the equation of the tangent is $y - \frac{3\sqrt{2}}{2} = (-\frac{3}{2})(x - \sqrt{2})$ or $3x + 2y = 6\sqrt{2}$.

7. (a) $x = 1 - t$, $y = 1 - t^2$; $(1, 1)$. $\frac{dy}{dt} = -2t$, $\frac{dx}{dt} = -1$, $\frac{dy}{dx} = \frac{dy/dt}{dx/dt} = 2t$. At $(1, 1)$, $t = 0$, so $\frac{dy}{dx} = 0$, and the tangent is $y - 1 = 0(x - 1)$ or $y = 1$.

 (b) $y = 1 - t^2 = 1 - (1 - x)^2 = 2x - x^2$, so $\left[\frac{dy}{dx}\right]_{x=1} = [2 - 2x]_{x=1} = 0$, and as in (a) the tangent is $y = 1$.

9. (a) $x = 5\cos t$, $y = 5\sin t$; $(3, 4)$. $\frac{dy}{dt} = 5\cos t$, $\frac{dx}{dt} = -5\sin t$, $\frac{dy}{dx} = \frac{dy/dt}{dx/dt} = -\cot t$. At $(3, 4)$, $t = \tan^{-1}(y/x) = \tan^{-1}(4/3)$, so $\frac{dy}{dx} = -3/4$, and the tangent is $y - 4 = -(\frac{3}{4})(x - 3)$, or $3x + 4y = 25$.

(b) $x^2 + y^2 = 25$, so $2x + 2y\dfrac{dy}{dx} = 0$, or $\dfrac{dy}{dx} = -x/y$. At $(3, 4)$, $\dfrac{dy}{dx} = -\dfrac{3}{4}$, and
as in (a) the tangent is $3x + 4y = 25$.

11. $x = t^2 + t$, $y = t^2 + 1$. $\dfrac{dy}{dx} = \dfrac{dy/dt}{dx/dt} = \dfrac{2t}{2t + 1} = 1 - \dfrac{1}{2t + 1}$. $\dfrac{d}{dt}\left(\dfrac{dy}{dx}\right) = \dfrac{2}{(2t + 1)^2}$

$\dfrac{d^2y}{dx^2} = \dfrac{d}{dx}\left(\dfrac{dy}{dx}\right) = \dfrac{d(dy/dx)/dt}{dx/dt} = \dfrac{2}{(2t + 1)^3}$

13. $x = \sqrt{t + 1}$, $y = t^2 - 3t$. $\dfrac{dy}{dx} = \dfrac{dy/dt}{dx/dt} = \dfrac{2t - 3}{1/\left[2\sqrt{t + 1}\right]} = 2(2t - 3)\sqrt{t + 1}$

$\dfrac{d}{dt}\left(\dfrac{dy}{dx}\right) = 4\sqrt{t + 1} + 2(2t - 3)\dfrac{1}{2\sqrt{t + 1}} = \dfrac{1}{\sqrt{t + 1}}(4t + 4 + 2t - 3) = \dfrac{6t + 1}{\sqrt{t + 1}}$

$\dfrac{d^2y}{dx^2} = \dfrac{d}{dx}\left(\dfrac{dy}{dx}\right) = \dfrac{d(dy/dx)/dt}{dx/dt} = 2\sqrt{t + 1}\dfrac{6t + 1}{\sqrt{t + 1}} = 2(6t + 1)$

15. $x = \sin \pi t$, $y = \cos \pi t$. $\dfrac{dy}{dx} = \dfrac{dy/dt}{dx/dt} = \dfrac{-\pi \sin \pi t}{\pi \cos \pi t} = -\tan \pi t$

$\dfrac{d^2y}{dx^2} = \dfrac{d}{dx}\left(\dfrac{dy}{dx}\right) = \dfrac{d(dy/dx)/dt}{dx/dt} = \dfrac{-\pi \sec^2 \pi t}{\pi \cos \pi t} = -\sec^3 \pi t$

17. $x = e^{-t}$, $y = te^{2t}$. $\dfrac{dy}{dx} = \dfrac{dy/dt}{dx/dt} = \dfrac{(2t + 1)e^{2t}}{-e^{-t}} = -(2t + 1)e^{3t}$

$\dfrac{d}{dt}\left(\dfrac{dy}{dx}\right) = -3(2t + 1)e^{3t} - 2e^{3t} = -(6t + 5)e^{3t}$

$\dfrac{d^2y}{dx^2} = \dfrac{d}{dx}\left(\dfrac{dy}{dx}\right) = \dfrac{d(dy/dx)/dt}{dx/dt} = \dfrac{-(6t + 5)e^{3t}}{-e^{-t}} = (6t + 5)e^{4t}$

19. $x = 1 - 2\cos t$, $y = 2 + 3\sin t$. $\dfrac{dx}{dt} = 2\sin t$, $\dfrac{dy}{dt} = 3\cos t$. $\dfrac{dy}{dt} = 0 \Leftrightarrow \cos t = 0$ (so that
$\sin t = \pm 1$) $\Leftrightarrow y = 5$ or $-1 \Leftrightarrow (x, y) = (1, 5)$ or $(1, -1)$. $\dfrac{dx}{dt} = 0 \Leftrightarrow \sin t = 0 \Leftrightarrow$
$y = 2$ and $x = 3$ or $-1 \Leftrightarrow (x, y) = (3, 2)$ or $(-1, 2)$. Thus the tangent is horizontal at
$(1, 5)$ and $(1, -1)$, and is vertical at $(3, 2)$ and $(-1, 2)$.

	$0 < t < \frac{\pi}{2}$	$\frac{\pi}{2} < t < \pi$	$\pi < t < \frac{3\pi}{2}$	$\frac{3\pi}{2} < t < 2\pi$
dx/dt	+	+	−	−
dy/dt	+	−	−	+
x	→	→	←	←
y	↑	↓	↓	↑
curve	↗	↘	↙	↖

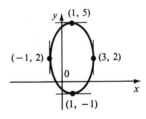

21. $x = t(t^2 - 3) = t^3 - 3t$, $y = 3(t^2 - 3)$.

$\dfrac{dx}{dt} = 3t^2 - 3 = 3(t - 1)(t + 1)$; $\dfrac{dy}{dt} = 6t$.

$\dfrac{dy}{dt} = 0 \Leftrightarrow t = 0 \Leftrightarrow (x, y) = (0, -9)$. $\dfrac{dx}{dt} = 0 \Leftrightarrow$

$t = \pm 1 \Leftrightarrow (x, y) = (-2, -6)$ or $(2, -6)$. So there is a horizontal tangent at $(0, -9)$ and there are vertical tangents at $(-2, -6)$ and $(2, -6)$. The graph is the reflection about the line $y = x$ of the graph in Exercise 9.1.27.

	$t < -1$	$-1 < t < 0$	$0 < t < 1$	$t < 1$
dx/dt	+	−	−	+
dy/dt	−	−	+	+
x	→	←	←	→
y	↓	↓	↑	↑
curve	↘	↙	↖	↗

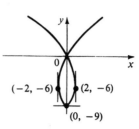

$(-2, -6)$ $(2, -6)$

$(0, -9)$

23. $x = \dfrac{3t}{1+t^3}$, $y = \dfrac{3t^2}{1+t^3}$. $\dfrac{dx}{dt} = \dfrac{(1+t^3)3 - 3t(3t^2)}{(1+t^3)^2} = \dfrac{3 - 6t^3}{(1+t^3)^2}$,

$\dfrac{dy}{dt} = \dfrac{(1+t^3)(6t) - 3t^2(3t^2)}{(1+t^3)^2} = \dfrac{6t - 3t^4}{(1+t^3)^2} = \dfrac{3t(2 - t^3)}{(1+t^3)^2}$. $\dfrac{dy}{dt} = 0 \Leftrightarrow t = 0$ or $\sqrt[3]{2} \Leftrightarrow$

$(x, y) = (0, 0)$ or $(\sqrt[3]{2}, \sqrt[3]{4})$. $\dfrac{dx}{dt} = 0 \Leftrightarrow t^3 = 1/2 \Leftrightarrow t = 2^{-\frac{1}{3}} \Leftrightarrow (x, y) = (\sqrt[3]{4}, \sqrt[3]{2})$.

There are horizontal tangents at $(0, 0)$ and $(\sqrt[3]{2}, \sqrt[3]{4})$, and there are vertical tangents

at $(\sqrt[3]{4}, \sqrt[3]{2})$ and $(0, 0)$. (The vertical tangent at $(0, 0)$ is undetectable by the methods of this

section because that tangent is the limiting position of the curve as $t \to \pm \infty$.)

	$t < -1$	$-1 < t < 0$	$0 < t < 1/\alpha$	$1/\alpha < t < \alpha$	$t > \alpha$
dx/dt	+	+	+	−	−
dy/dt	−	−	+	+	−
x	→	→	→	←	←
y	↓	↓	↑	↑	↓
curve	↘	↘	↗	↖	↙

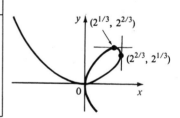

$(2^{1/3}, 2^{2/3})$

$(2^{2/3}, 2^{1/3})$

where $\alpha = \sqrt[3]{2}$

25. $x = \cos t$, $y = \sin t \cos t$. $\dfrac{dx}{dt} = -\sin t$,

$\dfrac{dy}{dt} = -\sin^2 t + \cos^2 t = \cos 2t$. $(x, y) = (0, 0)$

$\Leftrightarrow \cos t = 0 \Leftrightarrow t$ is an odd multiple of $\pi/2$.

When $t = \dfrac{\pi}{2}$, $\dfrac{dx}{dt} = -1$ and $\dfrac{dy}{dt} = -1$, so $\dfrac{dy}{dx} = 1$.

When $t = \dfrac{3\pi}{2}$, $\dfrac{dx}{dt} = 1$ and $\dfrac{dy}{dt} = -1$. So $\dfrac{dy}{dx} = -1$.

Thus $y = x$ and $y = -x$ are both tangent to the curve at $(0, 0)$.

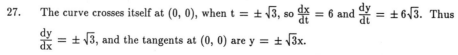

27.   The curve crosses itself at $(0, 0)$, when $t = \pm\sqrt{3}$, so $\dfrac{dx}{dt} = 6$ and $\dfrac{dy}{dt} = \pm 6\sqrt{3}$. Thus

$\dfrac{dy}{dx} = \pm\sqrt{3}$, and the tangents at $(0, 0)$ are $y = \pm\sqrt{3}x$.

29.   (a) $x = r\theta - d\sin\theta$, $y = r - d\cos\theta$; $\dfrac{dx}{d\theta} = r - d\cos\theta$, $\dfrac{dy}{d\theta} = d\sin\theta$. So $\dfrac{dy}{dx} = \dfrac{d\sin\theta}{r - d\cos\theta}$.

(b) If $0 < d < r$, then $|d \cos \theta| \le d < r$, so $r - d \cos \theta \ge r - d > 0$. This shows that $\frac{dx}{d\theta}$ never

vanishes, so the trochoid can have no vertical tangents if $d < r$.

31. The line with parametric equations $x = -7t$, $y = 12t - 5$ is $y = 12(-x/7) - 5$, which has

slope $-12/7$. The curve $x = t^3 + 4t$, $y = 6t^2$ has slope $\frac{dy}{dx} = \frac{dy/dt}{dx/dt} = \frac{12t}{3t^2 + 4}$.

This equals $-12/7 \iff 3t^2 + 4 = -7t \iff (3t + 4)(t + 1) = 0 \iff t = -1$ or $t = -4/3$

$\iff (x, y) = (-5, 6)$ or $(-208/27, 32/3)$.

33. By symmetry of the ellipse about the x- and y-axes,

$$A = 4 \int_0^a y \, dx = 4 \int_{\frac{\pi}{2}}^0 b \sin \theta (-a \sin \theta) \, d\theta = 4ab \int_0^{\frac{\pi}{2}} \sin^2 \theta \, d\theta = 4ab \int_0^{\frac{\pi}{2}} \frac{1 - \cos 2\theta}{2} \, d\theta$$

$$= 2ab \left[\theta - \frac{\sin 2\theta}{2} \right]_0^{\frac{\pi}{2}} = 2ab \left(\frac{\pi}{2} \right) = \pi ab.$$

35. $A = \int_0^1 (y - 1) \, dx = \int_{\frac{\pi}{2}}^0 (e^t - 1)(-\sin t) \, dt = \int_0^{\frac{\pi}{2}} (e^t \sin t - \sin t) \, dt$

$= \left[\frac{e^t}{2} (\sin t - \cos t) + \cos t \right]_0^{\frac{\pi}{2}} = \frac{1}{2}(e^{\frac{\pi}{2}} - 1)$ *[Formula 98]*

37. $A = \int_0^{2\pi r} y \, dx = \int_0^{2\pi} (r - d \cos \theta)(r - d \cos \theta) \, d\theta = \int_0^{2\pi} (r^2 - 2dr \cos \theta + d^2 \cos^2 \theta) \, d\theta$

$= \left[r^2 \theta - 2dr \sin \theta + \frac{d^2}{2} \left(\theta + \frac{\sin 2\theta}{2} \right) \right]_0^{2\pi} = 2\pi r^2 + \pi d^2$

39. The coordinates of T are $(r \cos \theta, r \sin \theta)$. Since TP

was unwound from arc TA, TP has length $r\theta$. Also

$\angle PTQ = \angle PTR - \angle QTR = \frac{1}{2}\pi - \theta$, so P has coordinates

$x = r \cos \theta + r\theta \cos \left(\frac{1}{2}\pi - \theta \right) = r(\cos \theta + \theta \sin \theta)$, and

$y = r \sin \theta - r\theta \sin \left(\frac{1}{2}\pi - \theta \right) = r(\sin \theta - \theta \cos \theta)$.

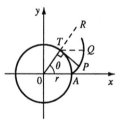

EXERCISES 9.3

1. $L = \int_0^1 \sqrt{\left(\frac{dx}{dt} \right)^2 + \left(\frac{dy}{dt} \right)^2} \, dt$ and $\frac{dx}{dt} = 3t^2$, $\frac{dy}{dt} = 4t^3 \Rightarrow$

$L = \int_0^1 \sqrt{9t^4 + 16t^6} \, dt = \int_0^1 t^2 \sqrt{9 + 16t^2} \, dt$.

3. $\frac{dx}{dt} = \sin t + t \cos t$ and $\frac{dy}{dt} = \cos t - t \sin t$

$\Rightarrow L = \int_0^{\pi/2} \sqrt{(\sin t + t \cos t)^2 + (\cos t - t \sin t)^2} \, dt = \int_0^{\pi/2} \sqrt{1 + t^2} \, dt$.

5. $x = 1 + 2\sin \pi t$, $y = 3 - 2\cos \pi t$, $0 \le t \le 1$. $\left(\frac{dx}{dt}\right)^2 + \left(\frac{dy}{dt}\right)^2 = (2\pi \cos \pi t)^2 + (2\pi \sin \pi t)^2$

$= 4\pi^2 \Rightarrow L = \int_0^1 \sqrt{(dx/dt)^2 + (dy/dt)^2}\, dt = \int_0^1 2\pi\, dt = 2\pi$

7. $x = 5t^2 + 1$, $y = 4 - 3t^2$, $0 \le t \le 2$. $\left(\frac{dx}{dt}\right)^2 + \left(\frac{dy}{dt}\right)^2 = (10t)^2 + (-6t)^2 = 136t^2 \Rightarrow$

$L = \int_0^2 \sqrt{136t^2}\, dt = \int_0^2 \sqrt{136}\, t\, dt = \left[\frac{1}{2} 2\sqrt{34}\, t^2\right]_0^2 = 4\sqrt{34}$

9. $x = e^t \cos t$, $y = e^t \sin t$, $0 \le t \le \pi$.

$\left(\frac{dx}{dt}\right)^2 + \left(\frac{dy}{dt}\right)^2 = \left[e^t(\cos t - \sin t)\right]^2 + \left[e^t(\sin t + \cos t)\right]^2$

$= e^{2t}(2\cos^2 t + 2\sin^2 t) = 2e^{2t} \Rightarrow L = \int_0^\pi \sqrt{2}\, e^t\, dt = \sqrt{2}(e^\pi - 1)$

11. $x = 2 - 3\sin^2 \theta$, $y = \cos 2\theta$, $0 \le \theta \le \pi/2$.

$\left(\frac{dx}{d\theta}\right)^2 + \left(\frac{dy}{d\theta}\right)^2 = (-6\sin \theta \cos \theta)^2 + (-2\sin 2\theta)^2 = (-3\sin 2\theta)^2 + (-2\sin 2\theta)^2$

$= 13\sin^2 2\theta \Rightarrow L = \int_0^{\pi/2} \sqrt{13} \sin 2\theta\, d\theta = \left[-\frac{\sqrt{13}}{2} \cos 2\theta\right]_0^{\frac{\pi}{2}} = -\frac{\sqrt{13}}{2}(-1 - 1) = \sqrt{13}$

13. $x = \ln t$ and $y = e^{-t} \Rightarrow \frac{dx}{dt} = \frac{1}{t}$ and $\frac{dy}{dt} = -e^{-t} \Rightarrow L = \int_1^2 \sqrt{t^{-2} + e^{-2t}}\, dt$. Using Simpson's

rule with $n = 10$, $\Delta x = (2 - 1)/10 = 0.1$ and $f(t) = \sqrt{t^{-2} + e^{-2t}}$ we get

$L \approx 0.1/3\left(f(1.0) + 4f(1.1) + 2f(1.2) + \cdots + 2f(1.8) + 4f(1.9) + f(2.0)\right) \approx 0.7314$.

15. $x = \sin^2 \theta$, $y = \cos^2 \theta$, $0 \le \theta \le 3\pi$. $\left(\frac{dx}{d\theta}\right)^2 + \left(\frac{dy}{d\theta}\right)^2$

$= (2\sin \theta \cos \theta)^2 + (-2\cos \theta \sin \theta)^2 = 8\sin^2 \theta \cos^2 \theta = 2\sin^2 2\theta \Rightarrow$

distance $= \int_0^{3\pi} \sqrt{2}|\sin 2\theta|\, d\theta = 6\sqrt{2}\int_0^{\frac{\pi}{2}} \sin 2\theta\, d\theta$ [by symmetry]

$= \left[-3\sqrt{2}\cos 2\theta\right]_0^{\frac{\pi}{2}} = -3\sqrt{2}(-1 - 1) = 6\sqrt{2}$

The full curve is traversed as θ goes from 0 to $\pi/2$, because the curve is the segment of

$x + y = 1$ that lies in the first quadrant (since $x, y \ge 0$), and this segment is completely

traversed as θ goes from 0 to $\pi/2$. Thus $L = \int_0^{\pi/2} \sin 2\theta\, d\theta = \sqrt{2}$, as above.

17. $x = a\sin \theta$, $y = b\cos \theta$, $0 \le \theta \le 2\pi$. $\left(\frac{dx}{d\theta}\right)^2 + \left(\frac{dy}{d\theta}\right)^2 = (a\cos \theta)^2 + (-b\sin \theta)^2$

$= a^2\cos^2 \theta + b^2\sin^2 \theta = a^2(1 - \sin^2 \theta) + b^2 \sin^2 \theta = a^2 - (a^2 - b^2)\sin^2 \theta$

$= a^2 - c^2 \sin^2 \theta = a^2\left(1 - \frac{c^2}{a^2} \sin^2 \theta\right) = a^2(1 - e^2 \sin^2 \theta)$ So $L = 4\int_0^{\frac{\pi}{2}} \sqrt{a^2(1 - e^2 \sin^2 \theta)}\, d\theta$

[by symmetry] $= 4a\int_0^{\frac{\pi}{2}} \sqrt{1 - e^2 \sin^2 \theta}\, d\theta$

19. $x = t^3$ and $y = t^4 \Rightarrow \frac{dx}{dt} = 3t^2$ and $\frac{dy}{dt} = 4t^3$. So

311

$$S = \int_0^1 2\pi t^4 \sqrt{9t^4 + 16t^6} \, dt = \int_0^1 2\pi t^6 \sqrt{9 + 16t^2} \, dt.$$

21. $\left(\dfrac{dx}{dt}\right)^2 + \left(\dfrac{dy}{dt}\right)^2 = (2t - 2/t^2)^2 + (4/\sqrt{t})^2 = 4t^2 + \dfrac{8}{t} + \dfrac{4}{t^4} = 4\left(t + \dfrac{1}{t^2}\right)^2$

$$S = \int_1^9 2\pi y \sqrt{(dx/dt)^2 + (dy/dt)^2} \, dt = 2\pi \int_1^9 (8\sqrt{t}) 2\left(t + \dfrac{1}{t^2}\right) dt = 32\pi \int_1^9 (t^{\frac{3}{2}} + t^{-\frac{3}{2}}) \, dt$$

$$= 32\pi \left[\dfrac{2}{5} t^{\frac{5}{2}} - 2t^{-\frac{1}{2}}\right]_1^9 = 32\pi \left[\left[\dfrac{2}{5}(243) - 2(\tfrac{1}{3})\right] - \left[\dfrac{2}{5}(1) - 2(1)\right]\right] = \dfrac{47104}{15}\pi$$

23. $x = e^t \cos t, \ y = e^t \sin t, \ 0 \le t \le \pi/2.$

$$\left(\dfrac{dx}{dt}\right)^2 + \left(\dfrac{dy}{dt}\right)^2 = \left[e^t(\cos t - \sin t)\right]^2 + \left[e^t(\sin t + \cos t)\right]^2$$

$$= e^{2t}(2\cos^2 t + 2\sin^2 t) = 2e^{2t}, \text{ so } S = \int_0^{\frac{\pi}{2}} 2\pi y \sqrt{(dx/dt)^2 + (dy/dt)^2} \, dt$$

$$= \int_0^{\frac{\pi}{2}} 2\pi e^t \sin t \ \sqrt{2} e^t \, dt = 2\sqrt{2}\pi \int_0^{\frac{\pi}{2}} e^{2t} \sin t \, dt = \left[2\sqrt{2}\pi \dfrac{e^{2t}}{5}(2\sin t - \cos t)\right]_0^{\frac{\pi}{2}} \ [Formula \ 98]$$

$$= \dfrac{2\sqrt{2}}{5}\pi \left[2e^\pi - (-1)\right] = \dfrac{2\sqrt{2}\pi}{5}(2e^\pi + 1)$$

25. $\left(\dfrac{dx}{d\theta}\right)^2 + \left(\dfrac{dy}{d\theta}\right)^2 = (-2\sin\theta + 2\sin 2\theta)^2 + (2\cos\theta - 2\cos 2\theta)^2$

$$= 4\left[(\sin^2\theta - 2\sin\theta\sin 2\theta + \sin^2 2\theta) + (\cos^2\theta - 2\cos\theta\cos 2\theta + \cos^2 2\theta)\right]$$

$$= 4\left[1 + 1 - 2(\cos 2\theta\cos\theta + \sin 2\theta\sin\theta)\right] = 8\left[1 - \cos(2\theta - \theta)\right] = 8(1 - \cos\theta)$$

Note that $x(2\pi - \theta) = x(\theta)$ and $y(2\pi - \theta) = -y(\theta)$, so the piece of the curve from $\theta = 0$ to

$\theta = \pi$ generates the same surface as the piece from $\theta = \pi$ to $\theta = 2\pi$. Also note that

$y = 2\sin\theta - \sin 2\theta = 2\sin\theta(1 - \cos\theta).$ So $S = \int_0^\pi 2\pi 2\sin\theta(1 - \cos\theta)2\sqrt{2}\sqrt{1 - \cos\theta} \, d\theta$

$$= 8\sqrt{2}\pi \int_0^\pi (1 - \cos\theta)^{\frac{3}{2}} \sin\theta \, d\theta = 8\sqrt{2}\pi \int_0^2 \sqrt{u^3} \, du \ [where \ u = 1 - \cos\theta, \ du = \sin\theta \, d\theta]$$

$$= \left[8\sqrt{2}\pi(\tfrac{2}{5}) u^{\frac{5}{2}}\right]_0^2 = \dfrac{128}{5}\pi.$$

27. $\left(\dfrac{dx}{dt}\right)^2 + \left(\dfrac{dy}{dt}\right)^2 = (6t)^2 + (6t^2)^2 = 36t^2(1 + t^2)$

$$\Rightarrow S = \int_0^5 2\pi x \sqrt{(dx/dt)^2 + (dy/dt)^2} \, dt = \int_0^5 2\pi(3t^2)6t\sqrt{1 + t^2} \, dt = 18\pi \int_0^5 t^2 \sqrt{1 + t^2} \, 2t \, dt$$

$$= 18\pi \int_1^{26} (u - 1)\sqrt{u} \, du \ [where \ u = 1 + t^2] = 18\pi \int_1^{26} (u^{3/2} - u^{1/2}) \, du$$

$$= 18\pi \left[\tfrac{2}{5} u^{5/2} - \tfrac{2}{3} u^{3/2}\right]_1^{26} = 18\pi \left[\left(\tfrac{2}{5}(676)\sqrt{26} - \tfrac{2}{3}(26)\sqrt{26}\right) - (\tfrac{2}{5} - \tfrac{2}{3})\right] = \dfrac{24}{5}\pi(949\sqrt{26} + 1)$$

29. $x = a\cos\theta, \ y = b\sin\theta, \ 0 \le \theta \le 2\pi.$ $\left(\dfrac{dx}{d\theta}\right)^2 + \left(\dfrac{dy}{d\theta}\right)^2 = (-a\sin\theta)^2 + (b\cos\theta)^2$

$$= a^2\sin^2\theta + b^2\cos^2\theta = a^2(1 - \cos^2\theta) + b^2\cos^2\theta = a^2 - (a^2 - b^2)\cos^2\theta$$

$$= a^2 - c^2\cos^2\theta = a^2\left(1 - \tfrac{c^2}{a^2}\cos^2\theta\right) = a^2(1 - e^2\cos^2\theta)$$

(a) $\displaystyle S = \int_0^\pi 2\pi b\sin\theta\, a\sqrt{1 - e^2\cos^2\theta}\, d\theta = 2\pi ab\int_{-e}^{e}\sqrt{1 - u^2}(\tfrac{1}{e})\,du$ [where $u = -e\cos\theta$,

$\qquad du = e\sin\theta\, d\theta] = \dfrac{4\pi ab}{e}\displaystyle\int_0^e (1 - u^2)^{1/2}\,du = \dfrac{4\pi ab}{e}\int_0^{\sin^{-1}e}\cos^2 v\, dv$ (where $u = \sin v$)

$\qquad = \dfrac{2\pi ab}{e}\displaystyle\int_0^{\sin^{-1}e}(1 + \cos 2v)\,dv = \dfrac{2\pi ab}{e}\Big[v + \tfrac{1}{2}\sin 2v\Big]_0^{\sin^{-1}e}$

$\qquad = \dfrac{2\pi ab}{e}\Big[v + \sin v\cos v\Big]_0^{\sin^{-1}e} = \dfrac{2\pi ab}{e}\Big(\sin^{-1}e + e\sqrt{1 - e^2}\Big)$. But

$\qquad \sqrt{1 - e^2} = \sqrt{1 - (c^2/a^2)} = \sqrt{(a^2 - c^2)/a^2} = \sqrt{b^2/a^2} = b/a$, so $S = \dfrac{2\pi ab}{e}\sin^{-1}e + 2\pi b^2$

(b) $\displaystyle S = \int_{-\frac{\pi}{2}}^{\frac{\pi}{2}} 2\pi a\cos\theta\, a\sqrt{1 - e^2\cos^2\theta}\, d\theta = 4\pi a^2\int_0^{\frac{\pi}{2}}\cos\theta\sqrt{(1 - e^2) + e^2\sin^2\theta}\, d\theta$

$\qquad = \dfrac{4\pi a^2(1 - e^2)}{e}\displaystyle\int_0^{\frac{\pi}{2}}\dfrac{e}{\sqrt{1 - e^2}}\cos\theta\sqrt{1 + \left[\dfrac{e\sin\theta}{\sqrt{1 - e^2}}\right]^2}\, d\theta$

$\qquad = \dfrac{4\pi a^2(1 - e^2)}{e}\displaystyle\int_0^{e/\sqrt{1-e^2}}\sqrt{1 + u^2}\, du$ [(where $u = e\sin\theta/\sqrt{1 - e^2}$]

$\qquad = \dfrac{4\pi a^2(1 - e^2)}{e}\displaystyle\int_0^{\sin^{-1}e}\sec^3 v\, dv$ [where $u = \tan v,\ du = \sec^2 v\, dv$]

$\qquad = \dfrac{2\pi a^2(1 - e^2)}{e}\Big[\sec v\tan v + \ln|\sec v + \tan v|\Big]_0^{\sin^{-1}e}$

$\qquad = \dfrac{2\pi a^2(1 - e^2)}{e}\left[\dfrac{1}{\sqrt{1 - e^2}}\dfrac{e}{\sqrt{1 - e^2}} + \ln\left|\dfrac{1}{\sqrt{1 - e^2}} + \dfrac{e}{\sqrt{1 - e^2}}\right|\right]$

$\qquad = 2\pi a^2 + \dfrac{2\pi a^2(1 - e^2)}{e}\ln\sqrt{\dfrac{1 + e}{1 - e}} = 2\pi a^2 + \dfrac{2\pi b^2}{e}\tfrac{1}{2}\ln\left(\dfrac{1 + e}{1 - e}\right)$ [since $1 - e^2 = b^2/a^2$]

$\qquad = 2\pi[a^2 + \dfrac{b^2}{2e}\ln\{(1 + e)/(1 - e)\}]$

31. (a) $\phi = \tan^{-1}\left(\dfrac{dy}{dx}\right) \Rightarrow \dfrac{d\phi}{dt} = \dfrac{d}{dt}\tan^{-1}\left(\dfrac{dy}{dx}\right) = \dfrac{1}{1 + (dy/dx)^2}\left[\dfrac{d}{dt}\left(\dfrac{dy}{dx}\right)\right]$.

$\qquad$ But $\dfrac{dy}{dx} = \dfrac{dy/dt}{dx/dt} = \dfrac{\dot{y}}{\dot{x}} \Rightarrow \dfrac{d}{dt}\left(\dfrac{dy}{dx}\right) = \dfrac{d}{dt}\left(\dfrac{\dot{y}}{\dot{x}}\right) = \dfrac{\ddot{y}\dot{x} - \ddot{x}\dot{y}}{\dot{x}^2}$

$\qquad \Rightarrow \dfrac{d\phi}{dt} = \dfrac{1}{1 + (\dot{y}/\dot{x})^2}\left[\dfrac{\ddot{y}\dot{x} - \ddot{x}\dot{y}}{\dot{x}^2}\right] = \dfrac{\dot{x}\ddot{y} - \ddot{x}\dot{y}}{\dot{x}^2 + \dot{y}^2}$. Using the chain rule, and the fact that

$\qquad s = \displaystyle\int_0^t \sqrt{\left(\dfrac{dx}{dt}\right)^2 + \left(\dfrac{dy}{dt}\right)^2}\, dt \Rightarrow \dfrac{ds}{dt} = \sqrt{\left(\dfrac{dx}{dt}\right)^2 + \left(\dfrac{dy}{dt}\right)^2} = (\dot{x}^2 + \dot{y}^2)^{1/2}$ we have that

$\qquad \dfrac{d\phi}{ds} = \dfrac{d\phi/dt}{ds/dt} = \left(\dfrac{\dot{x}\ddot{y} - \ddot{x}\dot{y}}{\dot{x}^2 + \dot{y}^2}\right)\dfrac{1}{(\dot{x}^2 + \dot{y}^2)^{1/2}} = \dfrac{\dot{x}\ddot{y} - \ddot{x}\dot{y}}{[\dot{x}^2 + \dot{y}^2]^{3/2}}$. So

$$\kappa = \left|\frac{d\phi}{ds}\right| = \left|\frac{\dot{x}\ddot{y} - \ddot{x}\dot{y}}{\left[\dot{x}^2 + \dot{y}^2\right]^{3/2}}\right| = \frac{|\dot{x}\ddot{y} - \ddot{x}\dot{y}|}{\left[\dot{x}^2 + \dot{y}^2\right]^{3/2}}.$$

(b) $x = x$ and $y = f(x) \Rightarrow \dot{x} = 1$, $\ddot{x} = 0$ and $\dot{y} = \frac{dy}{dx}$, $\ddot{y} = \frac{d^2y}{dx^2}$. So

$$\kappa = \frac{\left|1\cdot(d^2y/dx^2) - 0\cdot(dy/dx)\right|}{\left[1 + (dy/dx)^2\right]^{3/2}} = \frac{\left|d^2y/dx^2\right|}{\left[1 + (dy/dx)^2\right]^{3/2}}.$$

33. $x = \theta - \sin\theta \Rightarrow \dot{x} = 1 - \cos\theta \Rightarrow \ddot{x} = \sin\theta$ and $y = 1 - \cos\theta \Rightarrow \dot{y} = \sin\theta \Rightarrow \ddot{y} = \cos\theta$.

Therefore, $\kappa = \dfrac{\left|\cos\theta - \cos^2\theta - \sin^2\theta\right|}{\left[(1 - \cos\theta)^2 + \sin^2\theta\right]^{3/2}}$

$$= \frac{\left|\cos\theta - (\cos^2\theta + \sin^2\theta)\right|}{\left[1 - 2\cos\theta + \cos^2\theta + \sin^2\theta\right]^{3/2}} = \frac{\left|\cos\theta - 1\right|}{\left[2 - 2\cos\theta\right]^{3/2}}. \text{ The top of the arch is}$$

characterized by a horizontal tangent and from example 1 of section 9.2 the tangent is

horizontal when $\theta = (2n - 1)\pi$ so take n = 1 and substitute $\theta = \pi$ into the expression for κ:

$$\kappa = \frac{\left|\cos\pi - 1\right|}{\left[2 - 2\cos\pi\right]^{3/2}} = \frac{\left|-1 - 1\right|}{\left[2 - 2(-1)\right]^{3/2}} = \frac{1}{4}.$$

EXERCISES 9.4

1. $(1, \pi/2)$

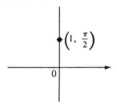

$(1, 5\pi/2), (-1, 3\pi/2)$

3. $(-1, \pi/5)$

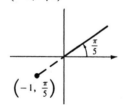

$(1, 6\pi/5), (-1, 11\pi/5)$

5. $(4, -2\pi/3)$

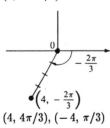

$(4, 4\pi/3), (-4, \pi/3)$

7. $(-1, \pi)$

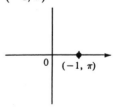

$(-1, 3\pi), (1, 0)$

9.

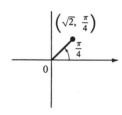

$x = \sqrt{2}\cos(\pi/4) = 1,$

$y = \sqrt{2}\sin(\pi/4) = 1$

11.

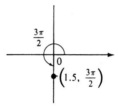

$(0, -3/2)$

13.

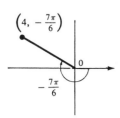

$x = 4\cos(-\frac{7\pi}{6}) = 4(-\frac{\sqrt{3}}{2}) = -2\sqrt{3}$

$y = 4\sin(-\frac{7\pi}{6}) = 4 \cdot \frac{1}{2} = 2$

15.

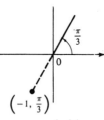

$x = -\cos(\pi/3) = -1/2$

$y = -\sin(\pi/3) = -\sqrt{3}/2$

17. $(x, y) = (-1, 1), r = \sqrt{(-1)^2 + 1^2} = \sqrt{2}, \tan\theta = y/x = -1$ and (x, y) is in quadrant II, so $\theta = 3\pi/4$. Coordinates $(\sqrt{2}, 3\pi/4)$.

19. $(x, y) = (2\sqrt{3}, -2), r = \sqrt{12 + 4} = 4, \tan\theta = y/x = -1/\sqrt{3}$ and (x, y) is in quadrant IV, so $\theta = 11\pi/6$. $(4, 11\pi/6)$

21. $r > 1$

23. $0 \le r \le 2, \pi/2 \le \theta \le \pi$

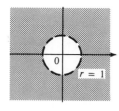

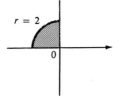

25. $3 < r < 4, -\pi/2 \le \theta \le \pi$

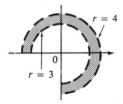

315

27. $(1, \pi/6)$ is $(\sqrt{3}/2, 1/2)$ Cartesian and $(3, 3\pi/4)$ is $(-3/\sqrt{2}, 3/\sqrt{2})$ Cartesian. The square

of the distance between them is $(\frac{\sqrt{3}}{2} + \frac{\sqrt{3}}{2})^2 + (\frac{1}{2} - \frac{3}{\sqrt{2}})^2 = \frac{1}{4}(40 + 6\sqrt{6} - 6\sqrt{2})$, so the distance

is $\frac{1}{2}\sqrt{40 + 6\sqrt{6} - 6\sqrt{2}}$.

29. Since $y = r\sin\theta$, the equation $r\sin\theta = 2$ becomes $y = 2$.

31. $r = \dfrac{1}{1 - \cos\theta} \Leftrightarrow r - r\cos\theta = 1 \Leftrightarrow r = 1 + r\cos\theta \Leftrightarrow r^2 = (1 + r\cos\theta)^2 \Leftrightarrow$

   $x^2 + y^2 = (1 + x)^2 = 1 + 2x + x^2 \Leftrightarrow y^2 = 1 + 2x$

33. $r^2 = \sin 2\theta = 2\sin\theta\cos\theta \Leftrightarrow r^4 = 2r\sin\theta\, r\cos\theta \Leftrightarrow (x^2 + y^2)^2 = 2yx$

35. $y = 5 \Leftrightarrow r\sin\theta = 5$

37. $x^2 + y^2 = 25 \Leftrightarrow r^2 = 25 \Leftrightarrow r = 5$

39. $2xy = 1 \Leftrightarrow 2r\cos\theta\, r\sin\theta = 1 \Leftrightarrow r^2\sin 2\theta = 1 \Leftrightarrow r^2 = \csc 2\theta$

41. $r = 5$ 43. $\theta = 3\pi/4$

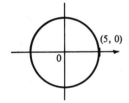
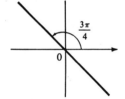

45. $r = 2\sin\theta \Leftrightarrow r^2 = 2r\sin\theta$ 47. $r = -\cos\theta \Leftrightarrow r^2 = -r\cos\theta$

    $\Leftrightarrow x^2 + y^2 = y \Leftrightarrow$ $\Leftrightarrow x^2 + y^2 = -x \Leftrightarrow$

    $x^2 + (y - 1)^2 = 1$ $(x + \frac{1}{2})^2 + y^2 = \frac{1}{4}$

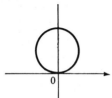
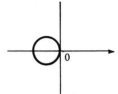

49. $r = \cos\theta - \sin\theta \Leftrightarrow$
    $r^2 = r\cos\theta - r\sin\theta \Leftrightarrow$
    $x^2 + y^2 = x - y \Leftrightarrow$
    $(x - \frac{1}{2})^2 + (y + \frac{1}{2})^2 = 1/2$

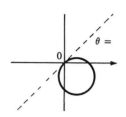

51. $r = 3(1 - \cos\theta)$

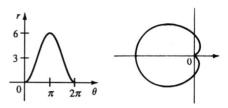

53. $r = \theta, \theta \geq 0$

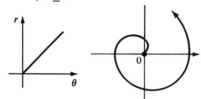

55. $r = 1/\theta$

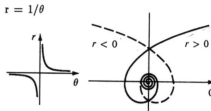

57. $r = 1 - 2\cos\theta$

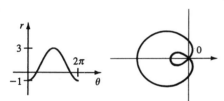

59. $r = 3 + 2\sin\theta$

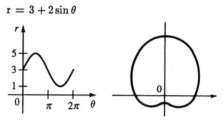

61. $r = -3\cos 2\theta$

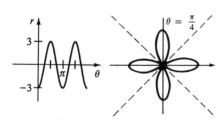

63. $r = \sin 3\theta$

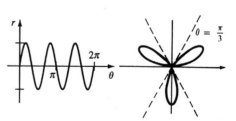

65. $r = 2\cos 4\theta$

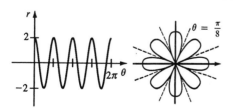

67. $r = \sin 5\theta$

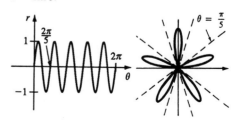

69. $r^2 = 4\cos 2\theta$

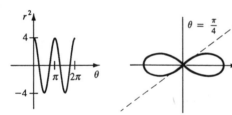

71. $r^2 = 3\cos(3\theta/2)$

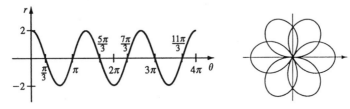

73. $x = r\cos\theta = 4\cos\theta + 2\sec\theta\cos\theta = 4\cos\theta + 2$. Now, $r \to \infty \Rightarrow (4 + 2\sec\theta) \to \infty$

     $\Rightarrow \theta \to \pi/2^-$ (since we need only consider $0 \le \theta < 2\pi$) so $\displaystyle\lim_{r\to\infty} x = \lim_{\theta\to\pi/2^-}(4\cos\theta + 2) = 2$.

     Also, $r \to -\infty \Rightarrow (4 + 2\sec\theta) \to -\infty \Rightarrow \theta \to \pi/2^+$ so

     $\displaystyle\lim_{r\to-\infty} x = \lim_{\theta\to\pi/2^+}(4\cos\theta + 2) = 2$. Therefore $\displaystyle\lim_{r\to\pm\infty} x = 2 \Rightarrow x = 2$ is a vertical asymptote.

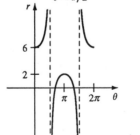 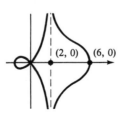

75.   To show that $x = 1$ is an asymptote we must prove $\displaystyle\lim_{r\to\pm\infty} x = 1$.

     $x = r\cos\theta = \sin\theta\tan\theta\cos\theta = \sin^2\theta$. Now, $r \to \infty \Rightarrow$

     $\sin\theta\tan\theta \to \infty \Rightarrow \theta \to \pi/2^-$ so $\displaystyle\lim_{r\to\infty} x = \lim_{\theta\to\pi/2^-}\sin^2\theta = 1$. And,

     $r \to -\infty \Rightarrow \sin\theta\tan\theta \to -\infty \Rightarrow \theta \to \pi/2^+$ so

     $\displaystyle\lim_{r\to-\infty} x = \lim_{\theta\to\pi/2^+}\sin^2\theta = 1$. Therefore $\displaystyle\lim_{r\to\pm\infty} x = 1 \Rightarrow x = 1$ is a

     vertical asymptote. Also notice $x = \sin^2\theta \ge 0$ for all θ and

     $x = \sin^2\theta \le 1$ for all θ. And $x \ne 1$ since the curve is not defined at odd

     multiples of $\pi/2$. Therefore the curve lies entirely within the vertical strip

     $0 \le x < 1$.

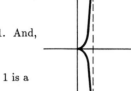

77. $\dfrac{dy}{dx} = \dfrac{dy/d\theta}{dx/d\theta} = \dfrac{(dr/d\theta)(\sin\theta) + r\cos\theta}{(dr/d\theta)(\cos\theta) - r\sin\theta} = \dfrac{-3\sin\theta\sin\theta + 3\cos\theta\cos\theta}{-3\sin\theta\cos\theta - 3\cos\theta\sin\theta}$

$$= \frac{3\cos 2\theta}{-3\sin 2\theta} = -\cot 2\theta = 1/\sqrt{3} \text{ when } \theta = \pi/3.$$

Alternate solution: $r = 3\cos\theta \Rightarrow x = r\cos\theta = 3\cos^2\theta$, $y = r\sin\theta = 3\sin\theta\cos\theta \Rightarrow$

$$\frac{dy}{dx} = \frac{dy/d\theta}{dx/d\theta} = \frac{-3\sin^2\theta + 3\cos^2\theta}{-6\cos\theta\sin\theta} = \frac{\cos 2\theta}{-\sin 2\theta} = -\cot 2\theta = 1/\sqrt{3} \text{ when } \theta = \pi/3$$

79. $r = \theta \Rightarrow x = r\cos\theta = \theta\cos\theta$, $y = r\sin\theta = \theta\sin\theta \Rightarrow$

$$\frac{dy}{dx} = \frac{dy/d\theta}{dx/d\theta} = \frac{\sin\theta + \theta\cos\theta}{\cos\theta - \theta\sin\theta} = -2/\pi \text{ when } \theta = \pi/2$$

81. $r = 1 + \cos\theta \Rightarrow x = r\cos\theta = \cos\theta + \cos^2\theta$, $y = r\sin\theta = \sin\theta + \sin\theta\cos\theta \Rightarrow$

$$\frac{dy}{dx} = \frac{dy/d\theta}{dx/d\theta} = \frac{\cos\theta + \cos^2\theta - \sin^2\theta}{-\sin\theta - 2\cos\theta\sin\theta} = \frac{\cos\theta + \cos 2\theta}{-\sin\theta - \sin 2\theta} = -1 \text{ when } \theta = \pi/6$$

83. $r = \sin 3\theta \Rightarrow x = r\cos\theta = \sin 3\theta\cos\theta$, $y = r\sin\theta = \sin 3\theta\sin\theta \Rightarrow$

$$\frac{dy}{dx} = \frac{dy/d\theta}{dx/d\theta} = \frac{3\sin\theta\cos 3\theta + \sin 3\theta\cos\theta}{3\cos\theta\cos 3\theta - \sin 3\theta\sin\theta} = \sqrt{3} \text{ when } \theta = \pi/3$$

85. $r = 3\cos\theta \Rightarrow x = r\cos\theta = 3\cos\theta\cos\theta$, $y = r\sin\theta = 3\cos\theta\sin\theta \Rightarrow$

$\frac{dy}{d\theta} = -3\sin^2\theta + 3\cos^2\theta = 3\cos 2\theta = 0 \Rightarrow 2\theta = \pi/2 \text{ or } 3\pi/2 \Leftrightarrow \theta = \pi/4 \text{ or } 3\pi/4$. So the tangent is horizontal at $(3/\sqrt{2}, \pi/4)$ and $(-3/\sqrt{2}, 3\pi/4)$ *(same as $(3/\sqrt{2}, -\pi/4)$)*.

$\frac{dx}{d\theta} = -6\sin\theta\cos\theta = -3\sin 2\theta = 0 \Rightarrow 2\theta = 0 \text{ or } \pi \Leftrightarrow \theta = 0 \text{ or } \pi/2$. So the tangent is vertical at $(3, 0)$ and $(0, \pi/2)$.

87. $r = \cos 2\theta \Rightarrow x = r\cos\theta = \cos 2\theta\cos\theta$, $y = r\sin\theta = \cos 2\theta\sin\theta \Rightarrow$

$\frac{dy}{d\theta} = -2\sin 2\theta\sin\theta + \cos 2\theta\cos\theta = -4\sin^2\theta\cos\theta + (\cos^3\theta - \sin^2\theta\cos\theta)$

$= \cos\theta(\cos^2\theta - 5\sin^2\theta) = \cos\theta(1 - 6\sin^2\theta) = 0 \Rightarrow \cos\theta = 0 \text{ or } \sin\theta = \pm 1/\sqrt{6} \Rightarrow$

$\theta = \pi/2, 3\pi/2, \alpha, \pi - \alpha, \pi + \alpha, \text{ or } 2\pi - \alpha$ [*where* $\alpha = \sin^{-1}(1/\sqrt{6})$]. So the tangent is horizontal at $(1, 3\pi/2)$, $(1, \pi/2)$, $(2/3, \alpha)$, $(2/3, \pi - \alpha)$, $(2/3, \pi + \alpha)$, and $(2/3, 2\pi - \alpha)$.

$\frac{dx}{d\theta} = -2\sin 2\theta\cos\theta - \cos 2\theta\sin\theta = -4\sin\theta\cos^2\theta - (2\cos^2\theta - 1)\sin\theta = \sin\theta(1 - 6\cos^2\theta) = 0 \Rightarrow \sin\theta = 0 \text{ or } \cos\theta = \pm 1/\sqrt{6} \Rightarrow \theta = 0, \pi, \pi/2 - \alpha, \pi/2 + \alpha, 3\pi/2 - \alpha, \text{ or } 3\pi/2 + \alpha$.

So the tangent is vertical at $(1, 0)$, $(1, \pi)$, $(2/3, 3\pi/2 - \alpha)$, $(2/3, 3\pi/2 + \alpha)$, $(2/3, \pi/2 - \alpha)$, and $(2/3, \pi/2 + \alpha)$.

89. $r = 1 + \cos\theta \Rightarrow x = r\cos\theta = \cos\theta(1 + \cos\theta)$, $y = r\sin\theta = \sin\theta(1 + \cos\theta) \Rightarrow$

$\frac{dy}{d\theta} = (1 + \cos\theta)\cos\theta - \sin^2\theta = 2\cos^2\theta + \cos\theta - 1 = (2\cos\theta - 1)(\cos\theta + 1) = 0 \Rightarrow$

$\cos\theta = 1/2 \text{ or } -1 \Rightarrow \theta = \pi/3, \pi, \text{ or } 5\pi/3 \Rightarrow$ horizontal tangent at $(3/2, \pi/3)$, $(0, \pi)$, and $(3/2, 5\pi/3)$.

$\frac{dx}{d\theta} = -(1 + \cos\theta)\sin\theta - \cos\theta\sin\theta = -\sin\theta(1 + 2\cos\theta) = 0 \Rightarrow \sin\theta = 0 \text{ or } \cos\theta = -1/2$

$\Rightarrow \theta = 0, \pi, 2\pi/3, \text{ or } 4\pi/3 \Rightarrow$ vertical tangent at $(2, 0)$, $(1/2, 2\pi/3)$, and $(1/2, 4\pi/3)$.

(Note that the tangent is horizontal, not vertical when $\theta = \pi$, since $\lim\limits_{\theta \to \pi} \frac{dy/d\theta}{dx/d\theta} = 0$.)

93. Following the hint, $\tan\psi = \tan(\phi - \theta) = \frac{\tan\phi - \tan\theta}{1 + \tan\phi\tan\theta} = \frac{(dy/dx) - \tan\theta}{1 + (dy/dx)\tan\theta}$

$$= \frac{\left[(dy/d\theta)/(dx/d\theta)\right] - \tan\theta}{1 + \left[(\tan\theta)(dy/d\theta)/(dx/d\theta)\right]} = \frac{(dy/d\theta) - ((\tan\theta)\,dx/d\theta)}{(dx/d\theta) + ((\tan\theta)\,dy/d\theta)}$$

$$= \frac{\left[\left[(dr/d\theta)\sin\theta\right] + r\cos\theta\right] - \tan\theta\left[\left[(dr/d\theta)\cos\theta\right] - r\sin\theta\right]}{\left[\left[(dr/d\theta)\cos\theta\right] - r\sin\theta\right] + \tan\theta\left[\left[(dr/d\theta)\sin\theta\right] + r\cos\theta\right]}$$

$$= \frac{r\cos^2\theta + r\sin^2\theta}{(dr/d\theta)\cos^2\theta + (dr/d\theta)\sin^2\theta} = \frac{r}{dr/d\theta}.$$

EXERCISES 9.5

1. $A = \displaystyle\int_0^\pi \tfrac{1}{2}r^2\,d\theta = \int_0^\pi \tfrac{1}{2}\theta^2\,d\theta = \left[\tfrac{1}{6}\theta^3\right]_0^\pi = \pi^3/6$

3. $A = \displaystyle\int_0^{\pi/6} \tfrac{1}{2}(2\cos\theta)^2\,d\theta = \int_0^{\pi/6}(1+\cos 2\theta)\,d\theta = \left[\theta + \tfrac{1}{2}\sin 2\theta\right]_0^{\pi/6} = \tfrac{\pi}{6} + \tfrac{\sqrt{3}}{4}$

5. $A = \displaystyle\int_{\pi/2}^{3\pi/2} \tfrac{1}{2}(\theta^2)^2\,d\theta = \left[\tfrac{1}{10}\theta^5\right]_{\pi/2}^{3\pi/2} = \tfrac{121}{160}\pi^5$

7. $A = \displaystyle\int_0^{\pi/6} \tfrac{1}{2}\sin^2 2\theta\,d\theta = \tfrac{1}{4}\int_0^{\pi/6}(1-\cos 4\theta)\,d\theta = \left[\tfrac{\theta}{4} - \tfrac{\sin 4\theta}{16}\right]_0^{\pi/6} = \tfrac{4\pi - 3\sqrt{3}}{96}$

9. $A = \displaystyle\int_0^\pi \tfrac{1}{2}(5\sin\theta)^2\,d\theta = \tfrac{25}{4}\int_0^\pi(1-\cos 2\theta)\,d\theta$

    $= \tfrac{25}{4}\left[\theta - \tfrac{1}{2}\sin 2\theta\right]_0^\pi = \tfrac{25}{4}\pi$

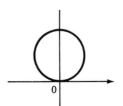

11. $A = \displaystyle 2\int_{-\pi/2}^{\pi/2} \tfrac{1}{2}(1+\sin\theta)^2\,d\theta$

    $= \displaystyle\int_{-\pi/2}^{\pi/2}(1 + 2\sin\theta + \sin^2\theta)\,d\theta$

    $= \left[\theta - 2\cos\theta\right]_{-\pi/2}^{\pi/2} + \displaystyle\int_{-\pi/2}^{\pi/2} \tfrac{1}{2}(1-\cos 2\theta)\,d\theta$

    $= \pi + \tfrac{1}{2}\left[\theta - \tfrac{1}{2}\sin 2\theta\right]_{-\pi/2}^{\pi/2} = \pi + \tfrac{\pi}{2} = 3\pi/2$

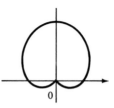

13. $A = 4\int_0^{\pi/4} \frac{1}{2}r^2\,d\theta = 8\int_0^{\pi/4} \cos 2\theta\,d\theta$

   $= \left[4\sin 2\theta\right]_0^{\pi/4} = 4$

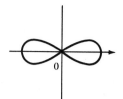

15. $A = 2\int_{-\pi/2}^{\pi/2} \frac{1}{2}(4 - \sin\theta)^2\,d\theta$

   $= \int_{-\pi/2}^{\pi/2} (16 - 8\sin\theta + \sin^2\theta)\,d\theta$

   $= 16\pi + 0 + \int_{-\pi/2}^{\pi/2} \sin^2\theta\,d\theta = 33\pi/2$

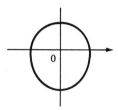

17. $A = 8\int_0^{\pi/4} \frac{1}{2}\sin^2 4\theta\,d\theta = 2\int_0^{\pi/4} (1 - \cos 8\theta)\,d\theta$

   $= \left[2\theta - \frac{1}{4}\sin 8\theta\right]_0^{\pi/4} = \pi/2$

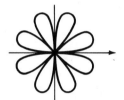

19. $A = 2\int_0^{\pi/6} \frac{1}{2}\cos^2 3\theta\,d\theta = \frac{1}{2}\int_0^{\pi/6} (1 + \cos 6\theta)\,d\theta = \frac{1}{2}\left[\theta + \frac{1}{6}\sin 6\theta\right]_0^{\pi/6} = \pi/12$

21. $A = \int_0^{\pi/5} \frac{1}{2}\sin^2 5\theta\,d\theta = \frac{1}{4}\int_0^{\pi/5} (1 - \cos 10\theta)\,d\theta = \frac{1}{4}\left[\theta - \frac{1}{10}\sin 10\theta\right]_0^{\pi/5} = \pi/20$

23. This is a limaçon, with inner loop traced out between $\theta = 7\pi/6$ and $11\pi/6$.

   $A = 2\int_{7\pi/6}^{3\pi/2} \frac{1}{2}(1 + 2\sin\theta)^2\,d\theta = \int_{7\pi/6}^{3\pi/2} (1 + 4\sin\theta + 4\sin^2\theta)\,d\theta$

   $= \left[\theta - 4\cos\theta + 2\theta - \sin 2\theta\right]_{7\pi/6}^{3\pi/2} = \pi - \frac{3\sqrt{3}}{2}$

25. $1 - \cos\theta = 3/2 \iff \cos\theta = -1/2 \Rightarrow \theta = 2\pi/3 \text{ or } 4\pi/3 \Rightarrow$

   $A = \int_{2\pi/3}^{4\pi/3} \frac{1}{2}\left[(1 - \cos\theta)^2 - (3/2)^2\right]d\theta$

   $= \frac{1}{2}\int_{2\pi/3}^{4\pi/3} \left((-5/4) - 2\cos\theta + \cos^2\theta\right)d\theta$

   $= \frac{1}{2}\left[-\frac{5}{4}\theta - 2\sin\theta\right]_{2\pi/3}^{4\pi/3} + \frac{1}{2}\int_{2\pi/3}^{4\pi/3} \frac{1 + \cos 2\theta}{2}\,d\theta$

   $= -\frac{5}{12}\pi + \sqrt{3} + \frac{1}{4}\left[\theta + \frac{1}{2}\sin 2\theta\right]_{2\pi/3}^{4\pi/3} = \frac{9\sqrt{3}}{8} - \frac{1}{4}\pi$

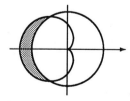

27. $4\sin\theta = 2 \Leftrightarrow \sin\theta = 1/2 \Rightarrow \theta = \pi/6$ or $5\pi/6 \Rightarrow$

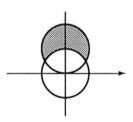

$$A = 2\int_{\pi/6}^{\pi/2} \tfrac{1}{2}\Big[(4\sin\theta)^2 - 2^2\Big]d\theta$$

$$= \int_{\pi/6}^{\pi/2}(16\sin^2\theta - 4)\,d\theta = \int_{\pi/6}^{\pi/2}\Big[8(1 - \cos 2\theta) - 4\Big]d\theta$$

$$= \Big[4\theta - 4\sin 2\theta\Big]_{\pi/6}^{\pi/2} = \tfrac{4}{3}\pi + 2\sqrt{3}$$

29. $3\cos\theta = 1 + \cos\theta \Leftrightarrow \cos\theta = 1/2 \Rightarrow \theta = \pi/3$

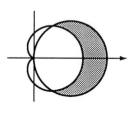

or $-\pi/3$. $\quad A = 2\int_{0}^{\pi/3}\tfrac{1}{2}\Big[(3\cos\theta)^2 - (1 + \cos\theta)^2\Big]d\theta$

$$= \int_{0}^{\pi/3}(8\cos^2\theta - 2\cos\theta - 1)\,d\theta$$

$$= \int_{0}^{\pi/3}\Big[4(1 + \cos 2\theta) - 2\cos\theta - 1\Big]d\theta$$

$$= \Big[3\theta + 2\sin 2\theta - 2\sin\theta\Big]_{0}^{\pi/3} = \pi + \sqrt{3} - \sqrt{3} = \pi$$

31. $\quad A = 2\int_{0}^{\pi/4}\tfrac{1}{2}\sin^2\theta\,d\theta = \int_{0}^{\pi/4}\dfrac{1 - \cos 2\theta}{2}\,d\theta$

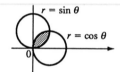

$r = \sin\theta$

$r = \cos\theta$

$$= \Big[\tfrac{1}{2}\theta - \tfrac{1}{4}\sin 2\theta\Big]_{0}^{\pi/4} = \tfrac{1}{8}\pi - \tfrac{1}{4}$$

33. $\sin 2\theta = \cos 2\theta \Rightarrow \tan 2\theta = 1 \Rightarrow 2\theta = \pi/4 \Rightarrow \theta = \pi/8$

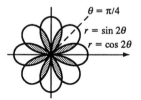

$\theta = \pi/4$

$r = \sin 2\theta$

$r = \cos 2\theta$

$$\Rightarrow A = 16\int_{0}^{\pi/8}\tfrac{1}{2}\sin^2 2\theta\,d\theta$$

$$= 4\int_{0}^{\pi/8}(1 - \cos 4\theta)\,d\theta = 4\Big[\theta - \tfrac{1}{4}\sin 4\theta\Big]_{0}^{\pi/8} = \tfrac{1}{2}\pi - 1$$

35. $\quad A = 2\left\{\int_{-\pi/2}^{-\pi/6}\tfrac{1}{2}(3 + 2\sin\theta)^2\,d\theta + \int_{-\pi/6}^{\pi/2}\tfrac{1}{2}2^2\,d\theta\right\}$

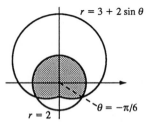

$r = 3 + 2\sin\theta$

$r = 2$

$\theta = -\pi/6$

$$= \int_{-\pi/2}^{-\pi/6}(9 + 12\sin\theta + 4\sin^2\theta)\,d\theta + \Big[4\theta\Big]_{-\pi/6}^{\pi/2}$$

$$= \Big[9\theta - 12\cos\theta + 2\theta - \sin 2\theta\Big]_{-\pi/2}^{-\pi/6} + 8\pi/3$$

$$= \tfrac{19}{3}\pi - \dfrac{11\sqrt{3}}{2}$$

37. $A = 2\left\{ \int_0^{2\pi/3} \frac{1}{2}\left(\frac{1}{2} + \cos\theta\right)^2 d\theta - \int_{2\pi/3}^{\pi} \frac{1}{2}\left(\frac{1}{2} + \cos\theta\right)^2 d\theta \right\}$

$= \left[\frac{\theta}{4} + \sin\theta + \frac{\theta}{2} + \frac{\sin 2\theta}{4}\right]_0^{2\pi/3} - \left[\frac{\theta}{4} + \sin\theta + \frac{\theta}{2} + \frac{\sin 2\theta}{4}\right]_{2\pi/3}^{\pi}$

$= \frac{1}{4}(\pi + 3\sqrt{3})$

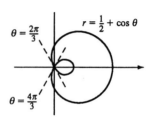

39. The two circles intersect at the pole since $(0, 0)$ satisfies the first equation and $(0, \pi/2)$ the second. The other intersection point $(1/\sqrt{2}, \pi/4)$ is where $\sin\theta = \cos\theta$.

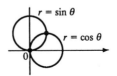

41. The curves intersect at the pole since $(0, \pi/2)$ satisfies the first equation and $(0, 0)$ the second.

$\cos\theta = 1 - \cos\theta \Rightarrow \cos\theta = 1/2 \Rightarrow \theta = \pi/3$ or $5\pi/3 \Rightarrow$ the other intersection points are $(1/2, \pi/3)$ and $(1/2, 5\pi/3)$.

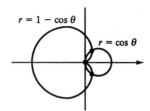

43. The pole is a point of intersection, $(0, 0)$ on both curves. $\sin\theta = \sin 2\theta = 2\sin\theta\cos\theta \Leftrightarrow$

$\sin\theta(1 - 2\cos\theta) = 0 \Leftrightarrow \sin\theta = 0$ or $\cos\theta = 1/2$

$\Rightarrow \theta = 0, \pi, \pi/3, -\pi/3 \Rightarrow (\sqrt{3}/2, \pi/3)$ and $(\sqrt{3}/2, 2\pi/3)$ are the other intersection points.

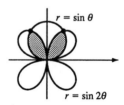

45. $L = \int_0^{3\pi/4} \sqrt{r^2 + (dr/d\theta)^2}\, d\theta = \int_0^{3\pi/4} \sqrt{(5\cos\theta)^2 + (-5\sin\theta)^2}\, d\theta$

$= 5\int_0^{3\pi/4} \sqrt{\cos^2\theta + \sin^2\theta}\, d\theta = 5\int_0^{3\pi/4} d\theta = \frac{15}{4}\pi$

47. $L = \int_0^{2\pi} \sqrt{(2^\theta)^2 + [(\ln 2)\, 2^\theta]^2}\, d\theta = \int_0^{2\pi} 2^\theta \sqrt{1 + \ln^2 2}\, d\theta = \left[\sqrt{1 + \ln^2 2}\, (2^\theta/\ln 2)\right]_0^{2\pi}$

$= \sqrt{1 + \ln^2 2}\,(2^{2\pi} - 1)/\ln 2$

49. $L = \int_0^{2\pi} \sqrt{(\theta^2)^2 + (2\theta)^2}\, d\theta = \int_0^{2\pi} \theta\sqrt{\theta^2 + 4}\, d\theta = \frac{1}{2}\cdot\frac{2}{3}\left[(\theta^2 + 4)^{3/2}\right]_0^{2\pi}$

$= \frac{8}{3}\left[(\pi^2 + 1)^{3/2} - 1\right]$

51. $L = 2\int_0^{2\pi} \sqrt{\cos^8(\theta/4) + \cos^6(\theta/4)\sin^2(\theta/4)}\, d\theta$

$$= 2 \int_0^{2\pi} |\cos^3(\theta/4)| \sqrt{\cos^2(\theta/4) + \sin^2(\theta/4)} \, d\theta = 2 \int_0^{2\pi} |\cos^3(\theta/4)| \, d\theta$$

$$= 8 \int_0^{\pi/2} \cos^3 u \, du \quad (where \ u = \theta/4) \ = 8 \left[\sin u - \tfrac{1}{3}\sin^3 u \right]_0^{\pi/2} = 16/3$$

(Note that the curve is retraced after every interval of length 4π.)

53.   From Figure 9.31 it is apparent that one loop lies between $\theta = -\pi/4$ and $\theta = \pi/4$. Therefore

$$L = \int_{-\pi/4}^{\pi/4} \sqrt{\cos^2 2\theta + (-2\sin 2\theta)^2} \, d\theta = 2 \int_0^{\pi/4} \sqrt{\cos^2 2\theta + 4\sin^2 2\theta} \, d\theta$$

$$= 2 \int_0^{\pi/4} \sqrt{1 + 3\sin^2 2\theta} \, d\theta. \quad \text{Using Simpson's rule with n} = 4, \ \Delta x = \frac{\pi/4}{4} = \frac{\pi}{16} \text{ and}$$

$$f(\theta) = \sqrt{1 + 3\sin^2 2\theta} \text{ we get}$$

$$L \approx 2 \cdot S_4 = 2 \frac{\pi}{16 \cdot 3} [f(0) + 4f(\pi/16) + 2f(\pi/8) + 4f(3\pi/16) + f(\pi/4)] \approx 2.4228.$$

Therefore the length of one loop of the four–leaved rose is approximately 2.42.

55.   (a)   From (9.9), $S = \int_a^b 2\pi y \sqrt{(dx/d\theta)^2 + (dy/d\theta)^2} \, d\theta = \int_a^b 2\pi y \sqrt{r^2 + (dr/d\theta)^2} \, d\theta$ [see the

derivation of (9.17)] $= \int_a^b 2\pi r \sin\theta \sqrt{r^2 + (dr/d\theta)^2} \, d\theta$

(b) $r^2 = \cos 2\theta \ \Rightarrow \ 2r\frac{dr}{d\theta} = -2\sin 2\theta \ \Rightarrow \ \left(\frac{dr}{d\theta}\right)^2 = \frac{\sin^2 2\theta}{r^2} = \frac{\sin^2 2\theta}{\cos 2\theta}$

$$S = 2 \int_0^{\pi/4} 2\pi \sqrt{\cos 2\theta} \sin\theta \sqrt{\cos 2\theta + ((\sin^2 2\theta)/\cos 2\theta)} \, d\theta = 4\pi \int_0^{\pi/4} \sin\theta \, d\theta$$

$$= \left[-4\pi \cos\theta \right]_0^{\pi/4} = -4\pi(\tfrac{1}{\sqrt{2}} - 1) = 2\pi(2 - \sqrt{2}).$$

EXERCISES 9.6

1. $x^2 = -8y$. $4p = -8$, so $p = -2$. The vertex is $(0, 0)$,
the focus is $(0, -2)$, and the directrix is $y = 2$.

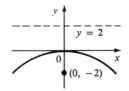

3. $y^2 = x$. $p = 1/4$ and the vertex is $(0, 0)$, so the focus is
$(1/4, 0)$, and the directrix is $x = -1/4$.

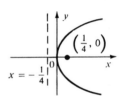

5. $x + 1 = 2(y - 3)^2 \Rightarrow (y - 3)^2 = \frac{1}{2}(x + 1) \Rightarrow p = 1/8$

    $\Rightarrow$ vertex $(-1, 3)$, focus $(-7/8, 3)$, directrix $x = -9/8$

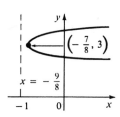

7. $2x + y^2 - 8y + 12 = 0 \Rightarrow (y - 4)^2 = -2(x - 2) \Rightarrow$

    $p = -1/2 \Rightarrow$ vertex $(2, 4)$, focus $(3/2, 4)$, directrix

    $x = 5/2$

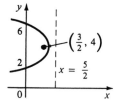

9. $\frac{x^2}{16} + \frac{y^2}{4} = 1 \Rightarrow a = 4, b = 2, c = \sqrt{16 - 4} = 2\sqrt{3} \Rightarrow$

    center $(0, 0)$, vertices $(\pm 4, 0)$, foci $(\pm 2\sqrt{3}, 0)$

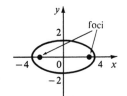

11. $25x^2 + 9y^2 = 225 \Leftrightarrow \frac{x^2}{9} + \frac{y^2}{25} = 1 \Rightarrow a = 5, b = 3,$

    $c = 4 \Rightarrow$ center $(0, 0)$, vertices $(0, \pm 5)$, foci $(0, \pm 4)$

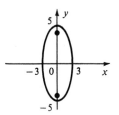

13. $\frac{x^2}{144} - \frac{y^2}{25} = 1 \Rightarrow a = 12, b = 5, c = \sqrt{144 + 25} = 13 \Rightarrow$

    center $(0, 0)$, vertices $(\pm 12, 0)$, foci $(\pm 13, 0)$, asymptotes

    $y = \pm 5x/12$

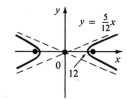

15. $9y^2 - x^2 = 9 \Rightarrow y^2 - \frac{x^2}{9} = 1 \Rightarrow a = 1, b = 3, c = \sqrt{10}$

    $\Rightarrow$ center $(0, 0)$, vertices $(0, \pm 1)$, foci

    $(0, \pm\sqrt{10})$, asymptotes $y = \pm x/3$

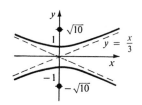

325

17. $9x^2 - 18x + 4y^2 = 27 \Leftrightarrow \dfrac{(x-1)^2}{4} + \dfrac{y^2}{9} = 1 \Rightarrow a = 3,$
$b = 2, c = \sqrt{5} \Rightarrow$ center $(1, 0)$, vertices $(1, \pm 3)$,
foci $(1, \pm\sqrt{5})$

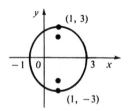

19. $2y^2 - 4y - 3x^2 + 12x = -8 \Leftrightarrow \dfrac{(x-2)^2}{6} - \dfrac{(y-1)^2}{9} = 1$
$\Rightarrow a = \sqrt{6}, b = 3, c = \sqrt{15} \Rightarrow$ center $(2, 1)$, vertices
$(2 \pm \sqrt{6}, 1)$, foci $(2 \pm \sqrt{15}, 1)$, asymptotes
$y - 1 = \pm\dfrac{3}{\sqrt{6}}(x-2)$

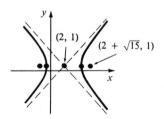

21. Vertex at $(0, 0)$, $p = 3$, opens upward $\Rightarrow x^2 = 4py = 12y$

23. Vertex at $(2, 0)$, $p = 1$, opens to right $\Rightarrow y^2 = 4p(x-2) = 4(x-2)$

25. Parabola must have equation $y^2 = 4px$, so $(-4)^2 = 4p(1) \Rightarrow p = 4 \Rightarrow y^2 = 16x.$

27. Center $(0, 0)$, $c = 1$, $a = 2 \Rightarrow b = \sqrt{2^2 - 1^2} = \sqrt{3} \Rightarrow \dfrac{x^2}{4} + \dfrac{y^2}{3} = 1$

29. Center $(3, 0)$, $c = 1$, $a = 3 \Rightarrow b = \sqrt{8} = 2\sqrt{2} \Rightarrow \dfrac{(x-3)^2}{8} + \dfrac{y^2}{9} = 1$

31. Center $(2, 2)$, $c = 2$, $a = 3 \Rightarrow b = \sqrt{5} \Rightarrow \dfrac{(x-2)^2}{9} + \dfrac{(y-2)^2}{5} = 1$

33. Center $(0, 0)$, vertical axis, $c = 3$, $a = 1 \Rightarrow b = \sqrt{8} = 2\sqrt{2} \Rightarrow y^2 - \dfrac{x^2}{8} = 1$

35. Center $(4, 3)$, horizontal axis, $c = 3$, $a = 2 \Rightarrow b = \sqrt{5} \Rightarrow \dfrac{(x-4)^2}{4} - \dfrac{(y-3)^2}{5} = 1$

37. Center $(0, 0)$, horizontal axis, $a = 3$, $\dfrac{b}{a} = 2 \Rightarrow b = 6 \Rightarrow \dfrac{x^2}{9} - \dfrac{y^2}{36} = 1$

39. In Figure 9.43, we see that the point on the ellipse closest to a focus is the closer vertex (which is a distance $a - c$ from it) while the farthest point is the other vertex (at a distance of $a + c$). So for this lunar orbit, $(a - c) + (a + c) = 2a = (1728 + 110) + (1728 + 314)$, or $a = 1940$; and $(a + c) - (a - c) = 2c = 314 - 110$, or $c = 102$. Thus $b^2 = a^2 - c^2 = 3{,}753{,}196$, and the equation is $\dfrac{x^2}{3{,}763{,}600} + \dfrac{y^2}{3{,}753{,}196} = 1.$

41. (a) Set up the coordinate system so that A is $(-200, 0)$ and B is $(200, 0)$.

$|PA| - |PB| = (1200)(980) = 1{,}176{,}000$ ft $= \dfrac{2450}{11}$ mi $= 2a \Rightarrow a = \dfrac{1225}{11}$, and $c = 200$

so $b^2 = c^2 - a^2 = \dfrac{3{,}339{,}375}{121} \Rightarrow \dfrac{121x^2}{1{,}500{,}625} + \dfrac{121y^2}{3{,}339{,}375} = 1$

(b)   Due north of B $\Rightarrow$ x = 200 $\Rightarrow$ $\dfrac{(121)(200)^2}{1,500,625} + \dfrac{121y^2}{3,339,375} = 1$ $\Rightarrow$ y = $\dfrac{133,575}{539} \approx 248$

mi.

43.   The function whose graph is the upper branch of this hyperbola is concave upward. The

function is

$y = f(x) = a\sqrt{1 + \dfrac{x^2}{b^2}} = \dfrac{a}{b}\sqrt{b^2 + x^2}$, so $y' = \dfrac{a}{b}x(b^2 + x^2)^{-1/2}$ and

$y'' = \dfrac{a}{b}\left[(b^2 + x^2)^{-1/2} - x^2(b^2 + x^2)^{-3/2}\right] = ab(b^2 + x^2)^{-3/2} > 0$ for all x, and so f is

concave upward.

45.   (a) ellipse (b) hyperbola (c) empty graph (no curve)

(d) In case (a), $a^2 = k$, $b^2 = k - 16$, and $c^2 = a^2 - b^2 = 16$, so the foci are at $(\pm 4, 0)$. In

case (b), $k - 16 < 0$, so $a^2 = k$, $b^2 = 16 - k$, and $c^2 = a^2 + b^2 = 16$, and so again the

foci are at $(\pm 4, 0)$.

47.   Use the parametrization $x = 2\cos t$, $y = \sin t$, $0 \le t \le 2\pi$ to get

$$L = 4\int_0^{\pi/2} \sqrt{(dx/dt)^2 + (dy/dt)^2}\, dt = 4\int_0^{\pi/2} \sqrt{4\sin^2 t + \cos^2 t}\, dt = 4\int_0^{\pi/2} \sqrt{3\sin^2 t + 1}\, dt.$$

Using Simpson's rule with n = 10,

$$L \approx \dfrac{4}{3}\left(\dfrac{\pi}{20}\right)\left\{f(0) + 4f(\pi/20) + 2f(\pi/10) + \ldots + 2f(2\pi/5) + 4f(9\pi/20) + f(\pi/2)\right\}, \text{ with}$$

$f(t) = \sqrt{3\sin^2 t + 1}$, so $L \approx 9.69$.

49. $\dfrac{x^2}{a^2} + \dfrac{y^2}{b^2} = 1$ $\Rightarrow$ $\dfrac{2x}{a^2} + \dfrac{2yy'}{b^2} = 0$ $\Rightarrow$ $y' = -\dfrac{b^2 x}{a^2 y}$ ($y \ne 0$). Thus the slope of the tangent

line at P is $-\dfrac{b^2 x_1}{a^2 y_1}$. The slope of $F_1 P$ is $\dfrac{y_1}{x_1 + c}$ and of $F_2 P$ is $\dfrac{y_1}{x_1 - c}$. By the formula in

Exercise 64(b) of Section 2 in Review and Preview, we have

$\tan \alpha = \dfrac{[y_1/(x_1 + c)] + [b^2 x_1/(a^2 y_1)]}{1 - (b^2 x_1 y_1)/[a^2 y_1(x_1 + c)]}$

$= \dfrac{a^2 y_1^2 + b^2 x_1(x_1 + c)}{a^2 y_1(x_1 + c) - b^2 x_1 y_1} = \dfrac{a^2 b^2 + b^2 c x_1}{c^2 x_1 y_1 + a^2 c y_1}$

$[\text{using } b^2 x_1^2 + a^2 y_1^2 = a^2 b^2 \text{ and } a^2 - b^2 = c^2] = \dfrac{b^2(c x_1 + a^2)}{c y_1(c x_1 + a^2)} = \dfrac{b^2}{c y_1}$

and $\tan \beta = \dfrac{[-y_1/(x_1 - c)] - [b^2 x_1/(a^2 y_1)]}{1 - (b^2 x_1 y_1)/[a^2 y_1(x_1 - c)]}$

$= \dfrac{-a^2 y_1^2 - b^2 x_1(x_1 - c)}{a^2 y_1(x_1 - c) - b^2 x_1 y_1} = \dfrac{-a^2 b^2 + b^2 c x_1}{c^2 x_1 y_1 - a^2 c y_1}$

$= \dfrac{b^2(c x_1 - a^2)}{c y_1(c x_1 - a^2)} = \dfrac{b^2}{c y_1}$.   So $\alpha = \beta$.

EXERCISES 9.7

1. $\quad r = \dfrac{ed}{1 + e\cos\theta} = \dfrac{(2/3)(3)}{1 + (2/3)\cos\theta} = \dfrac{6}{3 + 2\cos\theta}$

3. $\quad r = \dfrac{ed}{1 + e\sin\theta} = \dfrac{(1)(2)}{1 + \sin\theta} = \dfrac{2}{1 + \sin\theta}$

5. $\quad r = 5\sec\theta \Leftrightarrow x = r\cos\theta = 5$, so $r = \dfrac{ed}{1 + e\cos\theta} = \dfrac{(4)(5)}{1 + 4\cos\theta} = \dfrac{20}{1 + 4\cos\theta}$

7. $\quad$ Focus $(0, 0)$, vertex $(5, \pi/2) \Rightarrow$ directrix $y = 10 \Rightarrow r = \dfrac{ed}{1 + e\sin\theta} = \dfrac{10}{1 + \sin\theta}$

9. $\quad e = 3 \Rightarrow$ hyperbola; $ed = 4 \Rightarrow d = 4/3 \Rightarrow$ directrix
$x = 4/3$; vertices $(1, 0)$ and $(-2, \pi) = (2, 0)$; center
$(3/2, 0)$; asymptotes parallel to $\theta = \pm\cos^{-1}(-1/3)$

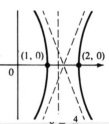

11. $\quad e = 1 \Rightarrow$ parabola; $ed = 2 \Rightarrow d = 2 \Rightarrow$ directrix
$x = -2$; vertex $(-1, 0) = (1, \pi)$

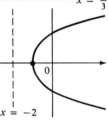

13. $\quad r = \dfrac{3}{1 + (1/2)\sin\theta} \Rightarrow e = 1/2 \Rightarrow$ ellipse; $ed = 3 \Rightarrow$
$d = 6 \Rightarrow$ directrix $y = 6$; vertices $(2, \pi/2)$ and
$(6, 3\pi/2)$; center $(2, 3\pi/2)$

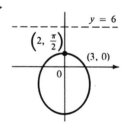

15. $\quad r = \dfrac{1/4}{1 - (3/4)\cos\theta} \Rightarrow e = 3/4 \Rightarrow$ ellipse; $ed = 1/4$
$\Rightarrow d = 1/3 \Rightarrow$ directrix $x = -1/3$; vertices $(1, 0)$ and
$(1/7, \pi)$; center $(3/7, 0)$; other focus $(6/7, 0)$.

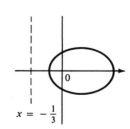

17. $r = \dfrac{7/2}{1 - (5/2)\sin\theta} \Rightarrow e = 5/2 \Rightarrow$ hyperbola;

ed $= 7/2 \Rightarrow d = 7/5 \Rightarrow$ directrix $y = -7/5$; center

$(5/3, 3\pi/2)$; vertices $(-7/3, \pi/2) = (7/3, 3\pi/2)$ and

$(1, 3\pi/2)$.

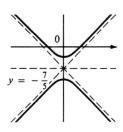

19. $r = \dfrac{5/2}{1 + \sin\theta} \Rightarrow e = 1 \Rightarrow$ parabola; ed $= 5/2 \Rightarrow$

$d = 5/2 \Rightarrow$ directrix $y = 5/2$; vertex $(5/4, \pi/2)$.

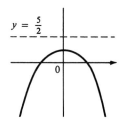

21. $|PF| = e|Pl| \Rightarrow r = e[d - r\cos(\pi - \theta)] = e(d + r\cos\theta) \Rightarrow$

$r(1 - e\cos\theta) = ed \Rightarrow r = \dfrac{ed}{1 - e\cos\theta}$.

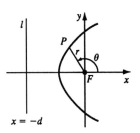

23. $|PF| = e|Pl| \Rightarrow r = e[d - r\sin(\theta - \pi)] = e(d + r\sin\theta)$

$\Rightarrow r(1 - e\sin\theta) = ed \Rightarrow r = \dfrac{ed}{1 - e\sin\theta}$

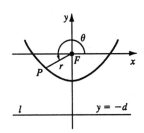

25. (a) If the directrix is $x = -d$ then $r = \dfrac{ed}{1 - e\cos\theta}$ [see figure 9.52 (b)], and, from (9.30),

$a^2 = \dfrac{e^2 d^2}{(1 - e^2)^2} \Rightarrow ed = a(1 - e^2)$. Therefore, $r = \dfrac{a(1 - e^2)}{1 - e\cos\theta}$.

(b) $e = 0.017$ and the major axis $= 2a = 2.99 \times 10^8 \Rightarrow a = 1.495 \times 10^8$. Therefore

$r = \dfrac{1.495 \times 10^8 [1 - (0.017)^2]}{1 - 0.017\cos\theta} \approx \dfrac{1.49 \times 10^8}{1 - 0.017\cos\theta}$.

27. The minimum distance is at perihelion where $4.6 \times 10^7 = r = a(1 - e) = a(1 - 0.206)$

$= a(0.794) \Rightarrow a = 4.6 \times 10^7/0.794$. So the maximum distance, which is at aphelion, is

$r = a(1+e) = (4.6 \times 10^7/0.794) \times 10^7 (1.206) \approx 7.0 \times 10^7$ km.

REVIEW EXERCISES FOR CHAPTER 9

1. $x = 1 - t^2, y = 1 - t \; (-1 \le t \le 1)$

    $x = 1 - (1-y)^2 = 2y - y^2$

    $(0 \le y \le 2)$

    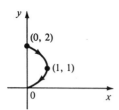

3. $x = 1 + \sin t, y = 2 + \cos t \;\Rightarrow$

    $(x-1)^2 + (y-2)^2 = \sin^2 t + \cos^2 t = 1$

    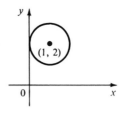

5. $r = 1 + 3\cos\theta$

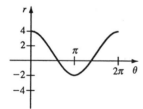
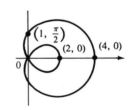

7. $r^2 = \sec 2\theta \;\Rightarrow\; r^2 \cos 2\theta = 1 \;\Rightarrow\; r^2(\cos^2\theta - \sin^2\theta) = 1 \;\Rightarrow\; x^2 - y^2 = 1$, a hyperbola

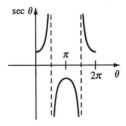
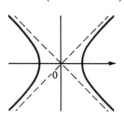

9. $r = 2\cos^2(\theta/2) = 1 + \cos\theta$

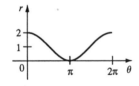
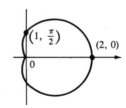

11. $r = \dfrac{1}{1 + \cos\theta} \Rightarrow e = 1 \Rightarrow$ parabola; $d = 1 \Rightarrow$

directrix $x = 1$ and vertex $(\frac{1}{2}, 0)$; y–intercepts are

$(1, \pi/2)$ and $(1, 3\pi/2)$.

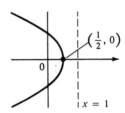

13. $x^2 + y^2 = 4x \Leftrightarrow r^2 = 4r\cos\theta \Leftrightarrow r = 4\cos\theta$

15. $x = t^2 + 2t$, $y = t^3 - t$. $\dfrac{dy}{dx} = \dfrac{dy/dt}{dx/dt} = \dfrac{3t^2 - 1}{2t + 2} = \dfrac{1}{2}$ when $t = 1$

17. $\dfrac{dy}{dx} = \dfrac{(dr/d\theta)\sin\theta + r\cos\theta}{(dr/d\theta)\cos\theta - r\sin\theta} = \dfrac{\sin\theta + \theta\cos\theta}{\cos\theta - \theta\sin\theta} = \dfrac{1/\sqrt{2} + (\pi/4)(1/\sqrt{2})}{1/\sqrt{2} - (\pi/4)(1/\sqrt{2})} = \dfrac{4 + \pi}{4 - \pi}$ when

$\theta = \dfrac{\pi}{4}$

19. $\dfrac{dy}{dx} = \dfrac{dy/dt}{dx/dt} = \dfrac{t\cos t + \sin t}{-t\sin t + \cos t}$. $\dfrac{d^2y}{dx^2} = \dfrac{\frac{d}{dt}\left(\frac{dy}{dx}\right)}{dx/dt}$, where

$\dfrac{d}{dt}\left(\dfrac{dy}{dx}\right) = \dfrac{(-t\sin t + \cos t)(-t\sin t + 2\cos t) - (t\cos t + \sin t)(-t\cos t - 2\sin t)}{(-t\sin t + \cos t)^2} = \dfrac{t^2 + 2}{(-t\sin t + \cos t)^2}$

$\Rightarrow \dfrac{d^2y}{dx^2} = \dfrac{t^2 + 2}{(-t\sin t + \cos t)^3}$

21. $\dfrac{dx}{dt} = -2a\sin t + 2a\sin 2t = 2a\sin t(2\cos t - 1) = 0 \Leftrightarrow \sin t = 0$ or $\cos t = \frac{1}{2} \Rightarrow t = 0$,

$\pi/3$, π, or $5\pi/3$.

$\dfrac{dy}{dt} = 2a\cos t - 2a\cos 2t = 2a(1 + \cos t - 2\cos^2 t) = 2a(1 - \cos t)(1 + 2\cos t) = 0 \Rightarrow t = 0$,

$2\pi/3$, or $4\pi/3$.

Thus the graph has vertical tangents where $t = \pi/3$, π and $5\pi/3$, and horizontal tangents where

$t = 2\pi/3$ and $4\pi/3$. To determine what the slope is where $t = 0$, we use l'Hospital's rule to

evaluate $\lim\limits_{t \to 0} \dfrac{dy/dt}{dx/dt} = 0$, so there is a horizontal tangent there.

t	x	y
0	a	0
$\pi/3$	$3a/2$	$\sqrt{3}a/2$
$2\pi/3$	$-a/2$	$3\sqrt{3}a/2$
π	$-3a$	0
$4\pi/3$	$-a/2$	$-3\sqrt{3}a/2$
$5\pi/3$	$3a/2$	$-\sqrt{3}a/2$

23. This curve will have 5 "petals", each petal corresponding to a segment of values of θ for which

$\cos 5\theta$ is positive (since $\cos 5\theta = r^2/9 \geq 0$).

$A = 5\int_{-\pi/10}^{\pi/10} \frac{1}{2} r^2 \, d\theta = \frac{5}{2}\int_{-\pi/10}^{\pi/10} 9 \cos 5\theta \, d\theta = 45\int_0^{\pi/10} \cos 5\theta \, d\theta = \left[9 \sin 5\theta\right]_0^{\pi/10} = 9$

25. The curves intersect where $4\cos\theta = 2$; that is, at $(2, \pi/3)$ and $(2, -\pi/3)$.

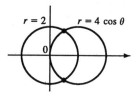

27. The curves intersect where $2\sin\theta = \sin\theta + \cos\theta \Rightarrow$ $\sin\theta = \cos\theta \Rightarrow \theta = \pi/4$, and also at the origin (at which $\theta = 3\pi/4$ on the second curve).

$A = \int_0^{\pi/4} \frac{1}{2}(2\sin\theta)^2 \, d\theta + \int_{\pi/4}^{3\pi/4} \frac{1}{2}(\sin\theta + \cos\theta)^2 \, d\theta$

$= \int_0^{\pi/4}(1 - \cos 2\theta)\, d\theta + \frac{1}{2}\int_{\pi/4}^{3\pi/4}(1 + \sin 2\theta)\, d\theta$

$= \left[\theta - \frac{1}{2}\sin 2\theta\right]_0^{\pi/4} + \left[\frac{1}{2}\theta - \frac{1}{4}\cos 2\theta\right]_{\pi/4}^{3\pi/4} = \frac{1}{2}(\pi - 1)$

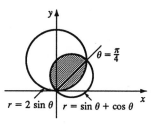

29. $x = 3t^2$, $y = 2t^3$. $L = \int_0^2 \sqrt{(dx/dt)^2 + (dy/dt)^2}\, dt = \int_0^2 \sqrt{(6t)^2 + (6t^2)^2}\, dt$

$= 6\int_0^2 t\sqrt{1 + t^2}\, dt = \left[2(1 + t^2)^{3/2}\right]_0^2 = 2(5\sqrt{5} - 1)$

31. $L = \int_\pi^{2\pi} \sqrt{r^2 + (dr/d\theta)^2}\, d\theta = \int_\pi^{2\pi} \sqrt{(1/\theta)^2 + (-1/\theta^2)^2}\, d\theta = \int_\pi^{2\pi} \frac{\sqrt{\theta^2 + 1}}{\theta^2}\, d\theta$

$= \left[-\frac{\sqrt{\theta^2 + 1}}{\theta} + \ln\left|\theta + \sqrt{\theta^2 + 1}\right|\right]_\pi^{2\pi} \quad [Formula\ 24] \quad = \frac{\sqrt{\pi^2 + 1}}{\pi} - \frac{\sqrt{4\pi^2 + 1}}{2\pi} + \ln\left|\frac{2\pi + \sqrt{4\pi^2 + 1}}{\pi + \sqrt{\pi^2 + 1}}\right|$

33. $S = \int_1^4 2\pi y\sqrt{(dx/dt)^2 + (dy/dt)^2}\, dt = \int_1^4 2\pi\left(\frac{1}{3}t^3 + \frac{1}{2}t^{-2}\right)\sqrt{(2/\sqrt{t})^2 + (t^2 - t^{-3})^2}\, dt$

$= 2\pi\int_1^4 \left(\frac{1}{3}t^3 + \frac{1}{2}t^{-2}\right)\sqrt{(t^2 + t^{-3})^2}\, dt = 2\pi\int_1^4 \left(\frac{1}{3}t^5 + \frac{5}{6} + \frac{1}{2}t^{-5}\right)dt$

$= 2\pi\left[\frac{1}{18}t^6 + \frac{5}{6}t - \frac{1}{8}t^{-4}\right]_1^4 = \frac{471295}{1024}\pi$

35. Ellipse, center $(0, 0)$, $a = 3$, $b = 2\sqrt{2}$, $c = 1$ $\Rightarrow$ foci $(\pm 1, 0)$, vertices $(\pm 3, 0)$.

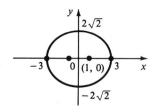

37. $6(y^2 - 6y + 9) = -(x + 1)$ $\Leftrightarrow$
$(y - 3)^2 = -\frac{1}{6}(x + 1)$ A parabola with vertex
$(-1, 3)$, opening to the left, $p = -\frac{1}{24}$ $\Rightarrow$ focus
$(-\frac{25}{24}, 3)$ and directrix $x = -\frac{23}{24}$.

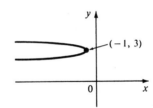

39. The parabola opens upward with vertex $(0, 4)$ and $p = 2$, so its equation is
$(x - 0)^2 = 4 \cdot 2(y - 4)$ $\Leftrightarrow$ $x^2 = 8(y - 4)$.

41. The hyperbola has center $(0, 0)$ and foci on the x–axis. $c = 3$ and $\frac{b}{a} = \frac{1}{2}$ [*from the*
asymptotes] $\Rightarrow$ $9 = c^2 = a^2 + b^2 = (2b)^2 + b^2 = 5b^2$ $\Rightarrow$ $b = \frac{3}{\sqrt{5}}$ $\Rightarrow$ $a = \frac{6}{\sqrt{5}}$ $\Rightarrow$
equation is $\frac{x^2}{36/5} - \frac{y^2}{9/5} = 1$ $\Leftrightarrow$ $5x^2 - 20y^2 = 36$.

43. $x^2 = -y + 100$ has its vertex at $(0, 100)$ so one of the vertices of the ellipse is $(0, 100)$. Another
form of the equation of a parabola is $x^2 = 4p(y - 100)$ so $4p(y - 100) = -y + 100$ $\Rightarrow$
$4py - 4p100 = 100 - y$ $\Rightarrow$ $4p = \frac{100 - y}{y - 100}$ $\Rightarrow$ $p = -0.25$. Therefore the shared focus is found
at $(0, 99.75)$ so $2c = 99.75 - 0$ $\Rightarrow$ $c = 399/8 = 49.875$ and the centre of the ellipse is
$(0, 49.875)$. So $a = 100 - 49.875 = 50.125 = 401/8$ and $b^2 = a^2 - c^2 = (401^2 - 399^2)/8^2$
$= 25$. Therefore the equation of the ellipse is $\frac{x^2}{b^2} + \frac{(y - 49.875)^2}{a^2} = 1$ $\Rightarrow$
$\frac{x^2}{25} + \frac{(y - 399/8)^2}{(401/8)^2} = 1$ or $\frac{x^2}{25} + \frac{(8y - 399)^2}{160,801} = 1$.

45. Directrix $x = 4$ $\Rightarrow$ $d = 4$, so $e = \frac{1}{3}$ $\Rightarrow$ $r = \frac{ed}{1 + e\cos\theta} = \frac{4}{3 + \cos\theta}$

47. In polar coordinates an equation for the circle is $r = 2a\sin\theta$. Thus the coordinates of Q are
$x = r\cos\theta = 2a\sin\theta\cos\theta$ and $y = r\sin\theta = 2a\sin^2\theta$. The coordinates of R are $x = 2a\cot\theta$
and $y = 2a$. Since P is the midpoint of QR, we use the midpoint formula to get
$x = a(\sin\theta\cos\theta + \cot\theta)$ and $y = a(1 + \sin^2\theta)$.

APPLICATIONS PLUS (page 554)

1. (a) While running from $(L, 0)$ to (x, y), the dog

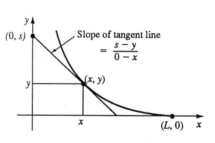

travels a distance $s = \int_x^L \sqrt{1 + \left(\dfrac{dy}{dx}\right)^2}\, dx.$

The dog and rabbit run at the same speed,

so the rabbit's position when the dog has

traveled a distance s is $(0, s)$. Since the dog

runs straight for the rabbit, $\dfrac{dy}{dx} = \dfrac{s - y}{0 - x}$ (see the figure). Thus $s = y - x\dfrac{dy}{dx}$ and

$$-\sqrt{1 + \left(\frac{dy}{dx}\right)^2} = \frac{ds}{dx} = \frac{dy}{dx} - x\frac{d^2y}{dx^2} - 1\frac{dy}{dx} = -x\frac{d^2y}{dx^2}. \quad \text{Hence } x\frac{d^2y}{dx^2} = \sqrt{1 + \left(\frac{dy}{dx}\right)^2} \text{ as}$$

claimed.

(b) Letting $z = \dfrac{dy}{dx}$, we obtain the differential equation $x\dfrac{dz}{dx} = \sqrt{1 + z^2}$, or $\dfrac{dz}{\sqrt{1 + z^2}} = \dfrac{dx}{x}.$

Integrating, we get $\ln x = \displaystyle\int \frac{dz}{\sqrt{1 + z^2}} \begin{bmatrix} z = tan\,\theta \\ dz = sec^2\theta\ d\theta \end{bmatrix} = \int (sec\,\theta)\,d\theta$

$$= \ln|sec\,\theta + tan\,\theta| + C = \ln\left|\sqrt{1 + z^2} + z\right| + C.$$

When $x = L$, $z = \dfrac{dy}{dx} = 0$, so $\ln L = \ln 1 + C$. Therefore $C = \ln L$, so

$$\ln x = \ln(\sqrt{1 + z^2} + z) + \ln L = \ln[L(\sqrt{1 + z^2} + z)] \text{ and } x = L(\sqrt{1 + z^2} + z).$$

$$\sqrt{1 + z^2} = \frac{x}{L} - z \Rightarrow 1 + z^2 = \left(\frac{x}{L}\right)^2 - \frac{2xz}{L} + z^2 \Rightarrow \left(\frac{x}{L}\right)^2 - 2z\left(\frac{x}{L}\right) - 1 = 0 \Rightarrow z = \frac{\left(\frac{x}{L}\right)^2 - 1}{2\left(\frac{x}{L}\right)}$$

$$= \frac{x^2 - L^2}{2Lx} = \frac{x}{2L} - \frac{L}{2}\frac{1}{x} \text{ (for } x > 0\text{). Since } z = \frac{dy}{dx}, y = \frac{x^2}{4L} - \frac{L}{2}\ln x + C'. \text{ But } y = 0 \text{ when}$$

$x = L$. Therefore $0 = \dfrac{L}{4} - \dfrac{L}{2}\ln L + C'$ and $C' = \dfrac{L}{2}\ln L - \dfrac{L}{4}$. Therefore

$$y = \frac{x^2}{4L} - \frac{L}{2}\ln x + \frac{L}{2}\ln L - \frac{L}{4} = \frac{x^2 - L^2}{4L} - \frac{L}{2}\ln\left(\frac{x}{L}\right).$$

(c) As $x \to 0$, $y \to \infty$, so the dog never catches the rabbit.

3. (a) The two spherical zones, whose surface areas we will call

S_1 and S_2, are generated by rotation about the y-axis of

circular arcs, as indicated in the figure. The arcs are the

upper and lower portions of the circle $x^2 + y^2 = r^2$ that

are obtained when the circle is cut with the line $y = d$.

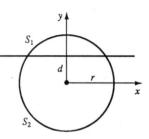

The portion of the upper arc in the first quadrant is sufficient to generate the upper spherical zone. That portion of the arc can be described by the relation $x = \sqrt{r^2 - y^2}$ for

$d \leq y \leq r$. Thus $\dfrac{dy}{dx} = \dfrac{-y}{\sqrt{r^2 - y^2}}$ and $ds = \sqrt{1 + \left(\dfrac{dx}{dy}\right)^2}\, dy = \sqrt{1 + \dfrac{y^2}{r^2 - y^2}}\, dy$

$= \sqrt{\dfrac{r^2}{r^2 - y^2}}\, dy = \dfrac{r\, dy}{\sqrt{r^2 - y^2}}$. From formula 8.26 we have $S_1 = \displaystyle\int_d^r 2\pi x \sqrt{1 + \left(\dfrac{dx}{dy}\right)^2}\, dy$

$= \displaystyle\int_d^r 2\pi \sqrt{r^2 - y^2}\, \dfrac{r\, dy}{\sqrt{r^2 - y^2}} = \int_d^r 2\pi r\, dy = 2\pi r(r - d)$. Similarly, we can compute

$S_2 = \displaystyle\int_{-r}^d 2\pi x \sqrt{1 + \left(\dfrac{dx}{dy}\right)^2}\, dy = \int_{-r}^d 2\pi r\, dy = 2\pi r(r + d)$. Note that $S_1 + S_2 = 4\pi r^2$, the

surface area of the entire sphere.

(b) $r = 3960\,\text{mi}$ and $d = r(\sin 75°) \approx 3825\,\text{mi}$,

so the surface area of the Arctic Ocean

$\approx 2\pi r(r - d) \approx 2\pi(3960)(135)$

$\approx 3.36 \times 10^6\ \text{mi}^2.$

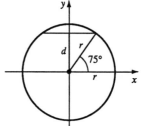

(c)   The area on the sphere lies between planes $y = y_1$ and $y = y_2$, where $y_2 - y_1 = h$. Thus we

compute the surface area on the sphere to be $S = \displaystyle\int_{y_1}^{y_2} 2\pi x \sqrt{1 + \left(\dfrac{dx}{dy}\right)^2}\, dy = \int_{y_1}^{y_2} 2\pi r\, dy$

$= 2\pi r(y_2 - y_1) = 2\pi r h$. This equals the lateral area of a cylinder of radius r and height h.

(The cylinder is obtained by rotating the line $x = r$

about the y-axis, so the surface area of the cylinder

between the planes $y = y_1$ and $y = y_2$ is

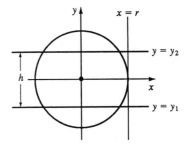

$\displaystyle\int_{y_1}^{y_2} 2\pi x \sqrt{1 + \left(\dfrac{dx}{dy}\right)^2}\, dy = \int_{y_1}^{y_2} 2\pi r \sqrt{1 + 0^2}\, dy$

$= 2\pi r y \big|_{y=y_1}^{y_2} = 2\pi r(y_2 - y_1) = 2\pi r h.)$

(d) $h = 2r \sin 23.45° \approx 3152\,\text{mi}$, so surface area of Torrid Zone $= 2\pi r h \approx 2\pi(3960)(3152)$

$\approx 7.84 \times 10^7\ \text{mi}^2.$

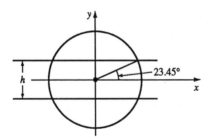

5. **(a)** $\frac{dy}{dt} = ky^{1+\epsilon} \Rightarrow y^{-1-\epsilon}\,dy = k\,dt \Rightarrow \frac{y^{-\epsilon}}{-\epsilon} = kt + C.$ Since $y(0) = y_0$, we have $C = \frac{y_0^{-\epsilon}}{-\epsilon}.$

Thus $\frac{y^{-\epsilon}}{-\epsilon} = kt + \frac{y_0^{-\epsilon}}{-\epsilon}$, or $y^{-\epsilon} = y_0^{-\epsilon} - \epsilon kt.$ Therefore $y^\epsilon = \frac{1}{y_0^{-\epsilon} - \epsilon kt} = \frac{y_0^\epsilon}{1 - \epsilon y_0^\epsilon kt}$ and

$$y(t) = \frac{y_0}{(1 - \epsilon y_0^\epsilon kt)^{1/\epsilon}}\,.$$

(b) $y(t) \to \infty$ as $1 - \epsilon y_0^\epsilon kt \to 0$, that is, as $t \to \frac{1}{\epsilon y_0^\epsilon k}$. Define $T = \frac{1}{\epsilon y_0^\epsilon k}$. Then

$$\lim_{t \to T^-} y(t) = \infty.$$

(c) According to the data given, we have $\epsilon = 0.01$, $y(0) = 2$, and $y(3) = 16$, where the time t

is given in months. Thus $y_0 = 2$ and $16 = y(3) = \frac{y_0}{(1 - \epsilon y_0^\epsilon k \cdot 3)^{1/\epsilon}}.$ We could solve for k,

but it is easier and more helpful to solve for $\epsilon y_0^\epsilon k$. [If you solved for k and want to check

your work, k turns out to be $\frac{1 - 8^{-.01}}{(.03)(2^{.01})} \approx .68125.$]

$16 = \frac{2}{(1 - 3\epsilon y_0^\epsilon k)^{100}}$, so $1 - 3\epsilon y_0^\epsilon k = \left(\frac{1}{8}\right)^{.01} = 8^{-.01}$ and $\epsilon y_0^\epsilon k = \frac{1}{3}(1 - 8^{-.01}).$ Thus

doomsday occurs when $t = T = \frac{1}{\epsilon y_0^\epsilon k} = \frac{3}{1 - 8^{-0.1}} \approx 145.77$ months or 12.15 years.

7. **(a)** If $\tan\theta = \sqrt{\frac{y}{C-y}}$, then $\tan^2\theta = \frac{y}{C-y}$, so $C\tan^2\theta - y\tan^2\theta = y$ and $y = \frac{C\tan^2\theta}{1 + \tan^2\theta}$

$= \frac{C\tan^2\theta}{\sec^2\theta} = C\tan^2\theta\cos^2\theta = C\sin^2\theta = \frac{C}{2}(1 - \cos 2\theta).$ Now $dx = \sqrt{\frac{y}{C-y}}\,dy$

$= \tan\theta \cdot \frac{C}{2} \cdot 2\sin 2\theta\,d\theta = C\tan\theta \cdot 2\sin\theta\cos\theta\,d\theta = 2C\sin^2\theta\,d\theta = C(1 - \cos 2\theta)\,d\theta.$ Thus

$x = C\left(\theta - \frac{\sin 2\theta}{2}\right) + K$ for some constant K. When $\theta = 0$, we have $y = 0$. We require that

$x = 0$ when $\theta = 0$ so that the curve passes through the origin when $\theta = 0$. This yields

$K = 0$. We now have $x = \frac{C}{2}(2\theta - \sin 2\theta)$, $y = \frac{C}{2}(1 - \cos 2\theta).$

(b) Setting $\phi = 2\theta$ and $r = \frac{C}{2}$, we get $x = r(\phi - \sin\phi)$, $y = r(1 - \cos\phi)$. Comparison with (9.2)

shows that the curve is a cycloid.

9. **(a)** Choose a vertical x-axis pointing downward with its origin at the surface. In order to

calculate the pressure at depth z, consider a partition P of the interval $[0, z]$ by points x_i

and choose a point $x_i^* \in [x_{i-1}, x_i]$ for each i. The thin layer of water lying between depth

x_{i-1} and depth x_i has a density of approximately $\rho(x_i^*)$, so the weight of a piece of that layer with unit cross-sectional area would be $\rho(x_i^*)g\,\Delta x_i$, where $\Delta x_i = x_i - x_{i-1}$. The total weight of a column of water extending from the surface to depth z (with unit cross-sectional area) would be approximately $\sum_i \rho(x_i^*)g\,\Delta x_i$. The estimate becomes exact if we take the limit as $\|P\| \to 0$: weight (or force) per unit area at depth $z = \lim\limits_{\|P\|\to 0} \sum_i \rho(x_i^*)g\,\Delta x_i$. In other words, $P(z) = \int_0^z \rho(x)g\,dx$. More generally, if we make no assumptions about the location of the origin of the x-axis, then

$P(z) = P_0 + \int_0^z \rho(x)g\,dx$, where P_0 is the pressure at $x = 0$. Differentiating, we get

$\dfrac{dP}{dz} = \rho(z)g$.

(b)

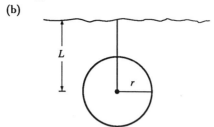

$$F = \int_{-r}^{r} P(L+x) \cdot 2\sqrt{r^2 - x^2}\,dx = \int_{-r}^{r}\left(P_0 + \int_0^{L+x} \rho_0 e^{z/H}g\,dz\right) \cdot 2\sqrt{r^2 - x^2}\,dx$$

$$= P_0 \int_{-r}^{r} 2\sqrt{r^2 - x^2}\,dx + \rho_0 gH \int_{-r}^{r}\left(e^{(L+x)/H} - 1\right) \cdot 2\sqrt{r^2 - x^2}\,dx$$

$$= (P_0 - \rho_0 gH)\int_{-r}^{r} 2\sqrt{r^2 - x^2}\,dx + \rho_0 gH \int_{-r}^{r} e^{(L+x)/H} \cdot 2\sqrt{r^2 - x^2}\,dx$$

$$= (P_0 - \rho_0 gH)(\pi r^2) + \rho_0 gH e^{L/H} \int_{-r}^{r} e^{x/H} \cdot 2\sqrt{r^2 - x^2}\,dx$$

Notice that the result of Exercise 21 in Section 8.5 does not apply here since pressure does not increase linearly with depth the way it does for a fluid of constant density.

11. **(a)** $\dfrac{d^2 y}{dx^2} = k\sqrt{1 + \left(\dfrac{dy}{dx}\right)^2}$. Setting $z = \dfrac{dy}{dx}$, we get $\dfrac{dz}{dx} = k\sqrt{1 + z^2} \Rightarrow \dfrac{dz}{\sqrt{1 + z^2}} = k\,dx$. Formula

25 in Appendix G gives $\ln(z + \sqrt{1 + z^2}) = kx + c \Rightarrow z + \sqrt{1 + z^2} = Ce^{kx}$ [*where* $C = e^c$] $\Rightarrow$

$\sqrt{1 + z^2} = Ce^{kx} - z \Rightarrow 1 + z^2 = C^2 e^{2kx} - 2Ce^{kx}z + z^2 \Rightarrow 2Ce^{kx}z = C^2 e^{2kx} - 1 \Rightarrow$

$z = \dfrac{C}{2}e^{kx} - \dfrac{1}{2C}e^{-kx}$. Now $\dfrac{dy}{dx} = \dfrac{C}{2}e^{kx} - \dfrac{1}{2C}e^{-kx} \Rightarrow y = \dfrac{C}{2k}e^{kx} + \dfrac{1}{2Ck}e^{-kx} + C'$.

From the diagram in the text, we see that $y(0) = a$ and $y(\pm b) = h$.

$a = y(0) = \dfrac{C}{2k} + \dfrac{1}{2Ck} + C' \Rightarrow C' = a - \dfrac{C}{2k} - \dfrac{1}{2Ck} \Rightarrow y = \dfrac{C}{2k}(e^{kx} - 1) + \dfrac{1}{2Ck}(e^{-kx} - 1) + a.$

From $h = y(\pm b)$, we find $h = \frac{C}{2k}(e^{kb} - 1) + \frac{1}{2Ck}(e^{-kb} - 1) + a$

and $h = \frac{C}{2k}(e^{-kb} - 1) + \frac{1}{2Ck}(e^{kb} - 1) + a$.

Subtracting the second equation from the first, we get $0 = \frac{C}{k}\frac{e^{kb} - e^{-kb}}{2} - \frac{1}{Ck}\frac{e^{kb} - e^{-kb}}{2}$

$= \frac{1}{k}\left(C - \frac{1}{C}\right)\sinh(kb)$. Now $k > 0$ and $b > 0$, so $\sinh(kb) > 0$ and $C = \pm 1$. If $C = 1$, then

$y = \frac{1}{2k}(e^{kx} - 1) + \frac{1}{2k}(e^{-kx} - 1) + a = \frac{1}{k}\frac{e^{kx} + e^{-kx}}{2} - \frac{1}{k} + a = a + \frac{1}{k}(\cosh kx - 1)$.

If $C = -1$, then $y = \frac{-1}{2k}(e^{kx} - 1) - \frac{1}{2k}(e^{-kx} - 1) + a = \frac{-1}{k}\frac{e^{kx} + e^{-kx}}{2} + \frac{1}{k} + a$

$= a - \frac{1}{k}(\cosh kx - 1)$. Since $k > 0$, $\cosh kx \geq 1$, and $y \geq a$, we conclude that $C = 1$ and

$y = a + \frac{1}{k}(\cosh kx - 1)$, where $h = y(b) = a + \frac{1}{k}(\cosh kb - 1)$. Since $\cosh(kb) = \cosh(-kb)$,

there is no further information to extract from the condition that $y(b) = y(-b)$.

However, we could replace a with the expression $h - \frac{1}{k}(\cosh kb - 1)$, obtaining

$y = h + \frac{1}{k}(\cosh kx - \cosh kb)$. It would be better still to keep a in the expression for y, and

use the expression for h to solve for k in terms of a, b, and h. That would enable us to

express y in terms of x and the given parameters a, b, and h. Unfortunately, it is not

possible to solve for k in closed form. That would have to be done by numerical methods

when specific values of a, b, and h are given.

(b) The length of the cable is $L = \int_{-b}^{b} \sqrt{1 + \left(\frac{dy}{dx}\right)^2}\, dx = \int_{-b}^{b} \sqrt{1 + \sinh^2 kx}\, dx$

$= \int_{-b}^{b} \cosh kx\, dx = \frac{1}{k}\sinh kx \Big|_{-b}^{b} = \frac{1}{k}[\sinh(kb) - \sinh(-kb)] = \frac{2}{k}\sinh(kb)$.

CHAPTER TEN

EXERCISES 10.1

1. $a_n = \frac{n}{2n+1}$ $\Rightarrow$ $\left\{\frac{1}{3}, \frac{2}{5}, \frac{3}{7}, \frac{4}{9}, \frac{5}{11}, \cdots\right\}$

3. $a_n = \frac{(-1)^{n-1}n}{2^n}$ $\Rightarrow$ $\left\{\frac{1}{2}, -\frac{1}{2}, \frac{3}{8}, -\frac{1}{4}, \frac{5}{32}, \cdots\right\}$

5. $a_n = \frac{1 \cdot 3 \cdot 5 \cdots (2n-1)}{n!}$ $\Rightarrow$ $\left\{1, \frac{3}{2}, \frac{5}{2}, \frac{35}{8}, \frac{63}{8}, \cdots\right\}$

7. $a_n = \sin\frac{n\pi}{2}$ $\Rightarrow$ $\{1, 0, -1, 0, 1, \ldots\}$

9. $a_1 = 1,\ a_{n+1} = \frac{1}{1+a_n}$ $\Rightarrow$ $\left\{1, \frac{1}{2}, \frac{2}{3}, \frac{3}{5}, \frac{5}{8}, \cdots\right\}$

11. $a_n = \frac{1}{2^n}$

13. $a_n = 3n - 2$

15. $a_n = (-1)^n\, n!$

17. $a_n = (-1)^{n+1}\left(\frac{n+1}{2n+1}\right)$

19.

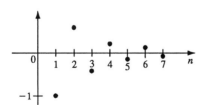

21. $\lim\limits_{n \to \infty} \frac{1}{4n^2} = \frac{1}{4}\ \lim\limits_{n \to \infty} \frac{1}{n^2} = \frac{1}{4} \cdot 0 = 0.$ Convergent.

23. $\lim\limits_{n \to \infty} \frac{n^2-1}{n^2+1} = \lim\limits_{n \to \infty} \frac{1-1/n^2}{1+1/n^2} = 1.$ Convergent.

25. $\{a_n\}$ diverges since $\frac{n^2}{n+1} = \frac{n}{1+1/n} \to \infty$ as $n \to \infty$.

27. $0 < |a_n| = \frac{n^2}{1+n^3} = \frac{1}{1/n^2 + n} < \frac{1}{n}$ and $\lim\limits_{n \to \infty} \frac{1}{n} = 0$, so by the Squeeze Theorem,

       $\lim\limits_{n \to \infty} \frac{n^2}{1+n^3} = 0$, and, by Theorem 10.6, $\lim\limits_{n \to \infty} (-1)^n\left(\frac{n^2}{1+n^3}\right) = 0$.

29. $\lim\limits_{n \to \infty} \frac{1}{5^n} = \lim\limits_{n \to \infty} \left(\frac{1}{5}\right)^n = 0$, by (10.8) with $r = \frac{1}{5}$.

31. $\{a_n\} = \{0, -1, 0, 1, 0, -1, 0, 1, \ldots\}$. This sequence oscillates among 0, -1, and 1 and so diverges.

33. $a_n = (\pi/3)^n$ so $\{a_n\}$ diverges by (10.8) with $r = \pi/3 > 1$.

35. $\lim\limits_{n \to \infty} \arctan 2n = \pi/2$ since $2n \to \infty$ as $n \to \infty$. Convergent.

37. $0 < \dfrac{3 + (-1)^n}{n^2} \le \dfrac{4}{n^2}$ and $\lim\limits_{n \to \infty} \dfrac{4}{n^2} = 0$, so $\left\{\dfrac{3 + (-1)^n}{n^2}\right\}$ converges to 0 by the Squeeze Theorem.

39. $\lim\limits_{x \to \infty} \dfrac{\ln(x^2)}{x} = \lim\limits_{x \to \infty} \dfrac{2 \ln x}{x} \overset{H}{=} \lim\limits_{x \to \infty} \dfrac{2/x}{1} = 0$, so by Theorem 10.2, $\left\{\dfrac{\ln(n^2)}{n}\right\}$ converges to 0.

41. $\sqrt{n+2} - \sqrt{n} = (\sqrt{n+2} - \sqrt{n})\dfrac{\sqrt{n+2} + \sqrt{n}}{\sqrt{n+2} + \sqrt{n}} = \dfrac{2}{\sqrt{n+2} + \sqrt{n}} < \dfrac{2}{2\sqrt{n}} = \dfrac{1}{\sqrt{n}} \to 0$ as $n \to \infty$. So by

    the Squeeze Theorem $\{\sqrt{n+2} - \sqrt{n}\}$ converges to 0.

43. $\lim\limits_{x \to \infty} \dfrac{x}{2^x} \overset{H}{=} \lim\limits_{x \to \infty} \dfrac{1}{(\ln 2)2^x} = 0$, so by Theorem 10.2 $\{a_n\}$ converges to 0.

45. Let $y = x^{-\frac{1}{x}}$. Then $\ln y = -(\ln x)/x$ and $\lim\limits_{x \to \infty} (\ln y) \overset{H}{=} \lim\limits_{x \to \infty} -(1/x)/1 = 0$, so

    $\lim\limits_{x \to \infty} y = e^0 = 1$, and so $\{a_n\}$ converges to 1.

47. $0 \le \dfrac{\cos^2 n}{2^n} \le \dfrac{1}{2^n}$ [since $0 \le \cos^2 n \le 1$], so since $\lim\limits_{n \to \infty} \dfrac{1}{2^n} = 0$, $\{a_n\}$ converges to 0 by the

    Squeeze Theorem.

49. The series converges, since $a_n = \dfrac{1 + 2 + 3 + \cdots + n}{n^2} = \dfrac{n(n+1)/2}{n^2}$ [*Theorem 5.3*]

    $= \dfrac{n+1}{2n} = \dfrac{1 + 1/n}{2} \to \dfrac{1}{2}$ as $n \to \infty$.

51. $a_n = \dfrac{1}{2} \cdot \dfrac{2}{2} \cdot \dfrac{3}{2} \cdots \dfrac{(n-1)}{2} \cdot \dfrac{n}{2} \ge \dfrac{1}{2} \cdot \dfrac{n}{2} = \dfrac{n}{4} \to \infty$ as $n \to \infty$, so $\{a_n\}$ diverges.

53. $0 < a_n = \dfrac{n^3}{n!} = \dfrac{n}{n} \cdot \dfrac{n}{(n-1)} \cdot \dfrac{n}{(n-2)} \cdot \dfrac{1}{(n-3)} \cdots \dfrac{1}{3} \cdot \dfrac{1}{2} \cdot \dfrac{1}{1} \le \dfrac{n^2}{(n-1)(n-2)(n-3)}$ [*for*

    $n \ge 4$] $= \dfrac{1/n}{(1 - 1/n)(1 - 2/n)(1 - 3/n)} \to 0$ as $n \to \infty$, so by the Squeeze Theorem, $\{a_n\}$

    converges to 0.

55. $0 < a_n = \dfrac{1 \cdot 3 \cdot 5 \cdots (2n-1)}{(2n)^n} = \dfrac{1}{2n} \cdot \dfrac{3}{2n} \cdot \dfrac{5}{2n} \cdots \dfrac{2n-1}{2n} \le \dfrac{1}{2n} \cdot (1) \cdot (1) \cdots (1) = \dfrac{1}{2n} \to 0$ as

    $n \to \infty$, so by the Squeeze Theorem $\{a_n\}$ converges to 0.

57. If $|r| \ge 1$, then $\{r^n\}$ diverges (10.8), so $\{nr^n\}$ diverges also since $|nr^n| = n|r^n| \ge |r^n|$. If $|r| < 1$

    then $\lim\limits_{x \to \infty} xr^x = \lim\limits_{x \to \infty} \dfrac{x}{r^{-x}} \overset{H}{=} \lim\limits_{x \to \infty} \dfrac{1}{(-\ln r)r^{-x}} = \lim\limits_{x \to \infty} \dfrac{r^x}{-\ln r} = 0$, so $\lim\limits_{n \to \infty} nr^n = 0$, and

    hence $\{nr^n\}$ converges whenever $|r| < 1$.

59. $3(n+1) + 5 > 3n + 5$ so $\dfrac{1}{3(n+1) + 5} < \dfrac{1}{3n + 5} \Leftrightarrow a_{n+1} < a_n$ so $\{a_n\}$ is decreasing.

340

61. $\left\{\frac{n-2}{n+2}\right\}$ is increasing since $a_n < a_{n+1} \Leftrightarrow \frac{n-2}{n+2} < \frac{(n+1)-2}{(n+1)+2} \Leftrightarrow (n-2)(n+3)$

$< (n+2)(n-1) \Leftrightarrow n^2+n-6 < n^2+n-2 \Leftrightarrow -6 < -2$, which is of course true.

63. $a_1 = 0 > a_2 = -1 < a_3 = 0$, so the sequence is not monotonic.

65. $\left\{\frac{n}{n^2+n-1}\right\}$ is decreasing since $a_{n+1} < a_n \Leftrightarrow \frac{n+1}{(n+1)^2+(n+1)-1} < \frac{n}{n^2+n-1} \Leftrightarrow$

$(n+1)(n^2+n-1) < n(n^2+3n+1) \Leftrightarrow n^3+2n^2-1 < n^3+3n^2+n \Leftrightarrow$

$0 < n^2+n+1 = (n+\frac{1}{2})^2 + \frac{3}{4}$, which is obviously true.

67. $a_1 = 2^{1/2}, a_2 = 2^{3/4}, a_3 = 2^{7/8}, \ldots, a_n = 2^{(2^n-1)/2^n} = 2^{(1-1/2^n)}$.

$\lim\limits_{n \to \infty} a_n = \lim\limits_{n \to \infty} 2^{(1-1/2^n)} = 2^1 = 2$

Alternate Solution: Let $L = \lim\limits_{n \to \infty} a_n$ (we could show the limit exists by showing $\{a_n\}$ is
bounded and increasing). So L must satisfy $L = \sqrt{2 \cdot L} \Rightarrow L^2 = 2L \Rightarrow L(L-2) = 0$
($L \neq 0$ since the sequence increases) so $L = 2$.

69. (a) We show by induction that $\{a_n\}$ is increasing and bounded above by 3. Let $P(n)$ be the
proposition that $a_{n+1} \geq a_n$ and $a_n \leq 3$. Clearly $P(1)$ is true. Assume $P(n)$ is true. Then

$a_{n+1} \geq a_n \Rightarrow \frac{1}{a_{n+1}} \leq \frac{1}{a_n} \Rightarrow -\frac{1}{a_{n+1}} \leq -\frac{1}{a_n} \Rightarrow a_{n+2} = 3 - \frac{1}{a_{n+1}} \geq 3 - \frac{1}{a_n} = a_{n+1} \Leftrightarrow$

$P(n+1)$. This proves that $\{a_n\}$ is increasing and bounded above by 3, so

$1 = a_1 \leq a_n \leq 3$, i.e. $\{a_n\}$ is bounded, and hence convergent by Theorem 10.11.

(b) If $L = \lim\limits_{n \to \infty} a_n$, then $\lim\limits_{n \to \infty} a_{n+1} = L$ also, so L must satisfy $L = 3 - 1/L$, so

$L^2 - 3L + 1 = 0$ and the quadratic formula gives $L = \frac{3 \pm \sqrt{5}}{2}$. But $L > 1$, so

$L = \frac{3 + \sqrt{5}}{2}$.

71. (a) Let a_n be the number of rabbit pairs in the nth month. Clearly $a_1 = 1 = a_2$. In the nth
month, each pair that is 2 or more months old (i.e. a_{n-2} pairs) will have a pair of
children to add to the a_{n-1} pairs already present. Thus $a_n = a_{n-1} + a_{n-2}$, so that
$\{a_n\} = \{f_n\}$, the Fibonacci sequence.

(b) $a_{n-1} = f_n/f_{n-1} = (f_{n-1} + f_{n-2})/f_{n-1} = 1 + f_{n-2}/f_{n-1} = 1 + 1/a_{n-2}$. If
$L = \lim\limits_{n \to \infty} a_n$, then L must satisfy $L = 1 + 1/L$ or $L^2 - L - 1 = 0$, so $L = \frac{1 + \sqrt{5}}{2}$

(since L must be non-negative).

73. If $\lim\limits_{n \to \infty} |a_n| = 0$ then $\lim\limits_{n \to \infty} -|a_n| = 0$, and since $-|a_n| \leq a_n \leq |a_n|$, we have that

$\lim\limits_{n \to \infty} a_n = 0$ by the Squeeze Theorem.

75. (a) First we show that $a > a_1 > b_1 > b$.

$$a_1 - b_1 = \frac{a+b}{2} - \sqrt{ab} = \tfrac{1}{2}(a - 2\sqrt{ab} + b) = \tfrac{1}{2}(\sqrt{a} - \sqrt{b})^2 > 0 \text{ (since } a > b) \Rightarrow a_1 > b_1.$$

Also $a - a_1 = a - \frac{a+b}{2} = \frac{a-b}{2} > 0$ and $b - b_1 = b - \sqrt{ab} = \sqrt{b}(\sqrt{b} - \sqrt{a}) < 0$,

so $a > a_1 > b_1 > b$. In the same way we can show that $a_1 > a_2 > b_2 > b_1$ and so the

given assertion is true for $n = 1$. Suppose it is true for $n = k$, i.e., $a_k > a_{k+1} > b_{k+1} > b_k$.

Then $a_{k+2} - b_{k+2} = \frac{a_{k+1} + b_{k+1}}{2} - \sqrt{a_{k+1}b_{k+1}} = \tfrac{1}{2}(a_{k+1} - 2\sqrt{a_{k+1}b_{k+1}} + b_{k+1})$

$= \tfrac{1}{2}(\sqrt{a_{k+1}} - \sqrt{b_{k+1}})^2 > 0$ and

$$a_{k+1} - a_{k+2} = a_{k+1} - \frac{a_{k+1} + b_{k+1}}{2} = \frac{a_{k+1} - b_{k+1}}{2} > 0$$

$$b_{k+1} - b_{k+2} = b_{k+1} - \sqrt{a_{k+1}b_{k+1}} = \sqrt{b_{k+1}}(\sqrt{b_{k+1}} - \sqrt{a_{k+1}}) < 0 \quad \Rightarrow$$

$a_{k+1} > a_{k+2} > b_{k+2} > b_{k+1}$ so the assertion is true for $n = k + 1$. Thus it is true for

all n by mathematical induction.

(b) From part (a) we have $a > a_n > a_{n+1} > b_{n+1} > b_n > b$, which shows that both

sequences are monotonic and bounded. So they are both convergent by Theorem 10.11.

(c) Let $\lim\limits_{n \to \infty} a_n = \alpha$ and $\lim\limits_{n \to \infty} b_n = \beta$. Then $\lim\limits_{n \to \infty} a_{n+1} = \lim\limits_{n \to \infty} \frac{a_n + b_n}{2} \Rightarrow$

$\alpha = \frac{\alpha + \beta}{2} \Rightarrow 2\alpha = \alpha + \beta \Rightarrow \alpha = \beta$.

EXERCISES 10.2

1. $\displaystyle\sum_{n=1}^{\infty} 4\left(\frac{2}{5}\right)^{n-1}$ is a geometric series with $a = 4$, $r = \frac{2}{5}$, so it converges to $\frac{4}{1 - 2/5} = \frac{20}{3}$.

3. $\displaystyle\sum_{n=1}^{\infty} \frac{2}{3}\left(-\frac{1}{3}\right)^{n-1}$ is geometric with $a = \frac{2}{3}$, $r = -\frac{1}{3}$, so it converges to $\frac{2/3}{1 - (-1/3)} = \frac{1}{2}$.

5. $\displaystyle\sum_{n=1}^{\infty} \frac{1}{36}\left(\frac{6}{5}\right)^{n-1}$ diverges since $r = \frac{6}{5} > 1$.

7. $a = 2$, $r = 3/4 < 1$, so series converges to $\frac{2}{1 - 3/4} = 8$.

9. $a = 5e/3$, $r = e/3 < 1$, so series converges to $\frac{5e/3}{1 - e/3} = \frac{5e}{3 - e}$.

11. $a = 1$, $r = 5/8 < 1$, so series converges to $\frac{1}{1 - 5/8} = \frac{8}{3}$.

13. $a = 64/3$, $r = 8/3 > 1$, so series diverges.

15. $a = \dfrac{(-2)^4}{5^{-1}} = 80$, $|r| = \left|-\frac{2}{5}\right| < 1$, series converges to $\dfrac{80}{1-(-2/5)} = \dfrac{400}{7}$.

17. $\lim\limits_{n\to\infty} \dfrac{n}{n+1} = 1 \neq 0$ so series diverges by the Test for Divergence.

19. This series diverges, since if it converged, so would [by Theorem 10.19(a)] $2 \cdot \sum\limits_{n=1}^{\infty} \dfrac{1}{2n} = \sum\limits_{n=1}^{\infty} \dfrac{1}{n}$, which we know diverges (Example 7).

21. Converges. $s_n = \sum\limits_{i=1}^{n} \dfrac{1}{(3i-2)(3i+1)} = \sum\limits_{i=1}^{n} \left[\dfrac{1/3}{3i-2} - \dfrac{1/3}{3i+1}\right]$ $\left\{\begin{array}{c} partial \\ fractions \end{array}\right\}$

$= \left[\frac{1}{3} \cdot 1 - \frac{1}{3} \cdot \frac{1}{4}\right] + \left[\frac{1}{3} \cdot \frac{1}{4} - \frac{1}{3} \cdot \frac{1}{7}\right] + \left[\frac{1}{3} \cdot \frac{1}{7} - \frac{1}{3} \cdot \frac{1}{10}\right] + \cdots$

$+ \left[\frac{1}{3} \cdot \dfrac{1}{3n-2} - \frac{1}{3} \cdot \dfrac{1}{3n+1}\right] = \frac{1}{3} - \dfrac{1}{3(3n+1)}$ [telescoping series]

$\lim\limits_{n\to\infty} s_n = \frac{1}{3} \Rightarrow \sum\limits_{n=1}^{\infty} \dfrac{1}{(3n-2)(3n+1)} = \frac{1}{3}$

23. Converges by Theorem 10.19.

$\sum\limits_{n=1}^{\infty} [2(0.1)^n + (0.2)^n] = 2\sum\limits_{n=1}^{\infty} (0.1)^n + \sum\limits_{n=1}^{\infty} (0.2)^n = 2\left(\dfrac{0.1}{1-0.1}\right) + \dfrac{0.2}{1-0.2} = \frac{2}{9} + \frac{1}{4} = \frac{17}{36}$

25. Diverges by the Test for Divergence. $\lim\limits_{n\to\infty} n/\sqrt{1+n^2} = \lim\limits_{n\to\infty} 1/\sqrt{1+1/n^2} = 1 \neq 0$.

27. Converges. $s_n = \sum\limits_{i=1}^{n} \dfrac{1}{i(i+2)} = \sum\limits_{i=1}^{n} \left[\dfrac{1/2}{i} - \dfrac{1/2}{i+2}\right]$ $\left\{\begin{array}{c} partial \\ fractions \end{array}\right\}$

$= \left[\frac{1}{2} - \frac{1}{6}\right] + \left[\frac{1}{4} - \frac{1}{8}\right] + \left[\frac{1}{6} - \frac{1}{10}\right] + \cdots + \left[\dfrac{1}{2n-2} - \dfrac{1}{2n+2}\right] + \left[\dfrac{1}{2n} - \dfrac{1}{2n+4}\right]$

$= \frac{1}{2} + \frac{1}{4} - \dfrac{1}{2n+2} - \dfrac{1}{2n+4}$ [telescoping series]

Thus $\sum\limits_{n=1}^{\infty} \dfrac{1}{n(n+2)} = \lim\limits_{n\to\infty} \left[\frac{1}{2} + \frac{1}{4} - \dfrac{1}{2n+2} - \dfrac{1}{2n+4}\right] = \frac{3}{4}$.

29. Converges. $\sum\limits_{n=1}^{\infty} \dfrac{3^n + 2^n}{6^n} = \sum\limits_{n=1}^{\infty} \left[\left(\frac{1}{2}\right)^n + \left(\frac{1}{3}\right)^n\right] = \dfrac{1/2}{1-1/2} + \dfrac{1/3}{1-1/3} = \frac{3}{2}$

31. Converges. $s_n = (\sin 1 - \sin 1/2) + (\sin 1/2 - \sin 1/3) + \cdots$

$+ [\sin(1/n) - \sin(1/(n+1))] = \sin 1 - \sin[1/(n+1)]$

$\sum\limits_{n=1}^{\infty} [\sin(1/n) - \sin(1/(n+1))] = \lim\limits_{n\to\infty} s_n = \sin 1 - \sin 0 = \sin 1$

33. Diverges since $\lim\limits_{n\to\infty} \arctan n = \frac{\pi}{2} \neq 0$.

35. $s_n = (\ln 1 - \ln 2) + (\ln 2 - \ln 3) + (\ln 3 - \ln 4) + \cdots + [\ln n - \ln(n+1)] = \ln 1 - \ln(n+1)$

$= -\ln(n+1)$ [telescoping series]. Thus $\lim\limits_{n\to\infty} s_n = -\infty$, so the series is divergent.

37. $0.\bar{5} = .5 + .05 + .005 + \cdots = \frac{.5}{1 - .1} = \frac{5}{9}$

39. $0.\overline{307} = .307 + .000307 + .000000307 + \cdots = \frac{.307}{1 - .001} = \frac{307}{999}$

41. $0.123\overline{456} = \frac{123}{1000} + \frac{.000456}{1 - .001} = \frac{123}{1000} + \frac{456}{999000} = \frac{123333}{999000} = \frac{41111}{333000}$

43. $\displaystyle\sum_{n=0}^{\infty} (x - 3)^n$ is a geometric series with $r = x - 3$, so converges whenever $|x - 3| < 1 \Rightarrow$

$-1 < x - 3 < 1 \Leftrightarrow 2 < x < 4$. The sum is $\dfrac{1}{1 - (x - 3)} = \dfrac{1}{4 - x}$.

45. $\displaystyle\sum_{n=2}^{\infty} \left(\frac{x}{5}\right)^n$ is a geometric series with $r = \frac{x}{5}$, so converges whenever $\left|\frac{x}{5}\right| < 1 \Leftrightarrow -5 < x < 5$.

The sum is $\dfrac{(x/5)^2}{1 - x/5} = \dfrac{x^2}{25 - 5x}$.

47. $\displaystyle\sum_{n=0}^{\infty} (2 \sin x)^n$ is geometric so converges whenever $|2 \sin x| < 1 \Leftrightarrow -\frac{1}{2} < \sin x < \frac{1}{2} \Leftrightarrow$

$n\pi - \frac{\pi}{6} < x < n\pi + \frac{\pi}{6}$, where the sum is $\dfrac{1}{1 - 2 \sin x}$.

49. Total distance $= 4 + 2 + 2 + 1 + 1 + \frac{1}{2} + \frac{1}{2} + \frac{1}{4} + \frac{1}{4} + \cdots$

$= 4 + 4 + 2 + 1 + \frac{1}{2} + \frac{1}{4} + \cdots = 10 + \displaystyle\sum_{n=0}^{\infty} \left(\frac{1}{2}\right)^n = 10 + \dfrac{1}{1 - 1/2} = 12$ m.

51. The series $1 - 1 + 1 - 1 + 1 - 1 + \cdots$ diverges (geometric series with $r = -1$) so we cannot

say $0 = 1 - 1 + 1 - 1 + 1 - 1 + \cdots$.

53. $\displaystyle\sum_{n=1}^{\infty} ca_n = \lim_{n \to \infty} \sum_{i=1}^{n} ca_i = \lim_{n \to \infty} c \sum_{i=1}^{n} a_i = c \lim_{n \to \infty} \sum_{i=1}^{n} a_i = c \sum_{n=1}^{\infty} a_n$, which exists by

hypothesis.

55. Suppose on the contrary that $\sum (a_n + b_n)$ converges. Then by Theorem 10.19 (c), so would

$\sum [(a_n + b_n) - a_n] = \sum b_n$, a contradiction.

57. (a) $s_1 = \text{lm} = 10 \, \frac{1}{1 \cdot 2} = \frac{1}{2}$, $s_2 = \frac{1}{2} + \frac{1}{1 \cdot 2 \cdot 3} = \frac{5}{6}$, $s_3 = \frac{5}{6} + \frac{3}{1 \cdot 2 \cdot 3 \cdot 4} = \frac{23}{24}$,

$s_4 = \frac{23}{24} + \frac{4}{1 \cdot 2 \cdot 3 \cdot 4 \cdot 5} = \frac{119}{120}$.

The denominators are $(n + 1)!$ so a guess would be $s_n = \dfrac{(n + 1)! - 1}{(n + 1)!}$.

(b) For $n = 1$, $s_1 = \frac{1}{2} = \frac{2! - 1}{2!}$, so the formula holds for $n = 1$. Assume $s_k = \dfrac{(k + 1)! - 1}{(k + 1)!}$.

Then $s_{k+1} = \dfrac{(k + 1)! - 1}{(k + 1)!} + \dfrac{k + 1}{(k + 2)!} = \dfrac{(k + 1)! - 1}{(k + 1)!} + \dfrac{k + 1}{(k + 1)!(k + 2)}$

$$= \frac{(k+2)! - (k+2) + k + 1}{(k+2)!} = \frac{(k+2)! - 1}{(k+2)!}. \text{ Thus the formula is true for } n = k+1.$$

So by induction the guess is correct.

(c) $\displaystyle \lim_{n \to \infty} s_n = \lim_{n \to \infty} \frac{(n+1)! - 1}{(n+1)!} = \lim_{n \to \infty} \left[1 - \frac{1}{(n+1)!} \right] = 1$ and so $\displaystyle \sum_{n=0}^{\infty} \frac{n}{(n+1)!} = 1.$

EXERCISES 10.3

1. $\displaystyle \sum_{n=1}^{\infty} 2/^3\sqrt{n} = 2 \sum_{n=1}^{\infty} \frac{1}{n^{1/3}}$, which is a p–series, $p = \frac{1}{3} < 1$, so it diverges.

3. $\displaystyle \sum_{n=5}^{\infty} \frac{1}{n^{1.0001}}$ is a p–series, $p = 1.0001 > 1$, so it converges.

5. $\displaystyle \sum_{n=5}^{\infty} \frac{1}{(n-4)^2} = \sum_{n=1}^{\infty} \frac{1}{n^2}$ is a p–series, $p = 2 > 1$, so it converges.

7. Since $\frac{1}{\sqrt{x}+1}$ is continuous, positive, and decreasing on $[0, \infty)$ we can apply the Integral Test.

$$\int_1^{\infty} \frac{1}{\sqrt{x}+1} dx = \lim_{t \to \infty} [2\sqrt{x} - 2\ln(\sqrt{x}+1)]\Big|_1^t \quad [using \ the \ substitution \ u = \sqrt{x}+1, \ so$$

$$x = (u-1)^2 \quad and \quad dx = 2(u-1)du] = \lim_{t \to \infty} [(2\sqrt{t} - 2\ln(\sqrt{t}+1)) - (2 - 2\ln 2)]. \quad \text{Now}$$

$$2\sqrt{t} - 2\ln(\sqrt{t}+1) = 2\ln\left(\frac{e^{\sqrt{t}}}{\sqrt{t}+1}\right) \text{ and so } \lim_{t \to \infty} (2\sqrt{t} - 2\ln(\sqrt{t}+1)) = \infty \ [using \ L'Hospital's$$

Rule] so both the integral and the original series diverge.

9. $f(x) = xe^{-x^2}$ is continuous and positive on $[1, \infty)$, and since $f'(x) = e^{-x^2}(1 - 2x^2) < 0$ for

$x > 1$, f is decreasing as well. We can use the Integral Test.

$\int_1^{\infty} xe^{-x^2} dx = \lim_{t \to \infty} [-\frac{1}{2} e^{-x^2}]\Big|_1^t = 0 - \left(-\frac{e^{-1}}{2}\right) = \frac{1}{2e}$ so the series converges.

11. $f(x) = \frac{x}{x^2+1}$ is continuous and positive on $[1, \infty)$, and since $f'(x) = \frac{1 - x^2}{(x^2+1)^2} < 0$ for $x > 1$,

f is also decreasing. Using the Integral Test,

$$\int_1^{\infty} \frac{x}{x^2+1} dx = \lim_{t \to \infty} \left[\frac{\ln(x^2+1)}{2} \right]_1^t = \infty, \text{ so the series diverges.}$$

13. $f(x) = \frac{1}{x\ln x}$ is continuous and positive on $[2, \infty)$, and also decreasing since

$f'(x) = -\frac{1 + \ln x}{x^2(\ln x)^2} < 0$ for $x > 2$, so we can use the Integral Test.

$$\int_2^\infty \frac{1}{x \ln x} \, dx = \lim_{t \to \infty} \left[\ln(\ln x) \right]_2^t = \lim_{t \to \infty} \left[\ln(\ln t) - \ln(\ln 2) \right] = \infty, \text{ so the series diverges.}$$

15. $f(x) = \frac{\arctan x}{1 + x^2}$ is continuous and positive on $[1, \infty)$. $f'(x) = \frac{1 - 2x \arctan x}{(1 + x^2)^2} < 0$ for $x > 1$,

since $2x \arctan x \geq \frac{\pi}{2} > 1$ for $x \geq 1$. So f is decreasing and we can use the Integral Test.

$$\int_1^\infty \frac{\arctan x}{1 + x^2} \, dx = \lim_{t \to \infty} \left[\frac{1}{2} \arctan^2 x \right]_1^t = \frac{(\pi/2)^2}{2} - \frac{(\pi/4)^2}{2} = \frac{3\pi^2}{32}, \text{ so the series converges.}$$

17. $f(x) = \frac{\ln x}{x^2}$ is continuous and positive for $x \geq 2$, and $f'(x) = \frac{1 - 2 \ln x}{x^3} < 0$ for $x \geq 2$ so f is

decreasing. $\int_2^\infty \frac{\ln x}{x^2} \, dx = \lim_{t \to \infty} \left[-\frac{\ln x}{x} - \frac{1}{x} \right]_1^t$ [*using integration by parts*] $= 1$ [*by*

L'Hospital's Rule]. Thus $\sum_{n=1}^\infty \frac{\ln n}{n^2} = \sum_{n=2}^\infty \frac{\ln n}{n^2}$ converges by the Integral Test.

19. $f(x) = \frac{\sin(1/x)}{x^2}$ is continuous and positive for $x \geq 1$, and

$f'(x) = -\frac{\cos(1/x) + 2x \sin(1/x)}{x^4} < 0$ if $x \geq 1$ [because then $0 < \frac{1}{x} \leq 1 < \frac{\pi}{2}$ so that both

$\cos(1/x)$ and $\sin(1/x)$ are positive] and so f is decreasing. Using the Integral Test,

$$\int_1^\infty \frac{\sin(1/x)}{x^2} \, dx = \lim_{t \to \infty} \left[\cos(1/x) \right]_1^t = 1 - \cos 1, \text{ and so the series converges.}$$

21. $f(x) = \frac{1}{x^2 + 2x + 2}$ is continuous and positive on $[1, \infty)$, and $f'(x) = -\frac{2x + 2}{(x^2 + 2x + 2)^2} < 0$ if

$x \geq 1$, so f is decreasing and we can use the Integral Test. $\int_1^\infty \frac{1}{x^2 + 2x + 2} \, dx$

$$= \int_1^\infty \frac{1}{(x+1)^2 + 1} \, dx = \lim_{t \to \infty} \left[\arctan(x + 1) \right]_1^t = \frac{\pi}{2} - \arctan 2,$$

so the series converges also.

23. We have already shown that when $p = 1$ the series diverges (in Exercise 13 above), so

assume $p \neq 1$. $f(x) = \frac{1}{x(\ln x)^p}$ is continuous and positive on $[2, \infty)$, and

$f'(x) = -\frac{p + \ln x}{x^2 (\ln x)^{p+1}} < 0$ if $x > e^{-p}$, so that f is eventually decreasing and we can use the

Integral Test. $\int_2^\infty \frac{1}{x(\ln x)^p} \, dx = \lim_{t \to \infty} \left[\frac{(\ln x)^{1-p}}{1 - p} \right]_2^t$ [*for $p \neq 1$*]

$$= \lim_{t \to \infty} \left(\frac{(\ln t)^{1-p}}{1 - p} \right) - \frac{(\ln 2)^{1-p}}{1 - p}. \text{ This limit exists whenever } 1 - p < 0 \Leftrightarrow p > 1,$$

so the series converges for $p > 1$.

25. Clearly the series cannot converge if $p \geq -\frac{1}{2}$, because then $\lim\limits_{n \to \infty} n(1+n^2)^p \neq 0$. Also, if

$p = -1$ the series diverges (see Exercise 11 above). So assume $p < -\frac{1}{2}$, $p \neq -1$. Then

$f(x) = x(1+x^2)^p$ is continuous, positive, and eventually decreasing on $[1, \infty)$, and

we can use the Integral Test. $\displaystyle\int_1^\infty x(1+x^2)^p \, dx = \lim\limits_{t \to \infty} \left[\frac{1}{2} \cdot \frac{(1+x^2)^{p+1}}{p+1} \right]_1^t =$

$\lim\limits_{t \to \infty} \frac{1}{2} \cdot \frac{(1+x^2)^{p+1}}{p+1} - \frac{2^p}{p+1}$. This limit exists and is finite $\Leftrightarrow p+1 < 0 \Leftrightarrow p < -1$, so the

series converges whenever $p < -1$.

27. Since this is a p–series with $p = x$, $\zeta(x)$ is defined when $x > 1$.

29. Using the proof of the Integral Test, in particular 10.21 and 10.22, we have $\int_1^n f \leq s_{n-1}$ and

$s_n \leq a_1 + \int_1^n f$. Letting $n \to \infty$ in both inequalities, we get

$$\int_1^\infty \frac{1}{x^{1.001}} \, dx \leq s \leq 1 + \int_1^\infty \frac{1}{x^{1.001}} \, dx, \quad \text{where } s = \sum_{n=1}^\infty \frac{1}{n^{1.001}}$$

But $\displaystyle\int_1^\infty \frac{1}{x^{1.001}} \, dx = \lim\limits_{t \to \infty} \int_1^t \frac{1}{x^{1.001}} \, dx = \lim\limits_{t \to \infty} \left[\frac{1}{-0.001} \left(\frac{1}{t^{.001}} - 1 \right) \right] = \frac{1}{0.001} = 1000.$

So $1000 \leq s \leq 1001$.

EXERCISES 10.4

1. $\dfrac{1}{n^3 + n^2} < \dfrac{1}{n^3}$ since $n^3 + n^2 > n^3$ for all n, and since $\displaystyle\sum_{n=1}^\infty \frac{1}{n^3}$ is a convergent p–series

$(p = 3 > 1)$, $\displaystyle\sum_{n=1}^\infty \frac{1}{n^3 + n^2}$ converges also by 10.24 (a).

3. $\dfrac{3}{n2^n} \leq \dfrac{3}{2^n}$. $\displaystyle\sum_{n=1}^\infty \frac{3}{2^n}$ is a geometric series with $|r| = \frac{1}{2} < 1$, and hence converges, so $\displaystyle\sum_{n=1}^\infty \frac{3}{n2^n}$

converges also by the Comparison Test.

5. $\dfrac{1+5^n}{4^n} > \dfrac{5^n}{4^n} = \left(\frac{5}{4}\right)^n$. $\displaystyle\sum_{n=0}^\infty \left(\frac{5}{4}\right)^n$ is a divergent geometric series $\left(|r| = \frac{5}{4} > 1\right)$ so $\displaystyle\sum_{n=0}^\infty \frac{1+5^n}{4^n}$

diverges by the Comparison Test.

7. $\dfrac{3}{n(n+3)} < \dfrac{3}{n^2}$. $\displaystyle\sum_{n=1}^\infty \frac{3}{n^2} = 3\sum_{n=1}^\infty \frac{1}{n^2}$ is a convergent p–series $(p = 2 > 1)$ so $\displaystyle\sum_{n=1}^\infty \frac{3}{n(n+3)}$

converges by the Comparison Test.

9. $\dfrac{1}{\sqrt{n^3+1}} < \dfrac{1}{\sqrt{n^3}} = \dfrac{1}{n^{3/2}}$. $\displaystyle\sum_{n=1}^{\infty} \dfrac{1}{n^{3/2}}$ is a convergent p–series $\left(p=\tfrac{3}{2}>1\right)$ so $\displaystyle\sum_{n=1}^{\infty}\dfrac{1}{\sqrt{n^3+1}}$

converges by the Comparison Test.

11. $\dfrac{\sqrt{n}}{n-1} > \dfrac{\sqrt{n}}{n} = \dfrac{1}{n^{1/2}}$. $\displaystyle\sum_{n=2}^{\infty}\dfrac{1}{n^{1/2}}$ is a divergent p–series $\left(p=\tfrac{1}{2}<1\right)$ so $\displaystyle\sum_{n=2}^{\infty}\dfrac{\sqrt{n}}{n-1}$

diverges by the Comparison Test.

13. $n^3+1 > n^3 \Rightarrow \dfrac{1}{n^3+1} < \dfrac{1}{n^3} \Rightarrow \dfrac{n}{n^3+1} < \dfrac{n}{n^3} \Rightarrow \dfrac{n-1}{n^3+1} < \dfrac{n}{n^3} = \dfrac{1}{n^2}$. $\displaystyle\sum_{n=1}^{\infty}\dfrac{1}{n^2}$ is a convergent

p–series $(p=2>1)$ so $\displaystyle\sum_{n=1}^{\infty}\dfrac{n-1}{n^3+1}$ converges by the Comparison Test.

15. $\dfrac{3+\cos n}{3^n} \le \dfrac{4}{3^n}$ since $\cos n \le 1$. $\displaystyle\sum_{n=1}^{\infty}\dfrac{4}{3^n}$ is a geometric series with $|r|=\tfrac{1}{3}<1$ so it converges,

and so $\displaystyle\sum_{n=1}^{\infty}\dfrac{3+\cos n}{3^n}$ converges by the Comparison Test.

17. $(n+1)(2n+1) > n \cdot 2n = 2n^2$ so $\dfrac{4}{(n+1)(2n+1)} < \dfrac{4}{2n^2} = \dfrac{2}{n^2}$. $\displaystyle\sum_{n=1}^{\infty}\dfrac{2}{n^2} = 2\sum_{n=1}^{\infty}\dfrac{1}{n^2}$ is a

convergent p–series $(p=2>1)$, so $\displaystyle\sum_{n=1}^{\infty}\dfrac{4}{(n+1)(2n+1)}$ converges by the

Comparison Test.

19. $\dfrac{n}{\sqrt{n^5+4}} < \dfrac{n}{\sqrt{n^5}} = \dfrac{1}{n^{3/2}}$. $\displaystyle\sum_{n=1}^{\infty}\dfrac{1}{n^{3/2}}$ is a convergent p–series $\left(p=\tfrac{3}{2}>1\right)$ so $\displaystyle\sum_{n=1}^{\infty}\dfrac{n}{\sqrt{n^5+4}}$

converges by the Comparison Test.

21. $\dfrac{2^n}{1+3^n} < \dfrac{2^n}{3^n} = \left(\tfrac{2}{3}\right)^n$. $\displaystyle\sum_{n=1}^{\infty}\left(\tfrac{2}{3}\right)^n$ is a convergent geometric series $\left(|r|=\tfrac{2}{3}<1\right)$, so

$\displaystyle\sum_{n=1}^{\infty}\dfrac{2^n}{1+3^n}$ converges by the Comparison Test.

23.   Let $a_n = \dfrac{1}{1+\sqrt{n}}$ and $b_n = \dfrac{1}{\sqrt{n}}$. Then $\displaystyle\lim_{n\to\infty}\dfrac{a_n}{b_n} = \lim_{n\to\infty}\dfrac{\sqrt{n}}{1+\sqrt{n}} = 1 > 0$. Since $\displaystyle\sum_{n=1}^{\infty}\dfrac{1}{\sqrt{n}}$ is a

divergent p–series $\left(p=\tfrac{1}{2}<1\right)$, $\displaystyle\sum_{n=1}^{\infty}\dfrac{1}{1+\sqrt{n}}$ also diverges by the Limit Comparison Test.

25.   Let $a_n = \dfrac{n^2+1}{n^4+1}$ and $b_n = \dfrac{1}{n^2}$. Then $\displaystyle\lim_{n\to\infty}\dfrac{a_n}{b_n} = \lim_{n\to\infty}\dfrac{n^4+n^2}{n^4+1} = 1$. Since $\displaystyle\sum_{n=1}^{\infty}\dfrac{1}{n^2}$ is a

convergent p–series $(p=2>1)$, so is $\displaystyle\sum_{n=1}^{\infty}\dfrac{n^2+1}{n^4+1}$ by the Limit Comparison Test.

27.   Let $a_n = \dfrac{n^2-n+2}{\sqrt[4]{n^{10}+n^5+3}}$ and $b_n = \dfrac{1}{\sqrt{n}}$. Then $\displaystyle\lim_{n\to\infty}\dfrac{a_n}{b_n}$

348

$$= \lim_{n \to \infty} \frac{n^{5/2} - n^{3/2} + 2n^{1/2}}{4\sqrt{n^{10} + n^5 + 3}} = \lim_{n \to \infty} \frac{1 - n^{-1} + 2n^{-2}}{4\sqrt{1 + n^{-5} + 3n^{-10}}} = 1. \quad \text{Since } \sum_{n=1}^{\infty} \frac{1}{\sqrt{n}} \text{ is a}$$

divergent p–series ($p = \frac{1}{2} < 1$), $\displaystyle\sum_{n=1}^{\infty} \frac{n^2 - n + 2}{4\sqrt{n^{10} + n^5 + 3}}$ diverges by the Limit Comparison

Test.

29. Let $a_n = \frac{n+1}{n2^n}$ and $b_n = \frac{1}{2^n}$. Then $\lim\limits_{n \to \infty} \frac{a_n}{b_n} = \lim\limits_{n \to \infty} \frac{n+1}{n} = 1$. Since $\sum\limits_{n=1}^{\infty} \frac{1}{2^n}$ is a

convergent geometric series ($|r| = \frac{1}{2} < 1$), $\displaystyle\sum_{n=1}^{\infty} \frac{n+1}{n2^n}$ converges by the Limit Comparison Test.

31. Let $a_n = \frac{\ln n}{n^3}$ and $b_n = \frac{1}{n^2}$. Then $\lim\limits_{n \to \infty} \frac{a_n}{b_n} = \lim\limits_{n \to \infty} \frac{\ln n}{n} = \lim\limits_{n \to \infty} \frac{1/n}{1} = 0$. So since $\sum\limits_{n=1}^{\infty} \frac{1}{n^2}$

converges (p–series, $p = 2 > 1$), so does $\displaystyle\sum_{n=1}^{\infty} \frac{\ln n}{n^3}$ by part (b) of the Limit Comparison Test.

33. Clearly $n! = n(n-1)(n-2) \cdots (3)(2) \geq 2 \cdot 2 \cdot 2 \cdots 2 \cdot 2 = 2^{n-1}$, so $\frac{1}{n!} \leq \frac{1}{2^{n-1}}$. $\displaystyle\sum_{n=1}^{\infty} \frac{1}{2^{n-1}}$ is

a convergent geometric series ($|r| = \frac{1}{2} < 1$) so $\displaystyle\sum_{n=1}^{\infty} \frac{1}{n!}$ converges by the Comparison Test.

35. $\frac{n!}{n^2} = \frac{n(n-1)(n-2) \cdots 1}{n^2} = \frac{(n-1)(n-2) \cdots 1}{n} \geq \frac{1}{n}$ and $\displaystyle\sum_{n=1}^{\infty} \frac{1}{n}$ diverges (harmonic series),

so $\displaystyle\sum_{n=1}^{\infty} \frac{n!}{n^2}$ diverges by the Comparison Test.

[OR: $\lim\limits_{n \to \infty} \frac{n!}{n^2} = \lim\limits_{n \to \infty} \frac{n(n-1)!}{n \cdot n} = \lim\limits_{n \to \infty} \frac{(n-1)!}{n} = \lim\limits_{n \to \infty} \left(1 - \frac{1}{n}\right)(n-2)! = \infty$, so the series

$\displaystyle\sum_{n=1}^{\infty} \frac{n!}{n^2}$ diverges by the Test for Divergence (10.18).]

37. Let $a_n = \sin\left(\frac{1}{n}\right)$ and $b_n = \frac{1}{n}$. Then $\lim\limits_{n \to \infty} \frac{a_n}{b_n} = \lim\limits_{n \to \infty} \frac{\sin(1/n)}{1/n} = \lim\limits_{\theta \to 0} \frac{\sin \theta}{\theta} = 1$, so since

$\displaystyle\sum_{n=1}^{\infty} b_n$ is the harmonic series (which diverges), $\displaystyle\sum_{n=1}^{\infty} \sin\left(\frac{1}{n}\right)$ diverges as well by the Limit

Comparison Test.

39. Since $\frac{d_n}{10^n} \leq \frac{9}{10^n}$ for each n, and since $\displaystyle\sum_{n=1}^{\infty} \frac{9}{10^n}$ is a convergent geometric series ($|r| = \frac{1}{10} < 1$),

$0.d_1 d_2 d_3 \cdots = \displaystyle\sum_{n=1}^{\infty} \frac{d_n}{10^n}$ will always converge by the Comparison Test.

41. Since $\sum a_n$ converges, $\lim\limits_{n \to \infty} a_n = 0$, so there exists N such that $|a_n - 0| < 1$ for all $n > N$

$\Rightarrow 0 \leq a_n < 1$ for all $n > N \Rightarrow 0 \leq a_n^2 \leq a_n$. Since $\sum a_n$ converges, so will $\sum a_n^2$ by the

Comparison Test.

43. We wish to prove that if $\lim\limits_{n \to \infty} \frac{a_n}{b_n} = \infty$ and $\sum b_n$ diverges, then so does $\sum a_n$. So suppose on

the contrary that $\sum a_n$ converges. Since $\lim\limits_{n \to \infty} \frac{a_n}{b_n} = \infty$, we have that $\lim\limits_{n \to \infty} \frac{b_n}{a_n} = 0$, so by

part (b) of the Limit Comparison Test (proved in Exercise 42), if $\sum a_n$ converges, so must

$\sum b_n$. But this contradicts our hypothesis, so $\sum a_n$ must diverge.

45. $\lim\limits_{n \to \infty} n a_n = \lim\limits_{n \to \infty} \frac{a_n}{1/n}$, so we apply the Limit Comparison Test with $b_n = \frac{1}{n}$. Since

$\lim\limits_{n \to \infty} n a_n > 0$ we know that either both series converge or both series diverge, and we also

know that $\sum\limits_{n=0}^{\infty} \frac{1}{n}$ diverges (p-series with $p = 1$). Therefore $\sum a_n$ must be divergent.

EXERCISES 10.5

1. $\sum\limits_{n=1}^{\infty} (-1)^{n-1} \frac{3}{n+4}$. $a_n = \frac{3}{n+4} > 0$ and $a_{n+1} < a_n$ for all n; $\lim\limits_{n \to \infty} a_n = 0$ so the

series converges by the Alternating Series Test.

3. $\sum\limits_{n=1}^{\infty} (-1)^n \frac{n}{n+1}$. $\lim\limits_{n \to \infty} \frac{n}{n+1} = 1$ so $\lim\limits_{n \to \infty} (-1)^n \frac{n}{n+1}$ does not exist and the series

diverges by the Test for Divergence.

5. $\sum\limits_{n=1}^{\infty} (-1)^{n-1} \frac{1}{n^2}$. $a_n = \frac{1}{n^2} > 0$ and $a_{n+1} < a_n$ for all n, and $\lim\limits_{n \to \infty} \frac{1}{n^2} = 0$, so the

series converges by the Alternating Series Test.

7. $\sum\limits_{n=1}^{\infty} (-1)^{n+1} \frac{n}{5n+1}$. $\lim\limits_{n \to \infty} \frac{n}{5n+1} = \frac{1}{5}$ so $\lim\limits_{n \to \infty} (-1)^{n+1} \frac{n}{5n+1}$ does not exist and

the series diverges by the Test for Divergence.

9. $\sum\limits_{n=1}^{\infty} (-1)^n \frac{n}{n^2+1}$. $a_n = \frac{n}{n^2+1} > 0$ for all n. $a_{n+1} < a_n \Leftrightarrow \frac{n+1}{(n+1)^2+1} < \frac{n}{n^2+1}$

$\Leftrightarrow (n+1)(n^2+1) < [(n+1)^2+1]n \Leftrightarrow n^3+n^2+n+1 < n^3+2n^2+2n \Leftrightarrow 0 < n^2+n-1$,

which is true for all $n \ge 1$. Also $\lim\limits_{n \to \infty} \frac{n}{n^2+1} = \lim\limits_{n \to \infty} \frac{1/n}{1+1/n^2} = 0$. Therefore the series

converges by the Alternating Series Test.

11. $\sum\limits_{n=1}^{\infty} (-1)^{n-1} \frac{\sqrt{n}}{n+4}$. $a_n = \frac{\sqrt{n}}{n+4} > 0$ for all n. Let $f(x) = \frac{\sqrt{x}}{x+4}$. Then

$f'(x) = \frac{4-x}{2\sqrt{x}(x+4)^2} < 0$ if $x > 4$, so $\{a_n\}$ is decreasing after $n = 4$.

$\lim\limits_{n \to \infty} \frac{\sqrt{n}}{n+4} = \lim\limits_{n \to \infty} \frac{1}{\sqrt{n}+4/\sqrt{n}} = 0$. So the series converges by the Alternating Series Test.

13. $\sum\limits_{n=2}^{\infty} (-1)^n \frac{n}{\ln n}$. $\lim\limits_{n\to\infty} \frac{n}{\ln n} = \lim\limits_{n\to\infty} \frac{1}{1/n} = \infty$ so the series diverges by the Test for Divergence.

15. $\sum\limits_{n=1}^{\infty} (-1)^{n+1} \frac{n+10}{n(n+1)}$. $\quad a_n = \frac{n+10}{n(n+1)} > 0$ for all n. Let $f(x) = \frac{x+10}{x(x+1)}$. Then

$f'(x) = -\frac{x^2 + 20x + 10}{(x^2+x)^2} < 0$ for $x \geq 1$, so $\{a_n\}$ is decreasing.

$\lim\limits_{n\to\infty} \frac{n+10}{n(n+1)} = \lim\limits_{n\to\infty} \frac{1/n + 10/n^2}{1 + 1/n} = 0$, so the series converges by the Alternating Series Test.

17. $\sum\limits_{n=1}^{\infty} \frac{\cos n\pi}{n^{3/4}} = \sum\limits_{n=1}^{\infty} \frac{(-1)^n}{n^{3/4}}$. $\quad a_n = \frac{1}{n^{3/4}}$ is decreasing and positive, and $\lim\limits_{n\to\infty} \frac{1}{n^{3/4}} = 0$ so

the series converges by the Alternating Series Test.

19. $\sum\limits_{n=1}^{\infty} (-1)^n \sin\left(\frac{\pi}{n}\right)$. $\quad a_n = \sin\left(\frac{\pi}{n}\right) > 0$ for $n \geq 2$ and $\sin\left(\frac{\pi}{n}\right) \geq \sin\left(\frac{\pi}{n+1}\right)$, and

$\lim\limits_{n\to\infty} \sin\left(\frac{\pi}{n}\right) = \sin 0 = 0$, so the series converges by the Alternating Series Test.

21. $\sum\limits_{n=1}^{\infty} (-1)^n \frac{1}{\sqrt[n]{n}}$. $\quad$ Let $L = \lim\limits_{n\to\infty} \frac{1}{\sqrt[n]{n}}$ (if it exists). Then $\ln L = \lim\limits_{n\to\infty} \ln\left(n^{-1/n}\right)$

$= \lim\limits_{n\to\infty} -\frac{\ln n}{n} = \lim\limits_{n\to\infty} -\frac{1/n}{1} = 0$, so $L = e^0 = 1$, so $\lim\limits_{n\to\infty} (-1)^n \frac{1}{\sqrt[n]{n}}$ does not exist

and the series diverges by the Test for Divergence.

23. First consider $f(x) = \frac{x^2 + x}{(x+2)^3} \Rightarrow f'(x) = \frac{(2x+1)(x+2)^3 - (x^2+x)3(x+2)^2}{(x+2)^6} = \frac{-x^2 + 2x + 2}{(x+2)^4}$

$\Rightarrow f'(x) < 0$ for $x > 1 + \sqrt{3}$, so f is decreasing on $(1 + \sqrt{3}, \infty) \Rightarrow \{a_n\}$ is decreasing for $n \geq 3$.

Also $\lim\limits_{n\to\infty} \frac{n(n+1)}{(n+2)^3} = \lim\limits_{n\to\infty} \frac{1 + 1/n}{n(1 + 2/n)^3} = 0$, so by the Alternating Series test the series is

convergent.

25. $\sum\limits_{n=1}^{\infty} \frac{(-3)^n}{n^2}$. $\quad \lim\limits_{n\to\infty} \frac{3^n}{n^2} = \infty$, so $\lim\limits_{n\to\infty} \frac{(-3)^n}{n^2}$ does not exist. Thus the given series is

divergent by the Test for Divergence.

27. Let $\sum b_n$ be the series for which $b_n = 0$ if n is odd and $b_n = \frac{1}{n^2}$ if n is even. Then

$\sum b_n = \sum \frac{1}{(2n)^2}$ clearly converges (by comparison with the p–series for $p = 2$). So suppose

that $\sum (-1)^{n-1} a_n$ converges. Then by Theorem 10.19 (b) so does $\sum [(-1)^{n-1} a_n + b_n]$

$= 1 + \frac{1}{3} + \frac{1}{5} + \cdots = \sum \frac{1}{2n-1}$. But this diverges (comparison with harmonic series), a

contradiction. $\sum (-1)^{n-1} a_n$ must diverge. The Alternating Series Test does not apply since

$\{a_n\}$ is not decreasing.

29. Clearly $a_n = \frac{1}{n+p}$ is decreasing and eventually positive and $\lim\limits_{n\to\infty} a_n = 0$ for any p. So the series will converge (by the Alternating Series Test) for any p for which all the a_n's are defined, i.e., $n + p \neq 0$ for $n \geq 1$, or p is not a negative integer.

31. If $a_n = \frac{1}{n^2}$, then $a_{11} = \frac{1}{121} < 0.01$, so by Theorem 10.28, $\sum\limits_{n=1}^{\infty} \frac{1}{n^2} \approx \sum\limits_{n=1}^{10} \frac{1}{n^2} \approx 0.82$.

33. $\sum\limits_{n=0}^{\infty} (-1)^n \frac{2^n}{n!}$ Since $\frac{2}{n} < \frac{2}{3}$ for $n \geq 4$, $0 < \frac{2^n}{n!} < \frac{2}{1} \cdot \frac{2}{2} \cdot \frac{2}{3} \cdot \left(\frac{2}{3}\right)^{n-3} \to 0$ as $n \to \infty$, so by the Squeeze Theorem, $\lim\limits_{n\to\infty} \frac{2^n}{n!} = 0$, and hence $\sum\limits_{n=0}^{\infty} (-1)^n \frac{2^n}{n!}$ is a convergent alternating series.

$\frac{2^8}{8!} = \frac{256}{40320} < 0.01$, so $\sum\limits_{n=0}^{\infty} (-1)^n \frac{2^n}{n!} \approx \sum\limits_{n=0}^{7} (-1)^n \frac{2^n}{n!} \approx 0.13$.

35. $\sum\limits_{n=1}^{\infty} \frac{(-1)^{n-1}}{(2n-1)!}$ $a_5 = \frac{1}{(2\cdot5-1)!} = \frac{1}{362880} < 0.00001$, so $\sum\limits_{n=1}^{\infty} \frac{(-1)^{n-1}}{(2n-1)!}$

$\approx \sum\limits_{n=1}^{4} \frac{(-1)^{n-1}}{(2n-1)!} \approx 0.8415$.

37. $\sum\limits_{n=0}^{\infty} \frac{(-1)^n}{2^n n!}$ $a_6 = \frac{1}{2^6 6!} = \frac{1}{46080} < 0.000022$, so $\sum\limits_{n=0}^{\infty} \frac{(-1)^n}{2^n n!} \approx \sum\limits_{n=0}^{5} \frac{(-1)^n}{2^n n!} \approx 0.6065$.

39. (a) We will prove this by induction. Let P(n) be the proposition that $s_{2n} = h_{2n} - h_n$. P(1) is true by an easy calculation. So suppose that P(n) is true. We will show that P(n + 1) must be true as a consequence.

$h_{2n+2} - h_{n+1} = (h_{2n} + \frac{1}{2n+1} + \frac{1}{2n+2}) - (h_n + \frac{1}{n+1}) = (h_{2n} - h_n) + \frac{1}{2n+1} - \frac{1}{2n+2} = s_{2n}$

$+ \frac{1}{2n+1} - \frac{1}{2n+2} = s_{2n+2}$, which is P(n + 1), and proves that $s_{2n} = h_{2n} - h_n$ for all n.

(b) We know that $h_{2n} - \ln(2n) \to \gamma$ and $h_n - \ln(n) \to \gamma$ as $n \to \infty$. So

$s_{2n} = h_{2n} - h_n = [h_{2n} - \ln(2n)] - [h_n - \ln(n)] + [\ln(2n) - \ln n]$, and

$\lim\limits_{n\to\infty} s_{2n} = \gamma - \gamma + \lim\limits_{n\to\infty} [\ln(2n) - \ln n] = \lim\limits_{n\to\infty} (\ln 2 + \ln n - \ln n) = \ln 2$.

EXERCISES 10.6

1. $\sum\limits_{n=1}^{\infty} \frac{1}{n\sqrt{n}} = \sum\limits_{n=1}^{\infty} \frac{1}{n^{3/2}}$ is a convergent p-series ($p = \frac{3}{2} > 1$), so the given series is absolutely convergent.

3. $\lim_{n\to\infty}\left|\frac{a_{n+1}}{a_n}\right| = \lim_{n\to\infty}\left|\frac{(-3)^{n+1}/(n+1)^3}{(-3)^n/n^3}\right| = 3\lim_{n\to\infty}\left(\frac{n}{n+1}\right)^3 = 3 > 1,$ so the series

diverges by the Ratio Test.

5. $\sum_{n=1}^{\infty}\frac{1}{2n+1}$ diverges (use the Integral Test or the Limit Comparison Test with $b_n = \frac{1}{n}$), but

since $\lim_{n\to\infty}\frac{1}{2n+1} = 0,$ $\sum_{n=1}^{\infty}\frac{(-1)^{n+1}}{2n+1}$ converges by the Alternating Series Test,

and so is conditionally convergent.

7. $\lim_{n\to\infty}\left|\frac{a_{n+1}}{a_n}\right| = \lim_{n\to\infty}\frac{1/(2n+1)!}{1/(2n-1)!} = \lim_{n\to\infty}\frac{1}{(2n+1)\,2n} = 0,$ so by the Ratio Test the

series is absolutely convergent.

9. $\sum_{n=1}^{\infty}\frac{n}{n^2+4}$ diverges (use the Limit Comparison Test with $b_n = \frac{1}{n}$). But since

$0 \le \frac{n+1}{(n+1)^2+4} < \frac{n}{n^2+4} \Leftrightarrow n^3+n^2+4n+4 < n^3+2n^2+5n \Leftrightarrow 0 < n^2+n-4$ (which is

true for $n \ge 2$), and since $\lim_{n\to\infty}\frac{n}{n^2+4} = 0,$ $\sum_{n=1}^{\infty}(-1)^n\frac{n}{n^2+4}$ converges by

the Alternating Series Test, and so converges conditionally.

11. $\lim_{n\to\infty}\frac{2n}{3n-4} = \frac{2}{3}$ so $\sum_{n=1}^{\infty}(-1)^n\frac{2n}{3n-4}$ diverges by the Test for Divergence.

13. $\left|\frac{\sin 2n}{n^2}\right| \le \frac{1}{n^2}$ and $\sum_{n=1}^{\infty}\frac{1}{n^2}$ converges (p–series, $p = 2 > 1$), so $\sum_{n=1}^{\infty}\frac{\sin 2n}{n^2}$ converges

absolutely (by the Comparison Test).

15. $\lim_{n\to\infty}\left|\frac{a_{n+1}}{a_n}\right| = \lim_{n\to\infty}\left|\frac{2^{n+1}/[(n+1)3^{n+2}]}{2^n/(n3^{n+1})}\right| = \frac{2}{3}\lim_{n\to\infty}\frac{n}{n+1} = \frac{2}{3} < 1$ so the series

converges absolutely by the Ratio Test.

17. $\lim_{n\to\infty}\left|\frac{a_{n+1}}{a_n}\right| = \lim_{n\to\infty}\frac{(n+2)5^{n+1}/[(n+1)3^{2(n+1)}]}{(n+1)5^n/(n3^{2n})} = \lim_{n\to\infty}\frac{5n(n+2)}{9(n+1)^2} = \frac{5}{9} < 1$ so the series

converges absolutely by the Ratio Test.

19. $\sum_{n=2}^{\infty}\frac{1}{\ln n}$ diverges (since $\frac{1}{\ln n} > \frac{1}{n}$ and $\sum_{n=1}^{\infty}\frac{1}{n}$ diverges), but $\sum_{n=2}^{\infty}\frac{(-1)^n}{\ln n}$ converges by the

Alternating Series Test (since $\frac{1}{\ln n}$ decreases to 0), so the series converges conditionally.

21. $\lim_{n\to\infty}\left|\frac{a_{n+1}}{a_n}\right| = \lim_{n\to\infty}\frac{(n+1)!/10^{n+1}}{n!/10^n} = \lim_{n\to\infty}\frac{n+1}{10} = \infty$ so the series diverges by the

Ratio Test.

23. $\dfrac{|\cos(n\pi/3)|}{n!} \le \dfrac{1}{n!}$ and $\sum\limits_{n=1}^{\infty} \dfrac{1}{n!}$ converges (Exercise 33, Section 10.4), so the given series

converges absolutely by the Comparison Test.

25. $\lim\limits_{n\to\infty} \left|\dfrac{a_{n+1}}{a_n}\right| = \lim\limits_{n\to\infty} \dfrac{(n+1)^{n+1}/5^{2n+5}}{n^n/5^{2n+3}} = \lim\limits_{n\to\infty} \dfrac{1}{25}\left(\dfrac{n+1}{n}\right)^n (n+1) = \infty$, so the series

diverges by the Ratio Test.

27. $\lim\limits_{n\to\infty} \sqrt[n]{|a_n|} = \lim\limits_{n\to\infty} \left|\dfrac{1-3n}{3+4n}\right| = \dfrac{3}{4} < 1$, so the series converges absolutely by the Root Test.

29. $\lim\limits_{n\to\infty} \left|\dfrac{a_{n+1}}{a_n}\right| = \lim\limits_{n\to\infty} \dfrac{(n+1)!/(1\cdot 3\cdot 5 \cdots (2n+1))}{n!/(1\cdot 3\cdot 5 \cdots (2n-1))} = \lim\limits_{n\to\infty} \dfrac{n+1}{2n+1} = \dfrac{1}{2} < 1$, so the

series converges absolutely by the Ratio Test.

31. $\sum\limits_{n=1}^{\infty} \dfrac{2\cdot 4\cdot 6 \cdots (2n)}{n!} = \sum\limits_{n=1}^{\infty} \dfrac{2^n n!}{n!} = \sum\limits_{n=1}^{\infty} 2^n$ which diverges by the Test for Divergence

since $\lim\limits_{n\to\infty} 2^n = \infty$.

33. $\lim\limits_{n\to\infty} \left|\dfrac{a_{n+1}}{a_n}\right| = \lim\limits_{n\to\infty} \dfrac{(n+3)!/[(n+1)!\,10^{n+1}]}{(n+2)!/(n!\,10^n)} = \dfrac{1}{10} \lim\limits_{n\to\infty} \dfrac{n+3}{n+1} = \dfrac{1}{10} < 1$ so the series

converges absolutely by the Ratio Test.

35. $\dfrac{|\sin 3n|\, n^2}{(1.1)^n} \le \dfrac{1\cdot n^2}{(1.1)^n}$ for all n, so we test the series $\sum\limits_{n=1}^{\infty} \dfrac{n^2}{(1.1)^n}$ using the Ratio Test.

$\lim\limits_{n\to\infty} \left|\dfrac{a_{n+1}}{a_n}\right| = \dfrac{(n+1)^2/(1.1)^{n+1}}{n^2/(1.1)^n} = \dfrac{1}{1.1} \lim\limits_{n\to\infty} \left(\dfrac{n+1}{n}\right)^2 = \dfrac{1}{1.1} < 1$ so $\sum\limits_{n=1}^{\infty} \dfrac{n^2}{(1.1)^n}$ converges

absolutely, and by the Comparison Test, so does $\sum\limits_{n=1}^{\infty} \dfrac{(\sin 3n)\, n^2}{(1.1)^n}$.

37. (a) $\lim\limits_{n\to\infty} \left|\dfrac{1/(n+1)^3}{1/n^3}\right| = \lim\limits_{n\to\infty} \dfrac{n^3}{(n+1)^3} = \lim\limits_{n\to\infty} \dfrac{1}{(1+1/n)^3} = 1$. So inconclusive.

(b) $\lim\limits_{n\to\infty} \left|\dfrac{(n+1)\cdot 2^n}{2^{n+1}\cdot n}\right| = \lim\limits_{n\to\infty} \dfrac{n+1}{2n} = \lim\limits_{n\to\infty}\left(\dfrac{1}{2}+\dfrac{1}{2n}\right) = \dfrac{1}{2}$. So conclusive (convergent).

(c) $\lim\limits_{n\to\infty} \left|\dfrac{(-3)^n}{\sqrt{n+1}}\cdot\dfrac{\sqrt{n}}{(-3)^{n-1}}\right| = 3\lim\limits_{n\to\infty}\sqrt{\dfrac{n}{n+1}} = 3\lim\limits_{n\to\infty}\sqrt{\dfrac{1}{1+1/n}} = 3$. So conclusive.

(Divergent)

(d) $\lim\limits_{n\to\infty} \left|\dfrac{\sqrt{n+1}}{1+(n+1)^2}\cdot\dfrac{1+n^2}{\sqrt{n}}\right| = \lim\limits_{n\to\infty}\left[\sqrt{1+\dfrac{1}{n}}\cdot\dfrac{(1/n^2)+1}{(1/n^2)+(1+1/n)^2}\right] = 1$. So inconclusive.

39. By the Triangle Inequality [see Exercise 4.1.46] we have

$$\left|\sum_{i=1}^{n} a_i\right| \le \sum_{i=1}^{n} |a_i| \;\Rightarrow\; -\sum_{i=1}^{n}|a_i| \le \sum_{i=1}^{n} a_i \le \sum_{i=1}^{n}|a_i| \;\Rightarrow$$

354

$$-\lim_{n\to\infty}\sum_{i=1}^{n}|a_i| \le \lim_{n\to\infty}\sum_{i=1}^{n}a_i \le \lim_{n\to\infty}\sum_{i=1}^{n}|a_i| \Rightarrow -\sum_{n=1}^{\infty}|a_n| \le \sum_{n=1}^{\infty}a_n \le \sum_{n=1}^{\infty}|a_n|$$

$$\Rightarrow \left|\sum_{n=1}^{\infty}a_n\right| \le \sum_{n=1}^{\infty}|a_n|.$$

41. (a) Since $\sum a_n$ is absolutely convergent, and since $\left|a_n^+\right| \le |a_n|$ and $|a_{\bar{n}}| \le |a_n|$ (because a_n^+ and $a_{\bar{n}}$ each equal either a_n or 0), we conclude by the Comparison Test that both $\sum a_n^+$ and $\sum a_{\bar{n}}$ must be absolutely convergent. [Or use Theorem 10.19.]

(b) We will show by contradiction that both $\sum a_n^+$ and $\sum a_{\bar{n}}$ must diverge. For suppose that $\sum a_n^+$ converged. Then so would $\sum (a_n^+ - \frac{1}{2}a_n)$ by Theorem 10.19. But

$$\sum (a_n^+ - \tfrac{1}{2}a_n) = \sum \left(\tfrac{a_n + |a_n|}{2} - \tfrac{a_n}{2}\right) = \tfrac{1}{2}\sum |a_n| \text{ which diverges because } \sum a_n \text{ is only}$$

conditionally convergent. Hence $\sum a_n^+$ can't converge. Similarly, neither can $\sum a_{\bar{n}}$.

EXERCISES 10.7

1. Use the Comparison Test, with $a_n = \dfrac{\sqrt{n}}{n^2+1}$ and $b_n = \dfrac{1}{n^{3/2}}$: $\dfrac{\sqrt{n}}{n^2+1} < \dfrac{\sqrt{n}}{n^2} = \dfrac{1}{n^{3/2}}$, and

$\displaystyle\sum_{n=1}^{\infty}\frac{1}{n^{3/2}}$ is a convergent p–series ($p = \frac{3}{2} > 1$), so $\displaystyle\sum_{n=1}^{\infty}a_n = \sum_{n=1}^{\infty}\frac{\sqrt{n}}{n^2+1}$ converges as well.

3. $\displaystyle\sum_{n=1}^{\infty}\frac{4^n}{3^{2n-1}} = 3\sum_{n=1}^{\infty}\left(\frac{4}{9}\right)^n$ which is a convergent geometric series ($|r| = \frac{4}{9} < 1$).

5. Converges by the Alternating Series Test, since $a_n = \dfrac{1}{(\ln n)^2}$ is decreasing ($\ln x$ is an increasing function) and $\displaystyle\lim_{n\to\infty}a_n = 0$.

7. $\displaystyle\sum_{k=1}^{\infty}\frac{1}{k^{1.7}}$ is a convergent p–series ($p = 1.7 > 1$).

9. $\displaystyle\lim_{n\to\infty}\left|\frac{a_{n+1}}{a_n}\right| = \lim_{n\to\infty}\frac{(n+1)/e^{n+1}}{n/e^n} = \frac{1}{e}\lim_{n\to\infty}\frac{n+1}{n} = \frac{1}{e} < 1$, so the series converges by the Ratio Test.

11. Use the Limit Comparison Test with $a_n = \dfrac{n^3+1}{n^4-1}$ and $b_n = \dfrac{1}{n}$. $\displaystyle\lim_{n\to\infty}\frac{a_n}{b_n}$

$= \displaystyle\lim_{n\to\infty}\frac{n^4+n}{n^4-1} = \lim_{n\to\infty}\frac{1+1/n^3}{1-1/n^4} = 1$, and since $\displaystyle\sum_{n=2}^{\infty}b_n$ diverges (harmonic series) so

does $\displaystyle\sum_{n=2}^{\infty}\frac{n^3+1}{n^4-1}$.

13. Let $f(x) = \dfrac{2}{x(\ln x)^3}$. $f(x)$ is clearly positive and decreasing for $x \geq 2$, so we apply the Integral

Test. $\displaystyle\int_2^\infty \dfrac{2}{x(\ln x)^3}\,dx = \lim_{t\to\infty} \left[\dfrac{-1}{(\ln x)^2}\right]_2^t = 0 - \dfrac{-1}{(\ln 2)^2}$ which is finite. So $\displaystyle\sum_{n=2}^\infty \dfrac{2}{n(\ln n)^3}$

converges.

15. $\displaystyle\lim_{n\to\infty}\left|\dfrac{a_{n+1}}{a_n}\right| = \lim_{n\to\infty} \dfrac{3^{n+1}(n+1)^2/(n+1)!}{3^n n^2/n!} = 3\lim_{n\to\infty}\dfrac{n+1}{n^2} = 0$, so the series converges

by the Ratio Test.

17. $\dfrac{3^n}{5^n + n} \leq \dfrac{3^n}{5^n} = \left(\dfrac{3}{5}\right)^n$. Since $\displaystyle\sum_{n=1}^\infty \left(\dfrac{3}{5}\right)^n$ is a convergent geometric series ($|r| = \dfrac{3}{5} < 1$),

$\displaystyle\sum_{n=1}^\infty \dfrac{3^n}{5^n + n}$ will converge by the Comparison Test.

19. $\displaystyle\lim_{n\to\infty}\left|\dfrac{a_{n+1}}{a_n}\right| = \lim_{n\to\infty}\dfrac{(n+1)!/(2\cdot5\cdot8\cdots(3n+5))}{n!/(2\cdot5\cdot8\cdots(3n+2))} = \lim_{n\to\infty}\dfrac{n+1}{3n+5} = \dfrac{1}{3} < 1$, so the series

converges by the Ratio Test.

21. Use the Limit Comparison Test with $a_i = \dfrac{1}{\sqrt{i(i+1)}}$ and $b_i = \dfrac{1}{i}$. $\displaystyle\lim_{i\to\infty}\dfrac{a_i}{b_i} = \lim_{i\to\infty}\dfrac{i}{\sqrt{i(i+1)}}$

$= \displaystyle\lim_{i\to\infty}\dfrac{1}{\sqrt{1+1/i}} = 1$. Since $\displaystyle\sum_{i=1}^\infty b_i$ diverges (harmonic series) so does $\displaystyle\sum_{i=1}^\infty \dfrac{1}{\sqrt{i(i+1)}}$.

23. $\displaystyle\lim_{n\to\infty} 2^{1/n} = 2^0 = 1$, so $\displaystyle\lim_{n\to\infty}(-1)^n 2^{1/n}$ does not exist and the series diverges by the Test

for Divergence.

25. Let $f(x) = \dfrac{\ln x}{\sqrt{x}}$. Then $f'(x) = \dfrac{2 - \ln x}{2x^{3/2}} < 0$ when $\ln x > 2$ or $x > e^2$ so $\dfrac{\ln n}{\sqrt{n}}$ is decreasing for

$n > e^2$. By l'Hospital's Rule, $\displaystyle\lim_{n\to\infty}\dfrac{\ln n}{\sqrt{n}} = \lim_{n\to\infty}\dfrac{1/n}{1/(2\sqrt{n})} = \lim_{n\to\infty}\dfrac{2}{\sqrt{n}} = 0$, so the series

converges by the Alternating Series Test.

27. The series diverges since it is a geometric series with $r = -\pi$ and $|r| = \pi > 1$. [Or use the

Test for Divergence.]

29. $\displaystyle\sum_{n=1}^\infty \dfrac{(-2)^{2n}}{n^n} = \sum_{n=1}^\infty \left(\dfrac{4}{n}\right)^n$. $\displaystyle\lim_{n\to\infty}\sqrt[n]{|a_n|} = \lim_{n\to\infty}\dfrac{4}{n} = 0$, so the series converges by the

Root Test.

31. Since $\dfrac{k\ln k}{(k+1)^3} < \dfrac{k\ln k}{k^3} = \dfrac{\ln k}{k^2}$, and since $\displaystyle\sum_{n=1}^\infty \dfrac{\ln n}{n^2}$ converges (Exercise 17, Section 10.3),

356

the given series converges by the Comparison Test.

33. $\lim\limits_{n \to \infty} \left| \dfrac{a_{n+1}}{a_n} \right| = \lim\limits_{n \to \infty} \dfrac{2^{n+1}/(2n+3)!}{2^n/(2n+1)!} = 2 \lim\limits_{n \to \infty} \dfrac{1}{(2n+3)(2n+2)} = 0,$ so the series

converges by the Ratio Test.

35. $0 < \dfrac{\tan^{-1} n}{n^{3/2}} < \dfrac{\pi/2}{n^{3/2}}. \displaystyle\sum_{n=1}^{\infty} \dfrac{\pi/2}{n^{3/2}} = \dfrac{\pi}{2} \sum_{n=1}^{\infty} \dfrac{1}{n^{3/2}}$ which is a convergent p-series ($p = \frac{3}{2} > 1$),

so $\displaystyle\sum_{n=1}^{\infty} \dfrac{\tan^{-1} n}{n^{3/2}}$ converges by the Comparison Test.

37. Since $\frac{3}{\pi} < 1$, $\lim\limits_{n \to \infty} \dfrac{1}{1 + (3/\pi)^n} = \dfrac{1}{1+0} = 1 \neq 0,$ so the series diverges by the Test for

Divergence.

39. $\lim\limits_{n \to \infty} \sqrt[n]{|a_n|} = \lim\limits_{n \to \infty} (2^{1/n} - 1) = 1 - 1 = 0,$ so the series converges by the Root Test.

EXERCISES 10.8

NOTE: "R" stands for "radius of convergence" and "I" stands for "interval of convergence" in

this section.

1. (a) We are given that the power series $\displaystyle\sum_{n=0}^{\infty} a_n x^n$ is convergent for $x = 4$. So by Theorem 10.38

it must converge at least for $-4 < x \leq 4$. In particular it converges when $x = -2$, that

is, $\displaystyle\sum_{n=0}^{\infty} a_n(-2)^n$ is convergent.

(b) But it does not follow that $\displaystyle\sum_{n=0}^{\infty} a_n(-4)^n$ is necessarily convergent. [See the comments after

Theorem 10.38. An example is $a_n = (-1)^n/(n4^n)$.]

3. If $u_n = \dfrac{x^n}{n+2}$, then $\lim\limits_{n \to \infty} \left| \dfrac{u_{n+1}}{u_n} \right| = \lim\limits_{n \to \infty} \left| \dfrac{x^{n+1}}{n+3} \cdot \dfrac{n+2}{x^n} \right| = |x| \lim\limits_{n \to \infty} \dfrac{n+2}{n+3} = |x| < 1$ for

convergence (by the Ratio Test). So $R = 1$. When $x = 1$, the series is $\displaystyle\sum_{n=0}^{\infty} \dfrac{1}{n+2}$ which

diverges (by the Integral Test or Comparison Test), and when $x = -1$, it is $\displaystyle\sum_{n=0}^{\infty} \dfrac{(-1)^n}{n+2}$

which converges (by the Alternating Series Test), so $I = [-1, 1)$.

5. If $u_n = nx^n$, then $\lim\limits_{n \to \infty} \left| \dfrac{u_{n+1}}{u_n} \right| = \lim\limits_{n \to \infty} \left| \dfrac{(n+1)x^{n+1}}{nx^n} \right| = |x| \lim\limits_{n \to \infty} \dfrac{n+1}{n} = |x| < 1$ for

convergence (by the Ratio Test). So $R = 1$. When $x = 1$ or -1, $\lim\limits_{n \to \infty} nx^n$ does not exist, so $\sum\limits_{n=0}^{\infty} nx^n$ diverges for these values. So $I = (-1, 1)$.

7. If $u_n = \dfrac{x^n}{n!}$, then $\lim\limits_{n \to \infty} \left| \dfrac{u_{n+1}}{u_n} \right| = \lim\limits_{n \to \infty} \left| \dfrac{x^{n+1}/(n+1)!}{x^n/n!} \right| = |x| \lim\limits_{n \to \infty} \dfrac{1}{n+1} = 0 < 1$ for all x.

So, by the Ratio Test, $R = \infty$, and $I = (-\infty, \infty)$.

9. If $u_n = \dfrac{(-1)^n x^n}{n2^n}$, then $\lim\limits_{n \to \infty} \left| \dfrac{u_{n+1}}{u_n} \right| = \lim\limits_{n \to \infty} \left| \dfrac{x^{n+1}/[(n+1)2^{n+1}]}{x^n/(n2^n)} \right| = \left| \dfrac{x}{2} \right| \lim\limits_{n \to \infty} \dfrac{n}{n+1}$

$= \left| \dfrac{x}{2} \right| < 1$ for convergence, so $|x| < 2$ and $R = 2$. When $x = 2$, $\sum\limits_{n=1}^{\infty} \dfrac{(-1)^n x^n}{n2^n}$

$= \sum\limits_{n=1}^{\infty} \dfrac{(-1)^n}{n}$ which converges by the Alternating Series Test. When $x = -2$

$\sum\limits_{n=1}^{\infty} \dfrac{(-1)^n x^n}{n2^n} = \sum\limits_{n=1}^{\infty} \dfrac{1}{n}$ which diverges (harmonic series), so $I = (-2, 2]$.

11. If $u_n = \dfrac{3^n x^n}{(n+1)^2}$, then $\lim\limits_{n \to \infty} \left| \dfrac{u_{n+1}}{u_n} \right| = \lim\limits_{n \to \infty} \left| \dfrac{3^{n+1} x^{n+1}}{(n+2)^2} \cdot \dfrac{(n+1)^2}{3^n x^n} \right|$

$= 3|x| \lim\limits_{n \to \infty} \left(\dfrac{n+1}{n+2} \right)^2 = 3|x| < 1$ for convergence, so $|x| < \dfrac{1}{3}$ and $R = \dfrac{1}{3}$. When $x = \dfrac{1}{3}$,

$\sum\limits_{n=0}^{\infty} \dfrac{3^n x^n}{(n+1)^2} = \sum\limits_{n=0}^{\infty} \dfrac{1}{(n+1)^2} = \sum\limits_{n=1}^{\infty} \dfrac{1}{n^2}$ which is a convergent p-series $(p = 2 > 1)$. When

$x = -\dfrac{1}{3}$, $\sum\limits_{n=0}^{\infty} \dfrac{3^n x^n}{(n+1)^2} = \sum\limits_{n=0}^{\infty} \dfrac{(-1)^n}{(n+1)^2}$ which converges by the Alternating Series Test, so

$I = [-1/3, 1/3]$.

13. If $u_n = \dfrac{x^n}{\ln n}$, then $\lim\limits_{n \to \infty} \left| \dfrac{u_{n+1}}{u_n} \right| = \lim\limits_{n \to \infty} \left| \dfrac{x^{n+1}}{\ln(n+1)} \cdot \dfrac{\ln n}{x^n} \right| = |x| \lim\limits_{n \to \infty} \dfrac{\ln n}{\ln(n+1)} = |x|$ (using

l'Hospital's Rule), so $R = 1$. When $x = 1$, $\sum\limits_{n=2}^{\infty} \dfrac{x^n}{\ln n} = \sum\limits_{n=2}^{\infty} \dfrac{1}{\ln n}$ which diverges because

$\dfrac{1}{\ln n} > \dfrac{1}{n}$ and $\sum\limits_{n=2}^{\infty} \dfrac{1}{n}$ is the divergent harmonic series. When $x = -1$, $\sum\limits_{n=2}^{\infty} \dfrac{x^n}{\ln n} = \sum\limits_{n=2}^{\infty} \dfrac{(-1)^n}{\ln n}$

which converges by the Alternating Series Test. So $I = [-1, 1)$.

15. If $u_n = \dfrac{n}{4^n}(2x-1)^n$, then $\left| \dfrac{u_{n+1}}{u_n} \right| = \left| \dfrac{(n+1)(2x-1)^{n+1}}{4^{n+1}} \cdot \dfrac{4^n}{n(2x-1)^n} \right| = \left| \dfrac{2x-1}{4} \left(1 + \dfrac{1}{n} \right) \right|$

$\to \dfrac{1}{2} \left| x - \dfrac{1}{2} \right|$ as $n \to \infty$. For convergence, $\dfrac{1}{2} \left| x - \dfrac{1}{2} \right| < 1 \Rightarrow \left| x - \dfrac{1}{2} \right| < 2 \Rightarrow R = 2$ and

$-2 < x - \dfrac{1}{2} < 2 \Rightarrow -\dfrac{3}{2} < x < \dfrac{5}{2}$. If $x = -\dfrac{3}{2}$, the series becomes $\sum\limits_{n=0}^{\infty} \dfrac{n}{4^n}(-4)^n = \sum\limits_{n=0}^{\infty} (-1)^n n$

which is divergent by the Test for Divergence. If $x = \dfrac{5}{2}$, the series is $\sum\limits_{n=0}^{\infty} \dfrac{n}{4^n} 4^n = \sum\limits_{n=0}^{\infty} n$ which is also divergent by the Test for Divergence. So $I = (-\frac{3}{2}, \frac{5}{2})$.

17.   If $u_n = \dfrac{(-1)^n(x-1)^n}{\sqrt{n}}$, then $\lim\limits_{n \to \infty} \left| \dfrac{u_{n+1}}{u_n} \right| = \lim\limits_{n \to \infty} \left| \dfrac{(x-1)^{n+1}}{\sqrt{n+1}} \cdot \dfrac{\sqrt{n}}{(x-1)^n} \right|$

$= |x-1| \lim\limits_{n \to \infty} \sqrt{\dfrac{n}{n+1}} = |x-1| < 1$ for convergence, or $0 < x < 2$, and $R = 1$. When

$x = 0$, $\displaystyle\sum_{n=1}^{\infty} \dfrac{(-1)^n(x-1)^n}{\sqrt{n}} = \sum_{n=1}^{\infty} \dfrac{1}{\sqrt{n}}$ which is a divergent p–series ($p = \frac{1}{2} < 1$). When

$x = 2$, the series is $\displaystyle\sum_{n=1}^{\infty} \dfrac{(-1)^n}{\sqrt{n}}$ which converges by the Alternating Series Test. So $I = (0, 2]$.

19.   If $u_n = \dfrac{(x-2)^n}{n^n}$, then $\lim\limits_{n \to \infty} \sqrt[n]{|u_n|} = \lim\limits_{n \to \infty} \dfrac{x-2}{n} = 0$, so the series converges for all x (by

the Root Test). $R = \infty$ and $I = (-\infty, \infty)$.

21.   If $u_n = \dfrac{2^n(x-3)^n}{n+3}$, then $\lim\limits_{n \to \infty} \left| \dfrac{u_{n+1}}{u_n} \right| = \lim\limits_{n \to \infty} \left| \dfrac{2^{n+1}(x-3)^{n+1}}{n+4} \cdot \dfrac{n+3}{2^n(x-3)^n} \right|$

$= 2|x-3| \lim\limits_{n \to \infty} \dfrac{n+3}{n+4} = 2|x-3| < 1$ for convergence, or $|x-3| < \frac{1}{2} \Leftrightarrow \frac{5}{2} < x < \frac{7}{2}$, and

$R = \frac{1}{2}$. When $x = \frac{5}{2}$, $\displaystyle\sum_{n=0}^{\infty} \dfrac{2^n(x-3)^n}{n+3} = \sum_{n=0}^{\infty} \dfrac{(-1)^n}{n+3}$ which converges by the Alternating

Series Test. When $x = \frac{7}{2}$, $\displaystyle\sum_{n=0}^{\infty} \dfrac{2^n(x-3)^n}{n+3} = \sum_{n=0}^{\infty} \dfrac{1}{n+3} = \sum_{n=3}^{\infty} \dfrac{1}{n}$, the harmonic series, which

diverges. So $I = [5/2, 7/2)$.

23.   If $u_n = \dfrac{n(x+10)^n}{(n^2+1)4^n}$, then $\lim\limits_{n \to \infty} \left| \dfrac{u_{n+1}}{u_n} \right|$

$= \lim\limits_{n \to \infty} \left| \dfrac{(n+1)(x+10)^{n+1}}{[(n+1)^2+1]\,4^{n+1}} \cdot \dfrac{(n^2+1)4^n}{n(x+10)^n} \right|$

$= \dfrac{|x+10|}{4} \lim\limits_{n \to \infty} \dfrac{n^3+n^2+n+1}{n^3+2n^2+2n} = \dfrac{|x+10|}{4} < 1$ for convergence, so $|x+10| < 4$,

$-14 < x < -6$, and $R = 4$. When $x = -14$, $\displaystyle\sum_{n=0}^{\infty} \dfrac{n(x+10)^n}{(n^2+1)4^n} = \sum_{n=0}^{\infty} \dfrac{(-1)^n n}{n^2+1}$ which

converges by the Alternating Series Test. When $x = -6$,

$\displaystyle\sum_{n=0}^{\infty} \dfrac{n(x+10)^n}{(n^2+1)4^n} = \sum_{n=0}^{\infty} \dfrac{n}{n^2+1}$ which diverges (by the Integral Test or the Limit

Comparison Test with $b_n = 1/n$). So $I = [-14, -6)$.

25.   If $u_n = \left(\dfrac{n}{2} \right)^n (x+6)^n$, then $\lim\limits_{n \to \infty} \sqrt[n]{|u_n|} = \lim\limits_{n \to \infty} \dfrac{n(x+6)}{2} = \infty$ unless $x = -6$, in which case

the limit is 0. So by the Root Test, the series converges only for $x = -6$. $R = 0$ and $I = \{-6\}$.

27. If $u_n = \dfrac{(2x-1)^n}{n^3}$, then $\lim\limits_{n \to \infty} \left| \dfrac{u_{n+1}}{u_n} \right| = |2x - 1| \lim\limits_{n \to \infty} \left(\dfrac{n}{n+1} \right)^3 = |2x - 1| < 1$ for

convergence, so $\left| x - 1/2 \right| < 1/2 \Leftrightarrow 0 < x < 1$, and $R = 1/2$. The series $\sum\limits_{n=1}^{\infty} \dfrac{(2x-1)^n}{n^3}$

converges both for $x = 0$ and $x = 1$ (in the first case because of the Alternating Series Test and

in the last case because we get a p-series with $p = 3 > 1$). So $I = [0, 1]$.

29. If $u_n = \dfrac{n}{\sqrt{n+1}}(x - e)^n$, then $\lim\limits_{n \to \infty} \left| \dfrac{u_{n+1}}{u_n} \right| = \lim\limits_{n \to \infty} \left| \dfrac{(n+1)(x-e)^{n+1}/\sqrt{n+2}}{n(x-e)^n/\sqrt{n+1}} \right|$

$= |x - e| \lim\limits_{n \to \infty} \left(\dfrac{n^3 + 3n^2 + 3n + 1}{n^3 + 2n^2} \right)^{1/2} = |x - e| < 1$ for convergence, so $e - 1 < x < e + 1$

and $R = 1$. When $x = e \pm 1$, $\sum\limits_{n=0}^{\infty} \dfrac{n}{\sqrt{n+1}}(x - e)^n$ will diverge by the Test for Divergence

since $\lim\limits_{n \to \infty} \dfrac{n}{\sqrt{n+1}} = \infty$. So $I = (e - 1, e + 1)$.

31. If $u_n = \dfrac{n! x^n}{(2n)!}$, then $\lim\limits_{n \to \infty} \left| \dfrac{u_{n+1}}{u_n} \right| = \lim\limits_{n \to \infty} \left| \dfrac{(n+1)! \, x^{n+1}}{(2n+2)!} \cdot \dfrac{(2n)!}{n! x^n} \right|$

$= \lim\limits_{n \to \infty} \dfrac{n+1}{(2n+1)(2n+2)} |x| = 0 < 1$ for all x, so $R = \infty$ and $I = (-\infty, \infty)$.

33. If $u_n = \dfrac{(-1)^n x^{2n+1}}{n!(n+1)! 2^{2n+1}}$, then $\lim\limits_{n \to \infty} \left| \dfrac{u_{n+1}}{u_n} \right| = \left(\dfrac{x}{2} \right)^2 \lim\limits_{n \to \infty} \dfrac{1}{(n+1)(n+2)} = 0$ for all x.

So $J_1(x)$ converges for all x; the domain is $(-\infty, \infty)$.

35. $s_{2n-1} = 1 + 2x + x^2 + 2x^3 + \cdots + x^{2n-2} + 2x^{2n-1} = (1 + 2x)(1 + x^2 + x^4 + \cdots + x^{2n-2})$

$= (1 + 2x) \dfrac{1 - x^{2n}}{1 - x^2} \to \dfrac{1 + 2x}{1 - x^2}$ as $n \to \infty$, when $|x| < 1$. Also $s_{2n} = s_{2n-1} + x^{2n} \to \dfrac{1 + 2x}{1 - x^2}$

since $x^{2n} \to 0$ for $|x| < 1$. Therefore $s_n \to \dfrac{1 + 2x}{1 - x^2}$ by Exercise 10.1.76(a).

Thus the interval of convergence is $(-1, 1)$ and $f(x) = \dfrac{1 + 2x}{1 - x^2}$.

37. We use the Root Test on the series $\sum a_n x^n$. $\lim\limits_{n \to \infty} \sqrt[n]{|a_n x^n|} = |x| \lim\limits_{n \to \infty} \sqrt[n]{|a_n|} = a|x| < 1$

for convergence, or $|x| < \dfrac{1}{a}$, so $R = \dfrac{1}{a}$.

EXERCISES 10.9

1. $f(x) = \cos x$ $f(0) = 1$

    $f'(x) = -\sin x$ $f'(0) = 0$

    $f''(x) = -\cos x$ $f''(0) = -1$

    $f^{(3)}(x) = \sin x$ $f^{(3)}(0) = 0$

    $f^{(4)}(x) = \cos x$ $f^{(4)}(0) = 1$

    $\cdots$ $\cdots$

$$\cos x = f(0) + f'(0)\,x + \frac{f''(0)}{2!}\,x^2 + \frac{f^{(3)}(0)}{3!}\,x^3 + \frac{f^{(4)}(0)}{4!}\,x^4 + \cdots = 1 - \frac{x^2}{2!} + \frac{x^4}{4!} - \cdots$$

$$= \sum_{n=0}^{\infty} \frac{(-1)^n x^{2n}}{(2n)!}. \quad \text{If } u_n = \frac{(-1)^n x^{2n}}{(2n)!}, \quad \text{then } \lim_{n\to\infty} \left|\frac{u_{n+1}}{u_n}\right| = x^2 \lim_{n\to\infty} \frac{1}{(2n+2)(2n+1)} = 0$$

< 1 for all x. So $R = \infty$.

3. $f(x) = \sin x$ $f(\pi/4) = \sqrt{2}/2$

    $f'(x) = \cos x$ $f'(\pi/4) = \sqrt{2}/2$

    $f''(x) = -\sin x$ $f''(\pi/4) = -\sqrt{2}/2$

    $f^{(3)}(x) = -\cos x$ $f^{(3)}(\pi/4) = -\sqrt{2}/2$

    $f^{(4)}(x) = \sin x$ $f^{(4)}(\pi/4) = \sqrt{2}/2$

$$\sin x = f(\pi/4) + f'(\pi/4)(x - \tfrac{\pi}{4}) + \frac{f''(\pi/4)}{2!}(x - \tfrac{\pi}{4})^2 + \frac{f^{(3)}(\pi/4)}{3!}(x - \tfrac{\pi}{4})^3 + \frac{f^{(4)}(\pi/4)}{4!}(x - \tfrac{\pi}{4})^4$$

$$+ \cdots = \frac{\sqrt{2}}{2}\left[1 + (x - \tfrac{\pi}{4}) - \tfrac{1}{2!}(x - \tfrac{\pi}{4})^2 - \tfrac{1}{3!}(x - \tfrac{\pi}{4})^3 + \tfrac{1}{4!}(x - \tfrac{\pi}{4})^4 + \cdots\right]$$

$$= \frac{\sqrt{2}}{2} \sum_{n=0}^{\infty} \frac{(-1)^{n(n-1)/2}(x - \pi/4)^n}{n!}. \quad \text{If } u_n = \frac{(-1)^{n(n-1)/2}(x - \pi/4)^n}{n!}, \quad \text{then } \lim_{n\to\infty} \left|\frac{u_{n+1}}{u_n}\right|$$

$$= \lim_{n\to\infty} \frac{|x - \pi/4|}{n+1} = 0 < 1 \text{ for all x, so } R = \infty.$$

5. $f(x) = (1+x)^{-2}$ $f(0) = 1$

    $f'(x) = -2(1+x)^{-3}$ $f'(0) = -2$

    $f''(x) = 2 \cdot 3(1+x)^{-4}$ $f''(0) = 2 \cdot 3$

    $f^{(3)}(x) = -2 \cdot 3 \cdot 4(1+x)^{-5}$ $f^{(3)}(0) = -2 \cdot 3 \cdot 4$

    $f^{(4)}(x) = 2 \cdot 3 \cdot 4 \cdot 5(1+x)^{-6}$ $f^{(4)}(0) = 2 \cdot 3 \cdot 4 \cdot 5$

    $\cdots$ $\cdots$

So $f^{(n)}(0) = (-1)^n(n+1)!$, and

$$\frac{1}{(1+x)^2} = \sum_{n=0}^{\infty} \frac{(-1)^n (n+1)!}{n!} x^n = \sum_{n=0}^{\infty} (-1)^n (n+1) x^n.$$

If $u_n = (-1)^n (n+1) x^n$, then $\lim\limits_{n \to \infty} \left| \frac{u_{n+1}}{u_n} \right| = |x|$ so $R = 1$.

7. $f(x) = x^{-1}$ $\qquad\qquad\qquad\qquad\qquad$ $f(1) = 1$

$f'(x) = -x^{-2}$ $\qquad\qquad\qquad\qquad$ $f'(1) = -1$

$f''(x) = 2x^{-3}$ $\qquad\qquad\qquad\qquad$ $f''(1) = 2$

$f^{(3)}(x) = -3 \cdot 2x^{-4}$ $\qquad\qquad\qquad$ $f^{(3)}(1) = -3 \cdot 2$

$f^{(4)}(x) = 4 \cdot 3 \cdot 2x^{-5}$ $\qquad\qquad\qquad$ $f^{(4)}(1) = 4 \cdot 3 \cdot 2$

$\ldots$ $\qquad\qquad\qquad\qquad\qquad\qquad\qquad$ $\ldots$

So $f^{(n)}(1) = (-1)^n n!$, and $\frac{1}{x} = \sum_{n=0}^{\infty} \frac{(-1)^n n!}{n!} (x-1)^n = \sum_{n=0}^{\infty} (-1)^n (x-1)^n$. If

$u_n = (-1)^n (x-1)^n$ then $\lim\limits_{n \to \infty} \left| \frac{u_{n+1}}{u_n} \right| = |x-1| < 1$ for convergence, so $0 < x < 2$

and $R = 1$.

9. Clearly $f^{(n)}(x) = e^x$, so $f^{(n)}(3) = e^3$ and $e^x = \sum_{n=0}^{\infty} \frac{e^3}{n!} (x-3)^n$. If $u_n = \frac{e^3}{n!} (x-3)^n$ then

$\lim\limits_{n \to \infty} \left| \frac{u_{n+1}}{u_n} \right| = \lim\limits_{n \to \infty} \frac{|x-3|}{n+1} = 0$ for all x, so $R = \infty$.

11. $f(x) = \sinh x$ $\qquad\qquad\qquad\qquad\qquad$ $f(0) = 0$

$f'(x) = \cosh x$ $\qquad\qquad\qquad\qquad\qquad$ $f'(0) = 1$

$f''(x) = \sinh x$ $\qquad\qquad\qquad\qquad\qquad$ $f''(0) = 0$

$f^{(3)}(x) = \cosh x$ $\qquad\qquad\qquad\qquad\qquad$ $f^{(3)}(0) = 1$

$f^{(4)}(x) = \sinh x$ $\qquad\qquad\qquad\qquad\qquad$ $f^{(4)}(0) = 0$

$\ldots$ $\qquad\qquad\qquad\qquad\qquad\qquad\qquad$ $\ldots$

So $f^{(n)}(0) = \begin{cases} 0 \text{ if } n \text{ is even} \\ 1 \text{ if } n \text{ is odd} \end{cases}$, and $\sinh x = \sum_{n=0}^{\infty} \frac{x^{2n+1}}{(2n+1)!}$. If $u_n = \frac{x^{2n+1}}{(2n+1)!}$ then

$\lim\limits_{n \to \infty} \left| \frac{u_{n+1}}{u_n} \right| = x^2 \lim\limits_{n \to \infty} \frac{1}{(2n+3)(2n+2)} = 0 < 1$ for all x, so $R = \infty$.

13. $f(x) = \frac{1}{1+x} = \frac{1}{1-(-x)} = \sum_{n=0}^{\infty} (-1)^n x^n$ with $|-x| < 1 \Leftrightarrow |x| < 1$ so $R = 1$.

15. $f(x) = \frac{1}{(1+x)^2} = -\frac{d}{dx}\left(\frac{1}{1+x}\right) = -\frac{d}{dx}\left(\sum_{n=0}^{\infty} (-1)^n x^n\right)$ [from Exercise 13]

$= \sum_{n=1}^{\infty} (-1)^{n+1} n x^{n-1} = \sum_{n=0}^{\infty} (-1)^n (n+1) x^n$ with $R = 1$.

17. $f(x) = \dfrac{1}{1+4x^2} = \displaystyle\sum_{n=0}^{\infty} (-1)^n(4x^2)^n$ [*substituting* $4x^2$ *for* x *in the series from Exercise* 13

$above] = \displaystyle\sum_{n=0}^{\infty} (-1)^n 4^n x^{2n}$, with $|4x^2| < 1$ so $x^2 < \frac{1}{4}$, $|x| < \frac{1}{2}$ and $R = \frac{1}{2}$.

19. $f(x) = \dfrac{1}{4+x^2} = \frac{1}{4}\left(\dfrac{1}{1+x^2/4}\right) = \frac{1}{4}\displaystyle\sum_{n=0}^{\infty} (-1)^n\left(\dfrac{x^2}{4}\right)^n$ [*using Exercise* 13]

$= \displaystyle\sum_{n=0}^{\infty} \dfrac{(-1)^n x^{2n}}{4^{n+1}}$, with $\left|\dfrac{x^2}{4}\right| < 1 \Leftrightarrow x^2 < 4 \Leftrightarrow |x| < 2$, so $R = 2$.

21. $f(x) = \dfrac{1}{1-x^2} = \displaystyle\sum_{n=0}^{\infty} (x^2)^n = \displaystyle\sum_{n=0}^{\infty} x^{2n}$, with $|x^2| < 1 \Leftrightarrow |x| < 1$ so $R = 1$.

23. $f(x) = \ln(1+x) - \ln(1-x) = \displaystyle\int \dfrac{dx}{1+x} + \int \dfrac{dx}{1-x} = \int\left[\displaystyle\sum_{n=0}^{\infty} (-1)^n x^n + \displaystyle\sum_{n=0}^{\infty} x^n\right] dx$

$= \displaystyle\int \sum_{n=0}^{\infty} 2x^{2n}\, dx = \displaystyle\sum_{n=0}^{\infty} \dfrac{2x^{2n+1}}{2n+1} + C$. But $f(0) = \ln 1 - \ln 1 = 0$, so $C = 0$ and we have

$f(x) = \displaystyle\sum_{n=0}^{\infty} \dfrac{2x^{2n+1}}{2n+1}$ with $R = 1$.

25. $e^{3x} = \displaystyle\sum_{n=0}^{\infty} \dfrac{(3x)^n}{n!} = \displaystyle\sum_{n=0}^{\infty} \dfrac{3^n x^n}{n!}$, with $R = \infty$.

27. $x^2 \cos x = x^2 \displaystyle\sum_{n=0}^{\infty} \dfrac{(-1)^n x^{2n}}{(2n)!} = \displaystyle\sum_{n=0}^{\infty} \dfrac{(-1)^n x^{2n+2}}{(2n)!}$, with $R = \infty$.

29. $x \sin\left(\dfrac{x}{2}\right) = x \displaystyle\sum_{n=0}^{\infty} \dfrac{(-1)^n (x/2)^{2n+1}}{(2n+1)!} = \displaystyle\sum_{n=0}^{\infty} \dfrac{(-1)^n x^{2n+2}}{(2n+1)!\,2^{2n+1}}$, with $R = \infty$.

31. $\sin^2 x = \frac{1}{2}[1 - \cos 2x] = \frac{1}{2}\left[1 - \displaystyle\sum_{n=0}^{\infty} \dfrac{(-1)^n (2x)^{2n}}{(2n)!}\right] = \frac{1}{2}\left[1 - 1 - \displaystyle\sum_{n=1}^{\infty} \dfrac{(-1)^n (2x)^{2n}}{(2n)!}\right]$

$= \displaystyle\sum_{n=1}^{\infty} \dfrac{(-1)^{n+1} 2^{2n-1} x^{2n}}{(2n)!}$, with $R = \infty$.

33. $\dfrac{\sin x}{x} = \dfrac{1}{x} \displaystyle\sum_{n=0}^{\infty} \dfrac{(-1)^n x^{2n+1}}{(2n+1)!} = \displaystyle\sum_{n=0}^{\infty} \dfrac{(-1)^n x^{2n}}{(2n+1)!}$ and this series also gives the required

value at $x = 0$, so $R = \infty$.

35. $f(x) = (1+x)^{1/2}$ $\qquad\qquad\qquad$ $f(0) = 1$

$f'(x) = \frac{1}{2}(1+x)^{-1/2}$ $\qquad\qquad\quad$ $f'(0) = \frac{1}{2}$

$f''(x) = -\frac{1}{4}(1+x)^{-3/2}$ $\qquad\qquad$ $f''(0) = -\frac{1}{4}$

$f^{(3)}(x) = \frac{3}{8}(1+x)^{-5/2}$ $\qquad$ $f^{(3)}(0) = \frac{3}{8}$

$f^{(4)}(x) = -\frac{15}{16}(1+x)^{-7/2}$ $\qquad$ $f^{(4)}(0) = -\frac{15}{16}$

$\cdots$ $\qquad\qquad\qquad$ $\cdots$

So $f^{(n)}(0) = \dfrac{(-1)^{n-1}1\cdot3\cdot5\,\cdots\,(2n-3)}{2^n}$ for $n \geq 2$, and $\sqrt{1+x} = 1 + \frac{x}{2}$

$+ \displaystyle\sum_{n=2}^{\infty} \dfrac{(-1)^{n-1}1\cdot3\cdot5\,\cdots\,(2n-3)}{2^n n!}\, x^n$. If $u_n = \dfrac{(-1)^{n+1}1\cdot3\cdot5\,\cdots\,(2n-3)}{2^n n!}\, x^n$ then

$\displaystyle\lim_{n\to\infty}\left|\frac{u_{n+1}}{u_n}\right| = \frac{|x|}{2}\lim_{n\to\infty}\frac{2n-1}{n+1} = |x| < 1$ for convergence, so $R = 1$.

37. $f(x) = (1-x)^{-1/3}$ $\qquad\qquad$ $f(0) = 1$

$f'(x) = \frac{1}{3}(1-x)^{-4/3}$ $\qquad\quad$ $f'(0) = \frac{1}{3}$

$f''(x) = \frac{4}{9}(1-x)^{-7/3}$ $\qquad\quad$ $f''(0) = \frac{4}{9}$

$f^{(3)}(x) = \frac{28}{27}(1-x)^{-10/3}$ $\qquad$ $f^{(3)}(0) = \frac{28}{27}$

$f^{(4)}(x) = \frac{280}{81}(1-x)^{-13/3}$ $\qquad$ $f^{(4)}(0) = \frac{280}{81}$

$\cdots$ $\qquad\qquad\qquad$ $\cdots$

So $f^{(n)}(0) = \dfrac{1\cdot4\cdot7\cdot10\cdot13\,\cdots\,(3n-2)}{3^n}$ for $n \geq 1$, and

$f(x) = 1 + \displaystyle\sum_{n=1}^{\infty} \dfrac{1\cdot4\cdot7\cdot10\cdot13\,\cdots\,(3n-2)}{3^n n!}\, x^n$. If $u_n = \dfrac{1\cdot4\cdot7\cdot10\cdot13\,\cdots\,(3n-2)}{3^n n!}\, x^n$

then $\displaystyle\lim_{n\to\infty}\left|\frac{u_{n+1}}{u_n}\right| = \frac{|x|}{3}\lim_{n\to\infty}\frac{3n+1}{n+1} = |x| < 1$ for convergence, so $R = 1$.

39. $f(x) = (1+x)^{-3} = -\frac{1}{2}\dfrac{d}{dx}\left[\dfrac{1}{(1+x)^2}\right] = -\frac{1}{2}\dfrac{d}{dx}\left[\displaystyle\sum_{n=0}^{\infty}(-1)^n(n+1)x^n\right]$ [*from Exercise 15*

above] $= -\frac{1}{2}\displaystyle\sum_{n=1}^{\infty}(-1)^n n(n+1)x^{n-1} = \displaystyle\sum_{n=0}^{\infty}\dfrac{(-1)^n(n+1)(n+2)x^n}{2}$, with $R = 1$ (since

that is the R in Exercise 15).

41. $\ln(5+x) = \ln[5(1+x/5)] = \ln(5) + \ln(1+x/5) = \ln(5) + \frac{1}{5}\displaystyle\int\dfrac{dx}{1+x/5}$

$= \ln(5) + \frac{1}{5}\displaystyle\int\sum_{n=0}^{\infty}(-1)^n\left(\frac{x}{5}\right)^n dx = \ln(5) + \displaystyle\sum_{n=0}^{\infty}\dfrac{(-1)^n x^{n+1}}{(n+1)5^{n+1}} = \ln(5) + \displaystyle\sum_{n=1}^{\infty}\dfrac{(-1)^{n-1}x^n}{n5^n}$,

with $R = 5$.

43. $\ln(1+x) = \displaystyle\int\dfrac{dx}{1+x} = \displaystyle\int\sum_{n=0}^{\infty}(-1)^n x^n\, dx = \displaystyle\sum_{n=1}^{\infty}\dfrac{(-1)^{n-1}x^n}{n}$ with $R = 1$, so

$\ln(1.1) = \sum\limits_{n=1}^{\infty} \dfrac{(-1)^{n-1}(0.1)^n}{n}$. This is an alternating series with $a_5 = \dfrac{(0.1)^5}{5} = .000002$, so

to 5 decimals, $\ln(1.1) \approx \sum\limits_{n=1}^{4} \dfrac{(-1)^{n-1}(0.1)^n}{n} \approx 0.09531$.

45. $\int \sin(x^2)\,dx = \int \sum\limits_{n=0}^{\infty} (-1)^n \dfrac{(x^2)^{2n+1}}{(2n+1)!}\,dx = \int \sum\limits_{n=0}^{\infty} \dfrac{(-1)^n x^{4n+2}}{(2n+1)!}\,dx$

$= C + \sum\limits_{n=0}^{\infty} \dfrac{(-1)^n x^{4n+3}}{(4n+3)(2n+1)!}$.

47. $\int \dfrac{dx}{1+x^4} = \int \sum\limits_{n=0}^{\infty} (-1)^n x^{4n}\,dx = C + \sum\limits_{n=0}^{\infty} \dfrac{(-1)^n x^{4n+1}}{4n+1}$.

49. Using the series we obtained in Exercise 35, we get

$$\sqrt{x^3+1} = 1 + \dfrac{x^3}{2} + \sum\limits_{n=2}^{\infty} \dfrac{(-1)^{n-1} 1\cdot 3\cdot 5\,\cdots\,(2n-3)}{2^n\,n!} x^{3n} \quad \text{so}$$

$$\int \sqrt{x^3+1}\,dx = \int \left(1 + \dfrac{x^3}{2} + \sum\limits_{n=2}^{\infty} \dfrac{(-1)^{n-1} 1\cdot 3\cdot 5\,\cdots\,(2n-3)}{2^n\,n!} x^{3n}\right) dx$$

$$= C + x + \dfrac{x^4}{8} + \sum\limits_{n=2}^{\infty} \dfrac{(-1)^{n-1} 1\cdot 3\cdot 5\,\cdots\,(2n-3)}{2^n\,n!\,(3n+1)} x^{3n+1}.$$

51. Using our series from Exercise 45, we get $\displaystyle\int_0^1 \sin(x^2)\,dx = \sum\limits_{n=0}^{\infty} \left[\dfrac{(-1)^n x^{4n+3}}{(4n+3)(2n+1)!}\right]_0^1$

$= \sum\limits_{n=0}^{\infty} \dfrac{(-1)^n}{(4n+3)(2n+1)!}$ and $|a_3| = \dfrac{1}{75600} < 0.000014$ so by Theorem 10.28, we have

$\sum\limits_{n=0}^{2} \dfrac{(-1)^n}{(4n+3)(2n+1)!} \approx \dfrac{1}{3} - \dfrac{1}{42} + \dfrac{1}{1320} \approx 0.310$.

53. $\displaystyle\int_0^{0.5} \dfrac{dx}{1+x^6} = \int_0^{0.5} \sum\limits_{n=0}^{\infty} (-1)^n x^{6n}\,dx = \sum\limits_{n=0}^{\infty} \left[\dfrac{(-1)^n x^{6n+1}}{6n+1}\right]_0^{1/2} = \sum\limits_{n=0}^{\infty} \dfrac{(-1)^n}{(6n+1)2^{6n+1}}$ and

$a_2 = \dfrac{1}{106496} < 0.00001$, so by Theorem 10.28, we use $\sum\limits_{n=0}^{1} \dfrac{(-1)^n}{(6n+1)2^{6n+1}} = \dfrac{1}{2} - \dfrac{1}{896} \approx 0.4989$.

55. $\displaystyle\int_0^{0.5} x^2 e^{-x^2}\,dx = \int_0^{0.5} \sum\limits_{n=0}^{\infty} \dfrac{(-1)^n x^{2n+2}}{n!}\,dx = \sum\limits_{n=0}^{\infty} \left[\dfrac{(-1)^n x^{2n+3}}{n!(2n+3)}\right]_0^{1/2}$

$= \sum\limits_{n=0}^{\infty} \dfrac{(-1)^n}{n!(2n+3)2^{2n+3}}$ and since $a_2 = \dfrac{1}{1792} < 0.001$ we use

$\sum\limits_{n=0}^{1} \dfrac{(-1)^n}{n!(2n+3)2^{2n+3}} = \dfrac{1}{24} - \dfrac{1}{160} \approx 0.0354$.

57. (a) $J_0(x) = \sum_{n=0}^{\infty} \frac{(-1)^n x^{2n}}{2^{2n}(n!)^2}$, $J_0'(x) = \sum_{n=1}^{\infty} \frac{(-1)^n 2n x^{2n-1}}{2^{2n}(n!)^2}$, and

$J_0''(x) = \sum_{n=1}^{\infty} \frac{(-1)^n 2n(2n-1) x^{2n-2}}{2^{2n}(n!)^2}$, so $x^2 J_0''(x) + x J_0'(x) + x^2 J_0(x)$

$= \sum_{n=1}^{\infty} \frac{(-1)^n 2n(2n-1) x^{2n}}{2^{2n}(n!)^2} + \sum_{n=1}^{\infty} \frac{(-1)^n 2n x^{2n}}{2^{2n}(n!)^2} + \sum_{n=0}^{\infty} \frac{(-1)^n x^{2n+2}}{2^{2n}(n!)^2}$

$= \sum_{n=1}^{\infty} \frac{(-1)^n 2n(2n-1) x^{2n}}{2^{2n}(n!)^2} + \sum_{n=1}^{\infty} \frac{(-1)^n 2n x^{2n}}{2^{2n}(n!)^2} + \sum_{n=1}^{\infty} \frac{(-1)^{n-1} x^{2n}}{2^{2n-2}((n-1)!)^2}$

$= \sum_{n=1}^{\infty} (-1)^n \left[\frac{2n(2n-1) + 2n - 2^2 n^2}{2^{2n}(n!)^2} \right] x^{2n} = \sum_{n=1}^{\infty} (-1)^n \left[\frac{4n^2 - 2n + 2n - 4n^2}{2^{2n}(n!)^2} \right] x^{2n}$

$= 0.$

(b) $\int_0^1 J_0(x)\,dx = \int_0^1 \left[\sum_{n=0}^{\infty} \frac{(-1)^n x^{2n}}{2^{2n}(n!)^2} \right] dx$

$= \int_0^1 dx + \int_0^1 -\frac{x^2}{4}\,dx + \int_0^1 \frac{x^4}{64}\,dx + \int_0^1 -\frac{x^6}{2304}\,dx + \cdots = \left[x - \frac{x^3}{3\cdot 4} + \frac{x^5}{5\cdot 64} - \frac{x^7}{7\cdot 2304} + \cdots \right]_0^1$

$= 1 - \frac{1}{12} + \frac{1}{320} - \frac{1}{16128} + \cdots.$ Since $\frac{1}{16128} \approx 0.000\,062$, it follows from Theorem 10.28

that, correct to 3 decimal places, $\int_0^1 J_0(x)\,dx \approx 1 - \frac{1}{12} + \frac{1}{320} \approx 0.920.$

59. $\sum_{n=0}^{\infty} (-1)^n \frac{x^{4n}}{n!} = \sum_{n=0}^{\infty} \frac{(-x^4)^n}{n!} = e^{-x^4}$ by (10.47).

61. $\sum_{n=0}^{\infty} \frac{(-1)^n \pi^{2n+1}}{4^{2n+1}(2n+1)!} = \sum_{n=0}^{\infty} \frac{(-1)^n (\pi/4)^{2n+1}}{(2n+1)!} = \sin\frac{\pi}{4} = \frac{1}{\sqrt{2}}$ by (10.49).

63. $\sum_{n=0}^{\infty} \frac{x^{n+1}}{(n+1)!} = \frac{x}{1!} + \frac{x^2}{2!} + \frac{x^3}{3!} + \cdots = \left(1 + \frac{x}{1!} + \frac{x^2}{2!} + \frac{x^3}{3!} + \cdots \right) - 1 = e^x - 1$ by (10.47).

65. As in Example 10(a), we have $e^{-x^2} = 1 - \frac{x^2}{1!} + \frac{x^4}{2!} + \frac{x^6}{3!} + \cdots$ and we know that

$\cos x = 1 - \frac{x^2}{2!} + \frac{x^4}{4!} - \cdots$ from Equation 10.50. Therefore

$e^{-x^2} \cos x = (1 - x^2 + \frac{1}{2}x^4 - \cdots)(1 - \frac{1}{2}x^2 + \frac{1}{24}x^4 - \cdots)$

$= 1 - \frac{1}{2}x^2 + \frac{1}{24}x^4 - x^2 + \frac{1}{2}x^4 + \frac{1}{2}x^4 + \cdots = 1 - \frac{3}{2}x^2 + \frac{25}{24}x^4 + \cdots$

67. From Example 2 we have $\ln(1-x) = -x - \frac{x^2}{2} - \frac{x^3}{3} - \cdots$, $|x| < 1$. Therefore

$$y = \frac{\ln(1-x)}{e^x} = \frac{-x - \frac{1}{2}x^2 - \frac{1}{3}x^3 - \cdots}{1 + x + \frac{1}{2}x^2 + \frac{1}{6}x^3 + \cdots}$$

Long division gives

$$\begin{array}{r}
-x + \frac{1}{2}x^2 \quad -\frac{1}{3}x^3 + \cdots \\
1 + x + \frac{1}{2}x^2 + \frac{1}{6}x^3 - \cdots \; \overline{)\; -x - \frac{1}{2}x^2 \quad -\frac{1}{3}x^3 - \cdots} \\
\underline{-x - \; x^2 \quad -\frac{1}{2}x^3 - \cdots} \\
\frac{1}{2}x^2 \quad +\frac{1}{6}x^3 + \cdots \\
\underline{\frac{1}{2}x^2 \quad +\frac{1}{2}x^3 + \cdots} \\
-\frac{1}{3}x^3 + \cdots \\
\underline{-\frac{1}{3}x^3 + \cdots}
\end{array}$$

So $\dfrac{\ln(1-x)}{e^x} = -x + \frac{1}{2}x^2 - \frac{1}{3}x^3 + \cdots$, $|x| < 1$.

69. If $u_n = \dfrac{x^n}{n^2}$, then $\lim\limits_{n\to\infty} \left|\dfrac{u_{n+1}}{u_n}\right| = |x|\lim\limits_{n\to\infty}\left(\dfrac{n}{n+1}\right)^2 = |x| < 1$ for convergence so $R = 1$.

When $x = \pm 1$, $\sum\limits_{n=1}^{\infty}\left|\dfrac{x^n}{n^2}\right| = \sum\limits_{n=1}^{\infty}\dfrac{1}{n^2}$ which is a convergent p–series $(p = 2 > 1)$, so the

interval of convergence for f is $[-1, 1]$. By Theorem 10.39, the radii of convergence of f' and

f'' are both 1, so we need only check the endpoints.

$f'(x) = \sum\limits_{n=1}^{\infty}\dfrac{nx^{n-1}}{n^2} = \sum\limits_{n=0}^{\infty}\dfrac{x^n}{n+1}$, and this series diverges for $x = 1$ (harmonic series)

and converges for $x = -1$ (Alternating Series Test), so the interval of convergence is

$[-1, 1)$. $f''(x) = \sum\limits_{n=1}^{\infty}\dfrac{nx^{n-1}}{n+1}$ diverges at both 1 and -1 (Test for Divergence) since

$\lim\limits_{n\to\infty}\dfrac{n}{n+1} = 1 \neq 0$, so its interval of convergence is $(-1, 1)$.

71. $f(x) = \begin{cases} e^{-1/x^2} & \text{if } x \neq 0 \\ 0 & \text{if } x = 0 \end{cases}$, so $f'(0) = \lim\limits_{x\to 0}\dfrac{f(x) - f(0)}{x - 0} = \lim\limits_{x\to 0}\dfrac{e^{-1/x^2}}{x} = \lim\limits_{x\to 0}\dfrac{1/x}{e^{1/x^2}}$

$= \lim\limits_{x\to 0}\dfrac{x}{2e^{1/x^2}}$ [*using l'Hospital's Rule and simplifying*] $= 0$. Similarly, we can use

the definition of the derivative and l'Hospital's Rule to show that $f''(0) = 0$, $f^{(3)}(0) = 0, \ldots$

$f^{(n)}(0) = 0$, so that the Maclaurin series for f consists entirely of zero terms. But since

$f(x) \neq 0$ except for $x = 0$, we see that f cannot equal its Maclaurin series except at $x = 0$.

EXERCISES 10.10

1. $(1+x)^{1/2} = \sum_{n=0}^{\infty} \binom{1/2}{n} x^n = 1 + \left(\frac{1}{2}\right)x + \frac{\left(\frac{1}{2}\right)\left(-\frac{1}{2}\right)}{2!} x^2 + \frac{\left(\frac{1}{2}\right)\left(-\frac{1}{2}\right)\left(-\frac{3}{2}\right)}{3!} x^3 + \cdots$

$= 1 + \frac{x}{2} - \frac{x^2}{2^2 \cdot 2!} + \frac{1 \cdot 3}{2^3 \cdot 3!} x^3 - \frac{1 \cdot 3 \cdot 5}{2^4 \cdot 4!} x^4 + \cdots$

$= 1 + \frac{x}{2} + \sum_{n=2}^{\infty} \frac{(-1)^{n-1} 1 \cdot 3 \cdot 5 \cdots (2n-3) \, x^n}{2^n \cdot n!}$. $R = 1$.

3. $[1 + (2x)]^{-4} = 1 + (-4)(2x) + \frac{(-4)(-5)}{2!}(2x)^2 + \frac{(-4)(-5)(-6)}{3!}(2x)^3 + \cdots$

$= 1 + \sum_{n=1}^{\infty} \frac{(-1)^n 2^n 4 \cdot 5 \cdot 6 \cdots (n+3)}{n!} x^n = \sum_{n=0}^{\infty} (-1)^n \frac{2^n (n+1)(n+2)(n+3)}{6} x^n$.

$|2x| < 1 \Leftrightarrow |x| < \frac{1}{2}$ so $R = \frac{1}{2}$.

5. $[1 + (-x)]^{-1/2} = \sum_{n=0}^{\infty} \binom{-1/2}{n}(-x)^n = 1 + \left(-\frac{1}{2}\right)(-x) + \frac{\left(-\frac{1}{2}\right)\left(-\frac{3}{2}\right)}{2!}(-x)^2 + \cdots$

$= 1 + \frac{x}{2} + \frac{1 \cdot 3}{2^2 2!} x^2 + \frac{1 \cdot 3 \cdot 5}{2^3 3!} x^3 + \frac{1 \cdot 3 \cdot 5 \cdot 7}{2^4 4!} x^4 + \cdots = 1 + \sum_{n=1}^{\infty} \frac{1 \cdot 3 \cdot 5 \cdots (2n-1)}{2^n \cdot n!} x^n$

so $\frac{x}{\sqrt{1-x}} = x + \sum_{n=1}^{\infty} \frac{1 \cdot 3 \cdot 5 \cdots (2n-1)}{2^n \cdot n!} x^{n+1}$ with $R = 1$.

7. $(8+x)^{-1/3} = \frac{1}{2}\left(1 + \frac{x}{8}\right)^{-1/3} = \frac{1}{2}\left[1 + \left(-\frac{1}{3}\right)\left(\frac{x}{8}\right) + \frac{\left(-\frac{1}{3}\right)\left(-\frac{4}{3}\right)}{2!}\left(\frac{x}{8}\right)^2 + \cdots \right]$

$= \frac{1}{2}\left[1 + \sum_{n=1}^{\infty} \frac{(-1)^n 1 \cdot 4 \cdot 7 \cdots (3n-2)}{3^n \, n! \, 8^n} x^n \right]$ and $\left|\frac{x}{8}\right| < 1 \Leftrightarrow |x| < 8$,   so $R = 8$.

9. $(1-x^4)^{1/4} = 1 + \left(\frac{1}{4}\right)(-x^4) + \frac{\left(\frac{1}{4}\right)\left(-\frac{3}{4}\right)}{2!}(-x^4)^2 + \frac{\left(\frac{1}{4}\right)\left(-\frac{3}{4}\right)\left(-\frac{7}{4}\right)}{3!}(-x^4)^3 + \cdots$

$= 1 - \frac{x^4}{4} - \sum_{n=2}^{\infty} \frac{3 \cdot 7 \cdot 11 \cdots (4n-5)}{4^n \, n!} x^{4n}$ with $R = 1$.

11. $(1-x)^{-5} = 1 + (-5)(-x) + \frac{(-5)(-6)}{2!}(-x)^2 + \frac{(-5)(-6)(-7)}{3!}(-x)^3 + \cdots$

$= 1 + \sum_{n=1}^{\infty} \frac{5 \cdot 6 \cdot 7 \cdots (n+4)}{n!} x^n = \sum_{n=0}^{\infty} \frac{(n+4)!}{4! \, n!} x^n \Rightarrow \frac{x^5}{(1-x)^5} = \sum_{n=0}^{\infty} \frac{(n+4)!}{4! \, n!} x^{n+5}$

$\left(\text{or } \sum_{n=0}^{\infty} \frac{(n+1)(n+2)(n+3)(n+4)}{24} x^{n+5} \right)$ with $R = 1$.

13. (a) $(1-x^2)^{-1/2} = 1 + \left(-\frac{1}{2}\right)(-x^2) + \frac{\left(-\frac{1}{2}\right)\left(-\frac{3}{2}\right)}{2!}(-x^2)^2 + \frac{\left(-\frac{1}{2}\right)\left(-\frac{3}{2}\right)\left(-\frac{5}{2}\right)}{3!}(-x^2)^3 + \cdots$

$$= 1 + \sum_{n=1}^{\infty} \frac{1 \cdot 3 \cdot 5 \cdots (2n-1)}{2^n \, n!} x^{2n}$$

(b) $\sin^{-1} x = \displaystyle\int \frac{1}{\sqrt{1-x^2}} \, dx = C + x + \sum_{n=1}^{\infty} \frac{1 \cdot 3 \cdot 5 \cdots (2n-1)}{(2n+1) \, 2^n \, n!} x^{2n+1}$

$$= x + \sum_{n=1}^{\infty} \frac{1 \cdot 3 \cdot 5 \cdots (2n-1)}{(2n+1) \, 2^n \, n!} x^{2n+1} \quad \text{since } 0 = \sin^{-1} 0 = C.$$

15. (a) $(1+x)^{-1/2} = 1 + \left(-\frac{1}{2}\right) x + \dfrac{\left(-\frac{1}{2}\right)\left(-\frac{3}{2}\right)}{2!} x^2 + \dfrac{\left(-\frac{1}{2}\right)\left(-\frac{3}{2}\right)\left(-\frac{5}{2}\right)}{3!} x^3 + \cdots$

$$= 1 + \sum_{n=1}^{\infty} \frac{(-1)^n 1 \cdot 3 \cdot 5 \cdots (2n-1)}{2^n \, n!} x^n$$

(b) Take $x = 0.1$ in the above series. $\dfrac{1 \cdot 3 \cdot 5 \cdot 7}{2^4 \, 4!} (0.1)^4 < 0.00003$, so

$$\frac{1}{\sqrt{1.1}} \approx 1 - \frac{0.1}{2} + \frac{1 \cdot 3}{2^2 \cdot 2!} (0.1)^2 - \frac{1 \cdot 3 \cdot 5}{2^3 \, 3!} (0.1)^3 \approx 0.953.$$

17. (a) $(1-x)^{-2} = 1 + (-2)(-x) + \dfrac{(-2)(-3)}{2!} (-x)^2 + \cdots = \displaystyle\sum_{n=0}^{\infty} (n+1) x^n$, so

$$\frac{x}{(1-x)^2} = \sum_{n=0}^{\infty} (n+1) x^{n+1} = \sum_{n=1}^{\infty} n x^n.$$

(b) With $x = \frac{1}{2}$ in part (a), we have $\displaystyle\sum_{n=1}^{\infty} \frac{n}{2^n} = \frac{1/2}{(1-1/2)^2} = 2.$

19. (a) $(1+x^2)^{1/2} = 1 + \left(\frac{1}{2}\right) x^2 + \dfrac{\left(\frac{1}{2}\right)\left(-\frac{1}{2}\right)}{2!} (x^2)^2 + \dfrac{\left(\frac{1}{2}\right)\left(-\frac{1}{2}\right)\left(-\frac{3}{2}\right)}{3!} (x^2)^3 + \cdots$

$$= 1 + \frac{x^2}{2} + \sum_{n=2}^{\infty} \frac{(-1)^{n-1} 1 \cdot 3 \cdot 5 \cdots (2n-3)}{2^n \, n!} x^{2n}$$

(b) The coefficient of x^{10} in the above Maclaurin series will be $\dfrac{f^{(10)}(0)}{10!}$, so

$$f^{(10)}(0) = 10! \left(\frac{1 \cdot 3 \cdot 5 \cdot 7}{2^5 \, 5!} \right) = 99{,}225.$$

21. (a) $g'(x) = \displaystyle\sum_{n=1}^{\infty} \binom{k}{n} n x^{n-1}.$ $(1+x) g'(x) = (1+x) \displaystyle\sum_{n=1}^{\infty} \binom{k}{n} n x^{n-1}$

$$= \sum_{n=1}^{\infty} \binom{k}{n} n x^{n-1} + \sum_{n=1}^{\infty} \binom{k}{n} n x^n = \sum_{n=0}^{\infty} \binom{k}{n+1} (n+1) x^n + \sum_{n=0}^{\infty} \binom{k}{n} n x^n$$

$$= \sum_{n=0}^{\infty} \left((n+1) \frac{k(k-1)(k-2)\cdots(k-n)}{(n+1)!} \right) x^n + \sum_{n=0}^{\infty} \left((n) \frac{k(k-1)(k-2)\cdots(k-n+1)}{n!} \right) x^n$$

$$= \sum_{n=0}^{\infty} \frac{(n+1) k (k-1)(k-2) \cdots (k-n+1)}{(n+1)!} [(k-n) + n] x^n$$

$$= \sum_{n=0}^{\infty} \frac{k^2 (k-1)(k-2) \cdots (k-n+1)}{n!} x^n = k \sum_{n=0}^{\infty} \binom{k}{n} x^n = k \, g(x).$$

So $g'(x) = \frac{kg(x)}{1+x}$.

(b) $h'(x) = -k(1+x)^{-k-1}g(x) + (1+x)^{-k}g'(x)$

$= -k(1+x)^{-k-1}g(x) + (1+x)^{-k}\left(\frac{kg(x)}{1+x}\right) = -k(1+x)^{-k-1}g(x) + k(1+x)^{-k-1}g(x)$

$= 0$

(c) From part (b) we see that h must be constant for $x \in (-1,1)$, so $h(x) = h(0) = 1$ for $x \in (-1,1)$. Thus $h(x) = 1 = (1+x)^{-k}g(x) \Leftrightarrow g(x) = (1+x)^k$ for $x \in (-1,1)$.

EXERCISES 10.11

1. $\quad f(x) = 1 + 2x + 3x^2 + 4x^3$ $\qquad\qquad f(-1) = -2$

$f'(x) = 2 + 6x + 12x^2$ $\qquad\qquad f'(-1) = 8$

$f''(x) = 6 + 24x$ $\qquad\qquad\qquad f''(-1) = -18$

$f^{(3)}(x) = 24$ $\qquad\qquad\qquad f^{(3)}(-1) = 24$

$f^{(4)}(x) = 0$ $\qquad\qquad\qquad f^{(4)}(-1) = 0$

$T_4(x) = \sum_{n=0}^{4} \frac{f^{(n)}(-1)}{n!}(x+1)^n = -2 + 8(x+1) - 9(x+1)^2 + 4(x+1)^3$

3. $\quad f(x) = \sin x$ $\qquad\qquad\qquad f(\frac{\pi}{6}) = \frac{1}{2}$

$f'(x) = \cos x$ $\qquad\qquad\qquad f'(\frac{\pi}{6}) = \frac{\sqrt{3}}{2}$

$f''(x) = -\sin x$ $\qquad\qquad\qquad f''(\frac{\pi}{6}) = -\frac{1}{2}$

$f^{(3)}(x) = -\cos x$ $\qquad\qquad\qquad f^{(3)}(\frac{\pi}{6}) = -\frac{\sqrt{3}}{2}$

$T_3(x) = \sum_{n=0}^{3} \frac{f^{(n)}(\pi/6)}{n!}(x - \frac{\pi}{6})^n = \frac{1}{2} + \frac{\sqrt{3}}{2}(x - \frac{\pi}{6}) - \frac{1}{4}(x - \frac{\pi}{6})^2 - \frac{\sqrt{3}}{12}(x - \frac{\pi}{6})^3$

$= -\frac{1}{2} - \frac{\sqrt{3}}{2}(x - \frac{2\pi}{3}) + \frac{1}{4}(x - \frac{2\pi}{3})^2 + \frac{\sqrt{3}}{12}(x - \frac{2\pi}{3})^3 - \frac{1}{48}(x - \frac{2\pi}{3})^4$

5. $\quad f(x) = \tan x$ $\qquad\qquad\qquad f(0) = 0$

$f'(x) = \sec^2 x$ $\qquad\qquad\qquad f'(0) = 1$

$f''(x) = 2\sec^2 x \tan x$ $\qquad\qquad f''(0) = 0$

$f^{(3)}(x) = 4\sec^2 x \tan^2 x + 2\sec^4 x$ $\qquad f^{(3)}(0) = 2$

$f^{(4)}(x) = 8\sec^2 x \tan^3 x + 16\sec^4 x \tan x$ $\qquad f^{(4)}(0) = 0$

$T_4(x) = \sum_{n=0}^{4} \frac{f^{(n)}(0)}{n!}x^n = x + \frac{2}{3!}x^3 = x + \frac{x^3}{3}$

7. $f(x) = e^x \sin x$ $f(0) = 0$

   $f'(x) = e^x (\sin x + \cos x)$ $f'(0) = 1$

   $f''(x) = 2e^x \cos x$ $f''(0) = 2$

   $f^{(3)}(x) = 2e^x (\cos x - \sin x)$ $f^{(3)}(0) = 2$

   $$T_3(x) = \sum_{n=0}^{3} \frac{f^{(n)}(0)}{n!} x^n = x + x^2 + \frac{x^3}{3}$$

9. $f(x) = x^{1/2}$ $f(9) = 3$

   $f'(x) = \frac{1}{2}x^{-1/2}$ $f'(9) = \frac{1}{6}$

   $f''(x) = -\frac{1}{4}x^{-3/2}$ $f''(9) = -\frac{1}{108}$

   $f^{(3)}(x) = \frac{3}{8}x^{-5/2}$ $f^{(3)}(9) = \frac{1}{648}$

   $$T_3(x) = \sum_{n=0}^{3} \frac{f^{(n)}(9)}{n!} (x-9)^n = 3 + \frac{1}{6}(x-9) - \frac{1}{216}(x-9)^2 + \frac{1}{3888}(x-9)^3.$$

11. $f(x) = \ln (\sin x)$ $f(\frac{\pi}{2}) = 0$

   $f'(x) = \cot x$ $f'(\frac{\pi}{2}) = 0$

   $f''(x) = -\csc^2 x$ $f''(\frac{\pi}{2}) = -1$

   $f^{(3)}(x) = 2\csc^2 x \cot x$ $f^{(3)}(\frac{\pi}{2}) = 0$

   $$T_3(x) = \sum_{n=0}^{3} \frac{f^{(n)}(\pi/2)}{n!}(x - \pi/2)^n = -\frac{1}{2}\left(x - \frac{\pi}{2}\right)^2$$

13. $f(x) = \cos x$ $f(0) = 1$

   $f'(x) = -\sin x$ $f'(0) = 0$

   $f''(x) = -\cos x$ $f''(0) = -1$

   $f^{(3)}(x) = \sin x$ $f^{(3)}(0) = 0$

   $f^{(4)}(x) = \cos x$ $f^{(4)}(0) = 1$

   $T_1(x) = 1$

   $T_2(x) = 1 - \frac{x^2}{2}$

   $T_3(x) = 1 - \frac{x^2}{2}$

   $T_4(x) = 1 - \frac{x^2}{2} + \frac{x^4}{24}$

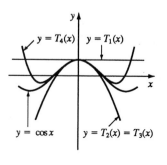

15. $f(x) = (1+x)^{1/2}$ $f(0) = 1$

   $f'(x) = \frac{1}{2}(1+x)^{-1/2}$ $f'(0) = \frac{1}{2}$

   $f''(x) = -\frac{1}{4}(1+x)^{-3/2}$

   (a) $(1+x)^{1/2} = 1 + \frac{1}{2}x + R_1(x)$ where $R_1(x) = \frac{f''(z)}{2!}x^2 = -\frac{1}{8(1+z)^{3/2}}x^2$ and z lies between 0 and x.

(b) $0 \le x \le 0.1 \Rightarrow 0 \le x^2 \le 0.01,$ and $0 < z < 0.1 \Rightarrow 1 < 1+z < 1.1$ so

$|R_1(x)| < \frac{0.01}{8.1} = 0.00125.$

17. $f(x) = \sin x$ $\qquad\qquad f(\frac{\pi}{4}) = \frac{\sqrt{2}}{2}$

$f'(x) = \cos x$ $\qquad\qquad f'(\frac{\pi}{4}) = \frac{\sqrt{2}}{2}$

$f''(x) = -\sin x$ $\qquad\qquad f''(\frac{\pi}{4}) = -\frac{\sqrt{2}}{2}$

$f^{(3)}(x) = -\cos x$ $\qquad\qquad f^{(3)}(\frac{\pi}{4}) = -\frac{\sqrt{2}}{2}$

$f^{(4)}(x) = \sin x$ $\qquad\qquad f^{(4)}(\frac{\pi}{4}) = \frac{\sqrt{2}}{2}$

$f^{(5)}(x) = \cos x$ $\qquad\qquad f^{(5)}(\frac{\pi}{4}) = \frac{\sqrt{2}}{2}$

$f^{(6)}(x) = -\sin x$

(a) $\sin x = \frac{\sqrt{2}}{2} + \frac{\sqrt{2}}{2}(x - \frac{\pi}{4}) - \frac{\sqrt{2}}{4}(x - \frac{\pi}{4})^2 - \frac{\sqrt{2}}{12}(x - \frac{\pi}{4})^3 + \frac{\sqrt{2}}{48}(x - \frac{\pi}{4})^4$

$+ \frac{\sqrt{2}}{240}(x - \frac{\pi}{4})^5 + R_5(x)$ where $R_5(x) = \frac{f^{(6)}(z)}{6!}(x - \frac{\pi}{4})^6 = \frac{-\sin z}{720}(x - \frac{\pi}{4})^6$ and z lies

between $\frac{\pi}{4}$ and x.

(b) Since $0 \le x \le \frac{\pi}{2}$, $-\frac{\pi}{4} \le x - \frac{\pi}{4} \le \frac{\pi}{4} \Rightarrow 0 \le (x - \frac{\pi}{4})^6 \le (\frac{\pi}{4})^6$, and since $0 < z < \frac{\pi}{2}$,

$0 < \sin z < 1$, so $|R_5(x)| < \frac{(\pi/4)^6}{720} \approx 0.00033.$

19. $f(x) = (1 + 2x)^{-4}$ $\qquad\qquad f(0) = 1$

$f'(x) = -8(1 + 2x)^{-5}$ $\qquad\qquad f'(0) = -8$

$f''(x) = 80(1 + 2x)^{-6}$ $\qquad\qquad f''(0) = 80$

$f^{(3)}(x) = -960(1 + 2x)^{-7}$ $\qquad\qquad f^{(3)}(0) = -960$

$f^{(4)}(x) = 13440(1 + 2x)^{-8}$

(a) $(1 + 2x)^{-4} = 1 - 8x + 40x^2 - 160x^3 + R_3(x)$ where

$R_3(x) = \frac{f^{(4)}(z)}{4!}x^4 = \frac{13440(1 + 2z)^{-8}}{4!}x^4 = \frac{560\,x^4}{(1 + 2z)^8}.$

(b) $|x| \le 0.1 \Rightarrow 0 \le x^4 \le 0.0001$ and $|z| < 0.1 \Rightarrow 0.8 < 1 + 2z < 1.2,$ so

$|R_3(x)| < \frac{560 \cdot (0.0001)}{(0.8)^8} < 0.34.$

21. $f(x) = \tan x$ $\qquad\qquad f(0) = 0$

$f'(x) = \sec^2 x$ $\qquad\qquad f'(0) = 1$

$f''(x) = 2\sec^2 x \tan x$ $\qquad\qquad f''(0) = 0$

$f^{(3)}(x) = 4\sec^2 x \tan^2 x + 2\sec^4 x$ $\qquad\qquad f^{(3)}(0) = 2$

$f^{(4)}(x) = 8\sec^2 x \tan^3 x + 16\sec^4 x \tan x$

(a) $\tan x = x + \dfrac{x^3}{3} + R_3(x)$ where $R_3(x) = \dfrac{f^{(4)}(z)}{4!}x^4$

$= \dfrac{8\sec^2 z \tan^3 z + 16\sec^4 z \tan z}{4!}x^4 = \dfrac{\sec^2 z \tan^3 z + 2\sec^4 z \tan z}{3}$ where z lies between 0 and x.

(b) $0 \le x^4 \le \left(\dfrac{\pi}{6}\right)^4$ and $0 < z < \dfrac{\pi}{6} \Rightarrow \sec^2 z < \dfrac{4}{3}$ and $\tan z < \dfrac{\sqrt{3}}{3}$ so

$$|R_3(x)| < \frac{(4/3)[1/(3\sqrt{3})] + 2(16/9)(1/\sqrt{3})}{3}\left(\frac{\pi}{6}\right)^4 = \frac{4\sqrt{3}}{9}\left(\frac{\pi}{6}\right)^4 < .06$$

23. $f(x) = e^{x^2}$ $\qquad\qquad\qquad\qquad$ $f(0) = 1$

$f'(x) = e^{x^2}(2x)$ $\qquad\qquad\qquad$ $f'(0) = 0$

$f''(x) = e^{x^2}(2 + 4x^2)$ $\qquad\qquad$ $f''(0) = 2$

$f^{(3)}(x) = e^{x^2}(12x + 8x^3)$ $\qquad$ $f^{(3)}(0) = 0$

$f^{(4)}(x) = e^{x^2}(12 + 48x^2 + 16x^4)$

(a) $e^{x^2} = 1 + x^2 + R_3(x)$ where $R_3(x) = \dfrac{f^{(4)}(z)}{4!}x^4 = \dfrac{e^{z^2}(3 + 12z^2 + 4z^4)}{6}x^4$ and z lies

between 0 and x.

(b) $0 \le x \le 0.1 \Rightarrow |R_3(x)| < \dfrac{e^{0.01}(3 + 0.12 + 0.0004)}{6}(0.0001) < 0.00006.$

25. $f(x) = x^{3/4}$ $\qquad\qquad\qquad\qquad$ $f(16) = 8$

$f'(x) = \dfrac{3}{4}x^{-1/4}$ $\qquad\qquad\qquad$ $f'(16) = \dfrac{3}{8}$

$f''(x) = \dfrac{-3}{16}x^{-5/4}$ $\qquad\qquad$ $f''(16) = \dfrac{-3}{512}$

$f^{(3)}(x) = \dfrac{15}{64}x^{-9/4}$ $\qquad\qquad$ $f^{(3)}(16) = \dfrac{15}{32768}$

$f^{(4)}(x) = -\dfrac{135}{256}x^{-13/4}$

(a) $x^{3/4} = 8 + \dfrac{3}{8}(x - 16) - \dfrac{3}{1024}(x - 16)^2 + \dfrac{5}{65536}(x - 16)^3 + R_3(x)$ where

$$R_3(x) = \frac{f^{(4)}(z)}{4!}(x - 16)^4 = -\frac{135(x - 16)^4}{256 \cdot 4! z^{13/4}}$$ and z lies between 16 and x.

(b) $|x - 16| \le 1$ and $z > 15 \Rightarrow |R_3(x)| < \dfrac{135}{256 \cdot 24 \cdot 15^{13/4}} < 0.0000034.$

27. $e^x = 1 + x + \dfrac{x^2}{2!} + \cdots + \dfrac{x^n}{n!} + R_n(x)$ where $R_n(x) = \dfrac{e^z}{(n + 1)!}x^{n+1}$ and z lies between 0 and x.

So taking $x = 0.1$, we have $z < 0.1 \Rightarrow e^z < 3^{0.1} < 2 \Rightarrow |R_3(0.1)| < \dfrac{2}{4!}(0.0001) < 0.00001$

and $e^{0.1} \approx 1 + 0.1 + \dfrac{(0.1)^2}{2} + \dfrac{(0.1)^3}{6} \approx 1.10517.$

29. $f(x) = (1 + x)^{1/5} = 1 + \dfrac{1}{5}x + \dfrac{\left(\frac{1}{5}\right)\left(-\frac{4}{5}\right)}{2!}x^2 + R_2(x) = 1 + \dfrac{1}{5}x - \dfrac{2}{25}x^2 + R_2(x)$ where

$|R_2(x)| = \frac{1 \cdot 4 \cdot 9}{5^3 \cdot 3!}|1 + z|^{-14/5}|x|^3$ and z lies between 0 and x, so $|R_2(0.1)|$

$< \frac{1 \cdot 4 \cdot 9}{5^3 \cdot 3!}(0.001) = 0.000048 < 0.00005.$ Thus $(1.1)^{1/5} \approx 1 + \frac{0.1}{5} - \frac{2}{25}(0.1)^2 = 1.0192$

(correct to four decimals).

31. If $f(x) = \ln(1 + x)$ then $f^{(n)}(x) = (-1)^n(n-1)!(1 + x)^{-n}$ so

$\ln(1 + x) = x - \frac{x^2}{2} + \frac{x^3}{3} - \frac{x^4}{4} + \frac{x^5}{5} + R_5(x)$ where $|R_5(x)| = \frac{|x|^6}{6|1 + z|^6}$ and z is between 0 and

x. So $|R_5(0.4)| < \frac{(0.4)^6}{6} \approx 0.00068 < 0.001$ and

$\ln(1.4) \approx 0.4 - \frac{(0.4)^2}{2} + \frac{(0.4)^3}{3} - \frac{(0.4)^4}{4} + \frac{(0.4)^5}{5} \approx 0.336.$

33. $\sin x = x - \frac{x^3}{3!} + \frac{x^5}{5!} + R_5(x)$, where $R_5(x) = \frac{-\sin z}{6!}x^6$ and z lies between 0 and x. So

$|R_5(0.5)| < \frac{1 \cdot (0.5)^6}{720} \approx 0.00002 < 0.0001$, and $\sin(0.5) \approx 0.5 - \frac{(0.5)^3}{3!} + \frac{(0.5)^5}{5!} \approx 0.4794.$

35. $\cos x = 1 - \frac{x^2}{2!} + \frac{x^4}{4!} + R_4(x)$ where $R_4(x) = \frac{\cos z}{5!}x^5$ and z is between 0 and x, so

$|R_4(10°)| = |R_4(\frac{\pi}{18})| < \frac{(\pi/18)^5}{5!} < 0.0000014$ so $\cos(10°) \approx 1 - \frac{(\pi/18)^2}{2!} + \frac{(\pi/18)^4}{4!} \approx 0.98481.$

37. Using the information of Exercise 3, above, we see that

$\sin x = \frac{1}{2} + \frac{\sqrt{3}}{2}(x - \frac{\pi}{6}) - \frac{1}{4}(x - \frac{\pi}{6})^2 - \frac{\sqrt{3}}{12}(x - \frac{\pi}{6})^3 + R_3(x)$, where $R_3(x) = \frac{\sin z}{4!}(x - \frac{\pi}{6})^4$ and z

lies between $\frac{\pi}{6}$ and x. Now 35° is $\frac{\pi}{6} + \frac{\pi}{36}$ in radian measure, so

$|R_3(\frac{\pi}{36})| < \frac{(\pi/36)^4}{4!} < 0.000003$ and $\sin 35° \approx \frac{1}{2} + \frac{\sqrt{3}}{2}(\frac{\pi}{36}) - \frac{1}{4}(\frac{\pi}{36})^2 - \frac{\sqrt{3}}{12}(\frac{\pi}{36})^3 \approx 0.57358.$

39. $\sin x = x - \frac{x^3}{6} + R_4(x)$ where $R_4(x) = \frac{\sin z}{5!}x^5$ and z is between 0 and x. So

$|R_4(x)| < \frac{|x|^5}{120} < 0.01 \Rightarrow |x|^5 < 1.2 \Rightarrow |x| < (1.2)^{1/5} \approx 1.037.$ This is certainly true if $|x| \leq 1.$

41. We will use Theorem 10.56. $R_n(x) = \frac{f^{(n+1)}(z)}{(n+1)!}x^{n+1}$ with $f^{(n+1)}(z) = \pm\sin z$ or $\pm\cos z$, and

with z between 0 and x. In every case $|f^{(n+1)}(z)| \leq 1.$ Thus $|R_n(x)| \leq \frac{|x|^{n+1}}{(n+1)!} \to 0$ as $n \to \infty$

by (10.57), so $\sin x$ is equal to its Taylor series by Theorem 10.56.

43. $R_n(x) = \frac{f^{(n+1)}(z)}{(n+1)!}x^{n+1}$ where $f^{(n+1)}(z) = \sinh z$ or $\cosh z.$ Since z lies between 0 and x,

$|\sinh z| < |\sinh x|$ and $|\cosh z| < |\cosh x|$, so in the case $f^{(n+1)}(z) = \sinh z$ we have

$|R_n(x)| < |\sinh z| \dfrac{|x|^{n+1}}{(n+1)!} \to 0$ as $n \to \infty$ by (10.57) (and similarly if $f^{(n+1)}(z) = \cosh z$). So by Theorem 10.56, $\sinh x$ is equal to its Taylor series.

45. From Section 10.10 $(1+x)^n = 1 + nx + \dfrac{n(n-1)}{2!}x^2 + \cdots > 1 + nx$ for all $x > 0$ and $n > 1$.

47. $\displaystyle\lim_{x \to 0} \frac{\sin x - x + \frac{1}{6}x^3}{x^5} = \lim_{x \to 0} \frac{(x - \frac{1}{6}x^3 + \frac{1}{5!}x^5 - \frac{1}{7!}x^7 + \cdots) - x + \frac{1}{6}x^3}{x^5}$

$= \displaystyle\lim_{x \to 0} \frac{\frac{1}{5!}x^5 - \frac{1}{7!}x^7 + \cdots}{x^5} = \lim_{x \to 0} (\tfrac{1}{5!} - \tfrac{1}{7!}x^2 + \tfrac{1}{9!}x^4 - \cdots) = \tfrac{1}{5!} = \tfrac{1}{120}$

since power series are continuous functions.

49. $y = T_1(x) = f(c) + f'(c)(x-c) \Leftrightarrow y - f(c) = f'(c)(x-c)$ is the equation of the line passing through $(c, f(c))$ with slope $f'(c)$, and this describes the tangent line.

51. Using Taylor's Formula (10.54) with $n = 1$, $c = x_n$, $x = r$, we get

$f(r) = f(x_n) + f'(x_n)(r - x_n) + R_2(x)$, where $R_2(x) = \frac{1}{2}f''(z)(r - x_n)^2$ and z lies between x_n and r. But r is a root, so $f(r) = 0$ and Taylor's Formula becomes

$$0 = f(x_n) + f'(x_n)(r - x_n) + \tfrac{1}{2}f''(z)(r - x_n)^2$$

Taking the first two terms to the left side and dividing by $f'(x_n)$, we have

$$x_n - r - \frac{f(x_n)}{f'(x_n)} = \frac{1}{2}\frac{f''(z)}{f'(x_n)}|x_n - r|^2$$

So the formula for Newton's method $\Rightarrow$

$$|x_{n+1} - r| = \left| x_n - \frac{f(x_n)}{f'(x_n)} - r \right| = \frac{1}{2}\frac{|f''(z)|}{|f'(x_n)|}|x_n - r|^2 \le \frac{M}{2K}|x_n - r|^2$$

since $|f''(z)| \le M$ and $|f'(x_n)| \ge K$.

53. $q!(e - s_q) = q!\left(\dfrac{p}{q} - 1 - \dfrac{1}{1!} - \dfrac{1}{2!} - \cdots - \dfrac{1}{q!}\right) = p(q-1)! - q! - q! - \dfrac{q!}{2!} - \cdots - 1$, which is

clearly an integer, and $q!(e - s_q) = q!\left[\dfrac{e^z}{(q+1)!}\right] = \dfrac{e^z}{q+1}$. We have

$0 < \dfrac{e^z}{q+1} < \dfrac{e}{q+1} < \dfrac{e}{3} < 1$ since $0 < z < 1$ and $q > 2$, and so $0 < q!(e - s_q) < 1$,

which is a contradiction since we have already shown $q!(e - s_q)$ must be an integer. So e cannot be rational.

REVIEW EXERCISES FOR CHAPTER 10

1. False. See the warning in Note 2 after Theorem 10.17.

3. False. For example, take $a_n = (-1)^n/(n6^n)$.

5. False, since $\lim\limits_{n \to \infty} \left| \dfrac{a_{n+1}}{a_n} \right| = \lim\limits_{n \to \infty} \left| \dfrac{n^3}{(n+1)^3} \right| = \lim\limits_{n \to \infty} \dfrac{1}{(1+1/n)^3} = 1.$

7. False. See the remarks after Example 3 in Section 10.4.

9. False. A power series has the form $a_0 + a_1 x + a_2 x^2 + a_3 x^3 + \cdots$.

11. True. See Example 8 in Section 10.1.

13. True. By Theorem 10.44 the coefficient of x^3 is $\dfrac{f'''(0)}{3!} = \frac{1}{3} \Rightarrow f'''(0) = 2.$

[Or use Theorem 10.39 to differentiate f three times.]

15. $\lim\limits_{n \to \infty} \dfrac{n}{2n+5} = \lim\limits_{n \to \infty} \dfrac{1}{2+5/n} = \frac{1}{2}$ and the sequence is convergent.

17. $\{2n+5\}$ is divergent since $2n+5 \to \infty$ as $n \to \infty.$

19. $\{\sin n\}$ is divergent since $\lim\limits_{n \to \infty} \sin n$ does not exist.

21. $\left\{ \left(1 + \dfrac{3}{n}\right)^{4n} \right\}$ is convergent. Let $y = \left(1 + \dfrac{3}{x}\right)^{4x}.$ Then $\lim\limits_{x \to \infty} \ln y$

$= \lim\limits_{x \to \infty} 4x \ln(1+3/x) = \lim\limits_{x \to \infty} \dfrac{\ln(1+3/x)}{1/(4x)} \overset{\text{H}}{=} \lim\limits_{x \to \infty} \dfrac{\frac{1}{1+3/x} \cdot (-3/x^2)}{-1/(4x^2)} = \lim\limits_{x \to \infty} \dfrac{12}{1+3/x}$

$= 12,$ so $\lim\limits_{x \to \infty} y = \lim\limits_{n \to \infty} \left(1 + \dfrac{3}{n}\right)^{4n} = e^{12}.$

23. Use the Limit Comparison Test with $a_n = \dfrac{n^2}{n^3+1}$ and $b_n = \frac{1}{n}.$

$\lim\limits_{n \to \infty} \dfrac{a_n}{b_n} = \lim\limits_{n \to \infty} \dfrac{n^2/(n^3+1)}{1/n} = \lim\limits_{n \to \infty} \dfrac{1}{1+1/n^3} = 1.$ Since $\sum\limits_{n=1}^{\infty} \frac{1}{n}$ (the harmonic series)

diverges, $\sum\limits_{n=1}^{\infty} \dfrac{n^2}{n^3+1}$ diverges also.

25. An alternating series with $a_n = \dfrac{1}{n^{1/4}},$ $a_n > 0$ for all n, and $a_n > a_{n+1}.$

$\lim\limits_{n \to \infty} a_n = \lim\limits_{n \to \infty} \dfrac{1}{n^{1/4}} = 0$ so the series converges by the Alternating Series Test.

27. $\lim\limits_{n \to \infty} \sqrt[n]{|a_n|} = \lim\limits_{n \to \infty} \dfrac{n}{3n+1} = \frac{1}{3} < 1,$ so series converges by the Root Test.

29. $\dfrac{|\sin n|}{1+n^2} \leq \dfrac{1}{1+n^2} < \dfrac{1}{n^2}$ and since $\sum\limits_{n=1}^{\infty} \dfrac{1}{n^2}$ converges (p–series with $p = 2 > 1$), so does

$\sum\limits_{n=1}^{\infty} \dfrac{|\sin n|}{1+n^2}$ by the Comparison Test.

31. $\lim\limits_{n \to \infty} \left| \dfrac{a_{n+1}}{a_n} \right| = \lim\limits_{n \to \infty} \dfrac{1 \cdot 3 \cdot 5 \cdots (2n-1)(2n+1)}{5^{n+1}(n+1)!} \cdot \dfrac{5^n n!}{1 \cdot 3 \cdot 5 \cdots (2n-1)} = \lim\limits_{n \to \infty} \dfrac{2n+1}{5(n+1)}$

376

$= \frac{2}{5} < 1$, so the series converges by the Ratio Test.

33. $\lim\limits_{n \to \infty} \left| \frac{a_{n+1}}{a_n} \right| = \lim\limits_{n \to \infty} \frac{4^{n+1}}{(n+1)3^{n+1}} \cdot \frac{n3^n}{4^n} = \frac{4}{3} \lim\limits_{n \to \infty} \frac{n}{n+1} = \frac{4}{3} > 1$ so the series diverges by the

   Ratio Test.

35. Consider the series of absolute values: $\sum\limits_{n=1}^{\infty} n^{-1/3}$ is a p-series with $p = \frac{1}{3} < 1$ and is therefore

   divergent. But if we apply the Alternating Series Test we see that $a_{n+1} < a_n$ and

   $\lim\limits_{n \to \infty} n^{-1/3} = 0$. Therefore $\sum\limits_{n=1}^{\infty} (-1)^{n-1} n^{-1/3}$ is conditionally convergent.

37. $\left| \frac{a_{n+1}}{a_n} \right| = \left| \frac{(-1)^{n+1}(n+2)3^{n+1}}{2^{2n+3}} \cdot \frac{2^{2n+1}}{(-1)^n(n+1)3^n} \right| = \frac{n+2}{n+1} \cdot \frac{3}{4} = \frac{1+(2/n)}{1+(1/n)} \cdot \frac{3}{4} \to \frac{3}{4} < 1$ as $n \to \infty$

   so by the Ratio Test $\sum\limits_{n=1}^{\infty} (-1)^n (n+1)3^n/2^{2n+1}$ is absolutely convergent.

39. Convergent geometric series. $\sum\limits_{n=1}^{\infty} \frac{2^{2n+1}}{5^n} = 2 \sum\limits_{n=1}^{\infty} \frac{4^n}{5^n} = 2 \left(\frac{4/5}{1-4/5} \right) = 8.$

41. $\sum\limits_{n=1}^{\infty} \left[\tan^{-1}(n+1) - \tan^{-1} n \right]$

   $= \lim\limits_{n \to \infty} [(\tan^{-1} 2 - \tan^{-1} 1) + (\tan^{-1} 3 - \tan^{-1} 2) + \cdots + (\tan^{-1}(n+1) - \tan^{-1} n)]$

   $= \lim\limits_{n \to \infty} [\tan^{-1}(n+1) - \tan^{-1} 1] = \frac{\pi}{2} - \frac{\pi}{4} = \frac{\pi}{4}$

43. $1.2 + 0.0\overline{345} = \frac{12}{10} + \frac{345/10000}{1 - 1/1000} = \frac{12}{10} + \frac{345}{9990} = \frac{4111}{3330}.$

45. $\sum\limits_{n=1}^{\infty} \frac{(-1)^{n+1}}{n^5} = 1 - \frac{1}{32} + \frac{1}{243} - \frac{1}{1024} + \frac{1}{3125} - \frac{1}{7776} + \frac{1}{16807} - \frac{1}{32768} + \cdots$ Since

   $\frac{1}{32768} < 0.000031, \quad \sum\limits_{n=1}^{\infty} \frac{(-1)^{n+1}}{n^5} \approx \sum\limits_{n=1}^{7} \frac{(-1)^{n+1}}{n^5} \approx 0.9721.$

47. Use the Limit Comparison Test. $\lim\limits_{n \to \infty} \left| \frac{\left(\frac{n+1}{n} \right) a_n}{a_n} \right| = \lim\limits_{n \to \infty} \frac{n+1}{n}$

   $= \lim\limits_{n \to \infty} \left(1 + \frac{1}{n} \right) = 1 > 0.$ Since $\sum |a_n|$ is convergent, so is $\sum \left| \left(\frac{n+1}{n} \right) a_n \right|$ by the Limit

   Comparison Test.

49. $\lim\limits_{n \to \infty} \left| \frac{u_{n+1}}{u_n} \right| = \lim\limits_{n \to \infty} \left| \frac{x^{n+1}}{3^{n+1}(n+1)^3} \cdot \frac{3^n n^3}{x^n} \right| = \frac{|x|}{3} \lim\limits_{n \to \infty} \left(\frac{n}{n+1} \right)^3 = \frac{|x|}{3} < 1$ for convergence

   (Ratio Test) $\Rightarrow |x| < 3$ and the radius of convergence is 3. When $x = \pm 3$, $\sum\limits_{n=1}^{\infty} |u_n| = \sum\limits_{n=1}^{\infty} \frac{1}{n^3}$

   which is a convergent p–series ($p = 3 > 1$), so the interval of convergence is $[-3, 3]$.

51. $\lim\limits_{n \to \infty} \left| \frac{u_{n+1}}{u_n} \right| = \lim\limits_{n \to \infty} \left| \frac{2^{n+1}(x-3)^{n+1}}{\sqrt{n+4}} \cdot \frac{\sqrt{n+3}}{2^n(x-3)^n} \right| = 2|x-3| \lim\limits_{n \to \infty} \sqrt{\frac{n+3}{n+4}} = 2|x-3| < 1$

$\Leftrightarrow |x - 3| < \frac{1}{2}$ so the radius of convergence is $1/2$. For $x = \frac{7}{2}$ the series becomes

$\sum_{n=0}^{\infty} \frac{1}{\sqrt{n+3}} = \sum_{n=3}^{\infty} \frac{1}{n^{1/2}}$ which diverges ($p = \frac{1}{2} < 1$), but for $x = \frac{5}{2}$ we get $\sum_{n=0}^{\infty} \frac{(-1)^n}{\sqrt{n+3}}$

which is a convergent alternating series, so the interval of convergence is $[5/2, 7/2)$.

53.

$f(x) = \sin x$ $\qquad\qquad\qquad\qquad f(\frac{\pi}{6}) = \frac{1}{2}$

$f'(x) = \cos x$ $\qquad\qquad\qquad\qquad f'(\frac{\pi}{6}) = \frac{\sqrt{3}}{2}$

$f''(x) = -\sin x$ $\qquad\qquad\qquad\qquad f''(\frac{\pi}{6}) = -\frac{1}{2}$

$f^{(3)}(x) = -\cos x$ $\qquad\qquad\qquad\qquad f^{(3)}(\frac{\pi}{6}) = -\frac{\sqrt{3}}{2}$

$f^{(4)}(x) = \sin x$ $\qquad\qquad\qquad\qquad f^{(4)}(\frac{\pi}{6}) = \frac{1}{2}$

$\cdots$ $\qquad\qquad\qquad\qquad\qquad\qquad \cdots$

$f^{(2n)}(\frac{\pi}{6}) = (-1)^n \cdot \frac{1}{2}$ and $f^{(2n+1)}(\frac{\pi}{6}) = (-1)^n \cdot \frac{\sqrt{3}}{2}$

$\sin x = \sum_{n=0}^{\infty} \frac{f^{(n)}(\pi/6)}{n!}(x - \frac{\pi}{6})^n = \sum_{n=0}^{\infty} \frac{(-1)^n}{2(2n)!}(x - \frac{\pi}{6})^{2n} + \sum_{n=0}^{\infty} \frac{(-1)^n \sqrt{3}}{2(2n+1)!}(x - \frac{\pi}{6})^{2n+1}$

55. $\frac{1}{1+x} = \frac{1}{1-(-x)} = \sum_{n=0}^{\infty} (-1)^n x^n$ for $|x| < 1 \Rightarrow \frac{x^2}{1+x} = \sum_{n=0}^{\infty} (-1)^n x^{n+2}$ with $R = 1$.

57. $\frac{1}{1-x} = \sum_{n=0}^{\infty} x^n$ for $|x| < 1 \Rightarrow \ln(1-x) = -\int \frac{dx}{1-x} = -\int \sum_{n=0}^{\infty} x^n \, dx = C - \sum_{n=0}^{\infty} \frac{x^{n+1}}{n+1}$.

$\ln(1-0) = C - 0 \Rightarrow C = 0 \Rightarrow \ln(1-x) = -\sum_{n=0}^{\infty} \frac{x^{n+1}}{n+1} = \sum_{n=1}^{\infty} \frac{-x^n}{n}$ with $R = 1$.

59. $\sin x = \sum_{n=0}^{\infty} \frac{(-1)^n x^{2n+1}}{(2n+1)!} \Rightarrow \sin(x^4) = \sum_{n=0}^{\infty} \frac{(-1)^n (x^4)^{2n+1}}{(2n+1)!} = \sum_{n=0}^{\infty} \frac{(-1)^n x^{8n+4}}{(2n+1)!}$ for all x,

so the radius of convergence is ∞.

61. $(16-x)^{-1/4} = \frac{1}{2}\left(1 - \frac{x}{16}\right)^{-1/4} = \frac{1}{2}\left[1 + \left(-\frac{1}{4}\right)\left(-\frac{x}{16}\right) + \frac{\left(-\frac{1}{4}\right)\left(-\frac{5}{4}\right)}{2!}\left(-\frac{x}{16}\right)^2 + \cdots\right]$

$= \sum_{n=0}^{\infty} \frac{1 \cdot 5 \cdot 9 \cdots (4n-3)}{2 \cdot 4^n \cdot n! \cdot 16^n} x^n = \sum_{n=0}^{\infty} \frac{1 \cdot 5 \cdot 9 \cdots (4n-3)}{2^{6n+1} n!} x^n$ for $\left|-\frac{x}{16}\right| < 1 \Rightarrow R = 16$.

63. $e^x = \sum_{n=0}^{\infty} \frac{x^n}{n!}$ so $\frac{e^x}{x} = \frac{1}{x} + \sum_{n=1}^{\infty} \frac{x^{n-1}}{n!}$ and $\int \frac{e^x}{x} dx = C + \ln|x| + \sum_{n=1}^{\infty} \frac{x^n}{n \cdot n!}$

65. $e^{-1/4} = \sum_{k=0}^{n} \frac{(-1/4)^k}{k!} + R_n(-1/4)$ where $R_n(-1/4) = \frac{e^z}{(n+1)!}\left(-\frac{1}{4}\right)^{n+1}$ and

$-\frac{1}{4} < z < 0 \Rightarrow |R_n(-1/4)| < \frac{e^0}{(n+1)!}\left(\frac{1}{4}\right)^{n+1} = \frac{1}{(n+1)!}\left(\frac{1}{4}\right)^{n+1} \Rightarrow |R_4(-1/4)| < \frac{1}{5! \cdot 4^5}$

$< 0.0000082 < 0.0001$ so $e^{-1/4} \approx 1 - \frac{1}{4} + \frac{1}{32} - \frac{1}{384} + \frac{1}{6144} \approx 0.7788$.

67. $f(x) = x^{1/2}$ $\qquad\qquad\qquad\qquad$ $f(1) = 1$

$\quad$ $f'(x) = \frac{1}{2}x^{-1/2}$ $\qquad\qquad\qquad$ $f'(1) = \frac{1}{2}$

$\quad$ $f''(x) = -\frac{1}{4}x^{-3/2}$ $\qquad\qquad\qquad$ $f''(1) = -\frac{1}{4}$

$\quad$ $f^{(3)}(x) = \frac{3}{8}x^{-5/2}$ $\qquad\qquad\qquad$ $f^{(3)}(1) = \frac{3}{8}$

$\quad$ $f^{(4)}(x) = -\frac{15}{16}x^{-7/2}$

$\quad$ $\sqrt{x} = 1 + \frac{1}{2}(x-1) - \frac{1}{8}(x-1)^2 + \frac{1}{16}(x-1)^3 + R_3(x)$ $\;$ where

$\quad$ $R_3(x) = \dfrac{f^{(4)}(z)}{4!}(x-1)^4 = -\dfrac{5(x-1)^4}{128z^{7/2}}$ $\;$ with z between x and 1.   If $0.9 \leq x \leq 1.1$ $\;$ then

$\quad$ $0 \leq |x-1| \leq 0.1$ and $z^{7/2} > (0.9)^{7/2}$ $\;$ so $|R_3(x)| < \dfrac{5(0.1)^4}{128(0.9)^{7/2}} < 0.000006.$

69. $e^x = \displaystyle\sum_{n=0}^{\infty} \frac{x^n}{n!} \;\Rightarrow\; e^{-1/x^2} = \sum_{n=0}^{\infty} \frac{(-1/x^2)^n}{n!} = 1 - \frac{1}{x^2} + \frac{1}{2x^4} - \cdots \;\Rightarrow$

$\quad$ $x^2(1 - e^{-1/x^2}) = x^2(\frac{1}{x^2} - \frac{1}{2x^4} + \cdots) = 1 - \frac{1}{2x^2} + \cdots \to 1$ $\;$ as $\;$ $x \to \infty.$

71. $e^x = \displaystyle\sum_{n=0}^{\infty} \frac{x^n}{n!} \;\Rightarrow\; e^{x^2} = \sum_{n=0}^{\infty} \frac{(x^2)^n}{n!} = \sum_{n=0}^{\infty} \frac{x^{2n}}{n!} = \sum_{k=0}^{\infty} \frac{f^{(k)}(0)}{k!}x^k \;\Rightarrow$

$\quad$ $\dfrac{f^{(2n)}(0)}{(2n)!} = \dfrac{1}{n!} \;\Rightarrow\; f^{(2n)}(0) = \dfrac{(2n)!}{n!}.$

PROBLEMS PLUS (page 627)

1. It would be far too much work to compute 15 derivatives of f. The key idea is to remember that $f^{(n)}(0)$ occurs in the coefficient of x^n in the Maclaurin series of f. We start with the Maclaurin series for sin: $\sin x = x - \frac{x^3}{3!} + \frac{x^5}{5!} - \cdots$. Then $\sin(x^3) = x^3 - \frac{x^9}{3!} + \frac{x^{15}}{5!} - \cdots$ and so the coefficient of x^{15} is $\frac{f^{(15)}(0)}{15!} = \frac{1}{5!}$. Therefore,

$$f^{(15)}(0) = \frac{15!}{5!} = 6 \cdot 7 \cdot 8 \cdot 9 \cdot 10 \cdot 11 \cdot 12 \cdot 13 \cdot 14 \cdot 15 = 10{,}897{,}286{,}400.$$

3. (a) At each stage, each side is replaced by 4 shorter sides, each of length $\frac{1}{3}$ of the side length at the preceding stage. Writing s_0 and ℓ_0 for the number of sides and the length of the side of the initial triangle, we have

$$s_0 = 3 \qquad\qquad \ell_0 = 1$$
$$s_1 = 3 \cdot 4 \qquad\qquad \ell_1 = \frac{1}{3}$$
$$s_2 = 3 \cdot 4^2 \qquad\qquad \ell_2 = \frac{1}{3^2}$$
$$s_3 = 3 \cdot 4^3 \qquad\qquad \ell_3 = \frac{1}{3^3}$$
$$\cdots \qquad\qquad\qquad \cdots$$

In general, we have $s_n = 3 \cdot 4^n$ and $\ell_n = 1/3^n$. Thus the length of the perimeter at the nth stage of construction is $p_n = s_n \ell_n = 3 \cdot 4^n \cdot (1/3^n) = 4^n/3^{n-1}$.

(b) $p_n = \frac{4^n}{3^{n-1}} = 4(\frac{4}{3})^{n-1}$. Since $\frac{4}{3} > 1$, $p_n \to \infty$ as $n \to \infty$.

(c) The area of each of the small triangles added at a given stage is $\frac{1}{9}$th of the area of the triangle added at the preceding stage. Let a be the area of the original triangle. Then the area a_n of each of the small triangles added at stage n is $a_n = a \cdot \frac{1}{9^n} = \frac{a}{9^n}$. Since a small triangle is added to each side at every stage, it follows that the total area A_n added to the figure at the nth stage is $A_n = s_{n-1} \cdot a_n = 3 \cdot 4^{n-1} \cdot \frac{a}{9^n} = a \cdot \frac{4^{n-1}}{3^{2n-1}}$. Then the total area enclosed by the snowflake curve is $A = a + A_1 + A_2 + A_3 + \cdots$

$$= a + a \cdot \frac{1}{3} + a \cdot \frac{4}{3^3} + a \cdot \frac{4^2}{3^5} + a \cdot \frac{4^3}{3^7} + \cdots.$$ After the first term, this is a geometric series with common ratio 4/9, so $A = a + \frac{a/3}{1 - (4/9)} = a + \frac{a}{3} \cdot \frac{9}{5} = \frac{8a}{5}$. But the area of the original equilateral triangle with side 1 is $a = \frac{1}{2} \cdot 1 \cdot \sin(\pi/3) = \frac{\sqrt{3}}{4}$. So the area enclosed by the snowflake curve is $\frac{8}{5} \cdot \frac{\sqrt{3}}{4} = \frac{2\sqrt{3}}{5}$.

5. (a) Using Equation 14a in Appendix B with $x = y = \theta$, we get $\tan 2\theta = \frac{2\tan\theta}{1 - \tan^2\theta}$, so

$$\cot 2\theta = \frac{1 - \tan^2\theta}{2\tan\theta} \;\Rightarrow\; 2\cot 2\theta = \frac{1 - \tan^2\theta}{\tan\theta} = \cot\theta - \tan\theta.$$ Replacing θ by $\frac{1}{2}x$, we get

$$2\cot x = \cot\tfrac{1}{2}x - \tan\tfrac{1}{2}x, \quad \text{or} \quad \tan\tfrac{1}{2}x = \cot\tfrac{1}{2}x - 2\cot x.$$

(b) From part (a) we have $\tan\frac{x}{2^n} = \cot\frac{x}{2^n} - 2\cot\frac{x}{2^{n-1}}$. So the partial sum of the given series

is $s_n = \frac{1}{2}\tan\frac{x}{2} + \frac{1}{4}\tan\frac{x}{4} + \frac{1}{8}\tan\frac{x}{8} + \cdots + \frac{1}{2^n}\tan\frac{x}{2^n}$

$= (\frac{1}{2}\cot\frac{x}{2} - \cot x) + (\frac{1}{4}\cot\frac{x}{4} - \frac{1}{2}\cot\frac{x}{2}) + (\frac{1}{8}\cot\frac{x}{8} - \frac{1}{4}\cot\frac{x}{4}) + \cdots + (\frac{1}{2^n}\cot\frac{x}{2^n} - \frac{1}{2^{n-1}}\cot\frac{x}{2^{n-1}})$

$= -\cot x + \frac{1}{2^n}\cot\frac{x}{2^n}$ [telescoping series]

Now $\frac{1}{2^n}\cot\frac{x}{2^n} = \frac{\cos(x/2^n)}{2^n\sin(x/2^n)} = \frac{1}{x}\cdot\frac{\cos(x/2^n)}{\sin(x/2^n)/(x/2^n)} \to \frac{1}{x}\cdot\frac{1}{1} = \frac{1}{x}$ as n $\to \infty$ since

$\frac{x}{2^n} \to 0$ for x $\neq$ 0. Therefore, if x $\neq$ 0 and x $\neq$ nπ, we have

$$\sum_{n=1}^{\infty}\frac{1}{2^n}\tan\frac{x}{2^n} = \lim_{n\to\infty}\left(-\cot x + \frac{1}{2^n}\cot\frac{x}{2^n}\right) = -\cot x + \frac{1}{x}.$$

If x = 0, then all terms in the series are 0, so the sum is 0.

7. $a_{n+1} = \frac{1}{2}(a_n + b_n)$, $b_{n+1} = \sqrt{b_n a_{n+1}}$. So $a_1 = \cos\theta$, $b_1 = 1 \Rightarrow a_2 = \frac{1}{2}(1 + \cos\theta) = \cos^2\frac{\theta}{2}$,

$b_2 = \sqrt{b_1 a_2} = \sqrt{\cos^2(\theta/2)} = \cos\frac{\theta}{2}$ since $-\frac{\pi}{2} \leq \theta \leq \frac{\pi}{2}$. Then $a_3 = \frac{1}{2}(\cos\frac{\theta}{2} + \cos^2\frac{\theta}{2})$

$= \cos\frac{\theta}{2}\cdot\frac{1}{2}(1 + \cos\frac{\theta}{2}) = \cos\frac{\theta}{2}\cos^2\frac{\theta}{4} \Rightarrow b_3 = \sqrt{b_2 a_3} = \sqrt{\cos\frac{\theta}{2}\cos\frac{\theta}{2}\cos^2\frac{\theta}{4}} = \cos\frac{\theta}{2}\cos\frac{\theta}{4}$

$\Rightarrow a_4 = \frac{1}{2}(\cos\frac{\theta}{2}\cos^2\frac{\theta}{4} + \cos\frac{\theta}{2}\cos\frac{\theta}{4}) = \cos\frac{\theta}{2}\cos\frac{\theta}{4}\cdot\frac{1}{2}(1 + \cos\frac{\theta}{4}) = \cos\frac{\theta}{2}\cos\frac{\theta}{4}\cos^2\frac{\theta}{8}$

$\Rightarrow b_4 = \sqrt{\cos\frac{\theta}{2}\cos\frac{\theta}{4}\cos\frac{\theta}{2}\cos\frac{\theta}{4}\cos^2\frac{\theta}{8}} = \cos\frac{\theta}{2}\cos\frac{\theta}{4}\cos\frac{\theta}{8}$. By now we see the pattern:

$b_n = \cos\frac{\theta}{2}\cos\frac{\theta}{2^2}\cos\frac{\theta}{2^3}\cdots\cdot\cos\frac{\theta}{2^{n-1}}$ and $a_n = b_n\cos\frac{\theta}{2^{n-1}}$. (This could be proved by

mathematical induction.) Note that $\sin\theta = 2\cos\frac{\theta}{2}\sin\frac{\theta}{2} = 2\cos\frac{\theta}{2}(2\cos\frac{\theta}{4}\sin\frac{\theta}{4})$

$= 4\cos\frac{\theta}{2}\cos\frac{\theta}{4}\sin\frac{\theta}{4} = 8\cos\frac{\theta}{2}\cos\frac{\theta}{4}\cos\frac{\theta}{8}\sin\frac{\theta}{8} = \cdots = 2^{n-1}\cos\frac{\theta}{2}\cos\frac{\theta}{4}\cdots\cos\frac{\theta}{2^{n-1}}\sin\frac{\theta}{2^{n-1}}$.

So $b_n = \cos\frac{\theta}{2}\cos\frac{\theta}{2^2}\cos\frac{\theta}{2^3}\cdots\cdot\cos\frac{\theta}{2^{n-1}} = \frac{\sin\theta}{2^{n-1}\sin(\theta/2^{n-1})}$. But $2^{n-1}\sin(\theta/2^{n-1})$

$= \theta\frac{\sin(\theta/2^{n-1})}{\theta/2^{n-1}} \to \theta$ as n $\to \infty$, so $b_n \to \frac{\sin\theta}{\theta}$ and $a_n = b_n\cos\frac{\theta}{2^{n-1}} \to \frac{\sin\theta}{\theta}\cdot 1 = \frac{\sin\theta}{\theta}$.

9. $u = 1 + \frac{x^3}{3!} + \frac{x^6}{6!} + \frac{x^9}{9!} + \cdots$, $v = x + \frac{x^4}{4!} + \frac{x^7}{7!} + \frac{x^{10}}{10!} + \cdots$, $w = \frac{x^2}{2!} + \frac{x^5}{5!} + \frac{x^8}{8!} + \cdots$.

The key idea is to differentiate: $\frac{du}{dx} = \frac{3x^2}{3!} + \frac{6x^5}{6!} + \frac{9x^8}{9!} + \cdots = \frac{x^2}{2!} + \frac{x^5}{5!} + \frac{x^8}{8!} + \cdots = w$.

Similarly, $\frac{dv}{dx} = 1 + \frac{x^3}{3!} + \frac{x^6}{6!} + \frac{x^9}{9!} + \cdots = u$, and $\frac{dw}{dx} = x + \frac{x^4}{4!} + \frac{x^7}{7!} + \frac{x^{10}}{10!} + \cdots = v$.

So $u' = w$, $v' = u$, and $w' = v$. Now differentiate the left hand side of the desired equation:

$\frac{d}{dx}(u^3 + v^3 + w^3 - 3uvw) = 3u^2u' + 3v^3v' + 3w^3w' - 3(u'vw + uv'w + uvw')$

$= 3u^2w + 3v^2u + 3w^2v - 3(vw^2 + u^2w + uv^2) = 0 \Rightarrow u^3 + v^3 + w^3 - 3uvw = C.$

To find the value of the constant C, we put $x = 1$ in the equation and get

$$1^3 + 0 + 0 - 3(1 \cdot 0 \cdot 0) = C \implies C = 1, \text{ so } u^3 + v^3 + w^3 - 3uvw = 1.$$

11. We start with the geometric series $\sum_{n=0}^{\infty} x^n = \frac{1}{1-x}$, $|x| < 1$, and differentiate:

$$\sum_{n=1}^{\infty} nx^{n-1} = \frac{d}{dx}\left[\sum_{n=0}^{\infty} x^n\right] = \frac{d}{dx}\left[\frac{1}{1-x}\right] = \frac{1}{(1-x)^2} \text{ for } |x| < 1 \implies$$

$$\sum_{n=1}^{\infty} nx^n = x \sum_{n=1}^{\infty} nx^{n-1} = \frac{x}{(1-x)^2} \text{ for } |x| < 1. \text{ Differentiate again:}$$

$$\sum_{n=1}^{\infty} n^2 x^{n-1} = \frac{d}{dx} \frac{x}{(1-x)^2} = \frac{(1-x)^2 - x \cdot 2(1-x)(-1)}{(1-x)^4} = \frac{x+1}{(1-x)^3} \implies \sum_{n=1}^{\infty} n^2 x^n = \frac{x^2+x}{(1-x)^3}$$

$$\implies \sum_{n=1}^{\infty} n^3 x^{n-1} = \frac{d}{dx} \frac{x^2+x}{(1-x)^3} = \frac{(1-x)^3(2x+1) - (x^2+x)3(1-x)^2(-1)}{(1-x)^6} = \frac{x^2+4x+1}{(1-x)^4}$$

$$\implies \sum_{n=1}^{\infty} n^3 x^n = \frac{x^3 + 4x^2 + x}{(1-x)^4}, \quad |x| < 1.$$

The radius of convergence is 1 because that is the radius of convergence for the geometric series we started with. If $x = \pm 1$, the series is $\sum_{n=1}^{\infty} n^3 (\pm 1)^n$, which diverges by the Test For Divergence, so the interval of convergence is $(-1, 1)$.

13. (a) Let $a = \arctan x$ and $b = \arctan y$. Then, by Equation 14(b) in Appendix B,

$$\tan(a - b) = \frac{\tan a - \tan b}{1 + \tan a \tan b} = \frac{\tan(\arctan x) - \tan(\arctan y)}{1 + \tan(\arctan x)\tan(\arctan y)} \implies \tan(a-b) = \frac{x-y}{1+xy}$$

$$\implies \arctan x - \arctan y = a - b = \arctan \frac{x-y}{1+xy} \text{ since } -\frac{\pi}{2} < \arctan x - \arctan y < \frac{\pi}{2}.$$

(b) From part (a) we have

$$\arctan \frac{120}{119} - \arctan \frac{1}{239} = \arctan \frac{\frac{120}{119} - \frac{1}{239}}{1 + \frac{120}{119} \cdot \frac{1}{239}} = \arctan \frac{\frac{28,561}{28,441}}{\frac{28,561}{28,441}} = \arctan 1 = \frac{\pi}{4}.$$

(c) Replacing y by $-y$ in the formula of part (a), we get $\arctan x + \arctan y = \arctan \frac{x+y}{1-xy}$.

So $4 \arctan \frac{1}{5} = 2(\arctan \frac{1}{5} + \arctan \frac{1}{5}) = 2 \arctan \frac{\frac{1}{5} + \frac{1}{5}}{1 - \frac{1}{5} \cdot \frac{1}{5}} = 2 \arctan \frac{5}{12}$

$$= \arctan \frac{5}{12} + \arctan \frac{5}{12} = \arctan \frac{\frac{5}{12} + \frac{5}{12}}{1 - \frac{5}{12} \cdot \frac{5}{12}} = \arctan \frac{120}{119}.$$

Thus, from part (b) we have $4 \arctan \frac{1}{5} - \arctan \frac{1}{239} = \arctan \frac{120}{119} - \arctan \frac{1}{239} = \frac{\pi}{4}$.

(d) From Example 9 in Section 10.9 we have $\arctan x = x - \frac{x^3}{3} + \frac{x^5}{5} - \frac{x^7}{7} + \frac{x^9}{9} - \frac{x^{11}}{11} + \cdots$, so

$$\arctan \frac{1}{5} = \frac{1}{5} - \frac{1}{3 \cdot 5^3} + \frac{1}{5 \cdot 5^5} - \frac{1}{7 \cdot 5^7} + \frac{1}{9 \cdot 5^9} - \frac{1}{11 \cdot 5^{11}} + \cdots. \text{ This is an alternating series}$$

and the size of the terms decreases to 0, so by Theorem 10.28, the sum lies between s_5 and s_6, i.e., $0.197395560 < \arctan \frac{1}{5} < 0.197395562$.

(e) From the series in part (d) we get $\arctan\frac{1}{239} = \frac{1}{239} - \frac{1}{3\cdot 239^3} + \frac{1}{5\cdot 239^5} - \cdots$. The third

term is less than 2.6×10^{-13}, so by Theorem 10.28 we have, to 9 decimal places,

$\arctan\frac{1}{239} \approx s_2 \approx 0.004184076$. Thus $0.004184075 < \arctan\frac{1}{239} < 0.004184077$.

(f) From part (c) we have $\pi = 16\arctan\frac{1}{5} - 4\arctan\frac{1}{239}$, so from parts (d) and (e) we have

$16(0.197395560) - 4(0.004184077) < \pi < 16(0.197395562) - 4(0.004184075) \Rightarrow$

$3.141592652 < \pi < 3.141592692$. So, to 7 decimal places, $\pi \approx 3.1415927$.

15. $(a^n + b^n + c^n)^{1/n} = \{c^n[(a/c)^n + (b/c)^n + 1]\}^{1/n} = c[(a/c)^n + (b/c)^n + 1]^{1/n}$. Since $0 \le a \le c$,

we have $0 \le a/c \le 1$, so $(a/c)^n \to 0$ or 1 as $n \to \infty$. Similarly, $(b/c)^n \to 0$ or 1 as $n \to \infty$.

Thus $(a/c)^n + (b/c)^n + 1 \to d$, where $d = 1, 2$, or 3 and so $[(a/c)^n + (b/c)^n + 1]^{1/n} \to 1$.

Therefore $\lim\limits_{n\to\infty} (a^n + b^n + c^n)^{1/n} = c$.

17. As in Section 10.9 we have to integrate the function x^x by integrating a series. Writing

$x^x = (e^{\ln x})^x = e^{x\ln x}$ and using the Maclaurin series for e^x, we have

$$x^x = e^{x\ln x} = \sum_{n=0}^{\infty} \frac{(x\ln x)^n}{n!} = \sum_{n=0}^{\infty} \frac{x^n(\ln x)^n}{n!}$$

As with power series, it turns out that we can integrate this series term-by-term:

$$\int_0^1 x^x\, dx = \sum_{n=0}^{\infty} \int_0^1 \frac{x^n(\ln x)^n}{n!}\, dx = \sum_{n=0}^{\infty} \frac{1}{n!}\int_0^1 x^n(\ln x)^n\, dx$$

We integrate by parts with $u = (\ln x)^n$, $dv = x^n dx$, so $du = \frac{n(\ln x)^{n-1}}{x}\, dx$ and $v = \frac{x^{n+1}}{n+1}$:

$$\int_0^1 x^n(\ln x)^n\, dx = \lim_{t\to 0^+}\int_t^1 x^n(\ln x)^n\, dx$$

$$= \lim_{t\to 0^+}\left[\frac{x^{n+1}}{n+1}(\ln x)^n\right]_t^1 - \lim_{t\to 0^+}\int_t^1 \frac{n}{n+1}x^n(\ln x)^{n-1}\, dx = 0 - \frac{n}{n+1}\int_0^1 x^n(\ln x)^{n-1}\, dx$$

(where l'Hospital's Rule was used to help evaluate the first limit). Further integration by parts

gives $\int_0^1 x^n(\ln x)^k\, dx = -\frac{k}{n+1}\int_0^1 x^n(\ln x)^{k-1}\, dx$ and, combining these steps, we get

$$\int_0^1 x^n(\ln x)^n\, dx = \frac{(-1)^n n!}{(n+1)^n}\int_0^1 x^n\, dx = \frac{(-1)^n n!}{(n+1)^{n+1}} \Rightarrow$$

$$\int_0^1 x^x\, dx = \sum_{n=0}^{\infty}\frac{1}{n!}\int_0^1 x^n(\ln x)^n\, dx = \sum_{n=0}^{\infty}\frac{1}{n!}\frac{(-1)^n n!}{(n+1)^{n+1}}$$

$$= \sum_{n=0}^{\infty}\frac{(-1)^n}{(n+1)^{n+1}} = \sum_{n=1}^{\infty}\frac{(-1)^{n-1}}{n^n}.$$

19. If L is the length of a side of the equilateral triangle, then the area is $A = \frac{1}{2}\cdot L\cdot \frac{\sqrt{3}}{2}L = \frac{\sqrt{3}}{4}L^2$

and so $L^2 = (4/\sqrt{3})A$. Let r be the radius of one of the circles when there are n rows of circles.

The figure shows that $L = \sqrt{3}r + r + (n-2)(2r) + r + \sqrt{3}r = r(2n - 2 + 2\sqrt{3})$,

so $r = \dfrac{L}{2(n + \sqrt{3} - 1)}$.

The number of circles is $1 + 2 + \cdots + n = \dfrac{n(n+1)}{2}$

and so the total area of the circles is

$$A_n = \frac{n(n+1)}{2}\pi r^2 = \frac{n(n+1)}{2}\pi \frac{L^2}{4(n+\sqrt{3}-1)^2}$$

$$= \frac{n(n+1)}{2}\pi \frac{(4/\sqrt{3})A}{4(n+\sqrt{3}-1)^2} = \frac{n(n+1)}{(n+\sqrt{3}-1)^2}\frac{\pi A}{2\sqrt{3}}$$

$$\Rightarrow \frac{A_n}{A} = \frac{n(n+1)}{(n+\sqrt{3}-1)^2}\frac{\pi}{2\sqrt{3}} = \frac{1+\frac{1}{n}}{\left(1+\frac{\sqrt{3}-1}{n}\right)^2}\frac{\pi}{2\sqrt{3}} \to \frac{\pi}{2\sqrt{3}} \quad \text{as } n \to \infty.$$

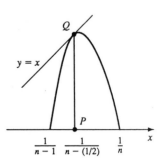

21. (a) f is clearly continuous when $x \neq 0$ since x, $\sin x$, and π/x are continuous when $x \neq 0$.

Also $|\sin(\pi/x)| \leq 1 \Rightarrow |x\sin(\pi/x)| \leq |x|$ and $\lim\limits_{x \to 0}|x| = 0$, so by the Squeeze Theorem

we have $\lim\limits_{x \to 0}|x\sin(\pi/x)| = 0$ and so $\lim\limits_{x \to 0} f(x) = \lim\limits_{x \to 0} x\sin(\pi/x) = 0 = f(0)$. This

shows that f is continuous at 0, so it is continuous on $(-1, 1)$.

(b) Note that $f(x) = 0$ when $x = 0$ and when

$\dfrac{\pi}{x} = n\pi \Rightarrow x = \dfrac{1}{n}$, n an integer. Since

$-1 \leq \sin(\pi/x) \leq 1$, the graph of f lies between

the lines $y = x$ and $y = -x$ and touches these

lines when $\dfrac{\pi}{x} = \dfrac{\pi}{2} + n\pi \Rightarrow x = \dfrac{1}{n+\frac{1}{2}}$.

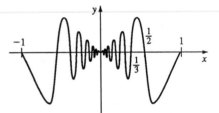

(c) The enlargement of the portion of the graph

between $x = \dfrac{1}{n}$ and $x = \dfrac{1}{n-1}$ (the case where n

is odd is illustrated) shows that the arc length

from $x = \dfrac{1}{n}$ to $x = \dfrac{1}{n-1}$ is greater than

$|PQ| = \dfrac{1}{n-\frac{1}{2}} = \dfrac{2}{2n-1}$. Thus the total length

of the graph is greater than $2\displaystyle\sum_{n=1}^{\infty}\dfrac{2}{2n-1}$.

This is a divergent series (by comparison with

the harmonic series), so the graph has infinite length.

EXERCISES A

1. $(-6ab)(0.5ac) = (-6)(0.5)(a \cdot abc) = -3a^2bc.$

3. $2x(x-5) = 2x \cdot x - 2x \cdot 5 = 2x^2 - 10x.$

5. $-2(4-3a) = -2 \cdot 4 + 2 \cdot 3a = -8 + 6a.$

7. $4(x^2 - x + 2) - 5(x^2 - 2x + 1) = 4x^2 - 4x + 8 - 5x^2 - 5(-2x) - 5$
   $= 4x^2 - 5x^2 - 4x + 10x + 8 - 5 = -x^2 + 6x + 3.$

9. $(4x-1)(3x+7) = 4x(3x+7) - (3x+7) = 12x^2 + 28x - 3x - 7 = 12x^2 + 25x - 7.$

11. $(2x-1)^2 = (2x)^2 - 2(2x)(1) + 1^2 = 4x^2 - 4x + 1.$

13. $y^4(6-y)(5+y) = y^4[6(5+y) - y(5+y)] = y^4(30 + 6y - 5y - y^2) = y^4(30 + y - y^2)$
    $= 30y^4 + y^5 - y^6.$

15. $(1+2x)(x^2 - 3x + 1) = 1(x^2 - 3x + 1) + 2x(x^2 - 3x + 1) = x^2 - 3x + 1 + 2x^3 - 6x^2 + 2x$
    $= 2x^3 - 5x^2 - x + 1.$

17. $\frac{2+8x}{2} = \frac{2}{2} + \frac{8x}{2} = 1 + 4x.$

19. $\frac{1}{x+5} + \frac{2}{x-3} = \frac{(1)(x-3) + 2(x+5)}{(x+5)(x-3)} = \frac{x-3+2x+10}{(x+5)(x-3)} = \frac{3x+7}{x^2 + 2x - 15}.$

21. $u + 1 + \frac{u}{u+1} = \frac{(u+1)(u+1) + u}{u+1} = \frac{u^2 + 2u + 1 + u}{u+1} = \frac{u^2 + 3u + 1}{u+1}.$

23. $\frac{x/y}{z} = \frac{x/y}{z/1} = \frac{1}{z} \cdot \frac{x}{y} = \frac{x}{yz}.$

25. $\left(\frac{-2r}{s}\right)\left(\frac{s^2}{-6t}\right) = \frac{-2rs^2}{-6st} = \frac{rs}{3t}.$

27. $\frac{1 + \frac{1}{c-1}}{1 - \frac{1}{c-1}} = \frac{\frac{c-1+1}{c-1}}{\frac{c-1-1}{c-1}} = \frac{\frac{c}{c-1}}{\frac{c-2}{c-1}} = \frac{c}{c-2} \cdot \frac{c}{c-1} = \frac{c}{c-2}.$

29. $2x + 12x^3 = 2x \cdot 1 + 2x \cdot 6x^2 = 2x(1 + 6x^2).$

31. The two integers that add to give 7 and multiply to give 6 are 6 and 1.
    Therefore $x^2 + 7x + 6 = (x+6)(x+1).$

33. The two integers that add to give -2 and multiply to give -8 are -4 and 2.
    Therefore $x^2 - 2x - 8 = (x-4)(x+2).$

35. $9x^2 - 36 = 9(x^2 - 4) = 9(x-2)(x+2)$ (*equation* (3) *with* $a = x$, $b = 2$).

37. $6x^2 - 5x - 6 = (3x+2)(2x-3).$

39. $t^3 + 1 = (t+1)(t^2 - t + 1)$ (*using equation* (5) *with* $a = t$, $b = 1$).

41. $4t^2 - 12t + 9 = (2t - 3)^2$ (*equation* (2) *with* $a = 2t$, $b = 3$).

43. $x^3 + 2x^2 + x = x(x^2 + 2x + 1) = x(x + 1)^2$ (*equation* (1) *with* $a = x$, $b = 1$).

45. Let $p(x) = x^3 + 3x^2 - x - 3$, and notice that $p(1) = 0$, so by the Factor Theorem, $(x - 1)$ is a factor. Using long division (*as in Example* 8):

$$
\begin{array}{r}
x^2 + 4x + 3 \\
x - 1 \overline{\smash{)}\, x^3 + 3x^2 - x - 3} \\
\underline{x^3 - x^2} \\
4x^2 - x \\
\underline{4x^2 - 4x} \\
3x - 3 \\
\underline{3x - 3}
\end{array}
$$

Therefore $x^3 + 3x^2 - x - 3 = (x - 1)(x^2 + 4x + 3) = (x - 1)(x + 1)(x + 3)$.

47. Let $p(x) = x^3 + 5x^2 - 2x - 24$, and notice that $p(2) = 2^3 + 5(2)^2 - 2(2) - 24 = 0$, so by the factor Theorem, $(x - 2)$ is a factor. Using long division (*as in Example* 8):

$$
\begin{array}{r}
x^2 + 7x + 12 \\
x - 2 \overline{\smash{)}\, x^3 + 5x^2 - 2x - 24} \\
\underline{x^3 - 2x^2} \\
7x^2 - 2x \\
\underline{7x^2 - 14x} \\
12x - 24 \\
\underline{12x - 24}
\end{array}
$$

Therefore $x^3 + 5x^2 - 2x - 24 = (x - 2)(x^2 + 7x + 12) = (x - 2)(x + 3)(x + 4)$.

49. $\dfrac{x^2 + x - 2}{x^2 - 3x + 2} = \dfrac{(x + 2)(x - 1)}{(x - 2)(x - 1)} = \dfrac{x + 2}{x - 2}$.

51. $\dfrac{x^2 - 1}{x^2 - 9x + 8} = \dfrac{(x - 1)(x + 1)}{(x - 8)(x - 1)} = \dfrac{x + 1}{x - 8}$.

53. $\dfrac{1}{x + 3} + \dfrac{1}{x^2 - 9} = \dfrac{1}{x + 3} + \dfrac{1}{(x - 3)(x + 3)} = \dfrac{1(x - 3) + 1}{(x - 3)(x + 3)} = \dfrac{x - 2}{x^2 - 9}$.

55. $x^2 + 2x + 5 = [x^2 + 2x] + 5 = [x^2 + 2x + (1)^2 - (1)^2] + 5 = (x + 1)^2 + 5 - 1$
$= (x + 1)^2 + 4$.

57. $x^2 - 5x + 10 = [x^2 - 5x] + 10 = [x^2 - 5x + (-\frac{5}{2})^2 - (-\frac{5}{2})^2] + 10 = (x - \frac{5}{2})^2 + 10 - \frac{25}{4}$
$= (x - \frac{5}{2})^2 + \frac{15}{4}$.

59. $4x^2 + 4x - 2 = 4[x^2 + x] - 2 = 4[x^2 + x + (\frac{1}{2})^2 - (\frac{1}{2})^2] - 2 = 4(x + \frac{1}{2})^2 - 2 - 4(\frac{1}{4})$
$= 4(x + \frac{1}{2})^2 - 3$.

61. $x^2 + 9x - 10 = 0 \Leftrightarrow (x + 10)(x - 1) = 0 \Leftrightarrow x + 10 = 0$ or $x - 1 = 0 \Leftrightarrow x = -10$ or $x = 1$.

63. Using the quadratic formula, $x^2 + 9x - 1 = 0 \Leftrightarrow x = \dfrac{-9 \pm \sqrt{9^2 - 4(1)(-1)}}{2(1)} = \dfrac{-9 \pm \sqrt{85}}{2}$.

65. Using the quadratic formula, $3x^2 + 5x + 1 = 0 \Leftrightarrow x = \dfrac{-5 \pm \sqrt{5^2 - 4(3)(1)}}{2(3)} = \dfrac{-5 \pm \sqrt{13}}{6}$.

67. Let $p(x) = x^3 - 2x + 1$, and notice that $p(1) = 0$, so by the **Factor Theorem**, $(x-1)$ is a factor. Using long division:

$$
\begin{array}{r}
x^2 + x - 1 \\
x - 1 \overline{\smash{\big)}\, x^3 + 0x^2 - 2x + 1} \\
\underline{x^3 - x^2} \\
x^2 - 2x \\
\underline{x^2 - x} \\
-x + 1 \\
\underline{-x + 1}
\end{array}
$$

Therefore $x^3 - 2x + 1 = (x-1)(x^2 + x - 1) = 0 \Leftrightarrow x - 1 = 0 \ \text{or} \ x^2 + x - 1 = 0$

$\Leftrightarrow x = 1$ or ($using\ the\ quadratic\ formula$) $x = \dfrac{-1 \pm \sqrt{1^2 - 4(1)(-1)}}{2(1)} = \dfrac{-1 \pm \sqrt{5}}{2}$.

69. $2x^2 + 3x + 4$ is irreducible because its discriminant is negative:

$b^2 - 4ac = 9 - 4(2)(4) = -23 < 0$.

71. $3x^2 + x - 6$ is not irreducible because its discriminant is nonnegative:

$b^2 - 4ac = 1 - 4(3)(-6) = 73 > 0$.

73. Using the Binomial Theorem with $k = 6$ we have,

$(a + b)^6 = a^6 + 6a^5b + \dfrac{6 \cdot 5}{1 \cdot 2}a^4b^2 + \dfrac{6 \cdot 5 \cdot 4}{1 \cdot 2 \cdot 3}a^3b^3 + \dfrac{6 \cdot 5 \cdot 4 \cdot 3}{1 \cdot 2 \cdot 3 \cdot 4}a^2b^4 + 6ab^5 + b^6$

$= a^6 + 6a^5b + 15a^4b^2 + 20a^3b^3 + 15a^2b^4 + 6ab^5 + b^6$.

75. Using the Binomial Theorem with $a = x^2$, $b = -1$, $k = 4$ we have,

$(x^2 - 1)^4 = [x^2 + (-1)]^4 = (x^2)^4 + 4(x^2)^3(-1) + \dfrac{4 \cdot 3}{1 \cdot 2}(x^2)^2(-1)^2$

$+ 4(x^2)(-1)^3 + (-1)^4 = x^8 - 4x^6 + 6x^4 - 4x^2 + 1$.

77. Using Equation 10, $\sqrt{32}\sqrt{2} = \sqrt{32 \cdot 2} = \sqrt{64} = 8$.

79. Using Equation 10, $\dfrac{\sqrt[4]{32x^4}}{\sqrt[4]{2}} = \dfrac{\sqrt[4]{32}\,\sqrt[4]{x^4}}{\sqrt[4]{2}} = \sqrt[4]{\dfrac{32}{2}}\,\sqrt[4]{x^4} = \sqrt[4]{16}|x| = 2|x|$.

81. Using Equation 10, $\sqrt{16a^4b^3} = \sqrt{16}\sqrt{a^4}\sqrt{b^3} = 4a^2b^{3/2} = 4a^2b\,b^{1/2} = 4a^2b\sqrt{b}$.

83. Using Laws 3 and 1 of Exponents respectively,

$3^{10} \times 9^8 = 3^{10} \times (3^2)^8 = 3^{10} \times 3^{2 \cdot 8} = 3^{10+16} = 3^{26}$.

85. Using Laws 4, 1 and 2 of Exponents respectively,

$\dfrac{x^9(2x)^4}{x^3} = \dfrac{x^9(2^4)x^4}{x^3} = \dfrac{16x^{9+4}}{x^3} = 16x^{9+4-3} = 16x^{10}$.

87. Using Law 2 of Exponents, $\dfrac{a^{-3}b^4}{a^{-5}b^5} = a^{-3-(-5)}b^{4-5} = a^2 b^{-1} = \dfrac{a^2}{b}$.

89. By definitions 3 and 4 for exponents respectively, $3^{-1/2} = \dfrac{1}{3^{1/2}} = \dfrac{1}{\sqrt{3}}$.

91. Using definition 4 for exponents, $125^{2/3} = [\sqrt[3]{125}]^2 = 5^2 = 25$.

93. $(2x^2 y^4)^{3/2} = 2^{3/2}(x^2)^{3/2}(y^4)^{3/2} = 2 \cdot 2^{1/2}[\sqrt{x^2}]^3[\sqrt{y^4}]^3 = 2\sqrt{2}|x|^3(y^2)^3 = 2\sqrt{2}|x|^3 y^6$.

95. $\sqrt[5]{y^6} = y^{6/5}$ by definition 4 for exponents.

97. $\dfrac{1}{(\sqrt{t})^5} = \dfrac{1}{(t^{1/2})^5} = \dfrac{1}{t^{5/2}} = t^{-5/2}$.

99. $\sqrt[4]{\dfrac{t^{1/2}\sqrt{st}}{s^{2/3}}} = \left(\dfrac{t^{1/2}s^{1/2}t^{1/2}}{s^{2/3}}\right)^{1/4} = \left(t^{(\frac{1}{2}+\frac{1}{2})}s^{(\frac{1}{2}-\frac{2}{3})}\right)^{1/4} = (ts^{-1/6})^{1/4}$

$= t^{1/4}s^{-\frac{1}{6}\cdot\frac{1}{4}} = \dfrac{t^{1/4}}{s^{1/24}}$.

101. $\dfrac{\sqrt{x}-3}{x-9} = \dfrac{\sqrt{x}-3}{x-9} \cdot \dfrac{\sqrt{x}+3}{\sqrt{x}+3} = \dfrac{(x-9)}{(x-9)(\sqrt{x}+3)} = \dfrac{1}{\sqrt{x}+3}$.

103. $\dfrac{x\sqrt{x}-8}{x-4} = \dfrac{x\sqrt{x}-8}{x-4} \cdot \dfrac{x\sqrt{x}+8}{x\sqrt{x}+8} = \dfrac{x^3-64}{(x-4)(x\sqrt{x}+8)} = \dfrac{(x-4)(x^2+4x+16)}{(x-4)(x\sqrt{x}+8)}$

$(equation\ 4\ with\ a = x,\ b = 4) = \dfrac{x^2+4x+16}{x\sqrt{x}+8}$.

105. $\dfrac{2}{3-\sqrt{5}} = \dfrac{2}{3-\sqrt{5}} \cdot \dfrac{3+\sqrt{5}}{3+\sqrt{5}} = \dfrac{2(3+\sqrt{5})}{9-5} = \dfrac{3+\sqrt{5}}{2}$.

107. $\sqrt{x^2+3x+4}-x = \left(\sqrt{x^2+3x+4}-x\right) \cdot \dfrac{\sqrt{x^2+3x+4}+x}{\sqrt{x^2+3x+4}+x}$

$= \dfrac{x^2+3x+4-x^2}{\sqrt{x^2+3x+4}+x} = \dfrac{3x+4}{\sqrt{x^2+3x+4}+x}$.

109. False. See Example 14(b).

111. True: $\dfrac{16+a}{16} = \dfrac{16}{16} + \dfrac{a}{16} = 1 + \dfrac{a}{16}$.

113. False.

115. False. Using Law 3 of Exponents, $(x^3)^4 = x^{3 \cdot 4} = x^{12} \neq x^7$.

117. Let S_n be the statement that $2^n > n$.

1. S_1 is true because $2^1 = 2 > 1$.

2. Assume S_k is true, that is $2^k > k$. Then $2^{k+1} = 2^k \cdot 2 > k \cdot 2 = 2k$ (since $2^k > k$).

But $k > 1 \Rightarrow k + k > k + 1 \Rightarrow 2k > k + 1$, so that $2^{k+1} > 2k > k + 1$

which shows that S_{k+1} is true.

3.   Therefore, by mathematical induction, $2^n > n$ for every positive integer n.

119.   Let S_n be the statement that $(1+x)^n \geq 1+nx$.

    1. S_1 is true because $(1+x)^1 = 1+(1)x$.

    2.   Assume S_k is true, that is, $(1+x)^k \geq 1+kx$.

        Then $(1+x)^{k+1} = (1+x)^k(1+x) \geq (1+kx)(1+x)$ $(since\ (1+z)^k \geq 1+kz)$

        $= 1+x+kx+kx^2 \geq 1+x+kx = 1+(k+1)x$ which shows that S_{k+1} is true.

    3.   Therefore, by mathematical induction, $(1+x)^n \geq 1+nx$ for every positive integer n.

121.   Let S_n be the statement that $7^n - 1$ is divisible by 6.

    1. S_1 is true because $7^1 - 1 = 6$ is divisible by 6.

    2.   Assume S_k is true, that is, $7^k - 1$ is divisible by 6; in other words $7^k - 1 = m \cdot 6$ for

        some positive integer m.   Then $7^{k+1} - 1 = 7^k \cdot 7 - 1 = (m \cdot 6 + 1) \cdot 7 - 1 = 6(m \cdot 7 + 1)$,

        which is divisible by 6, so S_{k+1} is true.

    3.   Therefore, by mathematical induction, $7^n - 1$ is divisible by 6 for every positive integer n.

123.   Let S_n be the statement that $1 + 3 + 5 + \cdots + (2n - 1) = n^2$.

    1. S_1 is true because $[2(1) - 1] = 1 = 1^2$.

    2.   Assume S_k is true, that is $1 + 3 + 5 + \cdots + (2k - 1) = k^2$.

        Then $1 + 3 + 5 + \cdots + [2(k + 1) - 1] = 1 + 3 + 5 + \cdots + (2k - 1) + (2k + 1)$

        $= k^2 + (2k + 1) = (k + 1)^2$ which shows that S_{k+1} is true.

    3.   Therefore, by mathematical induction, $1 + 3 + 5 + \cdots + (2n - 1) = n^2$ for every

        positive integer n.

125.   Let S_n be the statement that $\frac{1}{2} + \frac{1}{6} + \frac{1}{12} + \cdots + \frac{1}{n(n+1)} = \frac{n}{n+1}$.

    1. S_1 is true because $\frac{1}{1(1+1)} = \frac{1}{1+1}$.

    2.   Assume S_k is true, that is, $\frac{1}{2} + \frac{1}{6} + \frac{1}{12} + \cdots + \frac{1}{k(k+1)} = \frac{k}{k+1}$.

        Then $\frac{1}{2} + \frac{1}{6} + \frac{1}{12} + \cdots + \frac{1}{(k+1)[(k+1)+1]} = \frac{1}{2} + \frac{1}{6} + \frac{1}{12} + \cdots + \frac{1}{(k+1)(k+2)}$

        $= \frac{1}{2} + \frac{1}{6} + \frac{1}{12} + \cdots + \frac{1}{k(k+1)} + \frac{1}{(k+1)(k+2)} = \frac{k}{k+1} + \frac{1}{(k+1)(k+2)}$

        $= \frac{k(k+2)+1}{(k+1)(k+2)} = \frac{k^2 + 2k + 1}{(k+1)(k+2)} = \frac{(k+1)(k+1)}{(k+1)(k+2)} = \frac{(k+1)}{(k+1)+1}$ which shows that

        S_{k+1} is true.

    3.   Therefore, by mathematical induction, $\frac{1}{2} + \frac{1}{6} + \frac{1}{12} + \cdots + \frac{1}{n(n+1)} = \frac{n}{n+1}$ for every

        positive integer n.

EXERCISES B

1. $210° = 210\left(\frac{\pi}{180}\right) = \frac{7\pi}{6}$ rad.

3. $9° = 9\left(\frac{\pi}{180}\right) = \frac{\pi}{20}$ rad.

5. $900° = 900\left(\frac{\pi}{180}\right) = 5\pi$ rad.

7. 4π rad $= 4\pi\left(\frac{180}{\pi}\right) = 720°.$

9. $\frac{5\pi}{12}$ rad $= \frac{5\pi}{12}\left(\frac{180}{\pi}\right) = 75°.$

11. $-\frac{3\pi}{8}$ rad $= -\frac{3\pi}{8}\left(\frac{180}{\pi}\right) = -67.5°.$

13. Using formula 3, $a = r\theta = \frac{36\pi}{12} = 3\pi$ cm.

15. Using formula 3, $\theta = \frac{1}{1.5} = \frac{2}{3}$ rad $= \frac{2}{3}\left(\frac{180}{\pi}\right) = \left(\frac{120}{\pi}\right)°$

17.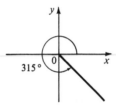

19.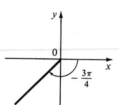

21.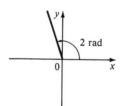

23.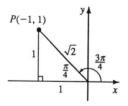

From the diagram we see that a point on the terminal line is $P(-1, 1)$. Therefore taking $x = -1, y = 1, r = \sqrt{2}$ in the definitions of the trigonometric ratios, we have

$\sin\frac{3\pi}{4} = \frac{1}{\sqrt{2}}, \cos\frac{3\pi}{4} = -\frac{1}{\sqrt{2}}, \tan\frac{3\pi}{4} = -1,$

$\csc\frac{3\pi}{4} = \sqrt{2}, \sec\frac{3\pi}{4} = -\sqrt{2}, \cot\frac{3\pi}{4} = -1.$

25.

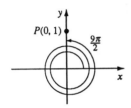

From the diagram we see that a point on the terminal line is $P(0,1)$. Therefore taking $x = 0$, $y = 1$, $r = 1$ in the definitions of the trigonometric ratios, we have $\sin\frac{9\pi}{2} = 1$, $\cos\frac{9\pi}{2} = 0$, $\tan\frac{9\pi}{2} = \frac{y}{x}$ is undefined since $x = 0$, $\csc\frac{9\pi}{2} = 1$, $\sec\frac{9\pi}{2} = \frac{r}{x}$ is undefined since $x = 0$, and $\cot\frac{9\pi}{2} = 0$.

27.

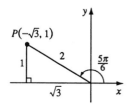

Using figure 8 we see that a point on the terminal line is $P(-\sqrt{3}, 1)$. Therefore taking $x = -\sqrt{3}$, $y = 1$, $r = 2$ in the definitions of the trigonometric ratios, we have $\sin\frac{5\pi}{6} = \frac{1}{2}$, $\cos\frac{5\pi}{6} = -\frac{\sqrt{3}}{2}$, $\tan\frac{5\pi}{6} = -\frac{1}{\sqrt{3}}$, $\csc\frac{5\pi}{6} = 2$, $\sec\frac{5\pi}{6} = -\frac{2}{\sqrt{3}}$, $\cot\frac{5\pi}{6} = -\sqrt{3}$.

29. $\sin\theta = \frac{y}{r} = \frac{3}{5} \Rightarrow y = 3$, $r = 5$, and $x = \sqrt{r^2 - y^2} = 4$ (since $0 < \theta < \pi/2$). Therefore taking $x = 4$, $y = 3$, $r = 5$ in the definitions of the trigonometric ratios, we have $\cos\theta = \frac{4}{5}$, $\tan\theta = \frac{3}{4}$, $\csc\theta = \frac{5}{3}$, $\sec\theta = \frac{5}{4}$, $\cot\theta = \frac{4}{3}$.

31. $\frac{\pi}{2} < \phi < \pi \Rightarrow \phi$ is in the second quadrant where x is negative and y is positive. Therefore $\sec\phi = \frac{r}{x} = -1.5 = -\frac{3}{2} \Rightarrow r = 3$, $x = -2$, and $y = \sqrt{r^2 - x^2} = \sqrt{5}$, Taking $x = -2$, $y = \sqrt{5}$, and $r = 3$ in the definitions of the trigonometric ratios, we have $\sin\phi = \frac{\sqrt{5}}{3}$, $\cos\phi = -\frac{2}{3}$, $\tan\phi = -\frac{\sqrt{5}}{2}$, $\csc\phi = \frac{3}{\sqrt{5}}$, and $\cot\theta = -\frac{2}{\sqrt{5}}$.

33. $\pi < \beta < 2\pi$ means that β is in the third or fourth quadrant where y is negative. Also since $\cot\beta = \frac{x}{y} = 3$ which is positive, x must also be negative. Therefore $\cot\beta = \frac{x}{y} = \frac{3}{1} \Rightarrow x = -3$, $y = -1$, and $r = \sqrt{x^2 + y^2} = \sqrt{10}$. Taking $x = -3$, $y = -1$ and $r = \sqrt{10}$ in the definitions of the trigonometric ratios we have, $\sin\beta = -\frac{1}{\sqrt{10}}$, $\cos\beta = -\frac{3}{\sqrt{10}}$, $\tan\beta = \frac{1}{3}$, $\csc\beta = -\sqrt{10}$, and $\sec\beta = -\frac{\sqrt{10}}{3}$.

35. $\sin 35° = \frac{x}{10} \Rightarrow x = 10\sin 35° \approx 5.73576$ cm.

37. $\tan\frac{2\pi}{5} = \frac{x}{8} \Rightarrow x = 8\tan\frac{2\pi}{5} \approx 24.62147$ cm.

39.

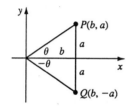

(a) From the diagram we see that $\sin\theta = \frac{y}{r} = \frac{a}{c}$, and $\sin(-\theta) = \frac{-a}{c} = -\sin\theta$.

(b) Again from the diagram we see that $\cos\theta = \frac{x}{r} = \frac{b}{c} = \cos(-\theta)$.

41. (a) Using (12a) and (13a) we have $\frac{1}{2}[\sin(x+y)+\sin(x-y)]$

$= \frac{1}{2}[\sin x\cos y + \cos x\sin y + \sin x\cos y - \cos x\sin y] = \frac{1}{2}(2\sin x\cos y) = \sin x\cos y$.

(b) This time using (12b) and (13b) we have $\frac{1}{2}[\cos(x+y)+\cos(x-y)]$

$= \frac{1}{2}[\cos x\cos y - \sin x\sin y + \cos x\cos y + \sin x\sin y] = \frac{1}{2}(2\cos x\cos y) = \cos x\cos y$.

(c) Again using (12b) and (13b) we have $\frac{1}{2}[\cos(x-y)-\cos(x+y)]$

$= \frac{1}{2}[\cos x\cos y + \sin x\sin y - \cos x\cos y + \sin x\sin y] = \frac{1}{2}(2\sin x\sin y) = \sin x\sin y$.

43. Using (12a), $\sin\left(\frac{\pi}{2}+x\right) = \sin\frac{\pi}{2}\cos x + \cos\frac{\pi}{2}\sin x = 1\cdot\cos x + 0\cdot\sin x = \cos x$.

45. Using (6), $\sin\theta\cot\theta = \sin\theta\,\frac{\cos\theta}{\sin\theta} = \cos\theta$.

47. $\sec y - \cos y \underset{(6)}{=} \frac{1}{\cos y} - \cos y = \frac{1-\cos^2 y}{\cos y} \underset{(7)}{=} \frac{\sin^2 y}{\cos y} = \frac{\sin y}{\cos y}\sin y \underset{(6)}{=} \tan y\sin y$.

49. $\cot^2\theta + \sec^2\theta \underset{(6)}{=} \frac{\cos^2\theta}{\sin^2\theta} + \frac{1}{\cos^2\theta} = \frac{\cos^2\theta\cos^2\theta + \sin^2\theta}{\sin^2\theta\cos^2\theta}$

$\underset{(7)}{=} \frac{(1-\sin^2\theta)(1-\sin^2\theta)+\sin^2\theta}{\sin^2\theta\cos^2\theta} = \frac{1-\sin^2\theta+\sin^4\theta}{\sin^2\theta\cos^2\theta}$

$\underset{(7)}{=} \frac{\cos^2\theta+\sin^4\theta}{\sin^2\theta\cos^2\theta} = \frac{1}{\sin^2\theta} + \frac{\sin^2\theta}{\cos^2\theta} \underset{(6)}{=} \csc^2\theta + \tan^2\theta$.

51. Using (14a) we have, $\tan 2\theta = \tan(\theta+\theta) = \frac{\tan\theta+\tan\theta}{1-\tan\theta\tan\theta} = \frac{2\tan\theta}{1-\tan^2\theta}$.

53. Using (15a) and (16a), $\sin x\sin 2x + \cos x\cos 2x = \sin x(2\sin x\cos x) + \cos x(2\cos^2 x - 1)$

$= 2\sin^2 x\cos x + 2\cos^3 x - \cos x \underset{(7)}{=} 2(1-\cos^2 x)\cos x + 2\cos^3 x - \cos x$

$= 2\cos x - 2\cos^3 x + 2\cos^3 x - \cos x = \cos x$.

55. $\frac{\sin\phi}{1-\cos\phi} = \frac{\sin\phi}{1-\cos\phi}\cdot\frac{1+\cos\phi}{1+\cos\phi} = \frac{\sin\phi(1+\cos\phi)}{1-\cos^2\phi} \underset{(7)}{=} \frac{\sin\phi(1+\cos\phi)}{\sin^2\phi} = \frac{1+\cos\phi}{\sin\phi}$

$= \frac{1}{\sin\phi} + \frac{\cos\phi}{\sin\phi} \underset{(6)}{=} \csc\phi + \cot\phi$.

57. Using (12a): $\sin 3\theta + \sin\theta = \sin(2\theta+\theta) + \sin\theta = \sin 2\theta\cos\theta + \cos 2\theta\sin\theta + \sin\theta$

$\underset{(16a)}{=} \sin 2\theta\cos\theta + (2\cos^2\theta - 1)\sin\theta + \sin\theta = \sin 2\theta\cos\theta + 2\cos^2\theta\sin\theta - \sin\theta + \sin\theta$

$\underset{(15a)}{=} \sin 2\theta\cos\theta + \sin 2\theta\cos\theta = 2\sin 2\theta\cos\theta$.

59. 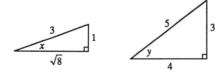 Since $\sin x = \frac{1}{3}$ we can label the opposite side as having length 1, the hypotenuse as having length 3, and use the Pythagorean Theorem to get that the adjacent side has length $\sqrt{8}$. Then, from the diagram, $\cos x = \frac{\sqrt{8}}{3}$. Similarly we have that $\sin y = \frac{3}{5}$.

Now use (12a): $\sin(x+y) = \sin x \cos y + \cos x \sin y = \left(\frac{1}{3}\right)\left(\frac{4}{5}\right) + \left(\frac{\sqrt{8}}{3}\right)\left(\frac{3}{5}\right) = \frac{4}{15} + \frac{3\sqrt{8}}{15}$

$= \frac{4 + 6\sqrt{2}}{15}.$

61. Using (13b) and the values for $\cos x$, and $\sin y$ obtained in Exercise 59 we have,

$\cos(x-y) = \cos x \cos y + \sin x \sin y = \left(\frac{\sqrt{8}}{3}\right)\left(\frac{4}{5}\right) + \left(\frac{1}{3}\right)\left(\frac{3}{5}\right) = \frac{8\sqrt{2}+3}{15}.$

63. Using (15a) and the value for $\sin y$ obtained in Exercise 59 we have,

$\sin 2y = 2\sin y \cos y = \frac{2\sin y}{\sec y} = 2\left(\frac{3}{5}\right)\left(\frac{4}{5}\right) = \frac{24}{25}.$

65. $2\cos x - 1 = 0 \Leftrightarrow \cos x = \frac{1}{2} \Rightarrow x = \frac{\pi}{3}, \frac{5\pi}{3}.$

67. $2\sin^2 x = 1 \Leftrightarrow \sin^2 x = \frac{1}{2} \Leftrightarrow \sin x = \pm\frac{1}{\sqrt{2}} \Rightarrow x = \frac{\pi}{4}, \frac{3\pi}{4}, \frac{5\pi}{4}, \frac{7\pi}{4}.$

69. Using (15a), $\sin 2x = \cos x \Rightarrow 2\sin x \cos x - \cos x = 0 \Leftrightarrow \cos x(2\sin x - 1) = 0$

$\Leftrightarrow \cos x = 0$ or $2\sin x - 1 = 0 \Rightarrow x = \frac{\pi}{2}, \frac{3\pi}{2}$ or $\sin x = \frac{1}{2} \Rightarrow x = \frac{\pi}{6}$ or $\frac{5\pi}{6}$. Therefore the

solutions are $x = \frac{\pi}{6}, \frac{\pi}{2}, \frac{5\pi}{6}, \frac{3\pi}{2}.$

71. $\sin x = \tan x \Leftrightarrow \sin x - \tan x = 0 \Leftrightarrow \sin x - \frac{\sin x}{\cos x} = 0 \Leftrightarrow \sin x\left(1 - \frac{1}{\cos x}\right) = 0 \Leftrightarrow \sin x = 0$

or $1 - \frac{1}{\cos x} = 0 \Rightarrow x = 0, \pi, 2\pi$ or $1 = \frac{1}{\cos x} \Rightarrow \cos x = 1 \Rightarrow x = 0, 2\pi$. Therefore the

solutions are $x = 0, \pi, 2\pi$

73. We know that $\sin x = \frac{1}{2}$ when $x = \frac{\pi}{6}$ or $\frac{5\pi}{6}$, and from Figure 13(a) we see that

$\sin x \leq \frac{1}{2} \Rightarrow 0 \leq x \leq \frac{\pi}{6}$ or $\frac{5\pi}{6} \leq x \leq 2\pi.$

75. $\tan x = -1$ when $x = \frac{3\pi}{4}, \frac{7\pi}{4}$, and $\tan x = 1$ when $x = \frac{\pi}{4}$ or $\frac{5\pi}{4}$. From Figure 14

we see that $-1 < \tan x < 1 \Rightarrow 0 \leq x < \frac{\pi}{4}, \frac{3\pi}{4} < x < \frac{5\pi}{4}$, and $\frac{7\pi}{4} < x \leq 2\pi.$

77. $y = \cos\left(x - \frac{\pi}{3}\right)$. We start with the graph of $y = \cos x$ and shift it $\frac{\pi}{3}$ units to the right.

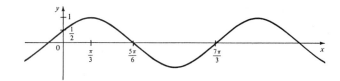

79. $y = \frac{1}{3}\tan\left(x - \frac{\pi}{2}\right)$. We start with the graph of $y = \tan x$, shift it $\frac{\pi}{2}$ units to the right and vertically compress it by a factor of $\frac{1}{3}$.

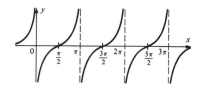

81. $y = |\sin x|$. We start with the graph of $y = \sin x$ and reflect the parts below the x-axis about the x-axis.

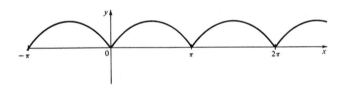

83. From the figure we see that $x = b \cos \theta$, $y = b \sin \theta$, and from the distance formula we have that the distance c from (x, y) to $(a, 0)$ is

$$c = \sqrt{(x-a)^2 + (y-0)^2} \Rightarrow c^2 = (b \cos \theta - a)^2 + (b \sin \theta)^2$$
$$= b^2 \cos^2 \theta - 2ab \cos \theta + a^2 + b^2 \sin^2 \theta = a^2 + b^2 (\cos^2 \theta + \sin^2 \theta) - 2ab \cos \theta$$
$$\underset{(7)}{=} a^2 + b^2 - 2ab \cos \theta.$$

85. Using the Law of Cosines, we have $c^2 = 1^2 + 1^2 - 2(1)(1) \cos (\alpha - \beta)$

$= 2[1 - \cos (\alpha - \beta)]$. Now, using the distance formula,

$c^2 = |AB|^2 = (\cos \alpha - \cos \beta)^2 + (\sin \alpha - \sin \beta)^2$. Equating these two expressions for c^2

we get $2[1 - \cos (\alpha - \beta)] = \cos^2 \alpha + \sin^2 \alpha + \cos^2 \beta + \sin^2 \beta - 2 \cos\alpha \cos \beta - 2 \sin \alpha \sin \beta$

$\Rightarrow 1 - \cos (\alpha - \beta) = 1 - \cos \alpha \cos \beta - \sin \alpha \sin \beta \Rightarrow \cos (\alpha - \beta) = \cos \alpha \cos \beta + \sin \alpha \sin \beta.$

87. In Exercise 86 we used the subtraction formula for cosine to prove the addition formula for cosine. Using that formula with $x = \frac{\pi}{2} - \alpha$, $y = \beta$, we get

$\cos [(\frac{\pi}{2} - \alpha) + \beta] = \cos (\frac{\pi}{2} - \alpha) \cos \beta - \sin (\frac{\pi}{2} - \alpha) \sin \beta$

$\Rightarrow \cos [\frac{\pi}{2} - (\alpha - \beta)] = \cos (\frac{\pi}{2} - \alpha) \cos \beta - \sin (\frac{\pi}{2} - \alpha) \sin \beta.$ Now we use the identities

given in the problem to get $\sin (\alpha - \beta) = \sin \alpha \cos \beta - \cos \alpha \sin \beta$.

EXERCISES D

1. Guessing: Calculator results (8 digits, 3 spare digits; don't forget to switch to radians!), followed by microcomputer results (16 digits, base 16).

   The last column shows the values of the sixth-degree Taylor polynomial for $f(x) = \csc^2 x - x^{-2}$ near $x = 0$ (note that the second arrangement of Taylor's polynomial is easier to use with a calculator):

   $$T_6(x) = \tfrac{1}{3} + \tfrac{1}{15}x^2 + \tfrac{2}{189}x^4 + \tfrac{1}{675}x^6 \equiv ((\tfrac{1}{675} x^2 + \tfrac{2}{189}) x^2 + \tfrac{1}{15}) x^2 + \tfrac{1}{3}$$

x	$f(x)_{\text{calculator}}$	$f(x)_{\text{computer}}$	$f(x)_{\text{Taylor}}$
0.1	0.33400107	.3340010596844678	0.33400106
0.01	0.333341	.3333400001065456	0.33334000
0.001	0.3334	.3333334000781178	0.33333340
0.0001	0.34	.3333333283662796	0.33333333
0.00001	2.0	.3333339691162109	0.33333333
0.000001	100 or 200	.33349609375	0.33333333
0.0000001	10000 or 20000	.328125	0.33333333
0.00000004	-230000	.5	0.33333333
0.00000001	1000000	0	0.33333333

We see that already at $x = 0.0001$, the results deteriorate, and for smaller x, they become meaningless. The different results "100 or 200" etc. depended on whether we calculated $((\sin x)^2)^{-1}$ or $((\sin x)^{-1})^2$.

With the microcomputer, the results become meaningless at 4×10^{-8}, when about half of the device's digits are used to store x (compare with the calculator result!) A detailed analysis reveals that the values of the function should be greater than $\tfrac{1}{3}$, but at $x = 0.0001$, the value is less than $\tfrac{1}{3}$.

The polynomial $T_6(x)$ was obtained by patient simplification of the expression for $f(x)$, starting with $\sin^2(x) = \tfrac{1}{2}(1 - \cos 2x)$, where
$$\cos 2x = 1 - (2x)^2/2! + (2x)^4/4! - \ldots - (2x)^{10}/10! + R_{12}(x).$$
Consequently, the exact value of the limit is $T_6(0) = \tfrac{1}{3}$. It can also be obtained by several applications of l'Hospital's Rule to the expression
$$f(x) = \frac{x^2 - \sin^2 x}{x^2 \sin^2 x}$$

with intermediate simplifications.

3. From $$f(x) = \frac{x^{25}}{(1.0001)^x} \quad \text{(we may assume } x > 0; \text{ why?)}$$

    we have $$\ln f(x) = 25 \ln x - x \ln (1.0001)$$

    and $$\frac{f'(x)}{f(x)} = \frac{25}{x} - \ln (1.0001).$$

    This derivative, as well as the derivative $f'(x)$ itself, is positive for
    $$0 < x < x_0 = \frac{25}{\ln (1.0001)} \approx 249{,}971.015, \text{ and negative for } x > x_0.$$

    Hence the maximum value of $f(x)$ is $f(x_0) = \dfrac{x_0^{25}}{(1.0001)^{x_0}}$, a number too large

    to be calculated directly. Using decimal logarithms,
    $$\log_{10} f(x_0) \approx 124.08987757, \text{ so that } f(x_0) \approx \underline{1.229922 \times 10^{124}}.$$

    The limit $\lim_{x \to \infty} f(x) = 0$; it would be wasteful and inelegant to use l'Hospital's Rule twenty-five times since we can transform $f(x)$ into
    $$f(x) = \left[\frac{x}{(1.0001)^{x/25}} \right]^{25}$$

    and the inside expression needs just one application of l'Hospital's Rule to yield limit 0.

5. For $f(x) = \ln \ln x$ with $x \in [a, b]$, $a = 10^9$, and $b = 10^9 + 1$, we need
    $$f'(x) = \frac{1}{x \ln x}, \qquad f''(x) = -\frac{\ln x + 1}{x^2 (\ln x)^2}$$

    In part (a): $f'(b) < D < f'(a)$, where

    $f'(a) \approx 4.82549\,42434 \times 10^{-11}$ $\qquad\qquad$ $f'(b) \approx 4.82549\,42383 \times 10^{-11}$

    In part (b): Let us estimate $f'(b) - f'(a) = (b - a)f''(c_1) = f''(c_1)$. Since f'' increases (its absolute value decreases), we have
    $$|f'(b) - f'(a)| < |f''(a)| \approx 5.0583 \times 10^{-20}.$$

7. (a) The 11-digit calculator value of $192 \sin \frac{\pi}{96}$ is $6.28206\,39018$, while the value (on the same device) of p before rationalization is $6.28206\,3885$, less than the trigonometric result by 1.68×10^{-8}.

    (b) The result is
    $$p = \frac{96}{\sqrt{2 + \sqrt{3}} \cdot \sqrt{2 + \sqrt{2 + \sqrt{3}}} \cdot \sqrt{2 + \sqrt{2 + \sqrt{2 + \sqrt{3}}}} \cdot \sqrt{2 + \sqrt{2 + \sqrt{2 + \sqrt{2 + \sqrt{3}}}}}}$$

    but of course we can avoid repetitious calculations by storing intermediate results in a memory:
    $$p_1 = \sqrt{2 + \sqrt{3}}, \quad p_2 = \sqrt{2 + p_1}, \quad p_3 = \sqrt{2 + p_2}, \quad p_4 = \sqrt{2 + p_3}, \quad p = \frac{96}{p_1 p_2 p_3 p_4}.$$

    According to this formula, a calculator gives $p \approx 6.28206\,39016$, in agreement with the trigonometric result within 2×10^{-10}.

A microcomputer program working with 16 digits gives $p \approx 6.28206\,39017\,81019$ with an error before rationalization about 4.1×10^{-14}, and after rationalization about 8.9×10^{-16}, also a gain of about 2 digits of accuracy.

9. (a) Let $A = \left(\dfrac{27q + \sqrt{729q^2 + 108p^3}}{2} \right)^{\frac{1}{3}}$ and $B = \left(\dfrac{27q - \sqrt{729q^2 + 108p^3}}{2} \right)^{\frac{1}{3}}.$

Then $A^3 + B^3 = 27q$ and $AB = \left(\dfrac{729q^2 - (729q^2 + 108p^3)}{4} \right)^{\frac{1}{3}} = -3p.$

Substitute into the formula $A + B = \dfrac{A^3 + B^3}{A^2 - AB + B^2}$ where we replace B by $-3p/A$:

$$ x = \tfrac{1}{3}(A + B) = \cfrac{27q/3}{\left(\dfrac{27q + \sqrt{729q^2 + 108p^3}}{2} \right)^{\frac{2}{3}} + 3p + 9p^2 \left(\dfrac{27q + \sqrt{729q^2 + 108p^3}}{2} \right)^{-\frac{2}{3}}} $$

which almost yields the given formula; since replacing q by $-q$ results in replacing x by $-x$, a simple discussion of the cases $q > 0$ and $q < 0$ allows us to replace q by $|q|$ in the denominator expression, so that it involves only positive numbers. The problems mentioned in the introduction to this exercise have disappeared.

(b) A direct attack works best here. To save space, let $\alpha = 2 + \sqrt{5}$, so we can rationalize, using $\alpha^{-1} = -2 + \sqrt{5}$ and $\alpha - \alpha^{-1} = 4$ (check it!):

$$ u = \frac{4}{\alpha^{2/3} + 1 + \alpha^{-2/3}} \cdot \frac{\alpha^{1/3} - \alpha^{-1/3}}{\alpha^{1/3} - \alpha^{-1/3}} = \frac{4(\alpha^{1/3} - \alpha^{-1/3})}{\alpha - \alpha^{-1}} = \alpha^{1/3} - \alpha^{-1/3} $$

and we cube the expression for u:

$$ u^3 = \alpha - 3\alpha^{1/3} + 3\alpha^{-1/3} - \alpha^{-1} = 4 - 3u, $$

$$ u^3 + 3u - 4 = (u - 1)(u^2 + u + 4) = 0, $$

so that the only real root is $u = 1$.

A check using the formula from part (a): $p = 3$, $q = -4$, so $729q^2 + 108p^3 = 14580 = 54^2 \times 5$, and

$$ x = \frac{36}{\left(54 + 27\sqrt{5} \right)^{\frac{2}{3}} + 9 + 81\left(54 + 27\sqrt{5} \right)^{-\frac{2}{3}}}, $$

which simplifies to the given form after reduction by 9.

11. Proof that $\lim_{n \to \infty} a_n = 0$:

From $1 \le e^{1-x} \le e$ it follows that $x^n \le e^{1-x} x^n \le x^n e$, and integration gives

$$ \tfrac{1}{n+1} = \int_0^1 x^n \, dx \le \int_0^1 e^{1-x} x^n \, dx \le \int_0^1 x^n e \, dx = \tfrac{e}{n+1}, $$

that is,
$$\tfrac{1}{n+1} \le a_n \le \tfrac{e}{n+1}$$

and since
$$\lim_{n \to \infty} \tfrac{1}{n+1} = \lim_{n \to \infty} \tfrac{e}{n+1} = 0,$$

it follows from the Squeeze Theorem that $\lim_{n \to \infty} a_n = 0$.

Of course, the expression $\tfrac{1}{n+1}$ on the left side could have been replaced by 0 and the proof would remain valid.

Calculations: An 11-digit scientific pocket calculator gives

n	a_n (red. formula)
0	1.71828 18284
1	0.71828 18284
2	0.43656 36568
3	0.30969 09704
4	0.23876 38816
5	0.19381 94080
6	0.16291 64480
7	0.14041 51360
8	0.12332 10880
9	0.10988 97920
10	0.09889 79200
11	0.08787 71200
12	0.05452 54400
13	$-0.29116\ 92800$
14	$-5.07636\ 992$
15	$-77.14554\ 88$
16	$-1235.32878\ 08$
17	$-21001.58927\ 4$
18	-378029.60693
19	-7182563.5317
20	-143651271.63

It is clear from step 13 that the values calculated from the direct reduction formula will diverge to $-\infty$.

13. We can start with expressing e^x and e^{-x} in terms of $E(x) = (e^x - 1)/x \ (x \ne 0)$, where $E(0) = 1$ to make E continuous at 0 (by L'Hospital's Rule). Namely,

$$e^x = 1 + xE(x), \qquad e^{-x} = 1 - xE(-x)$$

and
$$\sinh x = \frac{1 + xE(x) - (1 - xE(-x))}{2} = \tfrac{1}{2}x\left[E(x) + E(-x)\right]$$

where the addition involves only positive numbers $E(x)$ and $E(-x)$, thus presenting no loss of accuracy due to subtraction.

Another form which calls the function E only once: we write

$$\sinh x = \frac{(e^x)^2 - 1}{2e^x} = \frac{(1 + xE(x))^2 - 1}{2(1 + xE(x))} = \frac{x\left[1 + \tfrac{1}{2}|x|\,E(|x|)\right]E(|x|)}{1 + |x|\,E(|x|)},$$

taking advantage of the fact that $\frac{\sinh x}{x}$ is an even function, hence replacing x by $|x|$ does not change its value.

EXERCISES E

1. $X = 1 \cdot \cos 30° + 4 \sin 30° = 2 + \frac{\sqrt{3}}{2}$, $Y = -1 \cdot \sin 30° + 4 \cos 30° = 2\sqrt{3} - \frac{1}{2}$.

3. $X = -2 \cos 60° + 4 \sin 60° = -1 + 2\sqrt{3}$, $Y = 2 \sin 60° + 4 \cos 60° = \sqrt{3} + 2$.

5. $\cot 2\theta = \frac{A - C}{B} = 0 \Rightarrow 2\theta = \frac{\pi}{2} \Leftrightarrow \theta = \frac{\pi}{4}$

    $\Rightarrow$ (*by Equations* 2) $x = \frac{X - Y}{\sqrt{2}}$ and

    $y = \frac{X + Y}{\sqrt{2}}$. Substituting these into the

    curve equation gives $0 = (x - y)^2 - (x + y)$

    $= 2Y^2 - \sqrt{2}X$ or $Y^2 = \frac{X}{\sqrt{2}}$.

    (Parabola, vertex $(0,0)$, directrix

    $X = -1/(4\sqrt{2})$, focus $(1/(4\sqrt{2}), 0)$).

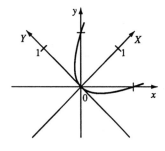

7. $\cot 2\theta = \frac{A - C}{B} = 0 \Rightarrow 2\theta = \frac{\pi}{2} \Leftrightarrow \theta = \frac{\pi}{4}$

    $\Rightarrow$ (*by Equations* 2) $x = \frac{X - Y}{\sqrt{2}}$ and $y = \frac{X + Y}{\sqrt{2}}$.

    Substituting these into the curve equation gives

    $1 = \frac{X^2 - 2XY + Y^2}{2} + \frac{X^2 - Y^2}{2} + \frac{X^2 + 2XY + Y^2}{2}$

    $\Rightarrow 3X^2 + Y^2 = 2 \Rightarrow \frac{X^2}{2/3} + \frac{Y^2}{2} = 1$.

    (An ellipse, center $(0,0)$, foci on Y–axis

    with $a = \sqrt{2}$, $b = \sqrt{6}/3$, $c = 2\sqrt{3}/3$.)

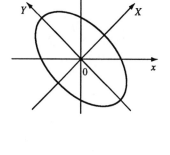

9. $\cot 2\theta = \frac{97 - 153}{192} = \frac{-7}{24} \Rightarrow \tan 2\theta = -\frac{24}{7}$

    $\Rightarrow \frac{\pi}{2} < 2\theta < \pi$ and $\cos 2\theta = \frac{-7}{25}$

    $\Rightarrow \frac{\pi}{4} < \theta < \frac{\pi}{2}$, $\cos \theta = \frac{3}{5}$, $\sin \theta = \frac{4}{5}$

    $\Rightarrow x = X \cos \theta - Y \sin \theta = \frac{3X - 4Y}{5}$

    and $y = X \sin \theta + Y \cos \theta = \frac{4X + 3Y}{5}$.

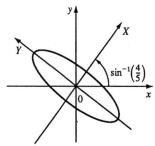

Substituting, we get $\frac{97}{25}(3X - 4Y)^2 + \frac{192}{25}(3X - 4Y)(4X + 3Y) + \frac{153}{25}(4X + 3Y)^2 = 225$, which

simplifies to $X^2 + \frac{Y^2}{9} = 1$ (an ellipse with foci on Y–axis, centered at origin, $a = 3$, $b = 1$).

11. $\cot 2\theta = \dfrac{A-C}{B} = \dfrac{1}{\sqrt{3}} \Rightarrow \theta = \dfrac{\pi}{6}$

$\Rightarrow x = \dfrac{\sqrt{3}X - Y}{2}, \; y = \dfrac{X + \sqrt{3}Y}{2}.$

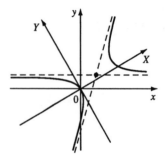

Substituting into the curve equation and

simplifying gives $4X^2 - 12Y^2 - 8X = 0$

$\Rightarrow (X-1)^2 - 3Y^2 = 1$ (a hyperbola with

foci on X–axis, centered at $(1,0)$, $a = 1$, $b = 1/\sqrt{3}$, $c = 2/\sqrt{3}$).

13. (a) $\cot 2\theta = \dfrac{A-C}{B} = \dfrac{-7}{24}$ so, as in Exercise 9, $x = \dfrac{3X - 4Y}{5}$ and $y = \dfrac{4X + 3Y}{5}$.

   Substituting and simplifying we get $100X^2 - 25Y + 25 = 0 \Rightarrow 4X^2 = Y - 1$, which is a

   parabola.

(b) The vertex is $(0,1)$ and $p = \dfrac{1}{16}$, so the XY–coordinates of the focus are $\left(0, \dfrac{17}{16}\right)$, and

   the xy–coordinates are $x = \dfrac{0\cdot 3}{5} - \left(\dfrac{17}{16}\right)\left(\dfrac{4}{5}\right) = -\dfrac{17}{20}$ and $y = \dfrac{0\cdot 4}{5} + \left(\dfrac{17}{16}\right)\left(\dfrac{3}{5}\right) = \dfrac{51}{80}$.

(c) The directrix is $Y = \dfrac{15}{16}$, so $-x\cdot\dfrac{4}{5} + y\cdot\dfrac{3}{5} = \dfrac{15}{16} \Rightarrow 64x - 48y + 75 = 0$.

15. A rotation through θ changes Equation 1 to

$A(X\cos\theta - Y\sin\theta)^2 + B(X\cos\theta - Y\sin\theta)(X\sin\theta + Y\cos\theta) + C(X\sin\theta + Y\cos\theta)^2$

$+ D(X\cos\theta - Y\sin\theta) + E(X\sin\theta + Y\cos\theta) + F = 0.$ Comparing this to Equation 4, we see that

$A' + C' = A(\cos^2\theta + \sin^2\theta) + C(\sin^2\theta + \cos^2\theta) = A + C.$

17. Choose θ so that $B' = 0$. Then $B^2 - 4AC = (B')^2 - 4A'C' = -4A'C'$. But $A'C'$ will be 0

   for a parabola, negative for a hyperbola (where the X^2 and Y^2 coefficients are of opposite sign),

   and positive for an ellipse (same sign for X^2 and Y^2 coefficients). So $B^2 - 4AC$: $= 0$ for a

   parabola, > 0 for a hyperbola, and < 0 for an ellipse. Note that the transformed equation

   takes the form $A'X^2 + C'Y^2 + D'X + E'Y + F = 0$, or by completing the square (assuming

   $A'C' \neq 0$), $A'(X')^2 + C'(Y')^2 = F'$, so that if $F' = 0$, the graph is either a pair of intersecting

   lines or a point depending on the signs of A' and C'. If $F' \neq 0$ and $A'C' > 0$, then the graph is

   either an ellipse, a point, or nothing, and if $A'C' < 0$, the graph is a hyperbola. If A' or C' is

   0, we cannot complete the square, so we get $A'(X')^2 + E'Y + F = 0$ or

   $C'(Y')^2 + D'X + F' = 0$. This is a parabola, a straight line (if only the second degree

   coefficient is nonzero), a pair of parallel lines (if the first degree coefficient is zero and the other

   two have opposite signs), or an empty graph (if the first degree coefficient is zero and the other

   two have the same sign).

EXERCISES F

1. $(3 + 2i) + (7 - 3i) = (3 + 7) + (2 - 3)i = 10 - i.$

3. $(3 - i)(4 + i) = 12 + 3i - 4i - (-1) = 13 - i.$

5. $\overline{12 + 7i} = 12 - 7i.$

7. $\dfrac{2 + 3i}{1 - 5i} = \dfrac{2 + 3i}{1 - 5i} \cdot \dfrac{1 + 5i}{1 + 5i} = \dfrac{2 + 10i + 3i + 15(-1)}{1 - 25(-1)} = \dfrac{-13 + 13i}{26} = -\tfrac{1}{2} + \tfrac{1}{2}i.$

9. $\dfrac{1}{1 + i} = \dfrac{1}{1 + i} \cdot \dfrac{1 - i}{1 - i} = \dfrac{1 - i}{1 - (-1)} = \dfrac{1 - i}{2} = \tfrac{1}{2} - \tfrac{1}{2}i.$

11. $i^3 = i^2 \cdot i = (-1)i = -i.$

13. $\sqrt{-25} = \sqrt{25}\,i = 5i.$

15. $\overline{3 + 4i} = 3 - 4i, |3 + 4i| = \sqrt{3^2 + 4^2} = \sqrt{25} = 5.$

17. $\overline{-4i} = \overline{0 - 4i} = 0 + 4i = 4i, |-4i| = \sqrt{0^2 + (-4)^2} = 4.$

19. $4x^2 + 9 = 0 \Leftrightarrow 4x^2 = -9 \Leftrightarrow x^2 = -\tfrac{9}{4} \Leftrightarrow x = \pm\sqrt{-\tfrac{9}{4}} = \pm\sqrt{\tfrac{9}{4}}\,i = \pm\tfrac{3}{2}i.$

21. By the quadratic formula,
$$x^2 - 8x + 17 = 0 \Leftrightarrow x = \frac{8 \pm \sqrt{8^2 - 4(1)(17)}}{2(1)} = \frac{8 \pm \sqrt{-4}}{2} = \frac{8 \pm 2i}{2} = 4 \pm i.$$

23. By the quadratic formula,
$$z^2 + z + 2 = 0 \Leftrightarrow z = \frac{-1 \pm \sqrt{1 - 4(1)(2)}}{2(1)} = \frac{-1 \pm \sqrt{-7}}{2} = -\tfrac{1}{2} \pm \tfrac{\sqrt{7}}{2}i.$$

25. $r = \sqrt{(-3)^2 + 3^2} = 3\sqrt{2}, \tan\theta = \tfrac{3}{-3} = -1 \Rightarrow \theta = \tfrac{3}{4}\pi$ (since the given number is in the second quadrant). Therefore $-3 + 3i = 3\sqrt{2}(\cos\tfrac{3\pi}{4} + i\sin\tfrac{3\pi}{4}).$

27. $r = \sqrt{3^2 + 4^2} = 5, \tan\theta = \tfrac{4}{3} \Rightarrow \theta = \tan^{-1}\tfrac{4}{3}$ (since the given number is in the second quadrant). Therefore $3 + 4i = 5[\cos(\tan^{-1}\tfrac{4}{3}) + i\sin(\tan^{-1}\tfrac{4}{3})].$

29. For $z = \sqrt{3} + i, r = \sqrt{[\sqrt{3}]^2 + 1^2} = 2,$ and $\tan\theta = \tfrac{1}{\sqrt{3}} \Rightarrow \theta = \tfrac{\pi}{6}$ so that $z = 2(\cos\tfrac{\pi}{6} + i\sin\tfrac{\pi}{6}).$
For $w = 1 + \sqrt{3}i, r = 2,$ and $\tan\theta = \sqrt{3} \Rightarrow \theta = \tfrac{\pi}{3}$ so that $w = 2(\cos\tfrac{\pi}{3} + i\sin\tfrac{\pi}{3}).$
Therefore $zw = 2 \cdot 2[\cos(\tfrac{\pi}{6} + \tfrac{\pi}{3}) + i\sin(\tfrac{\pi}{6} + \tfrac{\pi}{3})] = 4(\cos\tfrac{\pi}{2} + i\sin\tfrac{\pi}{2}),$
$z/w = \tfrac{2}{2}[\cos(\tfrac{\pi}{6} - \tfrac{\pi}{3}) + i\sin(\tfrac{\pi}{6} - \tfrac{\pi}{3})] = \cos(-\tfrac{\pi}{6}) + i\sin(-\tfrac{\pi}{6}),$ and
$1 = 1 + 0i = \cos 0 + i\sin 0 \Rightarrow 1/z = \tfrac{1}{2}[\cos(0 - \tfrac{\pi}{6}) + i\sin(0 - \tfrac{\pi}{6})] = \tfrac{1}{2}[\cos(-\tfrac{\pi}{6}) + i\sin(-\tfrac{\pi}{6})].$

31. For $z = 2\sqrt{3} - 2i, r = 4, \tan\theta = \tfrac{-2}{2\sqrt{3}} = -\tfrac{1}{\sqrt{3}} \Rightarrow \theta = -\tfrac{\pi}{6} \Rightarrow z = 4[\cos(-\tfrac{\pi}{6}) + i\sin(-\tfrac{\pi}{6})].$
For $w = -1 + i, r = \sqrt{2}, \tan\theta = \tfrac{1}{-1} = -1 \Rightarrow \theta = \tfrac{3\pi}{4} \Rightarrow z = \sqrt{2}(\cos\tfrac{3\pi}{4} + i\sin\tfrac{3\pi}{4}).$
Therefore $zw = 4\sqrt{2}[\cos(-\tfrac{\pi}{6} + \tfrac{3\pi}{4}) + i\sin(-\tfrac{\pi}{6} + \tfrac{3\pi}{4})] = 4\sqrt{2}(\cos\tfrac{7\pi}{12} + i\sin\tfrac{7\pi}{12}),$
$z/w = \tfrac{4}{\sqrt{2}}[\cos(-\tfrac{\pi}{6} - \tfrac{3\pi}{4}) + i\sin(-\tfrac{\pi}{6} - \tfrac{3\pi}{4})] = \tfrac{4}{\sqrt{2}}[\cos(-\tfrac{11\pi}{12}) + i\sin(-\tfrac{11\pi}{12})]$

$$= 2\sqrt{2}(\cos\tfrac{13\pi}{12} + i\sin\tfrac{13\pi}{12}), \text{ and } 1 = 1 + 0i = \cos 0 + i\sin 0$$
$$\Rightarrow 1/z = \tfrac{1}{4}[\cos(0 - (-\tfrac{\pi}{6})) + i\sin(0 - (-\tfrac{\pi}{6}))] = \tfrac{1}{4}(\cos\tfrac{\pi}{6} + i\sin\tfrac{\pi}{6}).$$

33. For $z = 1 + i$, $r = \sqrt{2}$, $\tan\theta = \tfrac{1}{1} = 1 \Rightarrow \theta = \tfrac{\pi}{4} \Rightarrow 1 + i = \sqrt{2}(\cos\tfrac{\pi}{4} + i\sin\tfrac{\pi}{4})$.

So by De Moivre's Theorem,

$$(1 + i)^{20} = [\sqrt{2}(\cos\tfrac{\pi}{4} + i\sin\tfrac{\pi}{4})]^{20} = [2^{1/2}]^{20}(\cos\tfrac{20\pi}{4} + i\sin\tfrac{20\pi}{4})$$
$$= 2^{10}(\cos 5\pi + i\sin 5\pi) = 2^{10}(-1 + i(0)) = -2^{10} = -1024.$$

35. For $z = 2\sqrt{3} + 2i$, $r = 4$, $\tan\theta = \tfrac{2}{2\sqrt{3}} = \tfrac{1}{\sqrt{3}} \Rightarrow \theta = \tfrac{\pi}{6} \Rightarrow 2\sqrt{3} + 2i = 4(\cos\tfrac{\pi}{6} + i\sin\tfrac{\pi}{6})$.

So by De Moivre's Theorem, $(2\sqrt{3} + 2i)^5 = [4(\cos\tfrac{\pi}{6} + i\sin\tfrac{\pi}{6})]^5 = 4^5(\cos\tfrac{5\pi}{6} + i\sin\tfrac{5\pi}{6})$

$$= 4^5(-\tfrac{\sqrt{3}}{2} + i(0.5)) = -512\sqrt{3} + 512i.$$

37. $1 = 1 + 0i = (\cos 0 + i\sin 0)$. Using equation (3) with $r = 1$, $n = 8$, and $\theta = 0$ we have

$$w_k = 1^{1/8}\left[\cos\left(\frac{0 + 2k\pi}{8}\right) + i\sin\left(\frac{0 + 2k\pi}{8}\right)\right] = \cos\tfrac{k\pi}{4} + i\sin\tfrac{k\pi}{4}, \text{ where } k = 0, 1, 2, \ldots, 7.$$

$w_0 = (\cos 0 + i\sin 0) = 1$

$w_1 = (\cos\tfrac{\pi}{4} + i\sin\tfrac{\pi}{4}) = \tfrac{1}{\sqrt{2}} + \tfrac{1}{\sqrt{2}}i$

$w_2 = (\cos\tfrac{\pi}{2} + i\sin\tfrac{\pi}{2}) = i$

$w_3 = (\cos\tfrac{3\pi}{4} + i\sin\tfrac{3\pi}{4}) = -\tfrac{1}{\sqrt{2}} + \tfrac{1}{\sqrt{2}}i$

$w_4 = (\cos\pi + i\sin\pi) = -1$

$w_5 = (\cos\tfrac{5\pi}{4} + i\sin\tfrac{5\pi}{4}) = -\tfrac{1}{\sqrt{2}} - \tfrac{1}{\sqrt{2}}i$

$w_6 = (\cos\tfrac{3\pi}{2} + i\sin\tfrac{3\pi}{2}) = -i$

$w_7 = (\cos\tfrac{7\pi}{4} + i\sin\tfrac{7\pi}{4}) = \tfrac{1}{\sqrt{2}} - \tfrac{1}{\sqrt{2}}i.$

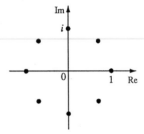

39. $0 = 0 + i = \cos\tfrac{\pi}{2} + i\sin\tfrac{\pi}{2}$. Using equation (3) with $r = 1$, $n = 3$, $\theta = \tfrac{\pi}{2}$ we have

$$w_k = 1^{1/3}\left[\cos\left(\frac{\tfrac{\pi}{2} + 2k\pi}{3}\right) + i\sin\left(\frac{\tfrac{\pi}{2} + 2k\pi}{3}\right)\right], \text{ where } k = 0, 1, 2.$$

$w_0 = (\cos\tfrac{\pi}{6} + i\sin\tfrac{\pi}{6}) = \tfrac{\sqrt{3}}{2} + \tfrac{1}{2}i$

$w_1 = (\cos\tfrac{5\pi}{6} + i\sin\tfrac{5\pi}{6}) = -\tfrac{\sqrt{3}}{2} + \tfrac{1}{2}i$

$w_2 = (\cos\tfrac{9\pi}{6} + i\sin\tfrac{9\pi}{6}) = -i.$

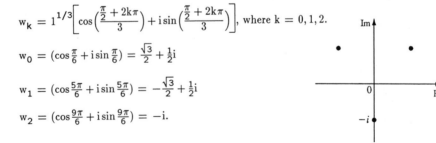

41. Using Euler's formula (6) with $y = \tfrac{\pi}{2}$, $e^{i\pi/2} = \cos\tfrac{\pi}{2} + i\sin\tfrac{\pi}{2} = i$.

43. Using Euler's formula with $y = \tfrac{3\pi}{4}$, $e^{i3\pi/4} = \cos\tfrac{3\pi}{4} + i\sin\tfrac{3\pi}{4} = -\tfrac{1}{\sqrt{2}} + \tfrac{1}{\sqrt{2}}i.$

45. Using Equation 7 with $x = 2$ and $y = \pi$,

$$e^{2+i\pi} = e^2 e^{i\pi} = e^2(\cos\pi + i\sin\pi) = e^2(-1 + 0) = -e^2.$$

47. Take r = 1 and n = 3 in De Moivre's Theorem to get

$$[1(\cos\theta + i\sin\theta)]^3 = 1^3(\cos 3\theta + i\sin 3\theta) \Rightarrow (\cos\theta + i\sin\theta)^3 = \cos 3\theta + i\sin 3\theta \Rightarrow$$

$$\cos^3\theta + 3(\cos^2\theta)(i\sin\theta) + 3(\cos\theta)(i\sin\theta)^2 + (i\sin\theta)^3 = \cos 3\theta + i\sin 3\theta$$

$$\Rightarrow (\cos^3\theta - 3\sin^2\theta\cos\theta) + (3\sin\theta\cos^2\theta - \sin^3\theta)i = \cos 3\theta + i\sin 3\theta.$$ Equating real and

imaginary parts gives $\cos 3\theta = \cos^3\theta - 3\sin^2\theta\cos\theta$ and $\sin 3\theta = 3\sin\theta\cos^2\theta - \sin^3\theta$.

49. $F(x) = e^{rx} = e^{(a + bi)x} = e^{ax+bxi} = e^{ax}(\cos bx + i\sin bx) = e^{ax}\cos bx + i(e^{ax}\sin bx)$

$$\Rightarrow F'(x) = (e^{ax}\cos bx)' + i(e^{ax}\sin bx)'$$

$$= (ae^{ax}\cos bx - be^{ax}\sin bx) + i(ae^{ax}\sin bx + be^{ax}\cos bx)$$

$$= a[e^{ax}(\cos bx + i\sin bx)] + b[e^{ax}(-\sin bx + i\cos bx)] = ae^{rx} + b[e^{ax}(i^2\sin bx + i\cos bx)]$$

$$= ae^{rx} + bi[e^{ax}(\cos bx + i\sin bx)] = ae^{rx} + bie^{rx} = (a + bi)e^{rx} = re^{rx}.$$